Analytic Methods in Geomechanics

Analytic Methods in Geomechanics

K. T. Chau

CRC Press
Taylor & Francis Group
Boca Raton London New York

CRC Press is an imprint of the
Taylor & Francis Group, an **informa** business

CRC Press
Taylor & Francis Group
6000 Broken Sound Parkway NW, Suite 300
Boca Raton, FL 33487-2742

International Standard Book Number: 978-1-4665-5584-6 (Hardback) 978-1-4665-5585-3 (Paperback)

Visit the Taylor & Francis Web site at
http://www.taylorandfrancis.com

and the CRC Press Web site at
http://www.crcpress.com

To

My wife Lim, son Magnum, and daughter Jaquelee

and

my inspirer in geomechanics, Prof. John W. Rudnicki

CONTENTS

PREFACE
THE AUTHOR

CHAPTER 1: ELEMENTARY TENSOR ANALYSIS 1

1.1 Introduction 1
1.2 General Tensors, Cartesian Tensors, and Tensor Rank 2
1.3 A Brief Review of Vector Analysis 2
1.4 Dyadic Form of Second-Order Tensors 5
1.5 Derivatives of Tensors 7
1.6 Divergence and Stokes Theorems 8
1.6.1 Divergence Theorem or Gauss Theorem 8
1.6.2 Stokes Theorem 9
1.7 Some Formulas in Cylindrical Coordinates 10
1.8 Some Formulas in Spherical Coordinates 12
1.9 Summary and Further Reading 13
1.10 Problems 13

CHAPTER 2: ELASTICITY AND ITS APPLICATIONS 17

2.1 Introduction 17
2.2 Basic Concepts for Stress Tensor 18
2.3 Piola–Kirchhoff Stresses 19
2.4 Coordinate Transformation of Stress 21
2.5 Basic Concepts for Strain Tensor 23
2.6 Rate of Deformation 25
2.7 Compatibility Equations 26
2.8 Hill's Work-Conjugate Stress Measures 26
2.9 Constitutive Relation 27
2.10 Isotropic Solids 27
2.11 Transversely Isotropic Solids 29
2.12 Equations of Motion and Equilibrium 30
2.13 Compatibility Equation in Terms of Stress Tensor 32
2.14 Strain Energy Density 33
2.15 Complementary Energy 34
2.16 Hyperelasticity and Hypoelasticity 34
2.17 Plane Stress, Plane Strain, and the Airy Stress Function 36
2.18 Stress Concentration at a Circular Hole 40
2.19 Force Acting at the Apex of a Wedge 43
2.20 Uniform Vertical Loading on Part of the Surface 45
2.21 Solution for Indirect Tensile Test (Brazilian Test) 46
2.22 Jaeger's Modified Brazilian Test 48
2.23 Edge Dislocation 49

2.24 Dislocation Pile-up and Crack 51
2.25 Screw Dislocation and Faulting 53
2.26 Mura Formula for Curved Dislocation 58
2.27 Summary and Further Reading 60
2.28 Problems 60

CHAPTER 3: COMPLEX VARIABLE METHODS FOR 2-D ELASTICITY 63

3.1 Introduction 63
3.2 Coordinate Transformation in Complex Variable Theory 66
3.3 Homogeneous Stresses in Terms Analytic Functions 67
3.4 A Borehole Subject to Internal Pressure 67
3.5 Kirsch Solution by Complex Variable Method 68
3.6 Definiteness and Uniqueness of the Analytic Function 69
3.7 Boundary Conditions for the Analytic Functions 70
3.8 Single-valued Condition for Multi-connected Bodies 72
3.9 Multi-connected Body of Infinite Extend 75
3.10 General Transformation of Quantities 76
3.11 Elastic Body with Holes 78
3.12 Stress Concentration at a Square Hole 82
3.13 Mapping Functions for Other Holes 87
3.14 Summary and Further Reading 88
3.15 Problems 89

CHAPTER 4: THREE-DIMENSIONAL SOLUTIONS IN ELASTICITY ... 93

4.1 Introduction 93
4.2 Displacement Formulation 94
4.2.1 Helmholtz Decomposition 94
4.2.2 Lamé's Strain Potential for Incompressible Solids 96
4.2.3 Galerkin Vector 97
4.2.4 Love's Displacement Potential for Cylindrical Solids 98
4.2.5 Papkovitch–Neuber Displacement Potential 99
4.2.6 2-D Papkovitch–Neuber vs. Kolosov–Muskhelisvili Methods 101
4.3 Stress Formulations 101
4.3.1 Beltrami and Beltrami–Schaefer Stress Functions 101
4.3.2 Maxwell Stress Functions 103
4.3.3 Morera Sress Function 104
4.3.4 Other Beltrami Stress Functions 104
4.4 Some 3-D Solutions in Geomechanics 106
4.4.1 Hollow Sphere Subject to Internal and External Pressures 106
4.4.2 Kelvin's Fundamental Solution 109
4.4.2.1 Papkovitch–Neuber Potential Method 109
4.4.2.2 Love's Displacement Potential Method 111

4.4.3	Boussinesq's Fundamental Solution	113
4.4.3.1	Love's and Lamé's Strain Potential Methods	114
4.4.3.2	Papkovitch–Neuber Potential Method	115
4.4.4	Cerruti's Fundamental Solution	118
4.4.5	Mindlin's Fundamental Solution in Half-space	122
4.4.6	Lorentz's Fundamental Solution	128
4.4.7	Melan's Fundamental Solution	130
4.5	Harmonic Functions and Indirect Method	132
4.6	Harmonic Functions in Spherical Coordinates	136
4.7	Harmonic Functions in Cylindrical Coordinates	137
4.8	Biharmonic Functions	138
4.9	Muki's Formulation in Cylindrical Coordinates	139
4.9.1	Muki's Vector Potentials	140
4.9.2	Method of Solution by Hankel Transform	141
4.9.3	Boussinesq Solution by Hankel Transform	144
4.10	Summary and Further Reading	147
4.10.1	Summary	147
4.10.2	Further Reading	148
4.10.2.1	General Method of Solutions for 3-D Elasticity	148
4.10.2.2	Integral Transform in Solving 3-D Problems	148
4.10.2.3	General Method of Solutions for Circular Cylinders	148
4.10.2.4	General Method of Solutions for Spheres	149
4.11	Problems	149

CHAPTER 5: PLASTICITY AND ITS APPLICATIONS 159

5.1	Introduction	159
5.2	Flow Theory and Deformation Theory	160
5.3	Yield Function and Plastic Potential	161
5.4	Elasto-plastic Constitutive Model	162
5.5	Rudnicki–Rice (1975) Model	163
5.6	Drucker's Postulate, PMPR, and Il'iushin's Postulate	163
5.7	Yield Vertex	165
5.8	Mohr–Coulomb Model	168
5.9	Lode Angle or Parameter	169
5.10	Yield Criteria on the π-Plane	171
5.11	Other Soil Yield Models	174
5.12	Cap Models	175
5.13	Physical Meaning of Cam-Clay Model	177
5.14	Modified Cam-Clay	178
5.15	A Cam-clay Model for Finite Strain	181
5.16	Plasticity by Internal Variables	183
5.17	Viscoplasticity	184
5.17.1	One-dimensional Modeling	184
5.17.2	Three-dimensional Models	186

5.17.3 Consistency Condition for Perzyna Model 187
5.17.4 Consistency Model of Wang et al. (1997) 188
5.17.5 Adachi-Oka (1984) Model 189
5.18 Summary and Further Reading 191
5.19 Problems 192

CHAPTER 6: FRACTURE MECHANICS AND ITS APPLICATIONS 197

6.1 Introduction 197
6.2 Stress Concentration at a Elliptical Hole 198
6.3 Stress Concentration at a Tensile Crack 202
6.4 Stress Field near a Shear Crack 205
6.5 The General Stress and Displacement Field for Mode I Cracks 207
6.6 The General Stress and Displacement Field for Mode II Cracks 211
6.7 The General Stress and Displacement Field for Mode III Cracks 212
6.8 The Energy Release Rate at Crack Tips 214
6.9 Fracture Toughness for Rocks 218
6.10 *J*-integral and the Energy Release Rate 219
6.11 Westergaard Stress Function and Superposition 223
6.12 Growth of Slip Surface in Slopes 227
6.13 Energy Release Rate for Earthquake 233
6.14 Wing Crack Model under Compressions 235
6.15 Bažant's Size Effect Law via *J*-integral 237
6.16 Continuum Damage Mechanics 240
6.17 Solids Containing Microcracks 242
6.17.1 Compliance Change due to a Single Crack 242
6.17.2 Effective Compliance for Cracked Bodies 243
6.17.3 Non-interacting Result for Planar Transverse Isotropy 243
6.17.4 Planar Transverse Isotropy by Self-consistent Method 244
6.17.5 Planar Transverse Isotropy by Differential Scheme 245
6.17.6 Non-interacting Result for Cylindrical Transverse Isotropy 245
6.17.7 Non-interacting Result for Isotropically Cracked Solids 246
6.18 Rudnicki–Chau (1996) Multiaxial Microcrack Model 247
6.19 Summary and Further Reading 249
6.20 Problems 250

CHAPTER 7: VISCOELASTICTY AND ITS APPLICATIONS 257

7.1 Introduction 257
7.2 Boltzmann's Integral Form of Stress and Strain 258
7.3 Stieltjes Convolution Notation 260
7.4 Stress-Strain Relation in Differential Equation Form 261
7.4.1 Maxwell Model 262
7.4.2 Kelvin–Voigt Model 263
7.4.3 Three-Parameter Models 263

7.4.4	Generalized Maxwell and Kelvin Models	265
7.5	Stress-strain Relation in Laplace Transform Space	265
7.5.1	Viscoelastic Solids with Elastic Bulk Modulus	267
7.5.2	Maxwell Solids	267
7.5.3	Kelvin–Voigt Solids	268
7.5.4	Standard Linear Solid and Three-Parameter Models	268
7.6	Correspondence Principle	270
7.6.1	Boussinesq Problem for Maxwell Half-space	271
7.6.2	Boussinesq Problem for Kelvin–Voigt Half-space	273
7.6.3	Boussinesq Problem for Three-Parameter Model A	274
7.7	Creeping and Relaxation Tests	275
7.7.1	Maxwell Material	275
7.7.1.1	Creeping Test	275
7.7.1.2	Relaxation Test	276
7.7.2	Kelvin–Voigt Material	277
7.7.2.1	Creeping Test	277
7.7.2.2	Relaxation Test	277
7.7.3	Three-parameter Model A or Standard Linear Solid	278
7.7.3.1	Creeping Test	278
7.7.3.2	Relaxation Test	280
7.7.3.3	Relaxation Test in Compression	281
7.8	Calibration of the Viscoelastic Model	281
7.9	Viscoelastic Crack Models for Steam Injection	284
7.9.1	Superposition of Auxiliary Problems I and II	284
7.9.2	Center of Dilatation in Two-dimensional Bimaterial	284
7.9.3	Stress Intensity Factor of Auxiliary Problem II	287
7.9.4	Inverse Laplace Transform	287
7.9.5	Numerical Results	288
7.10	Summary and Further Reading	289
7.11	Problems	290

CHAPTER 8: LINEAR ELASTIC FLUID-INFILTRATED SOLIDS AND POROELASTICITY 295

8.1	Introduction	295
8.2	Biot's Theory of Poroelasticity	298
8.2.1	McNamee and Gibson Cylindrical Form	298
8.2.2	Rice–Cleary (1976) Linearized Constitutive Relation	298
8.2.3	Rudnicki (1986) Constitutive Relation	301
8.2.4	Rudnicki's (1985) Anisotropic Diffusive Solids	302
8.3	Biot–Verruijt Displacement Function	304
8.4	McNamee–Gibson–Verruijt Displacement Function	306
8.5	Schiffman–Fungaroli–Verruijt Displacement Function	307
8.6	Schiffman–Fungaroli Displacement Function	308
8.7	Laplace–Hankel Transform Technique	309

8.8 Point Forces and Point Fluid Source in Half-space 310
8.8.1 Vertical Point Force Solution 310
8.8.2 Horizontal Point Force Solution 312
8.8.3 Fluid Point Source Solution 314
8.9 Cleary's Fundamental Solution of Point Forces in Full Space 315
8.9.1 Canonical Representation of Point Force Solution 315
8.9.2 Determination of Evolution Functions 316
8.9.3 Determination of Unknown Constant F_∞ 321
8.9.4 Final Solutions 322
8.10 Rudnicki's Fundamental Solutions in Full Space 323
8.10.1 Impulsive Fluid Source 323
8.10.2 Canonical Form of Displacement Solution 323
8.10.3 Error Function Representation 324
8.10.4 Suddenly Applied Fluid Mass Source 325
8.10.5 Equivalence of Fluid Mass Dipole and Body Force 327
8.10.6 Fluid Mass Dipoles 328
8.10.7 Point Force Solution by Rudnicki (1986) 328
8.11 Thermoelasticity vs. Poroelasticity 330
8.12 Summary and Further Reading 330
8.12.1 Summary 330
8.12.2 Further Reading 331
8.13 Problems 331

CHAPTER 9: DYNAMICS AND WAVES IN GEOMATERIALS 337

9.1 Introduction 337
9.2 Seismic Waves 338
9.3 Waves in Infinite Elastic Isotropic Solids 338
9.4 Helmholtz Theorem and Wave Speeds 340
9.5 Rayleigh Waves 341
9.5.1 Characteristics Equation for Rayleigh Wave Speed 341
9.5.2 Rayleigh Wave in Solids Satisfying Poisson Condition 342
9.5.3 Segel (1977) Method for Arbitrary Poisson's Ratio 345
9.6 Love Waves 346
9.6.1 Non-existence of SH-wave in Homogeneous Half-space 347
9.6.2 Love Waves in an Elastic Layer on a Half-space 347
9.6.3 Dispersion Characteristics of Love Waves 350
9.7 Stoneley Waves 351
9.8 Elastic-plastic Waves 353
9.8.1 Acceleration Waves in Solids 353
9.8.2 Shear Banding as Stationary Acceleration Wave 354
9.8.3 Acoustic Tensor for Geomaterials 355
9.8.4 Wave Speed Analysis 357
9.9 Waves in Viscoelastic Solids 357

9.9.1 Complex Moduli ... 357
9.9.2 Longitudinal and Transverse Waves Speeds ... 358
9.10 Dynamic Fracture Mechanics ... 359
9.10.1 Dynamic Solutions for a Stationary Crack ... 360
9.10.2 Asymptotic Fields near a Moving Crack-tip ... 361
9.10.3 Dynamic Energy Release Rate ... 362
9.10.4 Dynamic Fracture Toughness ... 363
9.11 Vibrations and Soil Dynamics ... 364
9.12 Summary and Further Reading ... 365
9.12.1 Summary ... 365
9.12.2 Further Reading ... 366
9.12.2.1 Waves in Solids and Elastodynamics ... 366
9.12.2.2 Seismic Waves on Earth ... 366
9.12.2.3 Waves in Porous Media ... 367
9.12.2.4 Dynamic Fracture Mechanics ... 367
9.12.2.5 Dynamic Fragmentation ... 367
9.13 Problems ... 367

Appendices ... 371
Appendix A: Nanson Formula ... 371
Appendix B: Laplace Transform ... 373
Appendix C: Legendre Transform and Work Increments ... 382
Selected Biographies ... 385
References ... 403
Author Index ... 425
Subject Index ... 433

PREFACE

Geomechanics deals with the deformation and failure process in geomaterials, which is literally defined as materials found on the surface of the Earth. Soil, rock, snow, and ice are typical geomaterials. Although in a broader sense in engineering mechanics geomaterials also include concrete, it will not be included explicitly in the discussion in this book. Ice and snow will not be covered either.

A general review article on geomechanics was given by Rudnicki (2000). There are at least two review articles using the title "Analytical methods in Geomechanics" (Booker, 1991; Selvadurai, 2007), and the name of the journal *International Journal of Numerical and Analytical Methods in Geomechanics* also includes Analytical Methods in Geomechanics. However, to date there has been no book devoted exclusively to such methods. Although numerical methods and tools (such as finite element method or distinct element method) have been widely adopted to solve practical problems in geomechanics, as Dr. Evert Hoek put it, "The answers (from the numerical methods) are only as good as the input information" (Hoek, 1986). In view of this, a sign that read "It is the duty of an engineer to judge soundly rather than to compute accurately" was put on the top of Dr. Hoek's computer (Hoek, 1986). Analytical solutions can often provide the order of magnitude of the solution and provide insight into the behavior and trend of the solutions in terms of the main controlling parameters of the problem, and thus they can provide the basis of "sound judgment."

In addition, although powerful numerical tools have been developed in the last few decades, it is not possible, and in many circumstances not particularly desirable, to conduct a full-scale investigation using finite element models, incorporating all fine details of geometric, materials, and loading conditions (Booker, 1991). In view of the uncertainties of the input data to finite element programs, analytical methods still have an important and valuable role to play in geomechanics.

Although Professor Y.H. Pao classified "Geomechanics" under "Earth Science" in his review article "Applied Mechanics in Science and Engineering" (Table 7 of Pao, 1998), we would rather consider it a multidisciplinary field encompassing both geophysics and civil engineering. Many postgraduate students and researchers who engage in geomechanics research are from different backgrounds, including geologists, seismologists, mining engineers, geophysicists, mathematicians, mechanical engineers, and civil engineers. Without a proper training in engineering mechanics or geomechanics, very often they find the terminology and mechanics techniques used in geomechanics journals or books incomprehensible. I am a civil engineer by training, with a master's degree in structural engineering. Under the supervision of Prof. John Rudnicki, my Ph.D. was, however, in theoretical and applied mechanics at Northwestern University, with a thesis related to bifurcation, localization, and inelastic deformation of pressure-sensitive dilatant materials (i.e., geomaterials). As a graduate student working in geomechanics, I had to take or sit in on a lot of fundamental courses in applied mathematics and mechanics before I could take on geomechanics research. If there had been a comprehensive book of the mathematical theory on geomaterials, I would have picked up the subject much more quickly.

I found that my graduate students are facing problems similar to those that I faced as a Ph.D. student. That is, the traditional bachelor's degree in engineering does not prepare them for geomechanics studies or for conducting geomechanics research. What makes the situation worse, however, is that there is no single textbook that covers the fundamentals of geomechanics, including tensor analysis, elasticity (essential to soils, rocks, and concrete), plasticity (essential to soils and rocks), fracture mechanics (essential to rocks and concrete) and viscoelasticity (essential to both soils and rocks), poroelasticity and wave propagation and dynamics, and puts them in the proper perspective relevant to the deformation behavior of geomaterials.

With this group of potential readers in mind, this book is written for scientists and engineers who have had some exposure to engineering mathematics and strength of materials. The text covers major topics in tensor analysis, elasticity (both 2-D and 3-D), plasticity, fracture mechanics, viscoelasticity, poroelasticity, and dynamics that are relevant to the modeling of geomaterials.

The text was developed and expanded from my course notes for a course called Analytic Methods in Geomechanics offered to graduate students at The Hong Kong Polytechnic University starting from 1995, under the encouragement of Prof. J.M. Ko, then the head of the Department of Civil & Structural Engineering (now Department of Civil & Environmental Engineering). I determined coverage of elasticity, plasticity, viscoelasticity, and fracture mechanics in such a way that selected topics are directly related to geomechanics, compared to typical textbooks. This book expands substantially and evolves from those lecture notes. Some contents and sections are of a more advanced nature and should provide useful reference material for researchers engaged in solving solid mechanics or geomechanics problems.

Chapter 1 summarizes the elements of tensor analysis, especially in dyadic forms. This constitutes an elegant basis for later chapters, especially dealing with messy analysis on polar cylindrical and spherical coordinates. Familiarity with materials in Chapter 1 will also help students to learn different branches of continuum mechanics. After a general discussion on the theory of elasticity and elastic dislocation theory in Chapter 2, Chapter 3 introduces the use of complex variable technique in solving two-dimensional elasticity problems. The treatise by Muskhelishvili on complex variable technique for solving 2-D problems is not easily accessible by engineering students and is not covered in most graduate courses. Chapter 3 serves as an introduction to this useful and powerful tool. In particular, complex variable technique has been found extremely useful in solving crack problems.

Although real geomaterials are three dimensional in nature, most graduate courses on elasticity only deal with 2-D analysis and do not include 3-D problems in their syllabi. I believe that 3-D elasticity is an important topic in geomechanics. Chapter 4 focuses on the methods of solution for three-dimensional elasticity. Solutions to be discussed include the Boussinesq solution, the Kelvin solution, and the Mindlin solution. These are three-dimensional solutions of fundamental importance in geomechanics. These solutions are discussed in detail. Muki's formalism with the Hankel transform is also discussed.

Chapter 5 introduces the framework of plasticity theory, with particular reference to the Rudnicki–Rice model for rocks and Cam-clay model for soils. The last few sections in Chapter 5 introduce the use of Helmholtz free energy in relating macroscopic deformation with microscopic damages using the internal variable approach of J.R. Rice and give an introduction to viscoplasticity. These topics are not covered in most plasticity books. Chapter 6 gives an overall introduction to fracture mechanics, including the use of *J*-integral on slope failure and earthquake energy release rate, and superposition technique involving fundamental solution in terms of the Westergaard stress function. The chapter concludes with discussions on continuum damage mechanics and microcrack models. Chapter 7 summarizes the essence of viscoelasticity and its application to geomechanics through the steam stimulation problem in oil sand extract. Chapter 8 discusses the use of displacement functions in poroelasticity. The constitutive forms of Biot's theory put forward by Rice and Cleary (1976) and by Rudnicki (1985, 1986) are discussed in detail. The use of the Laplace transform and the Hankel transform are included. The fundamental point force and point fluid source solutions are covered in detail. Chapter 9 covers the basics of wave propagations in half-spaces or layered half-spaces, as well as in viscoelastic and elastic-plastic solids. Essential results of dynamic fracture mechanics are summarized and serve as an introduction to dynamic fracture. A section on soil dynamics concludes the chapter. Brief biographies of about 70 selected mechanicians, scientists, and engineers whose works are used or described in the book are included to give brief historical developments of the mechanics and geomechanics topics covered in the book. Readers should find some of their scientific stories inspiring.

I have tried to keep the length of this book to an optimum such that I can strike a balance between breadth and depth, and between details and conciseness. In doing so, technical details have been kept to a minimum and therefore further readings are recommended for each chapter. To prevent this book project from becoming a never-ending endeavor, I had to stop somewhere and topics had to be selective. Suggestions are, however, welcome if some readers think that some important topics have been omitted.

I am indebted to many former professors, students, colleagues, friends, and authors, as interactions with them have shaped my thoughts and choices in writing this book either directly or indirectly. Chapter 1 was clearly influenced by the writing of E.L. Malvern and by the teaching of Continuum Mechanics of J.W. Rudnicki. Chapter 2 was influenced by the teachings of K.K. Koo, P. Karasudhi, J. Dundurs, J.W. Rudnicki, L. Keer, B. Moran, T. Mura, and J. Weertman. Chapter 3 was the result of my collaboration with Y.B. Wang, who is an expert in the Muskhelishvili method. My interest in three-dimensional elasticity started with my involvement in poroelasticity proposed by my former master's advisor W. Kanaok-Nukulchai and from the reading P. Karasudhi, A. Cheng, M. Cleary, J.W. Rudnicki, and J.R. Rice. This was further reinforced when I worked on diffuse mode bifurcations of solid cylinders as part of my doctorate, and vibrations of cylinders and spheres in my post-doctoral year. My former Ph.D. student X.X. Wei helped me continue to work on elasticity for cylinders and spheres, and S.Z. Wu helped with work on dynamic fragmentation on spheres. My interest in plasticity mainly arose from my study of Rudnicki–Rice model and from the teachings of T. Mura, J.W. Rudnicki, and B. Moran. My formal fracture mechanics training is from J. Achenbach, J. Dundurs, B. Moran, J.W. Rudnicki, L. Keer, J. Dundurs, and T. Mura, and my informal training from my collaboration with H. Muhlhaus during my visit to CSIRO in Australia in 1994. My former Ph.D. student R.H.C.

Wong and my colleague, Teng-fong Wong of SUNY, taught me about experimental fracture research. My knowledge of and interest in viscoelasticity result from my collaboration with R.C.K. Wong on delayed failure in shale overlying oil sand. My interests in wave propagations and dynamics are from my former teachers, K.K. Koo, P. Karasudhi, J. Rudnicki, E.L. Reiss, J. Achenbach, and M. Wieland.

I am indebted to discussions with numerous colleagues and friends in geomechanics or mechanics, including J.R. Rice, Z.P. Bažant, T.F. Wong, F. Oka, S. Sakurai, E.P. Chen, R.Y.S. Pak, R. Nova, R. Viesca, S. Kimoto, Y. Higo, A.N. Guz, G.C. Sih, C.T. Sun, W. Wu, R.L. Michalowski, F. Nicot, F. Darve, J.P. Bardet, A. Rechenmacher, J. Walsh, A.P.S. Selvadurai, Y.H. Pao, R.C.K. Wong, Y.B. Wang, D.M.Wood, E. Detourney, P.V. Lade, J. Labuz, S.K. Choi, I. Vardoulakis, R. Borja, J. Tejchman, J.F. Shao, R. Wan, Y.L. Li, C.S. Chang, T.T. Ng, P.Y. Hicher, A. Misra, N. Rajapakse, T. Senjuntichai, C.F. Lee, T.X. Yu, P. Lin, C.A. Tang, X. Yang, C. Yatomi, J.D. Zhao, G..S. Wang, X. Guo, J.J. Wu, J. Liu, W. Zhu, and S. Li, and many others.

My special appreciation goes to Dr. S.K. Yan of Hong Kong Baptist College (HKBC), who decided to give me a second change of academic life by creating an "extra" tutor and technician job for me at HKBC; Prof. P. Karasudhi of the Asian Institute of Technology for flying to Evanston to recommend me to Leon Keer and John Rudnicki; Prof. J.W. Rudnicki of Northwestern University for taking the chance of offering me an assistantship before meeting me. Without their special help, I would not have had a chance to write this book. Special thank also goes to Prof. J.G. Teng, Dean of the Faculty of Construction and Environment, who allowed me to step down from the Associate Deanship after serving for 5 years and to have more time to complete this book project.

The unfailing and continuous support of my wife Lim, my son Magnum and my daughter Jaquelee is what keeps me going when I face difficult times.

This book project was encouraged by Mr. Simon Bates of Spon Press (a subsidiary of Taylor & Francis) during one of his visits to Hong Kong. The expert helpings of Miss Laurie Schlags, my project coordinator of Taylor & Francis, and Miss Michele Dimont, my project editor of Taylor & Francis, are highly appreciated. A special thank goes to my former Ph.D. student, S.Z. Wu, who drew some of the diagrams in the second draft of my lecture notes in late 1990s. The cover of the book was conceptually designed under the technical guidance and advice of my son, Magnum, and my daughter, Jaquelee, and both of whom are excellent students and practitioners of visual arts.

K.T. Chau

THE AUTHOR

Professor K.T. Chau, Ph.D., is the Chair Professor of Geotechnical Engineering of the Department of Civil and Environmental Engineering at the Hong Kong Polytechnic University. He obtained his honors diploma with distinction from Hong Kong Baptist College (Hong Kong), his master of engineering in structural engineering from the Asian Institute of Technology (Thailand) where he was also awarded the Tim Kendall Memorial Prize (an academic prize for the best graduating student) with straight As, his Ph.D. in Theoretical and Applied Mechanics from Northwestern University (U.S.A.), and an Executive Certificate from the Graduate School of Business of Stanford University.

Dr. Chau worked as a full-time tutor/demonstrator/technician at Hong Kong Baptist College (1984–1985), as a research associate at the Asian Institute of Technology (summer of 1987), research assistant at Northwestern University (1987–1991), and as a post-doctoral fellow at Northwestern University (1991–1992). At Hong Kong Polytechnic University (PolyU), he has served as a lecturer, an assistant professor, an associate professor, a full professor and a chair professor since 1992. At PolyU, he served as the Associate Dean (Research and Development) of the Faculty of Construction and Environment, the Associated Head of the Department of Civil and Structural Engineering, the Chairman of the Appeals and Grievance Committee, the Alternate Chairman of the Academic Appeals Committee, and the Alternate Chairman of the University Staffing Committee.

Dr. Chau is a fellow of the Hong Kong Institution of Engineers (HKIE), the past Chairman of the Geomechanics Committee (2005–2010) of the Applied Mechanics Division (AMD) of ASME, the Chairman of the Elasticity Committee (2010–2013) of the Engineering Mechanics Institute (EMI) of ASCE, and Chairman of the TC103 of the ISSMGE. He is a recipient of the Distinguished Young Scholar Award of National Natural Science Foundation, China (2003), the France-Hong Kong Joint Research Scheme (2003–2004) of RGC of Hong Kong, and the Young Professor Overseas Placement Scheme of PolyU. He is a past President of the Hong Kong Society of Theoretical and Applied Mechanics (2004–2006) after serving as member-at-large and Vice President. He also served as a Scientific Advisor of the Hong Kong Observatory of HKSAR Government, a RGC Engineering Panel member of the HKSAR Government for 7 consecutive years, and served as the Vice President of the Hong Kong Institute of Science. He has delivered more than 12 keynote lectures at international/national conferences, served on advisory committee of 18 international conferences, and on organizing committees of 18 international conferences. He also held visiting positions at Harvard University (USA), Kyoto University (Japan), Polytech-Lille (France), Shandong University (China), Taiyuan University of Technology (China), the Rock Mechanics Research Center of CSIRO (Australia), and the University of Calgary (Canada).

Dr. Chau's research interests have included geomechanics and geohazards, including bifurcation and stability theories in geomaterials, rock mechanics, fracture and damage mechanics in brittle rocks, three-dimensional elasticity, earthquake engineering and mechanics, landslides and debris flows, tsunami and storm surges, and rockfalls and dynamic impacts, seismic pounding, vulnerability of tall buildings with transfer systems, and shaking table tests. He is the author of more than 100 journal papers and 200 conference publications.

In his leisure time, he enjoys swimming and takes part in master swimming competitions. He is the Honorable Manager of the Hong Kong Polytechnic University Swimming Team. Since 2001, he had competed in Hong Kong Masters Games, the Hong Kong Territory-wise Age-Group Swimming Competition, the Hong Kong Amateur Swimming Association (HKASA) Masters Swimming Championships, and District Swimming Meets of the Leisure and Cultural Services Department (LCSD). He has also participated in international masters swimming competitions, including the Macau Masters Swimming Championship, the Singapore National Masters Swimming, Standard Chartered Asia Pacific Masters Swim Meet, Wisdom-Act International Swimming Championship (Taiwan), Standard Chartered Singapore Masters Swim 2007, Japan Masters Long Distance Swim Meet 2008 (Aichi Meet and Machida Meet), Japan Short Course Masters Swimming Championship 2009 (Kyoto), Marblehead Sprint Classics (USA), The Masters Games Hamilton (New Zealand), Hawaii Senior Olympics, National China Masters Swimming Championships, the Third Annual Hawaii International Masters Swim Meet, and the Fifth Penang Invitational Masters Swimming Championship. By 2007, he had competed in all long course FINA events (i.e., 50 m, 100 m, 200 m, 400 m, 800 m and 1500 m freestyle; 50 m, 100 m and 200 m butterfly; 50 m, 100 m and 200 m breaststroke; 50 m, 100 m and 200 m backstroke; and 200 m and 400 m individual medley).

He also enjoys jogging and has completed four full marathons, including the Hong Kong International Marathon, the Chicago Oldstyle Marathon, and the China Coast Marathon with a personal best of 3 hours, 34 minutes and 15 seconds.

CHAPTER ONE

Elementary Tensor Analysis

1.1 INTRODUCTION

Continuum mechanics has been very successful in modeling physical phenomena in geomaterials, such as rock, soil, and concrete. Elasticity, plasticity, fracture mechanics, damage mechanics, viscoelasticity, and poroelasticity all can be considered branches of continuum mechanics that have found applications in geomechanics. Continuum mechanics, in fact, also includes fluid mechanics, but we will not deal with this aspect in this book. In order to understand and apply continuum mechanics more efficiently, tensor notation and analysis have been developed as the basic mathematical language for communication. This chapter deals only with elementary tensor analysis.

The physical laws, if they really describe the physical world, should be independent of the position and orientation of observers, that is, independent of the coordinates used in describing these phenomena. For this reason, physical laws are ideally written in *tensor equations* because tensor equations are invariant under coordinate transformation. If a tensor equation holds in one coordinate system, it also holds in any other coordinate system in the same reference frame. As we will see in later sections, many physical laws (such as the equation of equilibrium) in terms of a special coordinate system (such as a cylindrical or a spherical coordinate system) can be obtained by simply specializing the tensor equation to its component form. This coordinate-invariant property makes tensor analysis a very attractive technique for analysis of geomechanics problems. The physical quantities involved in the formulation of continuum mechanics, such as displacement, stress, strain, and modulus of elasticity, are more conveniently referred to as tensors. Mathematically, such tensors can either be expressed in polyadic or indicial forms. Tensor has its existence independent of any coordinate system, yet when it is specified in a particular coordinate system, it contains certain sets of quantities called *components*, identified by *free index* (or indices). Nowadays, technical papers in geomechanics or continuum mechanics are very often written in terms of tensors, taking advantage of their conciseness property. Tensor analysis, therefore, becomes a pre-requisite for any graduate student who wants access to the state-of-the-art information available in journal publications and advanced textbooks. This chapter will give a concise treatment of elementary tensor analysis for an orthogonal coordinate system, with particular reference to applications in geomechanics.

1.2 GENERAL TENSORS, CARTESIAN TENSORS, AND TENSOR RANK

Roughly speaking, *tensor* is a general term used for any physical quantity that may involve more than one physical component, all of which have directional sense. The term tensor in its modern meaning was introduced by German physicist W. Voigt in 1908 (see biography section). For *zeroth-rank or zeroth-order* tensors, there is only one component for a physical quantity, such as temperature and pressure; tensors of zeroth rank are normally called *scalars*, which are independent of direction. For *first-rank or first-order* tensors, there are three physical components in three-dimensional space, such as displacement. First-order tensors are normally referred to as *vectors*. For example, the components of a velocity are normally written as v_1, v_2, and v_3 along the x_1-, x_2-, and x_3-directions of a *Cartesian coordinate system*, respectively. In this case, v_i (i = 1,2,3) represents the physical components of the vector $\boldsymbol{v}$. These tensors are direction dependent, that is, components of different magnitude are observed along different directions. For *second-rank or second-order tensors*, there are nine physical components for a quantity in three-dimensional domain, such as stress and strain; second-order tensors are the most often encountered quantities. Again, all of these nine components are direction dependent. (As demonstrated in elementary elasticity textbooks, nine independent components are required to describe the stress at a point inside a body, although stress symmetry will normally lead to six independent components.) In general, for the N-th order tensors, there are 3^N components for a tensor in three-dimensional domain. For example, a fourth-order tensor has 81 components (note that a modulus tensor, which relates strains to stresses, is a fourth-order tensor). Therefore, tensors can be viewed as a generalization of an ordinary vector to incorporate more than one *free index* or to accommodate a quantity having more than three physical components.

The use of Riemann geometry (such as used in Einstein's theory of relativity) will not be discussed here. All vectors are assumed to be described in terms of Euclidean geometry.

1.3 A BRIEF REVIEW OF VECTOR ANALYSIS

Referring to Fig. 1.1, we let $\boldsymbol{e}_1$, $\boldsymbol{e}_2$, and $\boldsymbol{e}_3$ be the unit vectors along x_1-, x_2-, and x_3-directions, respectively. Then, it is well known that any vector $\boldsymbol{u}$ in the three-dimensional Euclidean space can be represented by the following linear combination:

$$\boldsymbol{u} = u_1\boldsymbol{e}_1 + u_2\boldsymbol{e}_2 + u_3\boldsymbol{e}_3 = \sum_{i=1}^{3} u_i\boldsymbol{e}_i = u_i\boldsymbol{e}_i \tag{1.1}$$

where u_i is also called *indicial notation or index notation of a tensor*. As mentioned earlier, u_i is also the physical component of the vector $\boldsymbol{u}$; physically, it is the length of the vector $\boldsymbol{u}$ projected along the i-th coordinate of the system. In more general tensor analysis, as shown in the last part of (1.1), the summation sign Σ is usually neglected; thus, repeated indices imply summation automatically. This is usually referred to as Einstein notation. The index i becomes a dummy index (i.e., it can be replaced arbitrarily by j, k, etc.), and it is no longer a free index (i.e., i cannot be set to 1, 2, or 3 arbitrarily).

For the curvilinear coordinate system, in which the base vectors are not necessarily orthogonal and are position dependent, there are two sets of physical components: the covariant and contravariant components, depending on whether the base vectors or reciprocal base vectors are used. The reciprocal base vectors are the orthogonal sets of the original base vectors. The term contravariant implies that the coordinate transformation rules for the contravariant tensor components and their base vectors are exactly the inverse of one another, while covariant components and their base vectors follow the same rule of coordinate transformation. When the base vectors are orthogonal (or perpendicular), this is called an orthogonal curvilinear coordinate. The tensor analysis for curvilinear coordinates is considerably more complicated than those for orthogonal coordinates. However, curvilinear coordinates can be very useful because it is sometimes more convenient to describe the boundary of a solid in certain curvilinear coordinates, such as cylinder and sphere problems. In this book, we will not discuss tensor analysis in general curvilinear coordinates, but two orthogonal curvilinear coordinates (cylindrical and spherical coordinates) will be discussed in Sections 1.7 and 1.8.

The addition rules for vectors are both associative and commutative, that is:

$$\boldsymbol{u}+\boldsymbol{v}=\boldsymbol{v}+\boldsymbol{u} \qquad (\boldsymbol{u}+\boldsymbol{v})+\boldsymbol{w}=\boldsymbol{u}+(\boldsymbol{v}+\boldsymbol{w}), \tag{1.2}$$

respectively. The *dot product* between two vectors $\boldsymbol{u}$ and $\boldsymbol{v}$, denoted by $\boldsymbol{u}\cdot\boldsymbol{v}$, is given by:

$$\boldsymbol{u}\cdot\boldsymbol{v}=|\boldsymbol{u}||\boldsymbol{v}|\cos\theta \qquad (0\le\theta\le\pi) \tag{1.3}$$

where θ is the angle between these two vectors. Since cosθ is always smaller than one, thus we have the following Schwarz inequality (Spiegel, 1968):

$$|\boldsymbol{u}\cdot\boldsymbol{v}|\le|\boldsymbol{u}||\boldsymbol{v}| \tag{1.4}$$

The *magnitude* of a vector $|\boldsymbol{u}|$ is defined as:

$$|\boldsymbol{u}|=\sqrt{u_1^2+u_2^2+u_3^2}=\sqrt{\boldsymbol{u}\cdot\boldsymbol{u}} \qquad \ge 0 \tag{1.5}$$

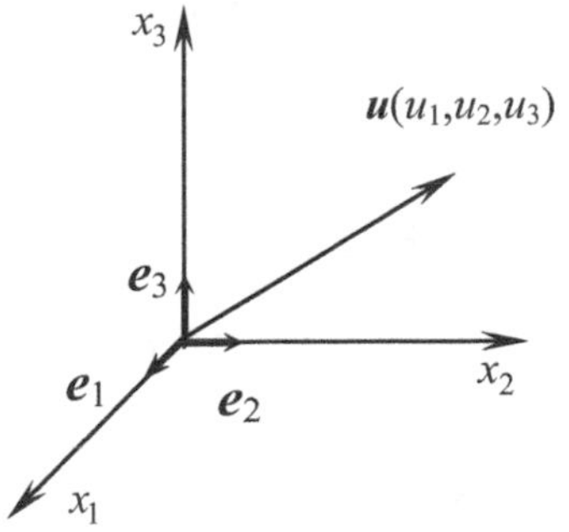

Figure 1.1 The Cartesian coordinates

and therefore $\boldsymbol{u} = \boldsymbol{0}$ if and only if $u_1 = u_2 = u_3 = 0$. Note that $\boldsymbol{e}_i \cdot \boldsymbol{e}_j = \delta_{ij}$, where $i, j = 1, 2, 3$; and δ_{ij} is the *Kronecker delta* which equals 1 if $i = j$ and 0 if $j \neq i$. Therefore, if we write

$$\boldsymbol{u} = u_1\boldsymbol{e}_1 + u_2\boldsymbol{e}_2 + u_3\boldsymbol{e}_3 \,, \quad \boldsymbol{v} = v_1\boldsymbol{e}_1 + v_2\boldsymbol{e}_2 + v_3\boldsymbol{e}_3 \tag{1.6}$$

then the dot product between $\boldsymbol{u}$ and $\boldsymbol{v}$ is clearly

$$\boldsymbol{u} \cdot \boldsymbol{v} = u_1v_1 + u_2v_2 + u_3v_3 = u_iv_i \tag{1.7}$$

Again, Einstein's notation for summing over all possible i is implied. The *cross product* of two vectors $\boldsymbol{u}$ and $\boldsymbol{v}$ produces another vector $\boldsymbol{w}$, which is usually denoted by $\boldsymbol{w} = \boldsymbol{u} \times \boldsymbol{v}$. The magnitude of $\boldsymbol{w}$ is defined as:

$$|\boldsymbol{w}| = |\boldsymbol{u} \times \boldsymbol{v}| = |\boldsymbol{u}||\boldsymbol{v}|\sin\theta \quad (0 \le \theta \le \pi/2) \tag{1.8}$$

In terms of tensor notation, the physical components w_i can be given in a right-handed coordinate system by

$$w_i = e_{ijk}u_jv_k \tag{1.9}$$

where e_{ijk} is in fact a third-order tensor called a permutation tensor. Note that the summation is done over both indices j and k. The magnitude is either -1, $+1$ or 0, which is determined by the following rules: (1) e_{ijk} equals 0 if any two indices are equal; (2) e_{ijk} equals $+1$ when i, j, k are 1, 2, 3 or an even permutation of 1, 2, 3; and (3) e_{ijk} equals -1 when i, j, k are 3, 2, 1, or an odd permutation of 1, 2, 3. The definition of even and odd permutation is illustrated in Fig. 1.2. Examples are $e_{123} = e_{231} = e_{312} = 1$, $e_{132} = e_{213} = e_{321} = -1$, and $e_{112} = e_{221} = e_{131} = 0$, etc. Mathematically, we can also write:

$$e_{ijk} = \frac{1}{2}(i-j)(j-k)(k-i) \tag{1.10}$$

Note, however, that (1.10) is not a tensor equation.

For vectors in Euclidean space, the cross products of vectors satisfy the following identities:

$$\boldsymbol{u} \times \boldsymbol{v} = -(\boldsymbol{v} \times \boldsymbol{u}) \tag{1.11}$$

$$\boldsymbol{u} \times (\boldsymbol{v} + \boldsymbol{w}) = \boldsymbol{u} \times \boldsymbol{v} + \boldsymbol{u} \times \boldsymbol{w} \tag{1.12}$$

$$\boldsymbol{u} \times \boldsymbol{u} = \boldsymbol{0} \tag{1.13}$$

$$\boldsymbol{e}_1 \times \boldsymbol{e}_2 = \boldsymbol{e}_3, \quad \boldsymbol{e}_2 \times \boldsymbol{e}_3 = \boldsymbol{e}_1, \quad \boldsymbol{e}_3 \times \boldsymbol{e}_1 = \boldsymbol{e}_2 \tag{1.14}$$

$$k\boldsymbol{u} \times \boldsymbol{v} = \boldsymbol{u} \times k\boldsymbol{v} = k(\boldsymbol{u} \times \boldsymbol{v}) \tag{1.15}$$

where k is a scalar.

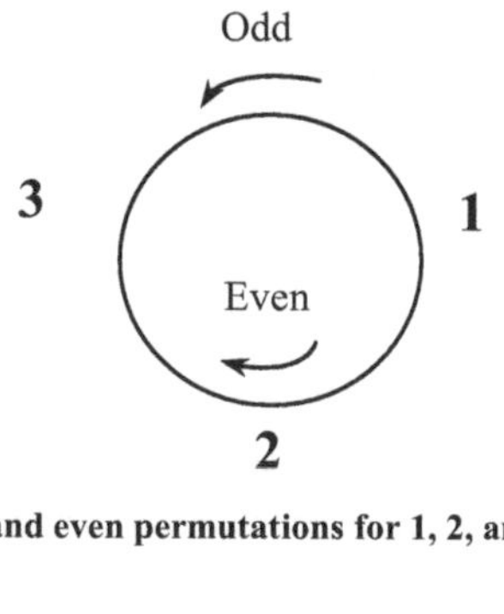

Figure 1.2 The odd and even permutations for 1, 2, and 3 in a permutation tensor

The cross product is sometimes easier to remember using the following expansion of determinant:

$$\boldsymbol{u}\times\boldsymbol{v}=\begin{vmatrix}\boldsymbol{e}_1 & \boldsymbol{e}_2 & \boldsymbol{e}_3\\ u_1 & u_2 & u_3\\ v_1 & v_2 & v_3\end{vmatrix}=(u_2v_3-u_3v_2)\boldsymbol{e}_1+(u_3v_1-u_1v_3)\boldsymbol{e}_2+(u_1v_2-u_2v_1)\boldsymbol{e}_3 \tag{1.16}$$

However, the determinant itself can also be written in tensor form:

$$\det\left|A_{ij}\right|=e_{ijk}A_{i1}A_{j2}A_{k3} \tag{1.17}$$

This can be proved by expanding the permutation tensor explicitly. The following *e-δ identity* has been found extremely useful in tensor analysis:

$$e_{ijk}e_{irs}=\delta_{jr}\delta_{ks}-\delta_{js}\delta_{kr} \tag{1.18}$$

This is proved in Problem 1.1 below.

1.4 DYADIC FORM OF SECOND-ORDER TENSORS

Instead of using physical components, vectors can be denoted by the *symbolic or Gibbs* notation. For example, velocity is sometimes written in terms of a bold-face letter as $\boldsymbol{v} = v_1\boldsymbol{e}_1 + v_2\boldsymbol{e}_2 + v_3\boldsymbol{e}_3$, where $\boldsymbol{e}_i$ is the *i*-th base vector in Cartesian coordinates. Similarly, this idea can be extended to second-order tensors. For example, stress can be represented by

$$\boldsymbol{\sigma}=\sigma_{ij}\boldsymbol{e}_i\boldsymbol{e}_j \tag{1.19}$$

where the indicial form is written in terms of Cartesian coordinates. Or more explicitly, we have

$$\boldsymbol{\sigma}=\sigma_{11}\boldsymbol{e}_1\boldsymbol{e}_1+\sigma_{22}\boldsymbol{e}_2\boldsymbol{e}_2+\sigma_{33}\boldsymbol{e}_3\boldsymbol{e}_3+\sigma_{13}\boldsymbol{e}_1\boldsymbol{e}_3+\sigma_{31}\boldsymbol{e}_3\boldsymbol{e}_1+\sigma_{12}\boldsymbol{e}_1\boldsymbol{e}_2+\sigma_{21}\boldsymbol{e}_2\boldsymbol{e}_1+\sigma_{23}\boldsymbol{e}_2\boldsymbol{e}_3+\sigma_{32}\boldsymbol{e}_3\boldsymbol{e}_2 \tag{1.20}$$

The symbolic form $\boldsymbol{\sigma}$ is most general and is independent of any coordinate system; the indicial form σ_{ij} is the physical component corresponding to a particular Cartesian coordinate system. Therefore, second order tensors can be written in terms of two vectors side-by-side called *dyads,* such as $\boldsymbol{e}_1\boldsymbol{e}_3$. Note that dyads are in general not commutative, i.e., $\boldsymbol{e}_1\boldsymbol{e}_3 \neq \boldsymbol{e}_3\boldsymbol{e}_1$. This dyadic form was proposed by Gibbs in the 1880s. A dyadic $\boldsymbol{D}$, which corresponds to a tensor of order two, may be represented as a finite sum of dyads

$$\boldsymbol{D}=\boldsymbol{a}_1\boldsymbol{b}_1+\boldsymbol{a}_2\boldsymbol{b}_2+\boldsymbol{a}_3\boldsymbol{b}_3=D_{ij}\boldsymbol{e}_i\boldsymbol{e}_j \tag{1.21}$$

which is, however, never unique. The specific form of $\boldsymbol{D}$ depends on the coordinate system used; for example, the physical component D_{ij} given in (1.21) refers to a particular Cartesian coordinate system. If each of the dyads is interchanged, the resulting dyadic is called the *conjugate dyadic* or *transpose* of $\boldsymbol{D}$:

$$\boldsymbol{D}^T=\boldsymbol{b}_1\boldsymbol{a}_1+\boldsymbol{b}_2\boldsymbol{a}_2+\boldsymbol{b}_3\boldsymbol{a}_3 \tag{1.22}$$

where the superscript T denotes transpose. A second-order tensor is said to be self-conjugate or symmetric if

$$\boldsymbol{D}=\boldsymbol{D}^T \tag{1.23}$$

and antisymmetric if

$$\boldsymbol{D} = -\boldsymbol{D}^T \tag{1.24}$$

Any second-order tensor can then be decomposed into symmetric and anti-symmetric tensors:

$$\boldsymbol{D} = \frac{1}{2}(\boldsymbol{D} + \boldsymbol{D}^T) + \frac{1}{2}(\boldsymbol{D} - \boldsymbol{D}^T) = \boldsymbol{G} + \boldsymbol{H} \tag{1.25}$$

where $\boldsymbol{G}$ and $\boldsymbol{H}$ are the symmetric and antisymmetric parts of the tensor $\boldsymbol{D}$. Note that double-transpose will yield the original tensor, i.e.,

$$\boldsymbol{D} = (\boldsymbol{D}^T)^T \tag{1.26}$$

The displacement gradient tensor $(\nabla \boldsymbol{u})^T$ is an example of decomposition that we often encounter in continuum mechanics. In particular, $(\nabla \boldsymbol{u})^T$ can always be decomposed into the strain tensor $\boldsymbol{\varepsilon}$ and the spin tensor $\boldsymbol{\Omega}$:

$$(\nabla \boldsymbol{u})^T = \frac{1}{2}[(\nabla \boldsymbol{u})^T + \nabla \boldsymbol{u}] + \frac{1}{2}[(\nabla \boldsymbol{u})^T - \nabla \boldsymbol{u}] = \boldsymbol{\varepsilon} + \boldsymbol{\Omega} \tag{1.27}$$

The dot product of a vector $\boldsymbol{v}$ and a second-order tensor $\boldsymbol{D}$ results in another vector defined by:

$$\boldsymbol{v} \bullet \boldsymbol{D} = (\boldsymbol{v} \bullet \boldsymbol{a}_1)\boldsymbol{b}_1 + (\boldsymbol{v} \bullet \boldsymbol{a}_2)\boldsymbol{b}_2 + (\boldsymbol{v} \bullet \boldsymbol{a}_3)\boldsymbol{b}_3 = \boldsymbol{u} \tag{1.28}$$

$$\boldsymbol{D} \bullet \boldsymbol{v} = \boldsymbol{a}_1(\boldsymbol{b}_1 \bullet \boldsymbol{v}) + \boldsymbol{a}_2(\boldsymbol{b}_2 \bullet \boldsymbol{v}) + \boldsymbol{a}_3(\boldsymbol{b}_3 \bullet \boldsymbol{v}) = \boldsymbol{w} \tag{1.29}$$

It is important to note that in general the order of dot product cannot be reversed. The role of $\boldsymbol{D}$ in (1.28) and (1.29) can be treated as a mapping or transformation between two vector functions (compared to ordinary functions between scalar quantities). This also provides a mathematical reason for the introduction of second-order tensors. Similarly, the dot product between two second-order tensors can also be defined:

$$\begin{aligned} \boldsymbol{D} \bullet \boldsymbol{E} &= (\boldsymbol{a}_1\boldsymbol{b}_1 + \boldsymbol{a}_2\boldsymbol{b}_2 + \boldsymbol{a}_3\boldsymbol{b}_3) \bullet (\boldsymbol{c}_1\boldsymbol{d}_1 + \boldsymbol{c}_2\boldsymbol{d}_2 + \boldsymbol{c}_3\boldsymbol{d}_3) \\ &= (\boldsymbol{b}_1 \bullet \boldsymbol{c}_1)\boldsymbol{a}_1\boldsymbol{d}_1 + (\boldsymbol{b}_2 \bullet \boldsymbol{c}_2)\boldsymbol{a}_2\boldsymbol{d}_2 + (\boldsymbol{b}_3 \bullet \boldsymbol{c}_3)\boldsymbol{a}_3\boldsymbol{d}_3 \end{aligned} \tag{1.30}$$

As expected, the result $\boldsymbol{G}$ is also a second-order tensor. Cross products $\boldsymbol{v} \times \boldsymbol{D}$ and $\boldsymbol{D} \times \boldsymbol{v}$ can be defined analogous to the cross product between vectors:

$$\boldsymbol{v} \times \boldsymbol{D} = (\boldsymbol{v} \times \boldsymbol{a}_1)\boldsymbol{b}_1 + (\boldsymbol{v} \times \boldsymbol{a}_2)\boldsymbol{b}_2 + (\boldsymbol{v} \times \boldsymbol{a}_3)\boldsymbol{b}_3 = \boldsymbol{F} \tag{1.31}$$

$$\boldsymbol{D} \times \boldsymbol{v} = \boldsymbol{a}_1(\boldsymbol{b}_1 \times \boldsymbol{v}) + \boldsymbol{a}_2(\boldsymbol{b}_2 \times \boldsymbol{v}) + \boldsymbol{a}_3(\boldsymbol{b}_3 \times \boldsymbol{v}) = \boldsymbol{G} \tag{1.32}$$

The results of these cross products $\boldsymbol{F}$ and $\boldsymbol{G}$ are again second-order tensors.

Another important operation between tensors is the *double dot product* or *scalar product of two tensors*, which is further subdivided into two types. In terms of a particular Cartesian coordinate system, we have

$$\begin{aligned} \boldsymbol{D} : \boldsymbol{E} &= (D_{ij}\boldsymbol{e}_i\boldsymbol{e}_j) : (E_{kl}\boldsymbol{e}_k\boldsymbol{e}_l) \\ &= D_{ij}E_{kl}(\boldsymbol{e}_i \bullet \boldsymbol{e}_k)(\boldsymbol{e}_j \bullet \boldsymbol{e}_l) = D_{ij}E_{ij} = \lambda_1 \end{aligned} \tag{1.33}$$

$$\begin{aligned} \boldsymbol{D} \bullet\bullet \boldsymbol{E} &= (D_{ij}\boldsymbol{e}_i\boldsymbol{e}_j) \bullet\bullet (E_{kl}\boldsymbol{e}_k\boldsymbol{e}_l) \\ &= D_{ij}E_{kl}(\boldsymbol{e}_i \bullet \boldsymbol{e}_l)(\boldsymbol{e}_j \bullet \boldsymbol{e}_k) = D_{ij}E_{ij} = \lambda_2 \end{aligned} \tag{1.34}$$

where λ_1 and λ_2 are scalars. For example, strain energy W in an elastic body can be written in terms of the scalar product of stress and strain tensors:

$$W = \frac{1}{2}\int_V \boldsymbol{\sigma} : \boldsymbol{\varepsilon}\, dV = \frac{1}{2}\int_V \sigma_{ij}\varepsilon_{ij}\, dV \tag{1.35}$$

where V is the volume of the solid.

1.5 DERIVATIVES OF TENSORS

In general, once a physical quantity, such as stress, is assigned to a tensor, we may write, say $\boldsymbol{\sigma} = \boldsymbol{\sigma}(\boldsymbol{x}, t)$ for a stress tensor, where $\boldsymbol{x}$ is the position vector of the point of consideration and t is the time of interest. This implies that a tensor may vary from point to point and thus represents a *tensor field*. In symbolic notation, the vector differential operator is expressed as $\boldsymbol{\nabla}$, or as a Cartesian tensor

$$\boldsymbol{\nabla} = \boldsymbol{e}_i \frac{\partial}{\partial x_i} \tag{1.36}$$

Frequently, the partial differentiation with respect to the variable x_i is represented by the *comma-subscript convention*, such as

$$\frac{\partial v_i}{\partial x_j} = v_{i,j}, \quad \frac{\partial \sigma_{ij}}{\partial x_k} = \sigma_{ij,k}, \quad \frac{\partial \varepsilon_{ij}}{\partial x_j} = \varepsilon_{ij,j} \tag{1.37}$$

It is obvious that the partial differentiation may raise the tensor order by one if one more free index is added [see the first two examples in (1.37)], but it may also decrease the tensor order by one if the partial differentiation makes j a repeated index [see the third example in (1.37)]. The following differential operators, which appear often in continuum mechanics, are given here for reference:

$$\text{grad}\,\varphi = \boldsymbol{\nabla}\varphi = \frac{\partial \varphi}{\partial x_i}\boldsymbol{e}_i = \varphi_{,i}\boldsymbol{e}_i \tag{1.38}$$

$$\text{div}\,\boldsymbol{v} = \boldsymbol{\nabla} \bullet \boldsymbol{v} = v_{i,i} \tag{1.39}$$

$$\text{curl}\,\boldsymbol{v} = \boldsymbol{\nabla} \times \boldsymbol{v} = e_{ijk} v_{k,j} \boldsymbol{e}_i \tag{1.40}$$

$$\nabla^2 \varphi = \boldsymbol{\nabla} \bullet \boldsymbol{\nabla}\varphi = \varphi_{,ii} \tag{1.41}$$

Note that all the indicial forms are for physical components in the Cartesian coordinate system only. A number of identities exist for the differential operator $\boldsymbol{\nabla}$ in the Cartesian coordinate system:

$$\boldsymbol{\nabla}(fg) = f\boldsymbol{\nabla}g + g\boldsymbol{\nabla}f \tag{1.42}$$

$$\nabla^2(fg) = f\nabla^2 g + 2(\boldsymbol{\nabla}f)\bullet(\boldsymbol{\nabla}g) + g\nabla^2 f \tag{1.43}$$

$$\boldsymbol{\nabla}\bullet(f\boldsymbol{v}) = (\boldsymbol{\nabla}f)\bullet\boldsymbol{v} + f\boldsymbol{\nabla}\bullet\boldsymbol{v} \tag{1.44}$$

$$\boldsymbol{\nabla}\bullet(f\boldsymbol{\nabla}g) = f\nabla^2 g + \boldsymbol{\nabla}f\bullet\boldsymbol{\nabla}g \tag{1.45}$$

$$\boldsymbol{\nabla}\times(\boldsymbol{\nabla}f) = 0 \tag{1.46}$$

$$\boldsymbol{\nabla}\bullet(\boldsymbol{\nabla}\times\boldsymbol{v}) = 0 \tag{1.47}$$

$$\boldsymbol{\nabla}\bullet(\boldsymbol{a}\times\boldsymbol{b}) = (\boldsymbol{\nabla}\times\boldsymbol{a})\bullet\boldsymbol{b} - \boldsymbol{a}\bullet(\boldsymbol{\nabla}\times\boldsymbol{b}) \tag{1.48}$$

$$\boldsymbol{\nabla}\times(f\boldsymbol{v}) = \boldsymbol{\nabla}f\times\boldsymbol{v} + f\boldsymbol{\nabla}\times\boldsymbol{v} \tag{1.49}$$

$$\boldsymbol{\nabla}\times(\boldsymbol{\nabla}\times\boldsymbol{v}) = \boldsymbol{\nabla}(\boldsymbol{\nabla}\bullet\boldsymbol{v}) - \nabla^2\boldsymbol{v} \tag{1.50}$$

$$\boldsymbol{\nabla}\bullet\boldsymbol{\nabla}\phi = \nabla^2\phi \tag{1.51}$$

$$\boldsymbol{\nabla}\bullet\nabla^2\boldsymbol{a} = \nabla^2(\boldsymbol{\nabla}\bullet\boldsymbol{a}) \tag{1.52}$$

$$\nabla^2(\boldsymbol{\nabla}\phi) = \boldsymbol{\nabla}(\nabla^2\phi) \tag{1.53}$$

$$\nabla^2(\boldsymbol{a}\bullet\boldsymbol{r}) = 2\nabla\bullet\boldsymbol{a} + \boldsymbol{r}\nabla^2\boldsymbol{a} \tag{1.54}$$

$$\nabla^2(\phi\boldsymbol{r}) = 2\nabla\phi + \boldsymbol{r}\nabla^2\phi \tag{1.55}$$

$$\nabla(\boldsymbol{a}\bullet\boldsymbol{r}) = \boldsymbol{a} + (\nabla\boldsymbol{a})\bullet\boldsymbol{r} \tag{1.56}$$

where $\boldsymbol{r}$ is a position vector. The divergence of a second-order tensor, say $\boldsymbol{\sigma}$, follows closely the definition for a vector field, and the result of such an operation will result in a vector field:

$$\nabla\bullet\sigma = (\boldsymbol{e}_k \frac{\partial}{\partial x_k})\bullet(\sigma_{ij}\boldsymbol{e}_i\boldsymbol{e}_j) = \sigma_{ij,i}\boldsymbol{e}_j \tag{1.57}$$

In cylindrical or spherical coordinate systems, the physical components will be more complicated since the base vectors change directions with the coordinate variables; they will be discussed a bit more in later sections. Analogously, the curls of a second-order tensor can be obtained by operating with the base vectors:

$$\nabla\times\sigma = (\boldsymbol{e}_i \frac{\partial}{\partial x_i})\times(\sigma_{jk}\boldsymbol{e}_j\boldsymbol{e}_k) = \sigma_{jk,i}(\boldsymbol{e}_i\times\boldsymbol{e}_j)\boldsymbol{e}_k = e_{ilj}\sigma_{jk,l}\boldsymbol{e}_i\boldsymbol{e}_k \tag{1.58}$$

So far, we have assumed all coordinates are orthogonal, i.e., the base vectors are perpendicular. For curvilinear coordinates, the vector differentiation is much more complicated; such an operation will involve the use of *Christoffel symbols of the first and second kinds* (Malvern, 1969). We will not discuss such a complication here.

1.6 DIVERGENCE AND STOKES THEOREMS

The formulation of problems in continuum mechanics always makes use of the *divergence theorem* of Gauss and *Stokes theorem*. The coordinate-invariant forms of these theorems in terms of tensors are given in this section without a proof. We emphasize that the derivations of these theorems do not put any restriction on the material response of the solids, thus these results apply regardless to any solid, elastic, plastic, or brittle, as long as the solid can be considered as a continuous medium and the variation of the tensor field is smooth.

1.6.1 Divergence Theorem or Gauss Theorem

Let V be the volume of a solid bounded by a smooth surface S and let $\boldsymbol{T}$ be a second-order tensor field (see Fig. 1.3). The divergence theorem is

$$\oint_S \boldsymbol{n}\bullet\boldsymbol{T}dS = \int_V \nabla\bullet\boldsymbol{T}dV \tag{1.59}$$

The divergence theorem relates volume integral to surface integral. This equation also applies to $\boldsymbol{T}$ being an n-th order tensor (see Segel, 1987). The proof for the ordinary divergence theorem can be found in standard textbooks for engineering mathematics (Kreyszig, 1996), and its extension for tensors can be found in Mal and Singh (1991) and Segel (1987).

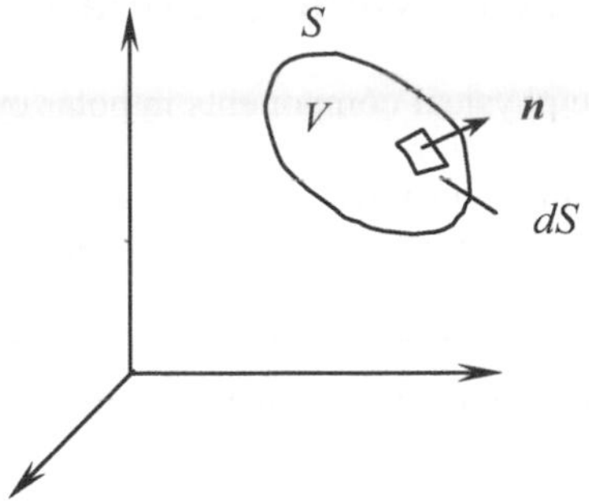

Figure 1.3 The domain for the divergence theorem

1.6.2 Stokes Theorem

Let S be a smooth surface bounded by a simple closed curve C, which does not intersect itself, and $\boldsymbol{T}$ be a tensor of arbitrary order. Then, the Stokes theorem states that

$$\int_C \boldsymbol{T} \bullet d\boldsymbol{s} = \int_S (\nabla \times \boldsymbol{T}) \bullet \boldsymbol{n} \; dS \tag{1.60}$$

where $d\boldsymbol{s}$ is an oriented element on C and $\boldsymbol{n}$ is the unit normal to S at dS along its positive curvature (see Fig. 1.4). Stokes theorem relates surface integral to surface integral.

There is an interesting story on the origin of the Stokes theorem, and a brief history is given in the biography of G.G. Stokes at the end of this book. In short, Lord Kelvin played a fundamental role in its development, so it is also known as the Kelvin–Stokes theorem.

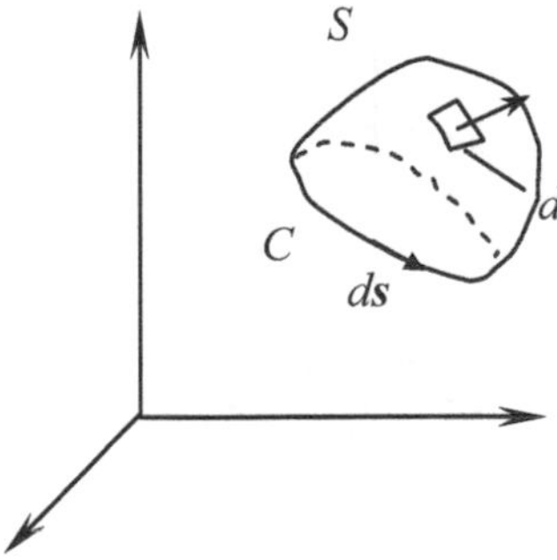

Figure 1.4 The domain for the Stokes theorem

1.7 SOME FORMULAS IN CYLINDRICAL COORDINATES

As mentioned previously, the physical components in polar cylindrical coordinates are more complicated. In this section, we provide some useful formulas, which appear regularly in geomechanical problems. Any position vector $\boldsymbol{r}$ in a Cartesian coordinate system can be written in terms of a cylindrical coordinate system (r, ϕ, z) as shown in Fig. 1.5:

$$\boldsymbol{r} = x_1\boldsymbol{e}_1 + x_2\boldsymbol{e}_2 + x_3\boldsymbol{e}_3 = r\cos\phi\,\boldsymbol{e}_1 + r\sin\phi\,\boldsymbol{e}_2 + z\boldsymbol{e}_3 \tag{1.61}$$

The new set of base vectors in cylindrical coordinates is defined by

$$\boldsymbol{e}_\alpha = \frac{1}{h_\alpha}\frac{\partial \boldsymbol{r}}{\partial x_\alpha} \tag{1.62}$$

where

$$h_\alpha = \left|\frac{\partial \boldsymbol{r}}{\partial \boldsymbol{x}_\alpha}\right| \tag{1.63}$$

and x_α equals r, ϕ, or z. In particular, we have

$$\begin{aligned} &\boldsymbol{e}_r = \frac{1}{h_r}\frac{\partial \boldsymbol{r}}{\partial r} = \cos\phi\,\boldsymbol{e}_1 + \sin\phi\,\boldsymbol{e}_2, \quad \boldsymbol{e}_\phi = \frac{1}{h_\phi}\frac{\partial \boldsymbol{r}}{\partial \phi} = -\sin\phi\,\boldsymbol{e}_1 + \cos\phi\,\boldsymbol{e}_2 \\ &\boldsymbol{e}_z = \frac{1}{h_z}\frac{\partial \mathbf{r}}{\partial z} = \boldsymbol{e}_3 \end{aligned} \tag{1.64}$$

Unlike the base vectors for the Cartesian coordinate system, these base vectors are not all constant with the change in coordinates; it can be shown that

$$\frac{\partial \boldsymbol{e}_r}{\partial \phi} = \boldsymbol{e}_\phi, \quad \frac{\partial \boldsymbol{e}_\phi}{\partial \phi} = -\boldsymbol{e}_r \tag{1.65}$$

while all other derivatives of the base vectors vanish. For example, the displacement gradient tensor can be formulated in dyadic notation as

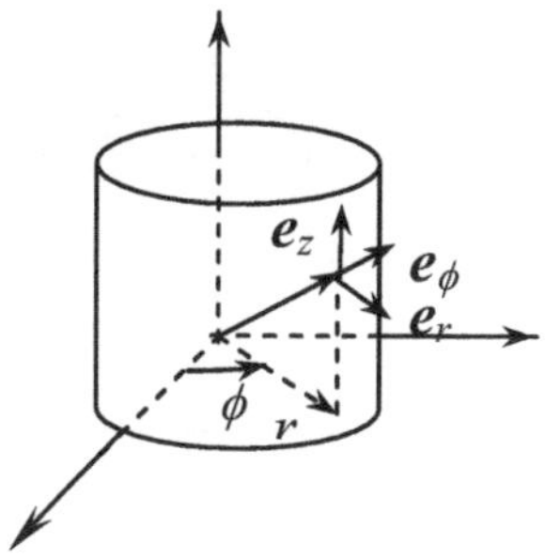

Figure 1.5 Cylindrical coordinates

$$\nabla \boldsymbol{u} = (\boldsymbol{e}_r \frac{\partial}{\partial r} + \boldsymbol{e}_\phi \frac{\partial}{r\partial\phi} + \boldsymbol{e}_z \frac{\partial}{\partial z})(u_r \boldsymbol{e}_r + u_\phi \boldsymbol{e}_\phi + u_z \boldsymbol{e}_z)$$

$$= \frac{\partial u_r}{\partial r}\boldsymbol{e}_r\boldsymbol{e}_r + \frac{1}{r}(u_r + \frac{\partial u_\phi}{\partial \phi})\boldsymbol{e}_\phi\boldsymbol{e}_\phi + \frac{\partial u_z}{\partial z}\boldsymbol{e}_z\boldsymbol{e}_z + \frac{1}{r}\frac{\partial u_z}{\partial \phi}\boldsymbol{e}_\phi\boldsymbol{e}_z + \frac{\partial u_\phi}{\partial z}\boldsymbol{e}_z\boldsymbol{e}_\phi \quad (1.66)$$

$$+ \frac{\partial u_r}{\partial z}\boldsymbol{e}_z\boldsymbol{e}_r + \frac{\partial u_z}{\partial r}\boldsymbol{e}_r\boldsymbol{e}_z + \frac{\partial u_\phi}{\partial r}\boldsymbol{e}_r\boldsymbol{e}_\phi + \frac{1}{r}(\frac{\partial u_r}{\partial \phi} - u_\phi)\boldsymbol{e}_\phi\boldsymbol{e}_r$$

In obtaining the above equation, we have already used the coordinate variation of the base vectors given in (1.65). This result can readily be used to obtain the strain tensor, which is defined as

$$\boldsymbol{\varepsilon} = \frac{1}{2}(\nabla \boldsymbol{u} + \boldsymbol{u}\nabla) \quad (1.67)$$

Combining (1.66) and (1.67), it is straightforward to see that the physical components are

$$\varepsilon_{zz} = \frac{\partial u_z}{\partial z}, \quad \varepsilon_{rr} = \frac{\partial u_r}{\partial r}, \quad \varepsilon_{\phi\phi} = \frac{u_r}{r} + \frac{1}{r}\frac{\partial u_\phi}{\partial \phi} \quad (1.68)$$

$$\varepsilon_{r\phi} = \frac{1}{2}(\frac{\partial u_\phi}{\partial r} + \frac{1}{r}\frac{\partial u_r}{\partial \phi} - \frac{u_\phi}{r}), \quad \varepsilon_{\phi z} = \frac{1}{2}(\frac{1}{r}\frac{\partial u_z}{\partial \phi} + \frac{\partial u_\phi}{\partial z}), \quad \varepsilon_{rz} = \frac{1}{2}(\frac{\partial u_r}{\partial z} + \frac{\partial u_z}{\partial r}) \quad (1.69)$$

This equations are the same as those obtained by Timoshenko and Goodier (1982), starting from the kinematics of compatibility in deformations. Thus, the tensor equation (1.67) provides a concise and elegant form for the strain-displacement relation, and more importantly it is independent of any coordinate system.

Following similar technique in specializing the above tensor equations, the following results can be obtained:

$$\nabla \bullet \boldsymbol{u} = \frac{\partial u_r}{\partial r} + \frac{1}{r}u_r + \frac{1}{r}\frac{\partial u_\phi}{\partial \phi} + \frac{\partial u_z}{\partial z} \quad (1.70)$$

$$\nabla \times \boldsymbol{u} = \boldsymbol{e}_r(\frac{1}{r}\frac{\partial u_z}{\partial \phi} - \frac{\partial u_\phi}{\partial z}) + \boldsymbol{e}_\phi(\frac{\partial u_r}{\partial z} - \frac{\partial u_z}{\partial r}) + \boldsymbol{e}_z(\frac{\partial u_\phi}{\partial r} + \frac{u_\phi}{r} - \frac{1}{r}\frac{\partial u_r}{\partial \phi}) \quad (1.71)$$

$$\nabla^2 f = \nabla \bullet \nabla f = \frac{\partial^2 f}{\partial r^2} + \frac{1}{r}\frac{\partial f}{\partial r} + \frac{1}{r^2}\frac{\partial^2 f}{\partial \phi^2} + \frac{\partial^2 f}{\partial z^2} \quad (1.72)$$

The three components of the equilibrium equations, $\nabla \cdot \boldsymbol{\sigma} = 0$, can be written explicitly as

$$\frac{\partial \sigma_{rr}}{\partial r} + \frac{\partial \sigma_{zr}}{\partial z} + \frac{1}{r}\frac{\partial \sigma_{r\phi}}{\partial \phi} + \frac{\sigma_{rr} - \sigma_{\phi\phi}}{r} = 0 \quad (1.73)$$

$$\frac{1}{r}\frac{\partial}{\partial r}(r\sigma_{rz}) + \frac{\partial \sigma_{zz}}{\partial r} + \frac{1}{r}\frac{\partial \sigma_{z\phi}}{\partial \phi} = 0 \quad (1.74)$$

$$\frac{1}{r}\frac{\partial \sigma_{\phi\phi}}{\partial \phi} + \frac{\partial \sigma_{r\phi}}{\partial r} + \frac{\partial \sigma_{z\phi}}{\partial z} + 2\frac{\sigma_{r\phi}}{r} = 0 \quad (1.75)$$

These equations are the same as those obtained by Timoshenko and Goodier (1982) by considering force equilibriums along r-, ϕ-, and z-directions for an infinitesimal small element. The proof of these equations is given as problems at the end of this chapter.

1.8 SOME FORMULAS IN SPHERICAL COORDINATES

The development of this section follows closely the discussion in the previous section. Any position vector $\boldsymbol{r}$ in a Cartesian coordinate system can be written in terms of a polar spherical coordinate system (r, φ, θ) as shown in Fig. 1.6:

$$\boldsymbol{r} = x_1\boldsymbol{e}_1 + x_2\boldsymbol{e}_2 + x_3\boldsymbol{e}_3 = r\sin\theta\cos\varphi\,\boldsymbol{e}_1 + r\sin\theta\sin\varphi\,\boldsymbol{e}_2 + r\cos\theta\,\boldsymbol{e}_3 \tag{1.76}$$

The base vectors can again be obtained using (1.62) and (1.63) as

$$\begin{aligned} \boldsymbol{e}_r &= (\sin\theta\cos\varphi)\boldsymbol{e}_1 + (\sin\theta\sin\varphi)\boldsymbol{e}_2 + \cos\theta\boldsymbol{e}_3 \\ \boldsymbol{e}_\theta &= (\cos\theta\cos\varphi)\boldsymbol{e}_1 + (\cos\theta\cos\varphi)\boldsymbol{e}_2 - \sin\theta\boldsymbol{e}_3 \\ \boldsymbol{e}_\varphi &= -\sin\varphi\boldsymbol{e}_1 + \cos\varphi\boldsymbol{e}_2 \end{aligned} \tag{1.77}$$

The variation of base vectors along coordinate directions is more complicated than that for cylindrical coordinates; in particular, the following nonzero terms are obtained:

$$\begin{aligned} &\frac{\partial \boldsymbol{e}_r}{\partial \theta} = \boldsymbol{e}_\theta\,, \quad \frac{\partial \boldsymbol{e}_\theta}{\partial \theta} = -\boldsymbol{e}_r\,, \quad \frac{\partial \boldsymbol{e}_\theta}{\partial \varphi} = \cos\theta\boldsymbol{e}_\varphi\,, \\ &\frac{\partial \boldsymbol{e}_\varphi}{\partial \varphi} = -\sin\theta\boldsymbol{e}_r - \cos\theta\boldsymbol{e}_\theta, \quad \frac{\partial \boldsymbol{e}_r}{\partial \varphi} = \sin\theta\boldsymbol{e}_\varphi \end{aligned} \tag{1.78}$$

while all other derivatives of the base vectors vanish. The differential operator in spherical polar coordinates is

$$\nabla = \left(\boldsymbol{e}_r \frac{\partial}{\partial r} + \boldsymbol{e}_\theta \frac{1}{r}\frac{\partial}{\partial \theta} + \boldsymbol{e}_\varphi \frac{1}{r\sin\theta}\frac{\partial}{\partial \varphi} \right) \tag{1.79}$$

Without showing the details here, we quote the following physical components for the strain-displacement by specializing (1.67) to the spherical coordinates:

$$\varepsilon_{\varphi\varphi} = \frac{1}{r\sin\theta}\frac{\partial u_\varphi}{\partial \varphi} + \frac{u_r}{r} + \frac{u_\theta}{r}\cot\theta, \quad \varepsilon_{rr} = \frac{\partial u_r}{\partial r}, \quad \varepsilon_{\theta\theta} = \frac{u_r}{r} + \frac{1}{r}\frac{\partial u_\theta}{\partial \theta}, \tag{1.80}$$

$$\varepsilon_{r\theta} = \frac{1}{2}\left(\frac{\partial u_\theta}{\partial r} + \frac{1}{r}\frac{\partial u_r}{\partial \theta} - \frac{u_\theta}{r}\right), \quad \varepsilon_{r\varphi} = \frac{1}{2}\left(\frac{1}{r\sin\theta}\frac{\partial u_r}{\partial \varphi} + \frac{\partial u_\varphi}{\partial r} - \frac{u_\varphi}{r}\right), \tag{1.81}$$

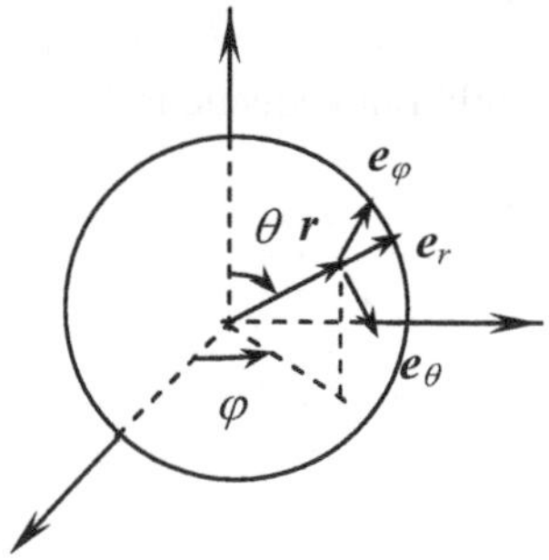

Figure 1.6 Spherical coordinates

$$\varepsilon_{\theta\varphi} = \frac{1}{2}\left(\frac{1}{r\sin\theta}\frac{\partial u_\theta}{\partial\varphi} + \frac{1}{r}\frac{\partial u_\varphi}{\partial\theta} - \frac{u_\varphi}{r}\cot\theta\right) \tag{1.82}$$

Again these results are the same as those obtained by considering the kinematics of compatibility in deformations.

Following the similar technique in specialization of tensor equations, the following results can be obtained:

$$\nabla\cdot\boldsymbol{u} = \frac{\partial u_r}{\partial r} + \frac{2}{r}u_r + \frac{1}{r}\left(\frac{\partial u_\theta}{\partial\theta} + \cot\theta u_\theta\right) + \frac{1}{r\sin\theta}\frac{\partial u_\varphi}{\partial\varphi} \tag{1.83}$$

$$\begin{aligned}\nabla\times\boldsymbol{u} = {} & \mathbf{e}_r\frac{1}{r}\left(\frac{\partial u_\varphi}{\partial\theta} + \cot\theta u_\varphi - \frac{1}{\sin\theta}\frac{\partial u_\theta}{\partial\varphi}\right) + \mathbf{e}_\theta\left(\frac{1}{r\sin\theta}\frac{\partial u_r}{\partial\varphi} - \frac{\partial u_\varphi}{\partial r} - \frac{u_\varphi}{r}\right) \\ & + \mathbf{e}_\varphi\left(\frac{\partial u_\theta}{\partial r} + \frac{u_\theta}{r} - \frac{1}{r}\frac{\partial u_r}{\partial\theta}\right)\end{aligned} \tag{1.84}$$

$$\nabla^2 f = \frac{1}{r^2}\frac{\partial}{\partial r}\left(r^2\frac{\partial f}{\partial r}\right) + \frac{1}{r^2\sin\theta}\frac{\partial}{\partial\theta}\left(\sin\theta\frac{\partial f}{\partial\theta}\right) + \frac{1}{r^2\sin^2\theta}\frac{\partial^2 f}{\partial\varphi^2} \tag{1.85}$$

In spherical coordinates, the three components of the equilibrium equations, $\nabla\cdot\boldsymbol{\sigma} = 0$, can be written explicitly as

$$\frac{\partial\sigma_{rr}}{\partial r} + \frac{1}{r}\frac{\partial\sigma_{r\theta}}{\partial\theta} + \frac{1}{r\sin\theta}\frac{\partial\sigma_{r\varphi}}{\partial\varphi} + \frac{1}{r}(2\sigma_{rr} - \sigma_{\theta\theta} - \sigma_{\varphi\varphi} + \sigma_{r\theta}\cot\theta) = 0 \tag{1.86}$$

$$\frac{\partial\sigma_{r\theta}}{\partial r} + \frac{1}{r}\frac{\partial\sigma_{\theta\theta}}{\partial\theta} + \frac{1}{r\sin\theta}\frac{\partial\sigma_{\theta\psi}}{\partial\varphi} + \frac{1}{r}[3\sigma_{r\theta} + (\sigma_{\theta\theta} - \sigma_{\varphi\varphi})\cot\theta] = 0 \tag{1.87}$$

$$\frac{\partial\sigma_{r\psi}}{\partial r} + \frac{1}{r}\frac{\partial\sigma_{\theta\psi}}{\partial\theta} + \frac{1}{r\sin\theta}\frac{\partial\sigma_{\psi\psi}}{\partial\varphi} + \frac{1}{r}(3\sigma_{r\varphi} + 2\sigma_{\theta\varphi}\cot\theta) = 0 \tag{1.88}$$

The proof of these equations is again given as problems at the end of this chapter.

1.9 SUMMARY AND FURTHER READING

The most comprehensive coverage of tensor analysis applying to solid mechanics is given by Malvern (1969). We also highly recommend the books by Chou and Pagano (1967) and Segel (1987). For a quick reference, Hughes and Gaylord (1964) summarized many useful formulas in tensor as well as component form. A good vector analysis is given in Chapter 1 of Wong (1991).

1.10 PROBLEMS

Problem 1.1 Prove the e-δ identity given in Eq. (1.18).

Problem 1.2 Show the following results: (a) $\delta_{ii} = 3$, (b) $\delta_{ij}\delta_{ij} = 3$, (c) $e_{ijk}e_{jki} = 6$, (d) $e_{ijk}A_jA_k = 0$, and (e) $e_{ijk}e_{pjk} = 2\delta_{ip}$.

Problem 1.3 Show that $(P_{ijk} + P_{jki} + P_{jik})x_ix_jx_k = 3P_{ijk}x_ix_jx_k$.

Problem 1.4 Use indicial notation to prove the following vector identities:

$$\boldsymbol{u}\times(\boldsymbol{v}\times\boldsymbol{w})=(\boldsymbol{u}\bullet\boldsymbol{w})\boldsymbol{v}-(\boldsymbol{u}\bullet\boldsymbol{v})\boldsymbol{w}\,,\quad (\boldsymbol{u}\times\boldsymbol{v})\bullet\boldsymbol{w}=0 \tag{1.89}$$

Problem 1.5 Prove Eq. (1.73) through (1.75), starting from specializing $\nabla\cdot\boldsymbol{\sigma}$ to the cylindrical coordinate system.

Problem 1.6 Prove Eq. (1.83) through (1.85) by specializing the appropriate tensor equation.

Problem 1.7 Prove Eq. (1.86) through (1.88) by specializing the appropriate tensor equation.

Problem 1.8 Show the validity of (1.46) and (1.47).

Problem 1.9 Prove (1.50) using Cartesian coordinates.

Problem 1.10 Using (1.65) to show the validity of the following equations:

$$\boldsymbol{e}_\phi\bullet\frac{1}{r}\frac{\partial}{\partial\phi}(\frac{\partial u_r}{\partial r}\boldsymbol{e}_r\boldsymbol{e}_r)=\frac{1}{r}\frac{\partial u_r}{\partial r}\boldsymbol{e}_r \tag{1.90}$$

$$\boldsymbol{e}_\phi\bullet\{\frac{1}{r}\frac{\partial}{\partial\phi}[\frac{1}{r}(u_r+\frac{\partial u_\phi}{\partial\phi})\boldsymbol{e}_\phi\boldsymbol{e}_\phi]\}=\frac{1}{r^2}(\frac{\partial u_r}{\partial\phi}+\frac{\partial^2 u_\phi}{\partial\phi^2})\boldsymbol{e}_\phi-\frac{1}{r^2}(u_r+\frac{\partial u_\phi}{\partial\phi})\boldsymbol{e}_r \tag{1.91}$$

$$\boldsymbol{e}_\phi\bullet\frac{1}{r}\frac{\partial}{\partial\phi}(\frac{\partial u_z}{\partial r}\boldsymbol{e}_r\boldsymbol{e}_z)=\frac{1}{r}\frac{\partial u_z}{\partial r}\boldsymbol{e}_z \tag{1.92}$$

$$\boldsymbol{e}_\phi\bullet\frac{1}{r}\frac{\partial}{\partial\phi}(\frac{\partial u_\phi}{\partial r}\boldsymbol{e}_r\boldsymbol{e}_\phi)=\frac{1}{r}\frac{\partial u_\phi}{\partial r}\boldsymbol{e}_\phi \tag{1.93}$$

$$\boldsymbol{e}_\phi\bullet\{\frac{1}{r}\frac{\partial}{\partial\phi}(\frac{1}{r}\frac{\partial u_z}{\partial\phi}\boldsymbol{e}_\phi\boldsymbol{e}_z)\}=\frac{1}{r^2}\frac{\partial^2 u_z}{\partial\phi^2}\boldsymbol{e}_z \tag{1.94}$$

$$\boldsymbol{e}_\phi\bullet\{\frac{1}{r}\frac{\partial}{\partial\phi}[\frac{1}{r}(\frac{\partial u_r}{\partial\phi}-u_\phi)]\boldsymbol{e}_\phi\boldsymbol{e}_r]\}=\frac{1}{r^2}(\frac{\partial^2 u_r}{\partial\phi^2}-\frac{\partial u_\phi}{\partial\phi})\boldsymbol{e}_r+\frac{1}{r^2}(\frac{\partial u_r}{\partial\phi}-u_\phi)\boldsymbol{e}_\phi \tag{1.95}$$

Problem 1.11 Use (1.66) and the results in Problem 1.10 to show the validity of the following equation:

$$\begin{aligned}\boldsymbol{e}_\phi(\frac{1}{r}\frac{\partial}{\partial\phi})\bullet(\nabla\boldsymbol{u}) &= \{\frac{1}{r}\frac{\partial u_r}{\partial r}-\frac{u_r}{r^2}+\frac{1}{r^2}\frac{\partial^2 u_r}{\partial\phi^2}-\frac{2}{r^2}\frac{\partial u_\phi}{\partial\phi}\}\boldsymbol{e}_r\\ &+\{\frac{2}{r}\frac{\partial u_r}{\partial\phi}+\frac{1}{r^2}\frac{\partial^2 u_\phi}{\partial\phi^2}+\frac{1}{r}\frac{\partial u_\phi}{\partial r}-\frac{u_\phi}{r^2}\}\boldsymbol{e}_\phi+\{\frac{1}{r}\frac{\partial u_z}{\partial r}+\frac{1}{r^2}\frac{\partial^2 u_z}{\partial\phi^2}\}\boldsymbol{e}_z\end{aligned} \tag{1.96}$$

Problem 1.12 Use the result of Problem 1.11 to show the validity of the following equation:

$$\nabla \cdot (\nabla \boldsymbol{u}) = \{\nabla^2 u_r - \frac{u_r}{r^2} - \frac{2}{r^2}\frac{\partial u_\phi}{\partial \phi}\}\boldsymbol{e}_r + \{\nabla^2 u_\phi - \frac{u_\phi}{r^2} + \frac{2}{r^2}\frac{\partial u_r}{\partial \phi}\}\boldsymbol{e}_\phi + \nabla^2 u_z \boldsymbol{e}_z \quad (1.97)$$

where the Laplacian operator is given in (1.72).

Problem 1.13 Substitute the result of Problem 1.12 and (1.70) into the following equation:

$$\nabla \cdot (\nabla \boldsymbol{u}) + \frac{1}{1-2\nu}\nabla(\nabla \cdot \boldsymbol{u}) + \frac{1}{\mu}\boldsymbol{F} = 0 \quad (1.98)$$

to show that

$$\nabla^2 u_r + \frac{1}{1-2\nu}\frac{\partial e}{\partial r} - \frac{u_r}{r^2} - \frac{2}{r^2}\frac{\partial u_\phi}{\partial \phi} + \frac{1}{\mu}F_r = 0 \quad (1.99)$$

$$\nabla^2 u_\phi + \frac{1}{1-2\nu}\frac{1}{r}\frac{\partial e}{\partial \phi} - \frac{u_\phi}{r^2} + \frac{2}{r^2}\frac{\partial u_r}{\partial \phi} + \frac{1}{\mu}F_\phi = 0 \quad (1.100)$$

$$\nabla^2 u_z + \frac{1}{1-2\nu}\frac{\partial e}{\partial r} + \frac{1}{\mu}F_z = 0 \quad (1.101)$$

$$e = \nabla \cdot \boldsymbol{u} = \frac{\partial u_r}{\partial r} + \frac{1}{r}u_r + \frac{1}{r}\frac{\partial u_\phi}{\partial \phi} + \frac{\partial u_z}{\partial z} \quad (1.102)$$

Note that these are the three-dimensional equilibrium equations in cylindrical coordinates in terms of displacements.

Problem 1.14 Adopting the procedures used in Problems 1.10–1.13, prove the following three-dimensional equilibrium equations in spherical coordinates in terms of displacements:

$$\nabla^2 u_r + \frac{1}{1-2\nu}\frac{\partial e}{\partial r} - \frac{2u_r}{r^2} - \frac{2}{r^2}\frac{\partial u_\theta}{\partial \theta} - \frac{2u_\theta \cot\theta}{r^2} - \frac{2}{r^2 \sin\theta}\frac{\partial u_\phi}{\partial \phi} + \frac{1}{\mu}F_r = 0 \quad (1.103)$$

$$\nabla^2 u_\theta + \frac{1}{1-2\nu}\frac{1}{r}\frac{\partial e}{\partial \theta} + \frac{2}{r^2}\frac{\partial u_r}{\partial \theta} - \frac{u_\theta}{r^2 \sin^2\theta} - \frac{2\cot\theta}{r^2 \sin\theta}\frac{\partial u_\varphi}{\partial \varphi} + \frac{1}{\mu}F_\theta = 0 \quad (1.104)$$

$$\nabla^2 u_\varphi + (\frac{1}{1-2\nu})\frac{1}{r\sin\theta}\frac{\partial e}{\partial \varphi} - \frac{u_\varphi}{r^2 \sin^2\theta} + \frac{2}{r^2 \sin^2\theta}\frac{\partial u_r}{\partial \varphi} + \frac{2\cot\theta}{r^2 \sin\theta}\frac{\partial u_\theta}{\partial \varphi} + \frac{1}{\mu}F_\varphi = 0$$

(1.105)

$$e = \nabla \cdot \boldsymbol{u} = \frac{\partial u_r}{\partial r} + \frac{2}{r}u_r + \frac{1}{r}(\frac{\partial u_\theta}{\partial \theta} + \cot\theta u_\theta) + \frac{1}{r\sin\theta}\frac{\partial u_\varphi}{\partial \varphi} \quad (1.106)$$

CHAPTER TWO

Elasticity and Its Applications

2.1 INTRODUCTION

Elasticity is perhaps the most successful theory ever developed to model the mechanical response of solids. Many important fields in solid mechanics, such as fracture mechanics and the theory of dislocation, are developed from the firm basis of the mathematical theory of elasticity. Thus, the theory of elasticity itself is a pre-requisite for and provides a fundamental background to any graduate student who wants to study more advanced topics in geomechanics or in engineering in general.

The mathematical theory of elasticity had occupied the minds of great scientists since the time of Galileo in the seventeenth century. Despite its development in the last 360 years, research on elasticity remains active today; it is fair to say that many theoretical and practical problems in elasticity remain to be solved.

Elasticity, in general, can be regarded as a branch of science that deals with the mechanical deformations of solids that deform under applied loads, such as applied traction, displacements, and temperature gradients, and then are able to recover their *original shape* upon unloading. The scope of elasticity can be extremely wide, depending on the type of elastic solids, the types of loading, and the form of deformation that we are interested in. There are about 100 textbooks directly devoted to or closely related to the theory of elasticity. The solids can be modeled as one, two or three dimensional although most analytic solutions exist only for one- or two-dimensional problems. The loading can be time dependent such that inertia effect may play a key role; for example, wave propagation is a typical phenomenon due to dynamic loads. Wave propagations and soil dynamics will be discussed in Chapter 9. Compared to the dimensions of the original solid, the deformations and strains in the body can be either infinitesimally small or finite. The deformation of solids may be either proportional to the applied loads (linear elasticity) or nonproportional to the loads (nonlinear elasticity). When nonlinear constitutive behavior sets in or when large deformation and strain are allowed to occur, the uniqueness theorem fails. The solution for elasticity problems becomes the result of solving nonlinear differential equations, which, in general, cannot be solved analytically. Approximate techniques, such as perturbation, have been developed to solve such problems. Numerical methods for solving these problems may yield unreliable solutions if no special care is taken. At certain critical loading conditions a trivial solution may yield to a nontrivial solution (this is normally called *bifurcation* in mathematical terms). Bifurcation problems in geomechanics have also been considered quite extensively (e.g., Rudnicki and Rice, 1975; Chau and Rudnicki, 1990, Chau, 1992, 1993, 1994a, 1995a–c, 1998a, 1999b; Chau and Choi, 1998; Muhlhaus et al., 1996;

Vardoulakis, 1979, 1983; Sulem and Vardoulakis, 1990; Vardoulakis and Sulem, 1996; Bigoni, 2012). In addition, the constitutive response of solids may be loading direction dependent; that is, the solid is not isotropic or it is *anisotropic* in response. For example, specimens taken horizontally from anisotropic solids will deform differently from those taken vertically. To date, most of the analytical solutions exist only for isotropic solids. Due to recent development of composite materials for the aerospace industry and other research, much effort has focused on the mechanical behavior of anisotropic elastic solids (e.g., Chau, 1994b, 1998b).

In terms of the existence of strain energy, elasticity can further be divided into two groups: *hyperelasticity*, for solids having an *elastic potential or strain energy function*; and *hypoelasticity*, for solids attaining a linear relationship between strain rate and stress rate. Strictly speaking, hyperelasticity is more than elasticity, which simply requires the recovery of strain and deformation upon unloading; on the other hand, hypoelasticity is less than elastic since it does not even require proportionality between stress and strain.

This chapter covers only problems with small deformations (i.e., linear elasticity). Both isotropic and anisotropic solids will be discussed, but the focus will mainly be on isotropic solids. Some practical examples will be used to illustrate the power of elasticity. The application of elasticity to model dislocation will also be introduced.

2.2 BASIC CONCEPTS FOR STRESS TENSORS

As we learned from elementary textbooks on the strength of material, the axial stress of a bar under tension is defined as the force per unit cross-sectional area. When we want to consider the stress of a particular surface in a three-dimensional solid, we need to use the concept of free body. Imagine if we cut a finite solid into two parts along a surface with unit normal $\boldsymbol{n}$ on which the magnitude of stress is of interest, then a resultant force has to be applied to this surface such that both parts of the cut body remain stationary as the solid was before the cutting process. The resultant amputated body is called a *free body*, such as the one shown in Fig. 2.1. *Traction* at a point P is defined as the force per unit area when the following limit is taken:

$$\boldsymbol{T} = \lim_{\Delta A \to 0} [\frac{\Delta \boldsymbol{F}}{\Delta A}] \tag{2.1}$$

where $\Delta \boldsymbol{F}$ is the resultant force on area ΔA. As shown in Fig. 2.2, this traction vector can be considered as the projection of the *Cauchy stress tensor* $\boldsymbol{\sigma}$ on the surface ΔA,

$$\boldsymbol{T} = \boldsymbol{n} \cdot \boldsymbol{\sigma} \tag{2.2}$$

Note that the Cauchy stress tensor is also called the *true stress tensor* since it is supposed to be determined on the deformed body. But for linear elasticity, it is assumed that there is no significant difference in shape and size between the original body and the deformed body; therefore, there is only one stress tensor, the Cauchy stress tensor.

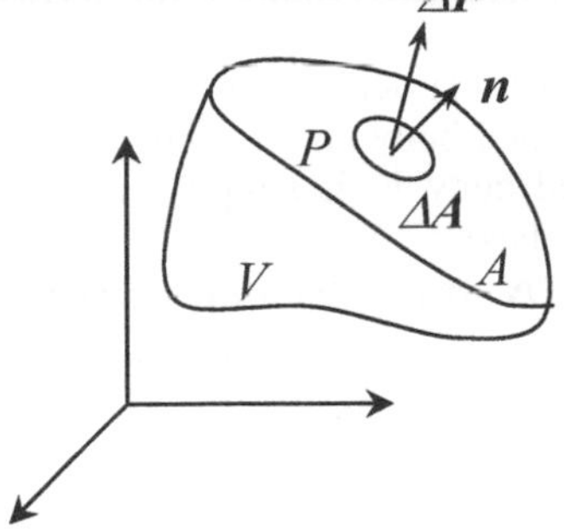

Figure 2.1 Surface force ΔF on ΔA at Point P

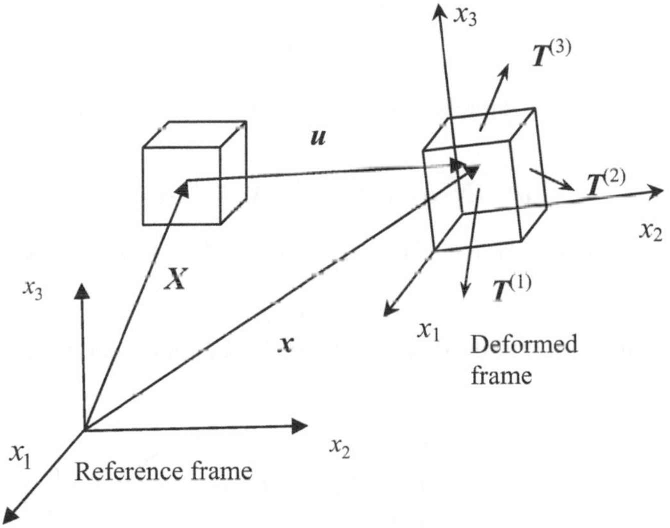

Figure 2.2 Traction vectors on three independent planes

2.3 PIOLA–KIRCHHOFF STRESSES

For a large deformation, the stress tensor determined on the undeformed (or reference) body can differ substantially from the stress tensor determined on the deformed body (or the Cauchy stress). For large deformation problems, the boundary conditions on the current body are in general functions of the deformation if Eulerian formulation is used or a current deformed body is used as a reference. Therefore, it is more convenient to formulate large deformation problems using Lagrangian formulation (i.e., problems formulated referring to the initial configuration). There two choices for stress tensor formulated on the reference state.

Physically, the first Piola–Kirchhoff (PK) stress is defined using the actual force $\boldsymbol{dP}$ on the undeformed body dS_0, as shown in Fig. 2.3. Note that the dotted force on the undeformed body V_0 is simply translated from the deformed body. Using this definition, we have

$$(\boldsymbol{N} \cdot \boldsymbol{\sigma}_{PK1})\, dS_0 = \boldsymbol{dP} = (\boldsymbol{n} \cdot \boldsymbol{\sigma})\, dS \tag{2.3}$$

The first Piola–Kirchhoff stress $\boldsymbol{\sigma}_{PK1}$ is also known as nominal stress, engineering stress, pseudo-stress tensor, or Lagrangian stress tensor. Since the deformation vector in the current configuration is defined by the following mapping

$$\boldsymbol{dx} = \boldsymbol{F} \cdot \boldsymbol{dX} \tag{2.4}$$

the second-order tensor $\boldsymbol{F}$ is called the deformation gradient. Further discussion of this deformation gradient tensor and strain will be given in the next section. The applied force vector also follows the same mapping:

$$\boldsymbol{dP} = \boldsymbol{F} \cdot \boldsymbol{dP^*} \quad \text{or} \quad \boldsymbol{dP^*} = \boldsymbol{F}^{-1} \cdot \boldsymbol{dP} \tag{2.5}$$

With Nanson formula, dS_0 can be related to dS as

$$\boldsymbol{n} dS = \frac{\rho_0}{\rho} \boldsymbol{N} \cdot \boldsymbol{F}^{-1} dS_0 \tag{2.6}$$

The proof of Nanson formula is given in Appendix A. Substitution of (2.6) into (2.3) gives

$$\boldsymbol{dP} = (\boldsymbol{N} \cdot \boldsymbol{\sigma}_{PK1}) dS_0 = \frac{\rho_0}{\rho} \boldsymbol{N} \cdot \boldsymbol{F}^{-1} \cdot \boldsymbol{\sigma} dS_0 \tag{2.7}$$

Therefore, the first Piola–Kirchhoff stress (or first PK stress) can be related to the Cauchy stress as

$$\boldsymbol{\sigma}_{PK1} = J \boldsymbol{G} \cdot \boldsymbol{\sigma} \tag{2.8}$$

where $\boldsymbol{G}$ is the inverse of the deformation gradient $\boldsymbol{F}$ and is defined as

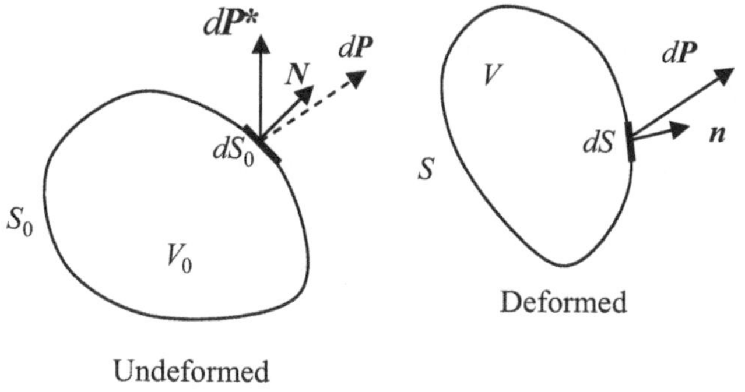

Figure 2.3 Forces on undeformed and deformed bodies

$$G = \frac{dX}{dx} = F^{-1} = \left(\frac{dx}{dX}\right)^{-1} \tag{2.9}$$

The Jacobian is the magnitude of the deformation gradient F or the volume ratio of the deformed to that of the reference body:

$$J = \frac{dV}{dV_0} = \frac{\rho_0}{\rho} \tag{2.10}$$

Since G is not symmetric, the first PK stress is not symmetric. It is awkward to use unsymmetric stress. Therefore, the second Piola–Kirchhoff stress is proposed.

The second Piola–Kirchhoff stress is defined as the transformed or mapped force dP^* on the undeformed body dS_0. In particular, we have

$$(N \bullet \sigma_{PK2})dS_0 = dP^* = F^{-1} \bullet (n \bullet \sigma)dS \tag{2.11}$$

Again, substitution of Nanson's formula (2.6) into (2.11) gives

$$\begin{aligned} dP^* &= (N \bullet \sigma_{PK2})dS_0 = F^{-1} \bullet \left(\frac{\rho_0}{\rho} N \bullet F^{-1} \bullet \sigma\right)dS_0 \\ &= \frac{\rho_0}{\rho} N \bullet F^{-1} \bullet \sigma \bullet (F^{-1})^T \, dS_0 \end{aligned} \tag{2.12}$$

Thus, the second Piola–Kirchhoff (2nd PK) stress can be related to the Cauchy stress as

$$\sigma_{PK2} = JG \bullet \sigma \bullet G^T \tag{2.13}$$

which is clearly symmetric. Thus, the second Piola–Kirchhoff stress is more convenient for numerical analysis for large deformations, such as the nonlinear finite element method. The disadvantage of using the first and second Piola–Kirchhoff stresses is that they do not physically relate to the traction on the surface of the solid or the force per unit deformed area, as shown in Fig. 2.3. More discussion on these stresses can be found in Malvern (1969).

2.4 COORDINATE TRANSFORMATION OF STRESS

We now restrict our discussion to small deformations, and for such cases all Cauchy, first PK, and second PK stresses are the same. A complete determination of the stress tensor σ requires the information of three independent traction vectors around the point P, obtained by cutting free surfaces along other orientations. For example, Fig. 2.4 illustrates the traction vectors on three independent planes for a particular Cartesian coordinate. The three components of $T^{(1)}$, $T^{(2)}$, and $T^{(3)}$ are (σ_{11}, σ_{12}, σ_{13}), (σ_{21}, σ_{22}, σ_{23}), and (σ_{31}, σ_{32}, σ_{33}). As shown in Fig. 2.4, note that the first subscript of σ_{ij} indicates the plane on which the stress component acts, and the second subscript j indicates the direction along which the stress component acts. It is obvious that the stress at point P has nine independent components, thus it is a second-order tensor (compare discussion in Chapter 1).

In addition to this surface traction, *body force* may also act on the body through "action-at-a-distance." Gravitational force is the most trivial example of body force, which is reckoned per unit mass or unit volume. Pore water pressure and thermal stress can also be considered body forces. The effect of pore pressure can be modeled by the theory of poroelasticity, a topic covered in Chapter 8. This body force will not be considered here but in the section of force equilibrium.

The transformation of stress tensor from one coordinate to the other can be done by multiplying the directional cosines, between the axes of the new and old systems, to the tensor components. For example, in index notation of the Cartesian tensor, it can be done as

$$\bar{\sigma}_{ij} = a_i^p a_j^q \sigma_{pq} \tag{2.14}$$

where a_i^p is the directional cosine between axis $\bar{x}_i$ of the new (barred) coordinate system and axis x_p of the old (unbarred) coordinate system, i.e.,

$$a_i^p = \cos(\bar{x}_i, x_p) \tag{2.15}$$

For example, the angle a_1^2 is illustrated in Fig. 2.5.

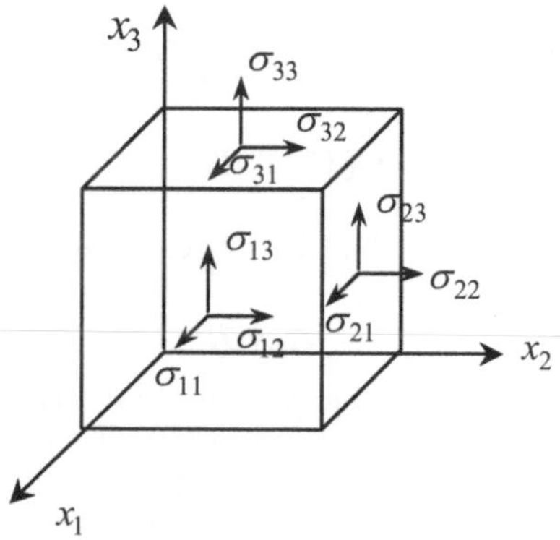

Figure 2.4 The stress components in Cartesian coordinates

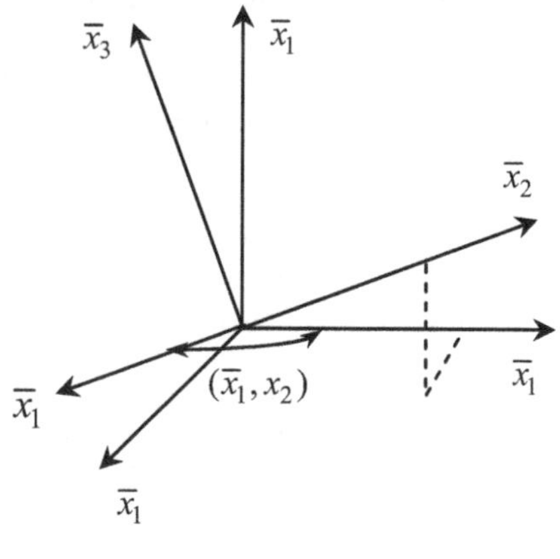

Figure 2.5 Rotation of axes

The Cauchy stress tensor $\boldsymbol{\sigma}$ can be shown to be symmetric by taking moments about the x_i- (i = 1, 2, 3) axis on the element shown in Fig. 2.4 shrunk to zero. That is, $\sigma_{ij} = \sigma_{ji}$ (This result will be shown later using a more rigorous approach). This conclusion is only valid when there are no distributed or surface couples on the solids. If that is not the case, a whole new theory called *micropolar elasticity* emerges. Such a medium is sometimes called the *Cosserate continuum*. Intensive research on micropolar elasticity is still being carried out. It also finds application in explaining the width of shear band observed in soils (Muhlhaus and Vardoulakis, 1987) and bifurcation in rock specimens (Sulem and Vardoulakis, 1990) under triaxial tests. But, we will not digress to such a theory since it is much more complicated than the theory for classical elastic solids.

2.5 BASIC CONCEPT FOR STRAIN TENSOR

The formulation of strain tensor in this section will not follow the traditional approach way of considering the kinematics of deformed elements (e.g., Timoshenko and Goodier, 1982). Instead, we use the exact approach called *Lagrangian formulation*, which is also valid for large deformation formulation. In this approach, the strain tensor is formulated on the undeformed body (now the deformation in the body is considered to be large such that there is a significant difference in position between the undeformed and deformed bodies). When the strain is formulated on the deformed body, we called it the *Eulerian formulation*. In elasticity formulation, however, the Lagrangian formulation is more suitable since there is a natural undeformed state to which all deformed states must return upon unloading.

Figure 2.6 depicts both the undeformed natural state of a body under unloaded condition and the deformed state of the same body under loading. It also shows the displacement, stretch, and rotation of a material vector $d\boldsymbol{X}$ in the undeformed body to $d\boldsymbol{x}$ in the deformed body. The initial point $\boldsymbol{X}$ in the body is displaced to a current position $\boldsymbol{x}$, by the following deformation:

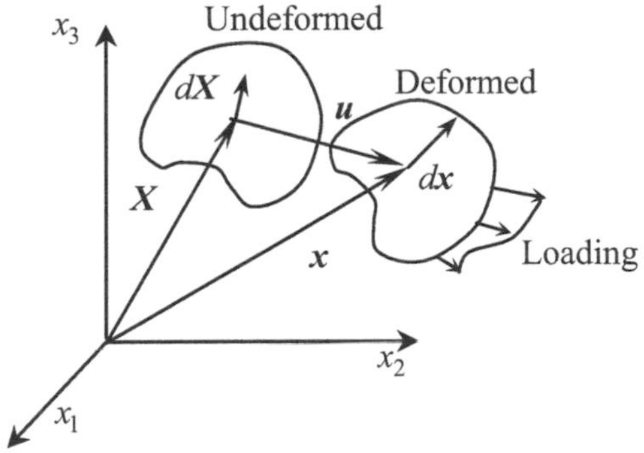

Figure 2.6 The deformation of dX at X to dx at x

$$\boldsymbol{x} = \boldsymbol{x}(\boldsymbol{X}, t), \quad \text{or} \quad x_i = x_i(X_1, X_2, X_3, t) \tag{2.16}$$

and inversely by

$$\boldsymbol{X} = \boldsymbol{X}(\boldsymbol{x}, t), \quad or \quad X_i = X_i(x_1, x_2, x_2, t) \tag{2.17}$$

We introduce here another useful second-order tensor called the *deformation gradient tensor*:

$$\boldsymbol{F} = (\nabla x)^T = x\nabla \tag{2.18}$$

where the vector differential operator ∇ is defined in (1.36) in Chapter 1. Then, an arbitrary infinitesimal material vector $d\boldsymbol{X}$ at $\boldsymbol{X}$ is associated with $d\boldsymbol{x}$ at the current position $\boldsymbol{x}$ as

$$d\boldsymbol{x} = \boldsymbol{F} \cdot d\boldsymbol{X}, \quad \textbf{or} \quad d\boldsymbol{x} = d\boldsymbol{X} \cdot \boldsymbol{F}^T \tag{2.19}$$

Let $dS = (d\boldsymbol{X} \cdot d\boldsymbol{X})^{1/2}$ be the length of the material vector $d\boldsymbol{X}$ in the original undeformed state, then the length of the material vector $d\boldsymbol{x}$ in the deformed state ds is given by

$$(ds)^2 = d\boldsymbol{x} \cdot d\boldsymbol{x} = (d\boldsymbol{X} \cdot \boldsymbol{F}^T) \cdot (\boldsymbol{F} \cdot d\boldsymbol{X}) = d\boldsymbol{X} \cdot (\boldsymbol{F}^T \cdot \boldsymbol{F}) \cdot d\boldsymbol{X} \tag{2.20}$$

By definition, the *Green's strain tensor* is

$$(ds)^2 - (dS)^2 = 2d\boldsymbol{X} \cdot \boldsymbol{E} \cdot d\boldsymbol{X} \tag{2.21}$$

which is one of the ways to define strain tensor. This is a strain measure introduced by George Green in 1838. Green was a self-taught, extraordinary scientist whose brief biography is given in the "Biography" section at the end of this book. Physically, Green's strain tensor is the change in the squared length of a line element in the reference configuration. Comparison of (2.20) and (2.21) yields:

$$\boldsymbol{E} = \tfrac{1}{2}(\boldsymbol{F}^T \cdot \boldsymbol{F} - \boldsymbol{I}) = \tfrac{1}{2}(\boldsymbol{C} - \boldsymbol{I}) \tag{2.22}$$

where $\boldsymbol{I}$ is the second-order unit tensor and $\boldsymbol{C}$ is the right Cauchy–Green tensor. Analogous to $\boldsymbol{C}$, we can define the left Cauchy–Green tensor $\boldsymbol{B}$ as

$$\boldsymbol{B} = \boldsymbol{F} \cdot \boldsymbol{F}^T \tag{2.23}$$

If the strain is formulated in the Eulerian or current configuration, the Eulerian or Alamani's strain tensor $\boldsymbol{e}$ is defined as

$$\boldsymbol{e} = \tfrac{1}{2}(\boldsymbol{I} - \boldsymbol{G}^T \cdot \boldsymbol{G}) = \tfrac{1}{2}(\boldsymbol{I} - \boldsymbol{c}) \tag{2.24}$$

where $\boldsymbol{c}$ is the Cauchy strain tensor and

$$\boldsymbol{G} = (\boldsymbol{F}^{-1})^T \cdot \boldsymbol{F}^{-1} \tag{2.25}$$

Referring to Fig. 2.6, we have

$$\boldsymbol{x} = \boldsymbol{X} + \boldsymbol{u} \tag{2.26}$$

Therefore, the deformation gradient tensor $\boldsymbol{F}$ is

$$\boldsymbol{F} = \boldsymbol{I} + (\nabla \boldsymbol{u})^T = (\delta_{ij} + \frac{\partial u_i}{\partial X_j})\boldsymbol{e}_i\boldsymbol{e}_j \tag{2.27}$$

Substitution of (2.27) into (2.22) gives

$$\begin{aligned} E_{ij} &= \tfrac{1}{2}(F_{ki}F_{kj} - \delta_{ij}) = \tfrac{1}{2}[(\delta_{ki} + \frac{\partial u_k}{\partial X_i})(\delta_{kj} + \frac{\partial u_k}{\partial X_j}) - \delta_{ij}] \\ &= \tfrac{1}{2}(\frac{\partial u_i}{\partial X_j} + \frac{\partial u_j}{\partial X_i}) + \tfrac{1}{2}\frac{\partial u_k}{\partial X_i}\frac{\partial u_k}{\partial X_j} \end{aligned} \tag{2.28}$$

For the case that the deformation is small (i.e., $X_i \approx x_i$), we have

$$e_{ij} = \tfrac{1}{2}(u_{i,j} + u_{j,i}) + \tfrac{1}{2}u_{k,i}u_{k,j} \tag{2.29}$$

For small strain problems, the last term on the right-hand side of (2.29) can be neglected and this results in the classical displacement-strain relationship (e.g., see Timoshenko and Goodier, 1982). Therefore for small strain and small deformation, the strain tensor can be defined as

$$\boldsymbol{\varepsilon} = \tfrac{1}{2}(\nabla \boldsymbol{u} + \boldsymbol{u}\nabla) \tag{2.30}$$

which applies to Cartesian coordinates, as well as to polar coordinates as demonstrated in Chapter 1.

2.6 RATE OF DEFORMATION

For later hypoelasticity discussion, the incremental forms of stress and strain in formulating constitutive law are introduced here. For such cases, the rate of deformation is needed. Let us start with considering the velocity change as

$$d\boldsymbol{v} = \boldsymbol{L} \cdot d\boldsymbol{x} \tag{2.31}$$

where $\boldsymbol{L}$ is the velocity gradient tensor. The physical meaning of $d\boldsymbol{v}$ is illustrated in Fig. 2.7. Since the deformation tensor is defined as

$$d\boldsymbol{x} = \boldsymbol{F} \cdot d\boldsymbol{X} \tag{2.32}$$

the time derivative of (2.32) gives

$$d\boldsymbol{v} = \dot{\boldsymbol{F}} \cdot d\boldsymbol{X} = \dot{\boldsymbol{F}} \cdot \boldsymbol{F}^{-1} \cdot d\boldsymbol{x} = \boldsymbol{L} \cdot d\boldsymbol{x} \tag{2.33}$$

This gives a relation between $\boldsymbol{L}$ and $\boldsymbol{F}$. The symmetric and asymmetric parts of $\boldsymbol{L}$ are defined as

$$\boldsymbol{D} = \frac{1}{2}(\boldsymbol{L} + \boldsymbol{L}^T), \quad \boldsymbol{\Omega} = \frac{1}{2}(\boldsymbol{L} - \boldsymbol{L}^T) \tag{2.34}$$

which are the rate of deformation and the rate of rotation tensor, respectively.

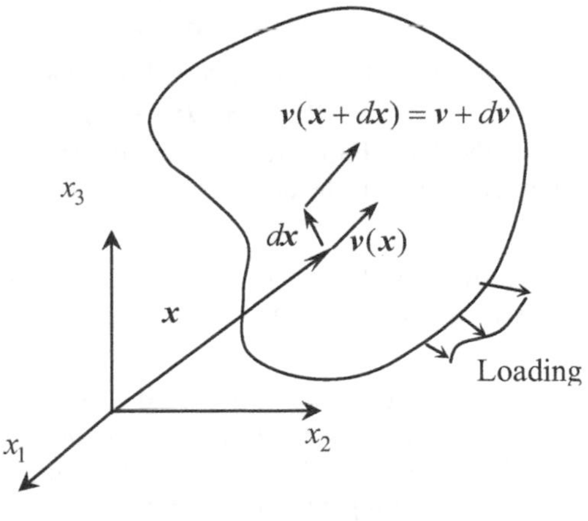

Figure 2.7 The definition of velocity gradient tensor

2.7 COMPATIBILITY EQUATIONS

Since (2.30) is symmetric with respect to its indices, there are only six independent strain components. If the strain in a body is given, then (2.30) provides six partial differential equations for three unknown displacement components. Therefore, the strain components cannot be prescribed arbitrarily; there must be some constraints or conditions that have to be satisfied among the strain components. The conditions for (2.30) to be integrable for $\boldsymbol{u}$ are called the *Saint–Venant's compatibility* equation, and can be obtained by applying the curl operator twice to $\boldsymbol{\varepsilon}$ given in (2.30), once from the left and once from the right:

$$\boldsymbol{S} = \nabla \times \boldsymbol{\varepsilon} \times \nabla = 0, \quad \text{or} \quad e_{ijl} e_{pkr} \varepsilon_{jk,ip} = 0 \tag{2.35}$$

This integrability condition can easily be seen if one remembers that the curl of a gradient is zero [see (1.46) of Chapter 1]. It can easily be shown that the index notation in (2.35) is symmetric, thus only six independent components remain. However, it is clear that we need only three conditions for (2.30) to be integrable. It can be further shown that the six components of $\boldsymbol{S}$ have to again satisfy three conditions called *Bianchi equations* (Malvern, 1969; Washizu, 1958); the details will not be discussed here.

It will be shown later that the compatibility equations have to be satisfied if stress is formulated as the unknown in boundary value problems, such as in the *Airy stress function* formulation in 2-D elasticity. If the displacement functions are used explicitly in the formulation, then the compatibility equations are not needed, such as the Galerkin vector for 3-D problems to be discussed in Chapter 4.

2.8 HILL'S WORK-CONJUGATE STRESS MEASURES

By now, we have introduced the concept of stress and strain. For large deformation, the pair of strain measure and its work-conjugate stress measure should be used in constitutive law formation or when the rate of stress working is considered (Hill, 1978). In particular, the rate of stress working per unit current volume is

$$\dot{W}_c = \boldsymbol{\sigma} : \boldsymbol{D} \tag{2.36}$$

Alternatively, the rate of stress working per unit reference volume is

$$\dot{W} = J\boldsymbol{\sigma} : \boldsymbol{D} = \boldsymbol{\tau} : \boldsymbol{D} \tag{2.37}$$

where $\boldsymbol{\tau}$ is called the Kirchhoff stress tensor. If the first Piola–Kirchhoff stress is adopted, the conjugate strain is the time derivative of the deformation gradient. That is,

$$\dot{W} = \boldsymbol{\sigma}_{PK1} : \dot{\boldsymbol{F}} \tag{2.38}$$

If the second Piola–Kirchhoff stress is adopted, the conjugate pair of strain is the time derivative of the Green's tensor defined in (2.22):

$$\dot{W} = \boldsymbol{\sigma}_{PK2} : \dot{\boldsymbol{E}} \tag{2.39}$$

Instead of rate of stress working, we can also determine the conjugate pair by considering the strain energy per unit initial volume. In particular, as discussed by Bažant and Cedolin (1991), the stress and strain used in constitutive law formulation must be a conjugate pair in doing work, which is the Helmholtz free energy if the condition is isothermal and the total energy if the condition is adiabatic.

2.9 CONSTITUTIVE RELATION

The deformation in a solid depends on how the solid responds to applied excitations, such as loads. Constitutive relation is an expression that relates stress and strain at any point inside the solid. In this section, we restrict our discussion to small deformation. The macroscopic constitutive response for a solid can be considered the overall effect of the individual microscopic constituents of the materials. It can be time or loading rate dependent. Most of the constitutive response is direction sensitive, i.e., anisotropic. If all time effect is neglected, we can simply assume

$$\boldsymbol{\sigma} = f(\boldsymbol{\varepsilon}) \tag{2.40}$$

Linearization of (2.40) leads to the following classical form:

$$\boldsymbol{\sigma} = \boldsymbol{C} : \boldsymbol{\varepsilon} \quad \text{or} \quad \sigma_{ij} = C_{ijkl}\varepsilon_{kl} \tag{2.41}$$

where $\boldsymbol{C}$ is a fourth-order tensor called the *elasticity tensor*. There are, in general, 81 components for $\boldsymbol{C}$, but due to the symmetric properties for both strain and stress (six independent components each) only 36 possible constants exist. If one further assumes that the potential energy function exists, they reduce to 21. This can be shown by assuming the *Clapeyron formula* for strain-energy function:

$$W = \frac{1}{2}\sigma_{ij}\varepsilon_{ij} = \frac{1}{2}C_{ijkl}\varepsilon_{ij}\varepsilon_{kl} \tag{2.42}$$

It is clear from the last part of (2.42) that $C_{ijkl} = C_{klij}$; this symmetric property reduces the elastic moduli to 21. In the terminology of material science, it is called *triclinic*. It can be shown that the number of constants reduces to 13 when there is one plane of symmetry; this kind of symmetry is also called *monoclinic*. When three mutual planes of symmetry exist, the number of independent constants reduces to 9. This is called *orthotropic* or *orthorhombic*. When there is a plane of isotropy (i.e., isotropic property within the plane), 5 independent constants remain and the solid becomes *transversely isotropic* or *hexagonal*. In *cubic* materials, there are 3 elastic constants. And, finally, for *isotropic* elastic solids, there are 2 independent constants. General discussion on anisotropic solids can be found in Hearmon (1961) and Lekhnitskii (1963).

To date, most of the analytic solutions are for isotropic solids. In this chapter, attention will be restricted to isotropic and transversely isotropic solids.

2.10 ISOTROPIC SOLIDS

The three-dimensional Hooke's law for an isotropic solid can easily be generalized from the one-dimensional Hooke's law. Consider a uniaxial tension σ_{11} applying on a parallelepiped along the x_1 direction, as shown in Case I of Fig. 2.8. The axial strain will be $\varepsilon_{11} = \sigma_{11}/E$, where E is called the Young's modulus; the lateral strains along both the x_2 and x_3 directions will be $-\nu\varepsilon_{11}$ (i.e., $\varepsilon_{22} = \varepsilon_{33} = -\nu\varepsilon_{11}$), where ν is called the Poisson's ratio. If the solid is isotropic, a uniaxial tension applied along the x_2 direction yields $\varepsilon_{22} = \sigma_{11}/E$ as the axial strain, and $-\nu\varepsilon_{22}$, as the lateral strains (this is illustrated in Case II of Fig. 2.8). That is, the same Young's modulus and Poisson's ratio apply irrespective of directions. A similar situation again applies to uniaxial tension along the x_3- direction, as shown in Case III of Fig. 2.8. For infinitesimal deformations, the relation between stress and strain is linear. Thus, the principle of superposition applies. If all normal stresses along x_1, x_2, and x_3 are applied simultaneously, the overall axial strains are simply the superposition of strains due to each stress component. For example, the axial strain along the x_1- direction is

$$\varepsilon_{11} = \frac{1}{E}[\sigma_{11} - \nu(\sigma_{22} + \sigma_{33})] = \frac{1}{E}[(1+\nu)\sigma_{11} - \nu\sigma_{kk}] \tag{2.43}$$

Combining all three normal strains, we can write

$$\boldsymbol{\varepsilon} = \frac{1}{E}[(1+\nu)\boldsymbol{\sigma} - \nu\sigma_{kk}\boldsymbol{I}] \tag{2.44}$$

For nondiagonal components, (2.44) implies

$$\varepsilon_{ij} = \frac{1+\nu}{E}\sigma_{ij} = \frac{1}{2G}\sigma_{ij}, \quad (i \neq j) \tag{2.45}$$

where G is called the shear modulus. This can easily be verified as the modulus governing deformation due to a shear stress applied on a parallelepiped, as covered in most textbooks on elasticity (e.g., Timoshenko and Goodier, 1982).
Taking the trace of (2.44) we get

$$\varepsilon_{kk} = \frac{1-2\nu}{E}\sigma_{kk} = \frac{1}{3K}\sigma_{kk} \tag{2.46}$$

where K is the bulk modulus relating the compressibility of solids. Note that the trace of a second-order tensor $\boldsymbol{\sigma}$ can be defined as $\mathrm{tr}(\boldsymbol{\sigma}) = \boldsymbol{I}:\boldsymbol{\sigma}$ or σ_{kk}. Rearranging (2.44) and applying (2.46), we have

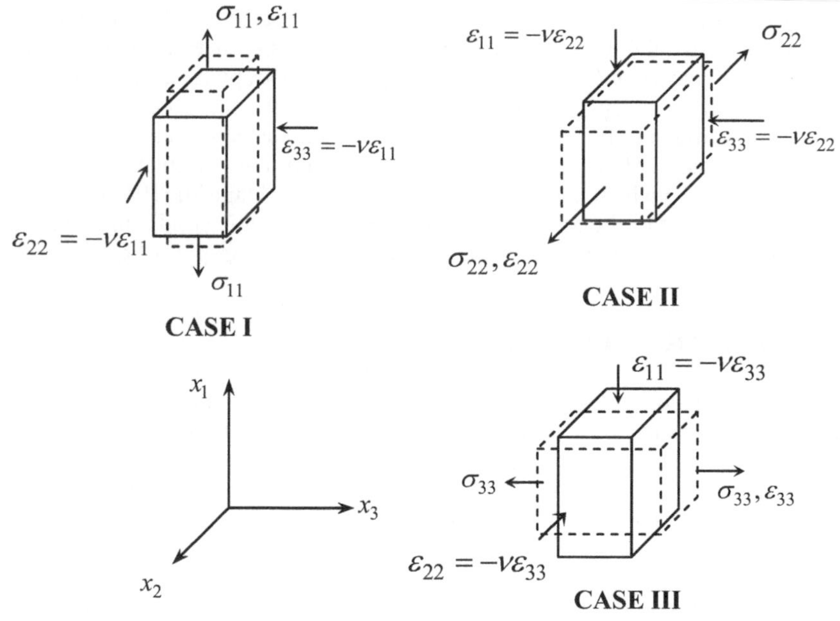

Figure 2.8 Uniaxial compression along three independent directions

$$\boldsymbol{\sigma} = \frac{E}{1+\nu}\boldsymbol{\varepsilon} + \frac{\nu E}{(1+\nu)(1-2\nu)}\varepsilon_{kk}\boldsymbol{I} = 2\mu\boldsymbol{\varepsilon} + \lambda\varepsilon_{kk}\boldsymbol{I} \tag{2.47}$$

where λ and μ are the Lamé constants. Note that μ is also called the shear modulus and denoted by G as defined in (2.45). In Cartesian components, the relationship between the stress $\boldsymbol{\sigma}$ and strain $\boldsymbol{\varepsilon}$ can be expressed as

$$\sigma_{ij} = \lambda\varepsilon_{kk}\delta_{ij} + 2\mu\varepsilon_{ij} \tag{2.48}$$

It can further be shown that the elastic tensor $\boldsymbol{C}$ for isotropic solids in Cartesian coordinates is

$$C_{ijkl} = \lambda\,\delta_{ij}\delta_{kl} + \mu(\delta_{ik}\delta_{jl} + \delta_{il}\delta_{jk}) \tag{2.49}$$

In isotropic solids, there are only two independent materials constants; all E, K, G, ν, λ, and μ can be shown to be related as

$$\lambda = \frac{2G\nu}{1-2\nu} = \frac{G(E-2G)}{3G-E} = K - \tfrac{2}{3}G = \frac{E\nu}{(1+\nu)(1-2\nu)} = \frac{3K\nu}{1+\nu} = \frac{3K(3K-E)}{9K-E} \tag{2.50}$$

$$\mu = G = \frac{\lambda(1-2\nu)}{2\nu} = \tfrac{3}{2}(K-\lambda) = \frac{E}{2(1+\nu)} = \frac{3K(1-2\nu)}{2(1+\nu)} = \frac{3KE}{9K-E} \tag{2.51}$$

$$\nu = \frac{\lambda}{2(\lambda+G)} = \frac{\lambda}{(3K-\lambda)} = \frac{E}{2G} - 1 = \frac{3K-2G}{2(3K+G)} = \frac{3K-E}{6K} \tag{2.52}$$

$$E = \frac{G(3\lambda+2G)}{\lambda+G} = \frac{\lambda(1+\nu)(1-2\nu)}{\nu} = \frac{9K(K-\lambda)}{2(3K+G)} = 2G(1+\nu) = \frac{9KG}{3K+G} = 3K(1-2\nu) \tag{2.53}$$

$$K = \lambda + \tfrac{2}{3}G = \frac{\lambda(1+\nu)}{3\nu} = \frac{2G(1+\nu)}{3(1-2\nu)} = \frac{GE}{3(3G-E)} = \frac{E}{3(1-2\nu)} \tag{2.54}$$

The method of solution for problems in isotropic elastic solids will be discussed later after the constitutive form of transversely isotropic solids is introduced.

Young's modulus and Poisson's ratio are normally determined from either uniaxial or triaxial compression tests. However, as discussed by Filon (1902) and Chau (1997), when friction exists between the loading platens and the end surfaces of the solid cylinder, a correction factor needs to be applied to get the true Young's modulus because of the nonuniform stresses. In the case of plane compression, the correction factor was considered by Chau (1999), using the hypercircle approach of Synge and Prager.

2.11 TRANSVERSELY ISOTROPIC SOLIDS

As mentioned in Section 2.9, there are five independent material constants for transversely isotropic solids. In explicit Cartesian component form, we have

$$\sigma_{11} = C_{11}\varepsilon_{11} + C_{12}\varepsilon_{22} + C_{13}\varepsilon_{33} \tag{2.55}$$

$$\sigma_{22} = C_{12}\varepsilon_{11} + C_{11}\varepsilon_{22} + C_{13}\varepsilon_{33} \tag{2.56}$$

$$\sigma_{33} = C_{13}(\varepsilon_{11} + \varepsilon_{22}) + C_{33}\varepsilon_{33} \tag{2.57}$$

$$\sigma_{12} = (C_{11} - C_{12})\varepsilon_{12}, \quad \sigma_{3i} = 2C_{44}\varepsilon_{3i} \quad (i = 1, 2) \tag{2.58}$$

Hooke's law for transversely isotropic solids can be written in terms of Cartesian tensors as, $\boldsymbol{C} = C_{iknm}\boldsymbol{e}_i\boldsymbol{e}_k\boldsymbol{e}_n\boldsymbol{e}_m$, where C_{iknm} is (Srinivasan and Nigam, 1969)

$$\begin{aligned} C_{iknm} &= \alpha\delta_{ik}\delta_{nm} + \lambda(\delta_{in}\delta_{km} + \delta_{im}\delta_{kn}) + \gamma\delta_{i3}\delta_{k3}\delta_{n3}\delta_{m3} \\ &+ \kappa(\delta_{i3}\delta_{k3}\delta_{nm} + \delta_{ik}\delta_{n3}\delta_{m3}) \\ &+ \beta(\delta_{im}\delta_{k3}\delta_{n3} + \delta_{i3}\delta_{m3}\delta_{kn} + \delta_{in}\delta_{k3}\delta_{m3} + \delta_{i3}\delta_{n3}\delta_{km}) \end{aligned} \tag{2.59}$$

in which we have assumed the x_3-axis is perpendicular to the plane of isotropy. Comparison of (2.59) to (2.55)–(2.58) yields the following definitions for α, λ, γ, κ, and β:

$$\begin{aligned} &\alpha = C_{12},\ \ \beta = C_{44} - \tfrac{1}{2}(C_{11} - C_{12}),\ \ \kappa = C_{13} - C_{12},\ \ \lambda = \tfrac{1}{2}(C_{11} - C_{12}) \\ &\gamma = C_{33} - 3C_{11} - 2(C_{13} - 2C_{12}) - 4C_{44} \end{aligned} \tag{2.60}$$

The physical meaning for C_{ij} $(i, j = 1, 2, 3)$ is not readily understandable. We invert Hooke's law such that

$$\varepsilon_{ik} = S_{iknm}\sigma_{nm} \tag{2.61}$$

where S_{iknm} is the compliance tensor and is defined as

$$\begin{aligned} S_{iknm} &= \bar{\alpha}\delta_{ik}\delta_{nm} + \bar{\lambda}(\delta_{in}\delta_{km} + \delta_{im}\delta_{kn}) + \bar{\gamma}\delta_{i3}\delta_{k3}\delta_{n3}\delta_{m3} \\ &+ \bar{\kappa}(\delta_{i3}\delta_{k3}\delta_{nm} + \delta_{ik}\delta_{n3}\delta_{m3}) \\ &+ \bar{\beta}(\delta_{im}\delta_{k3}\delta_{n3} + \delta_{i3}\delta_{m3}\delta_{kn} + \delta_{in}\delta_{k3}\delta_{m3} + \delta_{i3}\delta_{n3}\delta_{km}) \end{aligned} \tag{2.62}$$

These parameters are now related to the following physical parameters:

$$\bar{\alpha} = -\frac{\nu'}{E},\ \ \bar{\beta} = \frac{1}{G'} - \frac{1+\nu}{2E},\ \ \bar{\kappa} = -(\frac{\nu'}{E'} - \frac{\nu}{E}),\ \bar{\lambda} = \frac{1+\nu}{2E},\ \ \bar{\gamma} = \frac{1}{E'} - \frac{3}{E} + 2(\frac{\nu'}{E'} - \frac{2\nu}{E}) - \frac{4}{G'} \tag{2.63}$$

where E and E' are the Young's moduli in directions perpendicular and parallel to the plane of symmetry, respectively; ν and ν' are the Poisson's ratios for modeling the transverse deformation (compressive/tensile) in the directions perpendicular and parallel to the plane of symmetry, caused by perpendicular stresses (tensile/compressive).

The general theory for solving transversely isotropic solids was derived by Hu (1954) and is also found in Lekhnitskii (1963). For the application of the theory of transversely isotropic solids to rock mechanics, we refer to the analysis by Chau (1994b) and the related analyses on spherically isotropic solids by Wei and Chau (2002, 2009), Chau and Wei (1999), and Chau (1995a, 1998b).

2.12 EQUATIONS OF MOTION AND EQUILIBRIUM

Let us consider a body, either elastic or not, with volume V and surface S as shown in Fig. 2.9. Applying Newton's third law to the body V_i, which is an arbitrary volume inside the body V, we arrive at the following force equilibrium:

$$\int_V \boldsymbol{F} dV + \int_S \boldsymbol{n}\cdot\boldsymbol{\sigma} dS = \frac{d}{dt}\int_V \rho \frac{d\boldsymbol{u}}{dt} dV \tag{2.64}$$

where $\boldsymbol{u}$ is the displacement vector. In this equation, the body force is denoted by $\boldsymbol{F}dV$ and the surface force by $\boldsymbol{n}\cdot\boldsymbol{\sigma}\, dS$, where $\boldsymbol{n}$ is the unit normal to the surface S_i of body

V_i. The second term on the left-hand side of (2.64) can be transformed to volume integral by using the divergence theorem discussed in Section 1.6.1. Thus, we have:

$$\int_V (\nabla\bullet\boldsymbol{\sigma}+\boldsymbol{F})dV = \int_V \rho\frac{d^2\boldsymbol{u}}{dt^2}dV \tag{2.65}$$

Since the volume V_i can be taken arbitrarily as long as it is inside the body V, we must have

$$\nabla\bullet\boldsymbol{\sigma}+\boldsymbol{F} = \rho\frac{d^2\boldsymbol{u}}{dt^2} \tag{2.66}$$

at every point inside the body V. This is the equation of motions. When the force of inertia is absent or all loadings are *quasi-static*, the equation of equilibrium is obtained:

$$\nabla\bullet\boldsymbol{\sigma}+\boldsymbol{F} = 0 \tag{2.67}$$

In addition, the moment of momentum must also be in equilibrium, therefore we have

$$\int_V \boldsymbol{r}\times\boldsymbol{F}dV + \int_S \boldsymbol{r}\times(\boldsymbol{n}\bullet\boldsymbol{\sigma})dS = \frac{d}{dt}\int_V \boldsymbol{r}\times\rho\frac{d\boldsymbol{u}}{dt}dV \tag{2.68}$$

We now proceed with the Cartesian components such that (2.68) becomes

$$\int_V e_{rmn}x_mF_ndV + \int_S e_{rmn}x_mn_k\sigma_{kn}dS = \frac{d}{dt}\int_V e_{rmn}x_m\rho v_ndV \tag{2.69}$$

where $v_n = du_n/dt$ is the velocity component. We now apply the generalized Gauss theorem to the second term on the left-hand side to yield (Malvern, 1969)

$$\int_S e_{rmn}x_mn_k\sigma_{kn}dS = \int_V e_{rmn}\frac{\partial(x_m\sigma_{kn})}{\partial x_k}dV = \int_V e_{rmn}(\sigma_{mn}+x_m\frac{\partial\sigma_{kn}}{\partial x_k})dV \tag{2.70}$$

Back substitution of (2.70) into (2.69) and application of the equation of motion (2.66) again gives us

$$\int_V e_{rmn}(\sigma_{mn}+x_m\rho\frac{dv_n}{dt})dV = \int_V e_{rmn}(v_mv_n+x_m\rho\frac{dv_n}{dt})dV \tag{2.71}$$

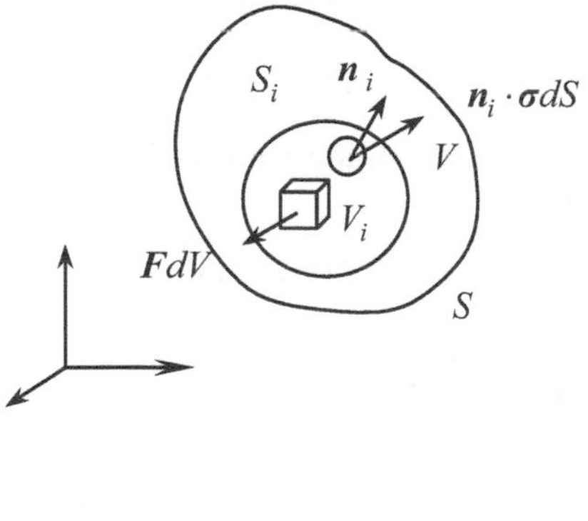

Figure 2.9 Momentum balance of body V_i

We further note that $e_{rmn}v_m v_n = 0$ since $v_m v_n$ is symmetric while e_{rmn} is antisymmetric in m and n. Thus, we must have $e_{rmn}\sigma_{mn} = 0$, which is satisfied if and only if $\boldsymbol{\sigma}$ is symmetric. Therefore, it can be concluded that the stress tensor must be symmetric if there is no concentrated or distributed moments inside the body. The theory of *micropolar elasticity* does allow such a possibility, but it is outside the scope of the present study.

For isotropic solids, substitution of (2.30) and (2.48) into (2.67) gives the equation of equilibrium in terms of displacement $\boldsymbol{u}$:

$$\mu\nabla\cdot\nabla\mathbf{u}+(\lambda+\mu)\nabla\nabla\cdot\boldsymbol{u}+\boldsymbol{F}=\mathbf{0} \tag{2.72}$$

In obtaining (2.72), we have used the identity $\nabla\cdot(\boldsymbol{u}\nabla)=\nabla\nabla\cdot\boldsymbol{u}$. Alternatively, using vector identity (1.50) of Chapter 1, (2.72) can be written as

$$(\lambda+2\mu)\nabla\nabla\cdot\boldsymbol{u}-\mu\nabla\times(\nabla\times\boldsymbol{u})+\boldsymbol{F}=\mathbf{0} \tag{2.73}$$

Equation (2.72) or (2.73) provides the equation of equilibrium in terms of displacement and is also called *Navier's equilibrium equation*, which is first obtained by Navier in 1827 (see Love, 1944). Similarly, the surface traction can also be written in terms of displacement. Substitution of (2.30) and (2.48) into (2.2) gives

$$\mu(\boldsymbol{n}\cdot\nabla)\boldsymbol{u}+\mu\nabla(\boldsymbol{n}\cdot\boldsymbol{u})+\lambda\boldsymbol{n}(\nabla\cdot\boldsymbol{u})=\boldsymbol{T} \tag{2.74}$$

Using the following identity

$$\boldsymbol{n}\times(\nabla\times\boldsymbol{u})=\nabla(\boldsymbol{n}\cdot\boldsymbol{u})-(\nabla\cdot\boldsymbol{n})\boldsymbol{u} \tag{2.75}$$

we find another form for the surface traction $\boldsymbol{T}$:

$$2\mu(\boldsymbol{n}\cdot\nabla)\boldsymbol{u}+\mu\boldsymbol{n}\times(\nabla\times\boldsymbol{u})+\lambda\boldsymbol{n}(\nabla\cdot\boldsymbol{u})=\boldsymbol{T} \tag{2.76}$$

When displacements are written explicitly as the unknown, the equation of compatibility is not required to be satisfied (it has been satisfied automatically).

For anisotropic solids, the equation of equilibrium will be much more complicated when it is written in terms of displacements.

2.13 COMPATIBILITY EQUATION IN TERMS OF STRESS TENSOR

Sometimes, it is more convenient to rewrite the compatibility in terms of stress tensor. We start by introducing a double cross product:

$$\boldsymbol{A}\overset{\times}{\underset{\times}{}}\boldsymbol{B}=\boldsymbol{e}_i\times\boldsymbol{e}_l\;\boldsymbol{e}_j\times\boldsymbol{e}_k A_{ij}B_{kl} \tag{2.77}$$

where $\boldsymbol{e}_i$ is the base vector of any Cartesian coordinate. We now apply this double cross product to (2.77) as

$$\boldsymbol{I}\overset{\times}{\underset{\times}{}}(\nabla\times\boldsymbol{\varepsilon}\times\nabla)=\boldsymbol{e}_i\boldsymbol{e}_i\overset{\times}{\underset{\times}{}}(\nabla\times\boldsymbol{e}_j\varepsilon_{jk}\boldsymbol{e}_k\times\nabla)=\boldsymbol{e}_i\times(\nabla\times\boldsymbol{e}_j)\boldsymbol{e}_i\times(\boldsymbol{e}_k\times\nabla)\varepsilon_{jk}=\boldsymbol{E} \tag{2.78}$$

Expanding the triple cross product using (2.75), we obtain

$$\begin{aligned}\boldsymbol{E}&=\boldsymbol{e}_i\cdot(\boldsymbol{e}_j\nabla-\nabla\boldsymbol{e}_j)\boldsymbol{e}_i\cdot(\nabla\boldsymbol{e}_k-\boldsymbol{e}_k\nabla)\varepsilon_{jk}\\&=\boldsymbol{e}_i\cdot\boldsymbol{e}_j\nabla\boldsymbol{e}_i\cdot\nabla\boldsymbol{e}_k\,\varepsilon_{jk}\nabla\\&\quad+\boldsymbol{e}_i\cdot\nabla\boldsymbol{e}_j\boldsymbol{e}_i\cdot\boldsymbol{e}_k\nabla\varepsilon_{jk}-\boldsymbol{e}_i\cdot\boldsymbol{e}_j\nabla\boldsymbol{e}_i\cdot\boldsymbol{e}_k\nabla\varepsilon_{jk}-\boldsymbol{e}_i\cdot\nabla\boldsymbol{e}_j\boldsymbol{e}_i\cdot\nabla\boldsymbol{e}_k\,\varepsilon_{jk}\end{aligned} \tag{2.79}$$

Each of these four terms in the second part of (2.79) can be interpreted again in symbolic form together with (2.35), and therefore we obtain

$$\nabla\nabla\cdot\boldsymbol{\varepsilon}+(\nabla\cdot\boldsymbol{\varepsilon})\nabla-\nabla\nabla\varepsilon_{kk}-\nabla\cdot\nabla\boldsymbol{\varepsilon}=\mathbf{0} \tag{2.80}$$

We now substitute (2.44) into (2.80), and the equation of compatibility can be written in terms of the stress tensor $\boldsymbol{\sigma}$.

$$(1+\nu)[\nabla\nabla\cdot\boldsymbol{\sigma}+(\nabla\cdot\boldsymbol{\sigma})\nabla-\nabla\cdot\nabla\boldsymbol{\sigma}]+\nu\nabla\cdot\nabla\sigma_{kk}\boldsymbol{I}-\nabla\nabla\sigma_{kk}=0 \tag{2.81}$$

Taking the trace of (2.81), we can show that

$$\nabla\cdot\nabla\sigma_{kk}=\frac{1+\nu}{1-\nu}\nabla\cdot(\nabla\cdot\boldsymbol{\sigma}) \tag{2.82}$$

Substituting (2.82) and the equilibrium equation (2.67) into (2.81), we finally obtain the following *Beltrami–Michell compatibility equation*

$$\frac{1}{1+\nu}\nabla\nabla\sigma_{kk}+\nabla\cdot\nabla\sigma=-\frac{\nu}{1-\nu}\nabla\cdot\boldsymbol{F}\boldsymbol{I}-(\nabla\boldsymbol{F}+\boldsymbol{F}\nabla) \tag{2.83}$$

which was obtained by Beltrami in 1892 and by Michell in 1900.

2.14 STRAIN ENERGY DENSITY

If a solid is purely elastic, it recovers all its deformation upon unloading. To put the statement into energy terms, all the external work done by applied loads on the body will be stored in the form of potential energy or the strain energy, and will be totally recuperated when the loads are removed. For most geomaterials, this is not true. In soils, frictional energy is dissipated when relative displacement between soil particles occurs under external loads; in rocks, growth of microcracks and frictional sliding between the surfaces of closed microcracks under external loads also result in energy loss. Therefore, energy loss due to these nonlinear processes must be incorporated in formulating the energy function in geomaterials if a more realistic prediction is desired. (This will be done in Chapter 6 on fracture mechanics.) Nevertheless, the existence of a recoverable strain energy does lead to some beautiful mathematical results and consequences.

Consider again an elastic body of volume V and surface S, as shown in Fig. 2.9. The total work done W on the elastic body by body force $\boldsymbol{F}$ and external traction $\boldsymbol{T}$ is

$$W=\int_V\left(\int_0^{u_f}\boldsymbol{F}\cdot d\boldsymbol{u}\right)dV+\int_S\left(\int_0^{u_f}\boldsymbol{T}\cdot d\boldsymbol{u}\right)dS \tag{2.84}$$

where $\boldsymbol{u}$ is the deformation and $\boldsymbol{u}_f$ is the current value of deformation field. Using $\boldsymbol{T}=\boldsymbol{n}\cdot\boldsymbol{\sigma}$ and applying the divergence theorem to the second term of (2.84), we obtain

$$W=\int_V\left(\int_0^{u_f}[\boldsymbol{F}\cdot d\boldsymbol{u}+\nabla\cdot(\boldsymbol{\sigma}\cdot d\boldsymbol{u})]\right)dV \tag{2.85}$$

Note that the last term can be expressed as two terms by using the following formula:

$$\nabla\cdot(\boldsymbol{\sigma}\cdot d\boldsymbol{u})=(\nabla\cdot\boldsymbol{\sigma})\cdot d\boldsymbol{u}+\boldsymbol{\sigma}:\nabla d\boldsymbol{u} \tag{2.86}$$

In Cartesian coordinates, the operators ∇ and d are commutative, and we have

$$\boldsymbol{\sigma}:\nabla d\boldsymbol{u}=\boldsymbol{\sigma}:d\nabla\boldsymbol{u}=\boldsymbol{\sigma}:d(\boldsymbol{\varepsilon}-\boldsymbol{\Omega})=\boldsymbol{\sigma}:d\boldsymbol{\varepsilon} \tag{2.87}$$

in which we have decomposed the displacement gradient tensor into the strain tensor $\boldsymbol{\varepsilon}$ and the rotation or spin tensor $\boldsymbol{\Omega}$. Note also that $\boldsymbol{\sigma}:d\boldsymbol{\Omega}=0$ since $\boldsymbol{\sigma}$ is symmetric while $\boldsymbol{\Omega}$ is antisymmetric. Therefore, the strain energy is

$$dW=\boldsymbol{\sigma}:d\boldsymbol{\varepsilon} \tag{2.88}$$

or

$$W = \int \boldsymbol{\sigma} : d\boldsymbol{\varepsilon} \tag{2.89}$$

Furthermore, if $\boldsymbol{\sigma}$ and $\boldsymbol{\varepsilon}$ are linearly related for elastic bodies under small deformations, we must have

$$\boldsymbol{\sigma} : d\boldsymbol{\varepsilon} = \tfrac{1}{2} d(\boldsymbol{\sigma} : \boldsymbol{\varepsilon}) \tag{2.90}$$

Combining (2.85) and (2.90), we finally get

$$W = \int_V \left[\int_0^{u_f} [(\boldsymbol{F} + \nabla \cdot \boldsymbol{\sigma}) \cdot d\boldsymbol{u} + \frac{1}{2} d(\boldsymbol{\sigma} : \boldsymbol{\varepsilon})] \right] dV = \frac{1}{2} \int_V \boldsymbol{\sigma} : \boldsymbol{\varepsilon} dV \tag{2.91}$$

since the first term inside the brackets vanishes by the equilibrium equation. Therefore, the strain energy density is

$$W_d = \tfrac{1}{2} \boldsymbol{\sigma} : \boldsymbol{\varepsilon} \tag{2.92}$$

Note that we have not made any assumption on the constitutive behavior of the solid in the derivation of (2.92), so it applies to isotropic as well as anisotropic solids.

2.15 COMPLEMENTARY ENERGY

To consider complementary energy, we first introduce the following Legendre transform as

$$\rho \Phi = W - \boldsymbol{\sigma} : \boldsymbol{\varepsilon} \tag{2.93}$$

Taking the differential form of (2.93) yields

$$\rho d\Phi = dW - \boldsymbol{\sigma} : d\boldsymbol{\varepsilon} - d\boldsymbol{\sigma} : \boldsymbol{\varepsilon} \tag{2.94}$$

Substituting (2.88) into (2.94), we have

$$\rho d\Phi = -\boldsymbol{\varepsilon} : d\boldsymbol{\sigma} \tag{2.95}$$

We now define the incremental complementary energy as

$$dW_c = -\rho d\Phi = \boldsymbol{\varepsilon} : d\boldsymbol{\sigma} \tag{2.96}$$

Thus, complementary energy can be obtained by integrating (2.96)

$$W_c = \int \boldsymbol{\varepsilon} : d\boldsymbol{\sigma} \tag{2.97}$$

The physical meaning and relation between strain and complementary energies are illustrated in Fig. 2.10. The horizontal hatched strip represents the increment of complementary energy (2.96), whereas the vertical hatched strip represents the increment of strain energy (2.88). For linear elastic solids, the stress strain relation becomes a straight line and in this case the complementary energy will be the same as the strain energy.

2.16 HYPERELASTICITY AND HYPOELASTICITY

As mentioned in the Introduction, if elastic potential or strain density function exists in an elastic body, the body is called *hyperelastic*. Elastic solids with the existence of strain energy function were first considered by George Green in 1839 and thus they are also called Green-elastic (Malvern, 1969). For example, substitution of (2.47) into (2.92) gives

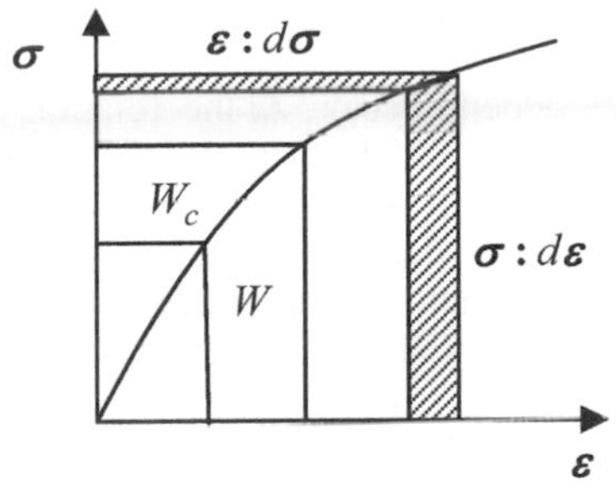

Figure 2.10 Strain energy and complementary energy

$$W_d = \tfrac{1}{2}\lambda(\text{tr}\,\boldsymbol{\varepsilon})^2 + \mu\boldsymbol{\varepsilon}:\boldsymbol{\varepsilon} \tag{2.98}$$

Inversely, substitution of (2.44) into (2.92) yields

$$W_d = \frac{1}{2E}[(1+\nu)\boldsymbol{\sigma}:\boldsymbol{\sigma} - \nu(tr\boldsymbol{\sigma})^2] \tag{2.99}$$

This elastic potential, of course, only applies to isotropic bodies under small strain and deformation. One main application of hyperelasticity has been on highly elastic bodies under finite deformation, such as rubber-like materials. The main issue is how to postulate the strain energy function in terms of invariants of the stress tensor, and subsequently how to determine the unknown coefficients involved in elastic potential. Most applied mathematicians working on elasticity follow such an approach. We will not explore such a theory in detail here. Some of the more commonly adopted models are

Neo–Hookean material:

$$W = c_1(I_B - 3) \tag{2.100}$$

Mooney–Rivlin material:

$$W = c_1(I_B - 3) + c_2(II_B - 3) \tag{2.101}$$

Rivlin–Saunders material:

$$W = c_1(I_B - 3) + f(II_B - 3) \tag{2.102}$$

where

$$I_B = B_{KK}\,, \quad II_B = \frac{1}{2}(B_{KK}B_{LL} - B_{KL}B_{KL}) \tag{2.103}$$

The stress tensor is given by

$$\boldsymbol{\sigma} = -p\boldsymbol{I} + 2\frac{\partial W}{\partial I_B}\boldsymbol{B} - 2\frac{\partial W}{\partial II_B}\boldsymbol{B}^{-1} \tag{2.104}$$

where p is the pressure, $\boldsymbol{I}$ is the unit tensor, and tensor $\boldsymbol{B}$ is the left Cauchy–Green tensor defined in (2.23).

As mentioned earlier, hypoelasticity deals with elastic solids in which the stress rate is proportional to the rate of deformation $\boldsymbol{D}$

$$\boldsymbol{\tau}^{\nabla} = \boldsymbol{C}:\boldsymbol{D} \tag{2.105}$$

where $\boldsymbol{\tau}^{\nabla}$, the Jaumann rate of Kirchhoff stress, which is a *frame-indifferent quantity*, and $\boldsymbol{D}$ are defined as

$$\boldsymbol{\tau}^{\nabla} = \dot{\boldsymbol{\tau}} - \boldsymbol{\Omega} \bullet \boldsymbol{\tau} + \boldsymbol{\tau} \bullet \boldsymbol{\Omega}, \quad \boldsymbol{D} = \frac{1}{2}(\nabla \boldsymbol{v} + \boldsymbol{v} \nabla), \tag{2.106}$$

respectively, and $\boldsymbol{v}$ is the velocity vector, $\boldsymbol{\Omega}$ the rotation tensor, and $\dot{\boldsymbol{\tau}}$ the material time derivative of Kirchhoff stress. If there is no abrupt change in the material response with respect to nonproportional load, the last two terms in the first part of (2.106) can be neglected. More detailed discussion on frame-indifference and the Jaumann stress rate can be found in Malvern (1969). Bažant and Cedolin (1991) showed that many other kinds of frame-indifferent stress rates (or objective stress rate) can be used, including Truesell's stress rate, Biot's stress rate, Green's stress rate, Oldroyd's stress rate, and Cotter–Rivlin stress rate. However, the choice of stress rate must be associated with work with some admissible finite strain as discussed in Section 2.8 (Section 11.3 of Bažant and Cedolin, 1991).

2.17 PLANE STRESS, PLANE STRAIN, AND THE AIRY STRESS FUNCTION

Three-dimensional problems in elasticity are difficult to solve. This is partly due to the fact that the geometrical shape of the boundary can, in general, be very irregular, and, partly because the conditions to be satisfied on the surface of the body can be *mixed* (i.e., displacement is prescribed on part of the surface while surface traction is imposed on the rest of the surface). In practice, two-dimensional (2-D) idealization is always used; two commonly used 2-D conditions are the plane strain and plane stress conditions. In particular, in plane strain condition the strain dyadic $\boldsymbol{\varepsilon}$ is reduced to

$$\boldsymbol{\varepsilon} = \varepsilon_{11}\boldsymbol{e}_1\boldsymbol{e}_1 + \varepsilon_{22}\boldsymbol{e}_2\boldsymbol{e}_2 + \varepsilon_{12}(\boldsymbol{e}_1\boldsymbol{e}_2 + \boldsymbol{e}_2\boldsymbol{e}_1) \tag{2.107}$$

In addition, it is assumed that all nonzero stress components are not functions of x_3. A typical example of plane strain condition is a slice of a dam section under water pressure, as shown in Fig. 2.11. Note that the displacement u_3 normal to the slice surface is identically zero, hence so is the strain component ε_{33}. Furthermore, from Fig. 2.11 it is obvious that u_1 and u_2 are functions of x_1 and x_2 only, but not of x_3. Therefore, the shear strain components ε_{13} and ε_{23} are identically zero. If we substitute (2.107) into Hooke's law, the nonzero stress components can easily to be shown as:

$$\boldsymbol{\sigma} = \sigma_{11}\boldsymbol{e}_1\boldsymbol{e}_1 + \sigma_{22}\boldsymbol{e}_2\boldsymbol{e}_2 + \sigma_{12}(\boldsymbol{e}_1\boldsymbol{e}_2 + \boldsymbol{e}_2\boldsymbol{e}_1) + \sigma_{33}\boldsymbol{e}_3\boldsymbol{e}_3 \tag{2.108}$$

As expected, a nonzero σ_{33} is required to prevent displacement along the x_3-direction. Plane strain condition does lead to a truly 2-D theory.

For *plane stress condition*, we consider a two-dimension body with thickness, say $2h$, much smaller than the other two dimensions, as shown in Fig. 2.12; and all loads are applied parallel to the x_1-x_2 plane. The stresses on the surface $x_3 = \pm h$ must vanish, i.e.,

$$\boldsymbol{e}_3 \bullet \boldsymbol{\sigma} = \sigma_{13}\boldsymbol{e}_1 + \sigma_{23}\boldsymbol{e}_2 + \sigma_{33}\boldsymbol{e}_3 = 0 \tag{2.109}$$

since all three stress components must vary from zero on $x_3 = -h$ to zero again on $x_3 = h$. We generally assume that all of them are identically zero. Without this assumption, the plane stress condition will not lead to two-dimensional theory. That is, the only nonzero stress components are

$$\boldsymbol{\sigma} = \sigma_{11}\boldsymbol{e}_1\boldsymbol{e}_1 + \sigma_{22}\boldsymbol{e}_2\boldsymbol{e}_2 + \sigma_{12}(\boldsymbol{e}_1\boldsymbol{e}_2 + \boldsymbol{e}_2\boldsymbol{e}_1) \tag{2.110}$$

where σ_{11}, σ_{22}, and σ_{12} are functions of x_1 and x_2 only, but not of x_3. Substitution of (2.110) into Hooke's law leads to

$$\boldsymbol{\varepsilon} = \varepsilon_{11}\boldsymbol{e}_1\boldsymbol{e}_1 + \varepsilon_{22}\boldsymbol{e}_2\boldsymbol{e}_2 + \varepsilon_{12}(\boldsymbol{e}_1\boldsymbol{e}_2 + \boldsymbol{e}_2\boldsymbol{e}_1) + \varepsilon_{33}\boldsymbol{e}_3\boldsymbol{e}_3 \tag{2.111}$$

in which the nonzero component ε_{33} indicating that the body is free to expand along the x_3-axis. As discussed by Timoshenko and Goodier (1982), this plane stress state satisfies the compatibility if and only if ε_{33} is a linear function of x_1 and x_2. This, of course, is too restrictive for general problems. Therefore, the solutions for plane stress problems are of approximate nature, but should closely resemble the actual solution if h is small (see Article 98 of Timoshenko and Goodier, 1982). In addition, the out-of-plane bucking in the x_3-direction is neglected in plane stress theory due to the assumption of small deformation.

For isotropic solids, it can easily be shown that the constitutive response for both plane strain and plane stress problems can be written in terms of a unified form (Karasudhi, 1991):

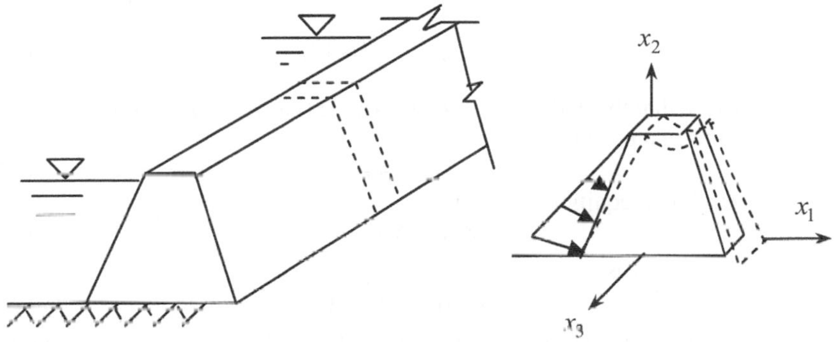

Figure 2.11 Example of plane strain condition

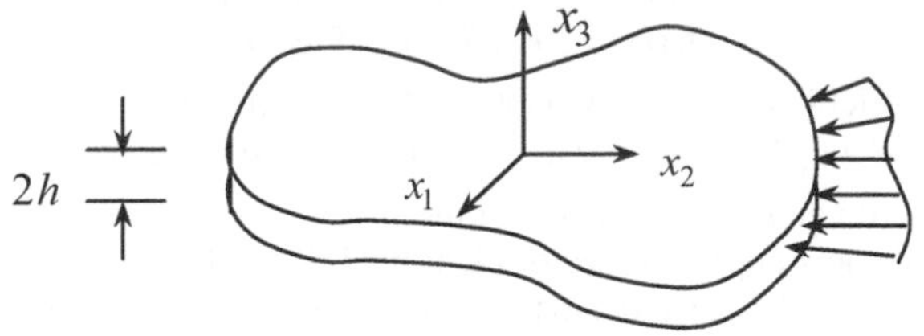

Figure 2.12 Example of plane stress condition

$$\varepsilon_{ij} = \frac{1}{2\mu}[\sigma_{ij} - \frac{3-\kappa}{4}(tr\,\boldsymbol{\sigma})\delta_{ij}] \tag{2.112}$$

where $i, j = 1, 2$ and $\boldsymbol{\sigma}$ is given in (2.110) and κ equals $(3-\nu)/(1+\nu)$ for the plane stress condition and $3-4\nu$ for the plane strain condition. Inversely, (2.112) can be written as

$$\sigma_{ij} = \frac{\mu}{\kappa - 1}[2(\kappa-1)\varepsilon_{ij} + (3-\kappa)(\mathrm{tr}\boldsymbol{\varepsilon})\delta_{ij}] \tag{2.113}$$

where, again, $i, j = 1, 2$.

For the plane strain condition, the equation of compatibility (2.35) is identically satisfied except for the following component:

$$\varepsilon_{11,22} + \varepsilon_{22,11} - 2\varepsilon_{12,12} = 0 \tag{2.114}$$

Substitution of (2.112) into (2.114) and elimination of the shear stress component by using the two-dimensional equilibrium equations yields the compatibility equation in terms of stress:

$$\nabla^2(tr\boldsymbol{\sigma}) = -\frac{4}{1+\kappa}(\nabla \cdot \boldsymbol{F}) \tag{2.115}$$

where

$$\boldsymbol{\nabla} = \boldsymbol{e}_i \frac{\partial}{\partial x_i}, \quad i = 1,2 \tag{2.116}$$

Now we assume that the body force is conservative (i.e., $F_{1,2} = F_{2,1}$), such that it can be written in terms of a potential V as

$$\boldsymbol{F} = -\boldsymbol{\nabla} V \tag{2.117}$$

Then the equation of equilibrium becomes

$$\boldsymbol{\nabla} \cdot \boldsymbol{\sigma} - \boldsymbol{\nabla} V = 0 \tag{2.118}$$

which can be written as

$$\boldsymbol{\nabla} \cdot [\boldsymbol{\sigma} - \boldsymbol{I}V] = 0 \tag{2.119}$$

where $\boldsymbol{I} = \boldsymbol{e}_1\boldsymbol{e}_1 + \boldsymbol{e}_2\boldsymbol{e}_2$. Since the divergence of a curl must be zero (see (1.47) of Chapter 1), $\boldsymbol{\sigma} - \boldsymbol{I}V$ must be the curl of some dyadic. In addition, $\boldsymbol{\sigma}$ is symmetric and with only nonzero components σ_{11}, σ_{22}, and σ_{12}, it is therefore natural to try the symmetric dyadic $\varphi(x_1, x_2)\,\boldsymbol{e}_3\boldsymbol{e}_3$ which is operated twice symmetrically with the curl operator. That is, we assume

$$\boldsymbol{\sigma} - \mathbf{I}V = -\boldsymbol{\nabla} \times (\boldsymbol{e}_3\boldsymbol{e}_3\varphi) \times \boldsymbol{\nabla}, \quad \textbf{or} \quad \boldsymbol{\sigma} = \mathbf{I}V - \boldsymbol{\nabla} \times (\boldsymbol{e}_3\boldsymbol{e}_3\varphi) \times \boldsymbol{\nabla} \tag{2.120}$$

We now consider the trace of the two-dimensional stress tensor:

$$\mathrm{tr}(\boldsymbol{\sigma}) = 2V + \boldsymbol{\nabla} \times \boldsymbol{e}_3 \bullet \boldsymbol{\nabla}\varphi \times \boldsymbol{e}_3 \tag{2.121}$$

Using the vector identity (1.48) of Chapter 1, the last term in (2.121) can be written as:

$$\boldsymbol{\nabla} \times \boldsymbol{e}_3 \bullet \boldsymbol{\nabla}\varphi \times \boldsymbol{e}_3 = \boldsymbol{\nabla} \bullet (\boldsymbol{e}_3 \times \boldsymbol{\nabla}\varphi \times \boldsymbol{e}_3) + \boldsymbol{e}_3 \bullet (\boldsymbol{\nabla} \times \boldsymbol{\nabla}\varphi \times \boldsymbol{e}_3) \tag{2.122}$$

where the last term vanishes as the curl of the grad of a scalar must be zero (see (1.45) of Chapter 1). We now apply the vector identity (1.48) of Chapter 1 to the first term on the right-hand side of (2.122) and substitute the result into (2.121), we get

$$\mathrm{tr}(\boldsymbol{\sigma}) = 2V + \boldsymbol{\nabla} \bullet [\boldsymbol{e}_3 \bullet (\boldsymbol{e}_3\boldsymbol{\nabla}\varphi - \boldsymbol{\nabla}\varphi\boldsymbol{e}_3)] = 2V + \boldsymbol{\nabla} \bullet \boldsymbol{\nabla}\varphi \tag{2.123}$$

Note that it is easy to see $\boldsymbol{e}_3 \cdot \boldsymbol{\nabla} \equiv 0$ in (2.123) if we recall the definition of $\boldsymbol{\nabla}$ given in (2.116). Substitution of (2.117) and (2.123) into (2.115) leads to the following governing equation for φ and V:

$$(\nabla\bullet\nabla)(\nabla\bullet\nabla)\varphi = \nabla^2\nabla^2\varphi = \nabla^4\varphi = -2(\frac{\kappa-1}{\kappa+1})\nabla^2 V \tag{2.124}$$

where ∇^4 is the biharmonic *operator* and φ is called the *Airy stress function,* which was originally proposed by G.B. Airy in 1863. This equation, of course, agrees with those given in Article 17 of Timoshenko and Goodier (1982). The above equation reduces to the usual biharmonic equation for the Airy stress function if body force is neglected. The stress is given in terms of Airy stress function in (2.120). If the body force is not conservative, we refer to the recent publications by Ho and Chau (1997, 1999) and Chau and Wei (2001b). A very detailed historical review on two-dimensional biharmonic functions was given by Meleshko (2003).

Although we have considered the Cartesian components in deriving (2.120) and (2.124), the procedure can easily be modified to cylindrical coordinate. That is, both of them remain valid for cylindrical coordinates. For example, in terms of Airy stress function and body force potential the stress tensor becomes

$$\boldsymbol{\sigma} = [(\frac{1}{r}\frac{\partial\varphi}{\partial r} + \frac{1}{r^2}\frac{\partial^2\varphi}{\partial\phi^2} + V)\boldsymbol{e}_r\boldsymbol{e}_r + (\frac{\partial^2\varphi}{\partial r^2} + V)\boldsymbol{e}_\phi\boldsymbol{e}_\phi - \frac{\partial}{\partial r}(\frac{1}{r}\frac{\partial\varphi}{\partial\phi})(\boldsymbol{e}_r\boldsymbol{e}_\phi + \boldsymbol{e}_\phi\boldsymbol{e}_r)] \tag{2.125}$$

That is, the stress components are

$$\sigma_{rr} = \frac{1}{r}\frac{\partial\varphi}{\partial r} + \frac{1}{r^2}\frac{\partial^2\varphi}{\partial\phi^2} + V \tag{2.126}$$

$$\sigma_{\phi\phi} = \frac{\partial^2\varphi}{\partial r^2} + V \tag{2.127}$$

$$\sigma_{r\phi} = -\frac{\partial}{\partial r}(\frac{1}{r}\frac{\partial\varphi}{\partial\phi}) \tag{2.128}$$

Now the solution for 2-D elasticity becomes the solution of (2.124) with appropriate boundary conditions. The general solution for (2.124) would be the homogenous solution for the biharmonic equation plus the particular solution, which depends on the form of the body force potential V.

We first consider the homogeneous solution. In explicit form, Airy stress function satisfies the following equation in polar coordinate:

$$(\frac{\partial^2}{\partial r^2} + \frac{1}{r}\frac{\partial}{\partial r} + \frac{1}{r^2}\frac{\partial^2}{\partial\phi^2})(\frac{\partial^2}{\partial r^2} + \frac{1}{r}\frac{\partial}{\partial r} + \frac{1}{r^2}\frac{\partial^2}{\partial\phi^2})\varphi = 0 \tag{2.129}$$

By using separation of variables, Michell (1899) presented a so-called "general" solution:

$$\begin{aligned}\varphi(r,\phi) &= (A + Br^2)\ln(\frac{r}{R}) + Cr^2 + D + (E + Fr^2)\phi + r\phi(G\cos\phi + H\sin\phi)\\ &+ (A_1 r^3 + B_1 r^{-1} + C_1 r\ln(\frac{r}{R}))(E_1\cos\phi + F_1\sin\phi)\\ &+ \sum_{n=2}^{\infty}[A_n r^{n+2} + B_n r^{-n} + C_n r^n + D_n r^{2-n}](E_n\cos n\phi + F_n\sin n\phi)\end{aligned} \tag{2.130}$$

where R is an arbitrary constant to normalize r. Note that the $r^2\phi$ term in (2.130) was actually added by Timoshenko and Goodier (1982). The validity of (2.130) can simply be checked by its direct substitution in (2.129). A comprehensive table for stress and displacement components due to each term of this homogeneous solution was compiled by Karasudhi (1991). For some given conservative body force potential V, the particular solutions were also tabulated (Karasudhi, 1991).

However, as summarized by Meleshko (2003), some terms were missing in Michell's solution. The most general solution form for separation of variables is (Filonenko-Borodich, 1965; Sadeh, 1967; Mann, 1949)

$$\begin{aligned}\varphi(r,\phi) &= A_0 r^2 \ln r\phi + B_0 \ln r\phi + C_0 r \ln r\phi \cos\phi + D_0 r \ln r\phi \sin\phi \\ &+(A+Br^2)\ln(\frac{r}{R}) + Cr^2 + D + (E+Fr^2)\phi + r\phi(G\cos\phi + H\sin\phi) \\ &+(A_1 r^3 + B_1 r^{-1} + C_1 r\ln(\frac{r}{R}))(E_1\cos\phi + F_1\sin\phi) \\ &+\sum_{n=2}^{\infty}[A_n r^{n+2} + B_n r^{-n} + C_n r^n + D_n r^{2-n}](E_n\cos n\phi + F_n\sin n\phi)\end{aligned} \tag{2.131}$$

Physically, the first four terms in (2.131) correspond to more general types of dislocations with discontinuous stresses (e.g., Mann, 1949). See also the discussion by Wan (1968), Bert (1968), and Hyman (1968).

2.18 STRESS CONCENTRATION AT A CIRCULAR HOLE

Consider an infinite two-dimensional elastic body containing a circular hole of radius R; and uniform compression is applied at infinity along the x_1-direction as shown in Fig. 2.13. This solution was obtained by Kirsch in 1898 and is also called the Kirsch solution (Hetnarski and Ignaczak, 2011). This simple problem can, however, be used to model the stress field around a tunnel (or a borehole) due to overburden pressure (or lateral earth pressure).

The far field stress tensor is $\boldsymbol{\sigma} = -T\boldsymbol{e}_1\boldsymbol{e}_1$ or in polar form

$$\begin{aligned}\boldsymbol{\sigma} &= -(\boldsymbol{e}_r\cos\phi - \boldsymbol{e}_\phi\sin\phi)(\boldsymbol{e}_r\cos\phi - \boldsymbol{e}_\phi)T \\ &= -\frac{T}{2}[(1+\cos 2\phi)\boldsymbol{e}_r\boldsymbol{e}_r + (1-\cos 2\phi)\boldsymbol{e}_\phi\boldsymbol{e}_\phi - \sin 2\phi(\boldsymbol{e}_r\boldsymbol{e}_\phi + \boldsymbol{e}_\phi\boldsymbol{e}_r)]\end{aligned} \tag{2.132}$$

as $r \to \infty$. On the contrary, the traction-free condition on the circular hole applies,

$$\boldsymbol{n}\cdot\boldsymbol{\sigma} = -\boldsymbol{e}_r\cdot\boldsymbol{\sigma} = 0 \tag{2.133}$$

on $r = R$. We now look for an Airy stress function that will yield the far field stress at infinity ($r \to \infty$) given by (2.132) but at the same time yield near field stress satisfying (2.133). An obvious choice for φ is

$$\varphi(r,\phi) = A\ln(\frac{r}{R}) + Cr^2 + (B_2 r^{-2} + C_1 r^2 + D_2)\cos 2\phi \tag{2.134}$$

Substitution of (2.134) into (2.125) yields

$$\begin{aligned}\boldsymbol{\sigma} &= \boldsymbol{e}_r\boldsymbol{e}_r[Ar^{-2} + 2C - (6B_2 r^{-4} + 2C_2 + 4D_2 r^{-2})\cos 2\phi] \\ &+ \boldsymbol{e}_\phi\boldsymbol{e}_\phi[-Ar^{-2} + 2C + (6B_2 r^{-4} + 2C_2)\cos 2\phi] \\ &+ (\boldsymbol{e}_r\boldsymbol{e}_\phi + \boldsymbol{e}_\phi\boldsymbol{e}_r)(-6B_2 r^{-4} + 2C_2 - 2D_2 r^{-2})\sin 2\phi\end{aligned} \tag{2.135}$$

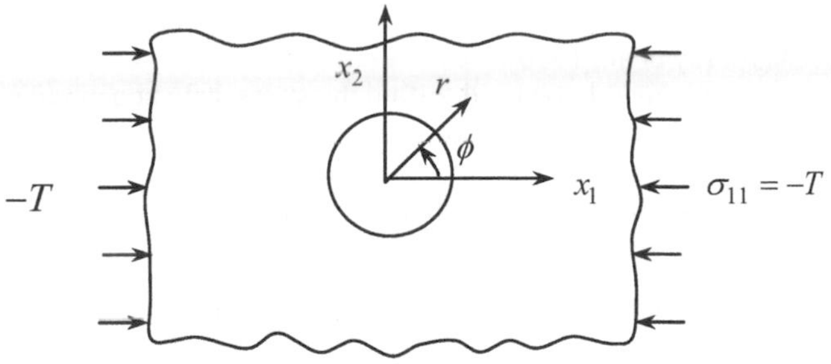

Figure 2.13 An infinite plane with a circular hole under compression

The far field condition (2.132) gives

$$2C = -\frac{T}{2}, \quad 2C_2 = \frac{T}{2} \tag{2.136}$$

The boundary condition on $r = R$ yields

$$A = \frac{TR^2}{2} \tag{2.137}$$

and the following coupled equations for B_2 and D_2:

$$\frac{T}{2} + 6B_2 R^{-4} + 4D_2 R^{-2} = 0, \tag{2.138}$$

$$\frac{T}{2} - 6B_2 R^{-4} - 2D_2 R^{-2} = 0 \tag{2.139}$$

The solutions for (2.138) and (2.139) are

$$B_2 = \frac{TR^4}{4}, \quad D_2 = -\frac{TR^2}{2} \tag{2.140}$$

Back substitution of all the constants into (2.135) gives

$$\begin{aligned} \boldsymbol{\sigma} &= -\frac{T}{2}\{\boldsymbol{e}_r\boldsymbol{e}_r\{1-(\frac{R}{r})^2+[1+3(\frac{R}{r})^4-4(\frac{R}{r})^2]\cos 2\phi\} \\ &+\boldsymbol{e}_\phi\boldsymbol{e}_\phi\{1+(\frac{R}{r})^2-[1+3(\frac{R}{r})^4]\cos 2\phi\} \\ &+(\boldsymbol{e}_r\boldsymbol{e}_\phi+\boldsymbol{e}_\phi\boldsymbol{e}_r)[-1+3(\frac{R}{r})^4-2(\frac{R}{r})^2]\sin 2\phi\} \end{aligned} \tag{2.141}$$

The stress concentration is expected to be the most severe on the circumference of the hole. On $r = R$, the stress dyadic reduces to

$$\boldsymbol{\sigma}_{r=R} = -T\boldsymbol{e}_\phi\boldsymbol{e}_\phi(1-2\cos 2\phi) \tag{2.142}$$

The hoop stress ($\sigma_{\phi\phi}$) at $\phi = 0$ is T (tensile) while the one at $\phi = \pi/2$ is $-3T$ (compression). For brittle rocks, the compressive strength to tensile strength is typically in the order of 10 to 40. Thus, if we assume that the required stresses for cracking and crushing are roughly proportional to the tensile and compressive strengths of the rock, respectively, one should expect cracking at $\phi = 0$ well before crushing takes place as shown in Fig. 2.14. More rigorous analysis including fracture mechanics of the vertical cracks emanating from the hole did show the same conclusion (Gharpuray et al., 1990).

This solution can readily be used to find the stress state at a circular tunnel under a more practical loading—biaxial compression, as shown in Fig. 2.15. We can simply add the stress dyadic given in (2.141) to another stress dyadic, which is obtained by replacing T and ϕ by αT and $\phi+\pi/2$ in (2.141), respectively. The following stress dyadic for the biaxial compression is obtained:

$$\begin{aligned}\boldsymbol{\sigma} &= \boldsymbol{e}_r\boldsymbol{e}_r\{-\frac{T}{2}(1+\alpha)[1-(\frac{R}{r})^2]-\frac{T}{2}(1-\alpha)[1+3(\frac{R}{r})^4-4(\frac{R}{r})^2]\cos 2\phi\} \\ &+\boldsymbol{e}_\phi\boldsymbol{e}_\phi\{-\frac{T}{2}(1+\alpha)[1+(\frac{R}{r})^2]+\frac{T}{2}(1-\alpha)[1+3(\frac{R}{r})^4]\cos 2\phi\} \\ &-\frac{T}{2}(1-\alpha)(\boldsymbol{e}_r\boldsymbol{e}_\phi+\boldsymbol{e}_\phi\boldsymbol{e}_r)[-1+3(\frac{R}{r})^4-2(\frac{R}{r})^2]\sin 2\phi\end{aligned} \tag{2.143}$$

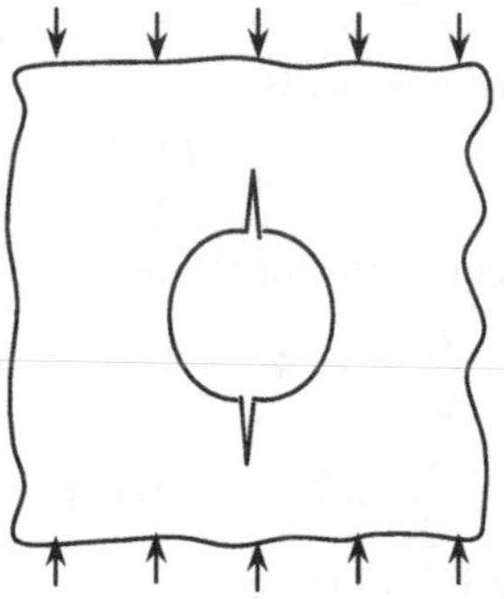

Figure 2.14 The position of crack nucleation at a circular hole under compression

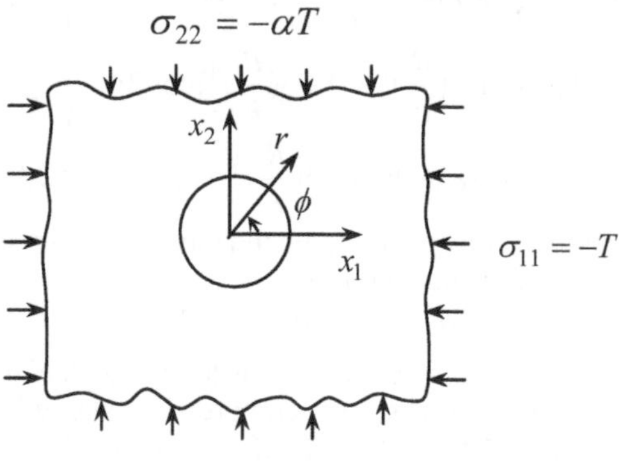

Figure 2.15 A tunnel subject to biaxial stresses

This solution can be applied to modeling the stress field around either a circular tunnel or a borehole. For example, Terzaghi and Richart (1952) applied this solution to examine the stress around a tunnel under geostatic stress state. We assume that σ_{22} is the vertical stress due to the overburden stress (i.e., $\sigma_{22} = -\gamma z$, where z is the depth of the tunnel and γ the unit weight of rock) and σ_{11} is the lateral earth pressure, which can be approximated by

$$\sigma_{11} = -\sigma_h = -K\gamma z = -(\frac{\nu}{1-\nu})\gamma z \tag{2.144}$$

where ν is the Poisson's ratio of the rock. For a typical Poisson's ratio of 0.2 for rock, K equals 0.25. In applying this to (2.144), we can set $T = K\gamma z$ and $\alpha = 1/K$. Similar to the uniaxial compression case, it is straightforward to see that tensile hoop stress of magnitude $0.25\gamma z$ develops at the roof of the tunnel.

2.19 FORCE ACTING AT THE APEX OF A WEDGE

In this section, we consider a thin plate in the form of a wedge which is bounded by the planes $x_3 = \pm b$, and by the planes $\phi = \pm\alpha$. As shown in Fig. 2.16, the wedge is assumed to be fixed at a great distance from the origin, and subject to a concentrated force $\boldsymbol{F}$ applied at the origin. This problem was first considered by Michell (1902). The boundary condition on the planes $\phi = \pm\alpha$ is

$$\boldsymbol{e}_\phi \bullet \boldsymbol{\sigma} = 0, \qquad (\phi = \pm\alpha) \tag{2.145}$$

Substitution of (2.125) into (2.145) gives:

$$[\boldsymbol{e}_\phi \frac{\partial^2 \varphi}{\partial r^2} - \boldsymbol{e}_r \frac{\partial}{\partial r}(\frac{1}{r}\frac{\partial \varphi}{\partial \phi})]_{\phi=\pm\alpha} = 0 \tag{2.146}$$

It can be shown that the following form of the Airy stress function can be used:

$$\varphi(r,\phi) = r\phi(G\cos\phi + H\sin\phi) \tag{2.147}$$

With the Airy stress function given in (2.147), (2.125) yields the stress dyadic

$$\boldsymbol{\sigma} = \boldsymbol{e}_r\boldsymbol{e}_r \frac{2}{r}(H\cos\phi - G\sin\phi) \tag{2.148}$$

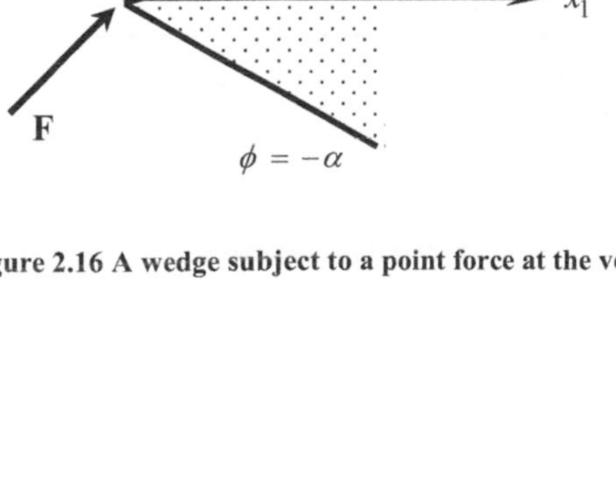

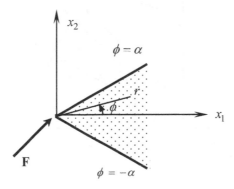

Figure 2.16 A wedge subject to a point force at the vertex

This expression predicts that the stress becomes infinity as r tends toward zero. To remove this singularity, we exclude the origin from the physical wedge and assume the force $\boldsymbol{F}$ being distributed on a small circular arc of radius δ. That is,

$$\int (-\boldsymbol{e}_r \bullet \boldsymbol{\sigma})_{r=\delta}\, dS = -2\int_{-\alpha}^{+\alpha} \boldsymbol{e}_r (H\cos\phi - G\sin\phi) d\phi = \frac{\boldsymbol{F}}{2b} \tag{2.149}$$

If we write $\boldsymbol{F}$ in terms of its Cartesian components (i.e., $\boldsymbol{F} = \boldsymbol{e}_1 F_1 + \boldsymbol{e}_2 F_2$ and $\boldsymbol{e}_r = \boldsymbol{e}_1 \cos\phi + \boldsymbol{e}_2 \sin\phi$) we have

$$H = -\frac{F_1/(2b)}{2\alpha + \sin 2\alpha}, \quad G = \frac{F_2/(2b)}{2\alpha - \sin 2\alpha} \tag{2.150}$$

The stress dyadic thus becomes

$$\boldsymbol{\sigma} = -\boldsymbol{e}_r \boldsymbol{e}_r \frac{1}{rb}\left[\frac{F_1 \cos\phi}{2\alpha + \sin 2\alpha} + \frac{F_2 \sin\phi}{2\alpha - \sin 2\alpha}\right] \tag{2.151}$$

The stress field for a semi-infinite plane subject to an inclined point force can readily be obtained as a special case of (2.151) by setting $\alpha = \pi/2$. Therefore, the stress dyadic becomes

$$\boldsymbol{\sigma} = -\boldsymbol{e}_r \boldsymbol{e}_r \frac{1}{\pi b r}[F_1 \cos\phi + F_2 \sin\phi] \tag{2.152}$$

which can further be simplified by letting the magnitude of the point force be $F = (F_1^2 + F_2^2)^{1/2}$. Thus, as shown in Fig. 2.17, if β is the angle between the normal and the direction of $\boldsymbol{F}$, we can write

$$\boldsymbol{\sigma} = -\boldsymbol{e}_r \boldsymbol{e}_r \frac{F}{\pi b r}\cos(\beta - \phi) \tag{2.153}$$

Now an interesting conclusion can be drawn if we draw a circle with diameter d such that $r = d\cos(\beta - \phi)$ as shown in Fig. 2.17, then the stress on such a circle becomes constant and equals

$$\boldsymbol{\sigma} = -\boldsymbol{e}_r \boldsymbol{e}_r \frac{F}{\pi b d} \tag{2.154}$$

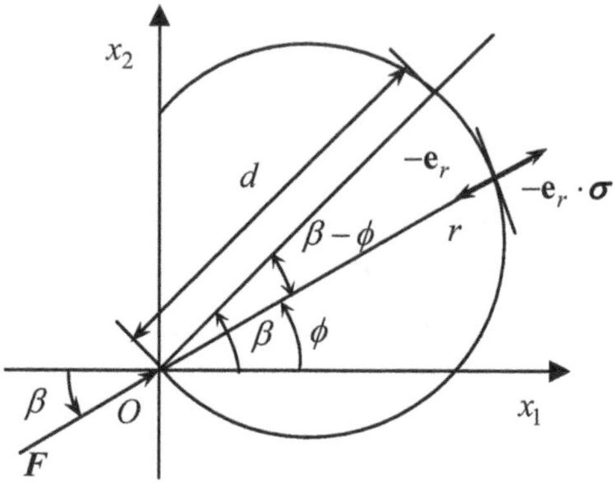

Figure 2.17 The circle of uniform radial stress near the surface

The traction on this circle is $-\boldsymbol{e}_r\cdot\boldsymbol{\sigma}$ or $\boldsymbol{e}_r F/(\pi bd)$, which is always radial and inversely proportional to the diameter d. This solution can be used to synthesize the solutions for many other loadings or other two-dimensional problems by the *principle of superposition*. Poulos and Davis (1974) have compiled a rather comprehensive list of solutions for various types of loading on the surface of a half plane.

2.20 UNIFORM VERTICAL LOADING ON PART OF THE SURFACE

The stress field under a patch of two-dimensional load can easily be obtained by integrating the solution for line load found in the previous section. We first set $F_2 = 0$ in (2.152) and rewrite the stress dyadic in Cartesian coordinates:

$$\begin{aligned}\boldsymbol{\sigma} &= -\boldsymbol{e}_r\boldsymbol{e}_r\frac{F_1\cos\phi}{\pi br} = -\frac{F_1\cos\phi}{\pi br}(\boldsymbol{e}_1\cos\phi+\boldsymbol{e}_2\sin\phi)(\boldsymbol{e}_1\cos\phi+\boldsymbol{e}_2\sin\phi)\\ &= -\frac{F_1}{\pi br}[\boldsymbol{e}_1\boldsymbol{e}_1\cos^3\phi+\boldsymbol{e}_2\boldsymbol{e}_2\sin^2\phi\cos\phi+(\boldsymbol{e}_1\boldsymbol{e}_2+\boldsymbol{e}_2\boldsymbol{e}_1)\sin\phi\cos^2\phi]\end{aligned} \tag{2.155}$$

We further write r and ϕ in terms of x_1 and x_2 as:

$$\boldsymbol{\sigma} = -\frac{F_1}{\pi b}[\boldsymbol{e}_1\boldsymbol{e}_1\frac{x_1^3}{(x_1^2+x_2^2)^2}+\boldsymbol{e}_2\boldsymbol{e}_2\frac{x_2^2x_1}{(x_1^2+x_2^2)^2}+(\boldsymbol{e}_1\boldsymbol{e}_2+\boldsymbol{e}_2\boldsymbol{e}_1)\frac{x_2x_1^2}{(x_1^2+x_2^2)^2}] \tag{2.156}$$

This line force solution can now be used to estimate the patch load solution. We replace F_1 by qdx_2, and carry out the integration along the surface. This can be done more conveniently in terms of $d\theta$. Referring to Fig. 2.18, we find

$$dx_2 = \frac{rd\theta}{\cos(\beta+\theta)} = \frac{x_1 d\theta}{\cos^2(\beta+\theta)} \tag{2.157}$$

In obtaining the second part of (2.157), we used $r\cos(\theta+\beta) = x_1$. The stress dyadic at point P becomes

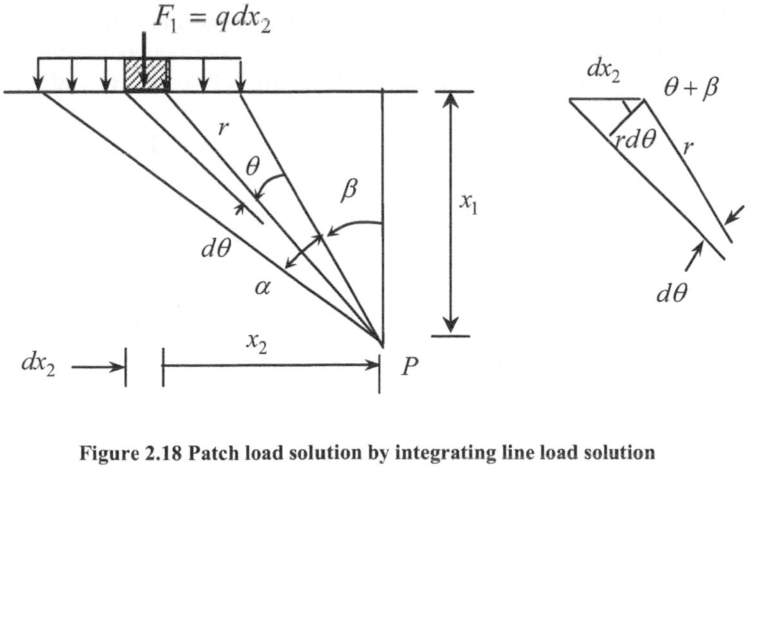

Figure 2.18 Patch load solution by integrating line load solution

$$\boldsymbol{\sigma} = -\frac{q}{\pi b}\int_0^{\alpha}[\boldsymbol{e}_1\boldsymbol{e}_1\frac{x_1^3}{(x_1^2+x_2^2)^2}+\boldsymbol{e}_2\boldsymbol{e}_2\frac{x_2^2 x_1}{(x_1^2+x_2^2)^2}+(\boldsymbol{e}_1\boldsymbol{e}_2+\boldsymbol{e}_2\boldsymbol{e}_1)\frac{x_2 x_1^2}{(x_1^2+x_2^2)^2}]\frac{x_1 d\theta}{\cos^2(\beta+\theta)} \tag{2.158}$$

Note that x_1 is independent of θ while $x_2 = x_1\tan(\theta + \beta)$. Using this information and after integration, we obtain

$$\begin{aligned}\boldsymbol{\sigma} = -\frac{q}{2\pi b}\{&\boldsymbol{e}_1\boldsymbol{e}_1[\alpha+\sin\alpha\cos(\alpha+2\beta)]+\boldsymbol{e}_2\boldsymbol{e}_2[\alpha-\sin\alpha\cos(\alpha+2\beta)]\\ &+(\boldsymbol{e}_1\boldsymbol{e}_2+\boldsymbol{e}_2\boldsymbol{e}_1)\sin\alpha\sin(\alpha+2\beta)\}\end{aligned} \tag{2.159}$$

This solution, of course, agrees with the formulas given by Poulos and Davis (1974).

2.21 SOLUTION FOR INDIRECT TENSILE TEST (BRAZILIAN TEST)

Both concrete and rocks are very strong under compression but extremely weak in tension. Experimental observations show that even when specimens are loaded under uniaxial compression, the actual failure is due to tensile crack growth. This issue will be discussed in more detail in a later chapter. Therefore, the tensile strength is a very important parameter for brittle geomaterials. Since it is very difficult to apply uniaxial tension on rock specimens without inducing any stress concentration or unwanted bending stress due to eccentricity, indirect tensile tests are more commonly used than uniaxial tensile tests. Among these indirect tensile tests, the *Brazilian test* remains the most popular test procedure for obtaining the tensile stress for rocks and concrete. Although this test is called the Brazilian test, it was proposed independently by Carneiro in Brazil and by Akazawa in Japan about 1943 (Fairbairn and Ulm, 2002).

Figure 2.19 illustrates a typical experimental set-up for the Brazilian test: a rock core of length $2b$ subjected to concentrated diametral forces $\boldsymbol{F}$ and $-\boldsymbol{F}$ acting on the circumference of the rock specimen at points O and O_1, respectively.

The results of Section 2.20 showed that the contour of constant stress $\boldsymbol{e}_r \cdot \boldsymbol{\sigma} \cdot \boldsymbol{e}_r$ is a circle through point O of the boundary of the infinite plate and with its center on the line of action of the point force. As shown in Fig. 2.19, our boundary for the indirect tensile test is exactly the circle of constant stress for both points O and O_1, so it is natural to try the following stress dyadic:

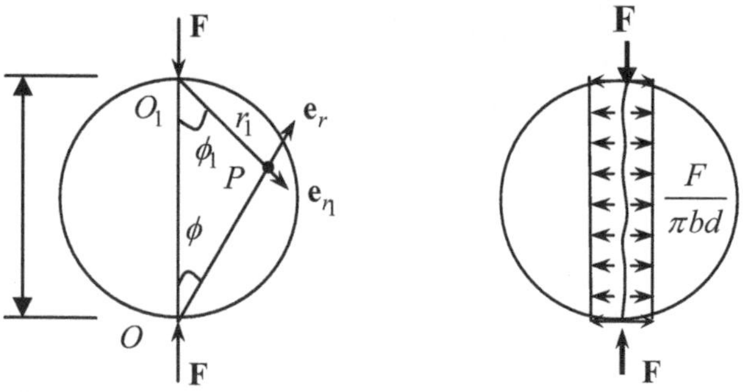

Figure 2.19 Rock core subject to diametral point forces

$$\boldsymbol{\sigma} = -\frac{F}{\pi b}[\boldsymbol{e}_r\boldsymbol{e}_r\frac{\cos\phi}{r} + \boldsymbol{e}_{r1}\boldsymbol{e}_{r1}\frac{\cos\phi_1}{r_1}] \tag{2.160}$$

On the circular boundary this stress dyadic, however, does not yield a traction-free boundary. In particular, we have $d = r/\cos\phi = r_1/\cos\phi_1$, and $\boldsymbol{e}_r$ and $\boldsymbol{e}_{r1}$ must be orthogonal. Therefore, on the boundary (2.160) becomes

$$-\frac{F}{\pi bd}[\boldsymbol{e}_r\boldsymbol{e}_r + \boldsymbol{e}_\phi\boldsymbol{e}_\phi] \tag{2.161}$$

To remove this unwanted stress dyadic on the circular boundary, we must add a stress tensor of different sign. Therefore, the solution for the stress field in a Brazilian test is

$$\boldsymbol{\sigma} = \frac{F}{\pi b}[(\boldsymbol{e}_r\boldsymbol{e}_r + \boldsymbol{e}_\phi\boldsymbol{e}_\phi)\frac{1}{d} - \boldsymbol{e}_r\boldsymbol{e}_r\frac{\cos\phi}{r} - \boldsymbol{e}_{r1}\boldsymbol{e}_{r1}\frac{\cos\phi_1}{r_1}] \tag{2.162}$$

Along the direction of the action of force (i.e., diameter O-O_1), we have $\phi = \phi_1 = 0$ and $r_1 = d - r$. Consequently, the stress becomes

$$\boldsymbol{\sigma} = \frac{F}{\pi b}[(\frac{rd - r^2 - d^2}{rd(d-r)})\boldsymbol{e}_r\boldsymbol{e}_r + \frac{1}{d}\boldsymbol{e}_\phi\boldsymbol{e}_\phi] \tag{2.163}$$

along the diameter O-O_1. The radial stress is always compressive with a minimum at the center, $(\sigma_{rr})^{\min} = 3F/(\pi bd)$, and approaches infinity at both points O and O_1. More importantly, the circumferential stress $\sigma_{\phi\phi}$ along O-O_1 is always tensile and equals $F/(\pi bd)$. This is the reason why the Brazilian test is suitable as an indirect tensile test.

In practice, the point load is actually recommended to be disturbed over a distance of $d/12$ at both points O and O_1. As shown in Fig. 2.20, this situation can be modeled by a uniform radial pressure p acting over an angle of 2α at both ends of the diameter. This loading can be modeled mathematically by Fourier series expansion, and the stress dyadic along O-O_1 is found to be (Hondros, 1959)

$$\begin{aligned}\boldsymbol{\sigma} = \frac{2p}{\pi}\{&-[\frac{(1-\rho^2)\sin 2\alpha}{1-2\rho^2\cos 2\alpha+\rho^4} + \tan^{-1}(\frac{1+\rho^2}{1-\rho^2})\tan\alpha]\boldsymbol{e}_r\boldsymbol{e}_r \\ &+[\frac{(1-\rho^2)\sin 2\alpha}{1-2\rho^2\cos 2\alpha+\rho^4} - \tan^{-1}(\frac{1+\rho^2}{1-\rho^2})\tan\alpha]\boldsymbol{e}_\phi\boldsymbol{e}_\phi\}\end{aligned} \tag{2.164}$$

where $\rho = r/R$ is the normalized radial distance.

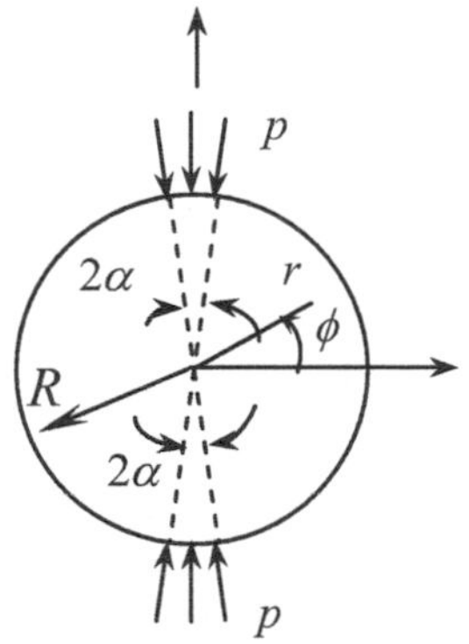

Figure 2.20 Brazilian test loaded by pressure over an angle of 2α at both ends

This theoretical analysis was also extended to transversely isotropic materials by Wang and Chong (1989); they concluded that the Brazilian test is valid for transversely isotropic solids but the detailed stress distribution was not examined.

Another related indirect strength test is the point load strength test, and the theory of elasticity has also been found essential in estimating the tensile stress inside the tested rock specimens (e.g., Wei and Chau,1998; Chau,1998c; Chau and Wong, 1996). The mathematical details will, however, not be discussed here.

2.22 JAEGER'S MODIFIED BRAZILIAN TEST

Jaeger and Cook (1967) proposed a modification to the traditional Brazilian test by applying three line loads instead of just two in the conventional one. Figure 2.21 shows a rock core of radius R being loaded by the three-line load test. Each of the line loads is W per unit length of the specimen, and the angle between each line load is $2\pi/3$. The stress field can again be modeled by Fourier series expansion, and the resulting series in the stress components can be summed exactly to give

$$\begin{aligned}\boldsymbol{\sigma} = \frac{3W}{2\pi R}\{&\boldsymbol{e}_r\boldsymbol{e}_r\left[\frac{1-\rho^6}{1-2\rho^3\cos 3\phi+\rho^6}-\frac{3\rho(\rho^2-1)[(1+\rho^6)\cos 3\phi-2\rho^3]}{[1-2\rho^3\cos 3\phi+\rho^6]^2}\right] \\ &+\boldsymbol{e}_\phi\boldsymbol{e}_\phi\left[\frac{1-\rho^6}{1-2\rho^3\cos 3\phi+\rho^6}+\frac{3\rho(\rho^2-1)[(1+\rho^6)\cos 3\phi-2\rho^3]}{[1-2\rho^3\cos 3\phi+\rho^6]^2}\right] \\ &+(\boldsymbol{e}_r\boldsymbol{e}_\phi+\boldsymbol{e}_\phi\boldsymbol{e}_r)\frac{3\rho(\rho^2-1)[(1-\rho^6)\sin 3\phi]}{[1-2\rho^3\cos 3\phi+\rho^6]^2}\}\end{aligned} \tag{2.165}$$

where $\rho = r/R$.

If the line loads are distributed over an angular width of 2α, the stress dyadic acting along $\phi = 0$ is given by (Jaeger and Cook, 1976):

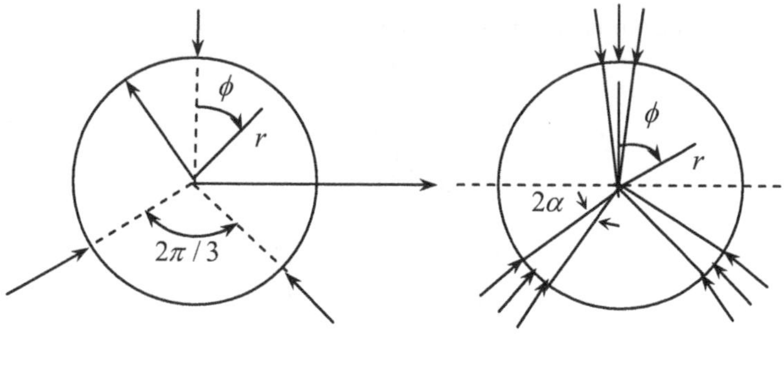

Figure 2.21 Jaeger's modified Brazilian test

$$\boldsymbol{\sigma} = \frac{3W}{2\pi R}\{\mathbf{e_r e_r}[3+\frac{2}{\alpha}\tan^{-1}(\frac{\rho^3\sin 3\alpha}{1-\rho^3\cos 3\phi}) - \frac{3\rho(\rho^2-1)\sin 3\phi}{\alpha(1-2\rho^3\cos 3\phi+\rho^6)}]$$

$$+\mathbf{e_\phi e_\phi}[3+\frac{2}{\alpha}\tan^{-1}(\frac{\rho^3\sin 3\alpha}{1-\rho^3\cos 3\phi}) + \frac{3\rho(\rho^2-1)\sin 3\phi}{\alpha(1-2\rho^3\cos 3\phi+\rho^6)}]\} \tag{2.166}$$

It can be shown that this stress does not lead to uniform tensile stress in the cylinder. Tensile stress only appears near the center of the cylinder.

2.23 EDGE DISLOCATION

In this section, we will discuss the theory of edge and screw dislocation based on two-dimensional elasticity. Dislocation was first proposed by G.I. Taylor in 1934 and E. Orowan in 1934 to explain defects in lattices. However, the mathematical theory of dislocation had been considered earlier in 1907 by V. Volterra who considered a thick circular cylinder subject to a cut parallel to the axis of the cylinder, as shown in Fig. 2.22.

The relative displacement b_x and b_y are called edge dislocations, whereas b_z is called screw dislocation. These are now called Burgers vectors. After relative displacement of Burgers vectors is imposed, the cuts are healed by welding back the cylinder material. The rough edges of the cylinder are then grinned and polished to get back the original shape. The cylinder looks the same, but residual stress is trapped in the cylinders because of the imposed Burgers vectors. These stresses are in self-equilibrium.

For pure edge dislocation as shown in Fig. 2.23, we have defined the Burgers vector to be positive if we go around a dislocation line (i.e., the direction z going out of the page in Fig. 2.23) clockwise, and the positive directions of b_x and b_y are shown. It can be shown that we can set the Airy stress function as

$$\varphi = \frac{2b_y\mu}{\pi(1+\kappa)} r\ln r\cos\theta \tag{2.167}$$

Note that logarithmic function of a dimensional number is meaningless. In order to have a physically meaningful solution, we must have normalized all distance respective to some arbitrary constant. For simplicity, we would not write out this dimensionless form explicitly. That is, r is understood hereafter as normalized distance from the dislocation. The corresponding stress components are

$$\sigma_{xx} = \frac{2b_y\mu}{\pi(1+\kappa)}[\frac{x(x^2-y^2)}{r^4}] \tag{2.168}$$

$$\sigma_{xy} = -\frac{2b_y\mu}{\pi(1+\kappa)}[\frac{y(y^2-x^2)}{r^4}] \tag{2.169}$$

$$\sigma_{yy} = \frac{2b_y\mu}{\pi(1+\kappa)}[\frac{x(x^2+3y^2)}{r^4}] \tag{2.170}$$

We can then integrate this to get the displacement field as

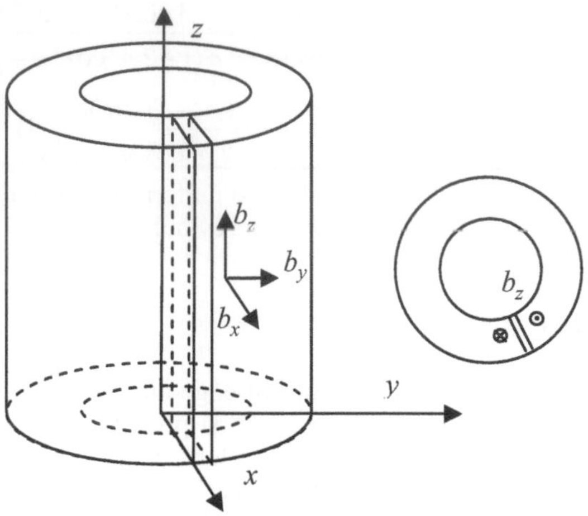

Figure 2.22 Volterra's dislocation on thick cylinder

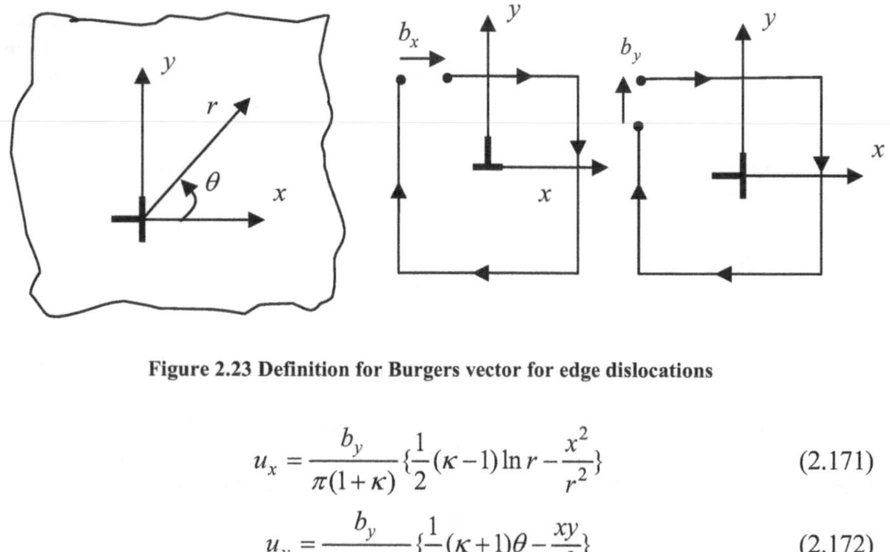

Figure 2.23 Definition for Burgers vector for edge dislocations

$$u_x = \frac{b_y}{\pi(1+\kappa)}\{\frac{1}{2}(\kappa-1)\ln r - \frac{x^2}{r^2}\} \tag{2.171}$$

$$u_y = \frac{b_y}{\pi(1+\kappa)}\{\frac{1}{2}(\kappa+1)\theta - \frac{xy}{r^2}\} \tag{2.172}$$

It is obvious from (2.172) that if we go around the dislocation line in the clockwise direction, θ will increase by 2π such that u_y attains a jump of displacement of b_y. All other stress and displacement components are, however, continuous and suffer no jump. Therefore, this solution is indeed the solution for edge dislocation.

For the case of the nonzero Burgers vector b_x, the solution has been given by Weertman and Weertman (1964):

$$u_x = \frac{b_x}{2\pi}[\theta + \frac{\lambda+\mu}{\lambda+2\mu}(\frac{xy}{x^2+y^2})] \tag{2.173}$$

$$u_y = \frac{b_x}{2\pi}[-\frac{\mu}{2(\lambda+2\mu)}\ln r + \frac{\lambda+\mu}{\lambda+2\mu}(\frac{y^2}{x^2+y^2})] \tag{2.174}$$

This result was first obtained by Koehler (1941). The corresponding stress components in cylindrical coordinates are:

$$\sigma_{rr} = \sigma_{\theta\theta} = -\frac{\mu b_x}{2\pi(1-\nu)}(\frac{\sin\theta}{r}) \tag{2.175}$$

$$\sigma_{zz} = -\frac{\mu\nu b_x}{\pi(1-\nu)}(\frac{\sin\theta}{r}), \quad \sigma_{r\theta} = \frac{\mu b_x}{2\pi(1-\nu)}(\frac{\cos\theta}{r}) \tag{2.176}$$

2.24 DISLOCATION PILE-UP AND CRACK

So far, we have assumed a single or discrete dislocation. To solve crack problems, we can actually superimpose an appropriate distribution of dislocation density. Figure 2.24 illustrates the superposition of discrete edge dislocation to model opening. The gap displacement $g(x)$ and normal traction $N(x)$ on $y = 0$ are also sketched in the figure.

For the shift dislocation shown in Fig. 2.24, we have from (2.172) and (2.170) that

$$g(x) = u_y(x,0^+) - u_y(x,0^-) = -b_y H(x) \tag{2.177}$$

$$\sigma_{yy}(x,0) = N(x) = -\frac{2b_y\mu}{\pi(\kappa+1)}\frac{1}{\xi - x} \tag{2.178}$$

For the distributed edge dislocation shown, we have

$$g(x) = -\int_{-a}^{a} B_y(\xi)H(x-\xi)d\xi = -\int_{-a}^{x} B_y(\xi)d\xi \tag{2.179}$$

$$N(x) = -\frac{2\mu}{\pi(\kappa+1)}\int_{-a}^{a}\frac{B_y(\xi)d\xi}{\xi - x}, \quad -a < x < a \tag{2.180}$$

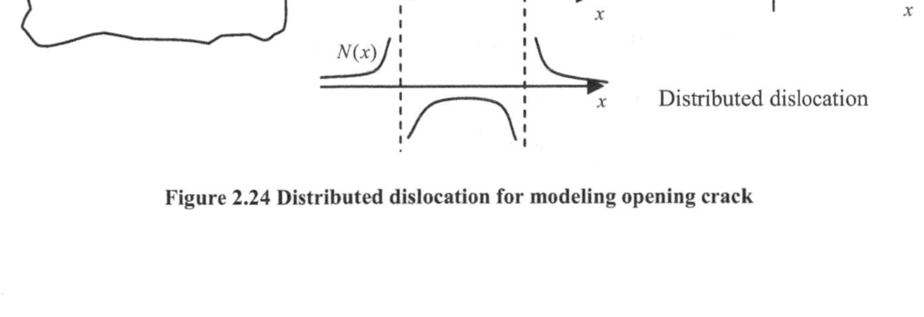

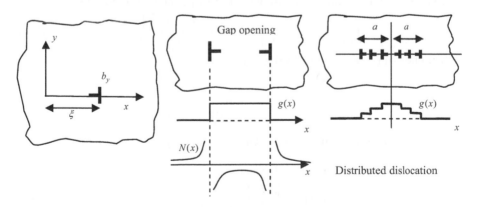

Figure 2.24 Distributed dislocation for modeling opening crack

This equation can be interpreted as a Hilbert transform as illustrated by Weertman (1996). Therefore, to solve (2.180) Hilbert transform tables can be used to obtain solutions for various crack problems. For example, Erdelyi (1954) and Appendix A of Weertman (1996) provide useful results for the solution of (2.180). In addition, since there is no permanent gap opening outside the distributed dislocation, we must have

$$\int_{-a}^{a} B_y(\xi)d\xi = 0 \tag{2.181}$$

Consider the case of a uniform pressure $-T$ within a crack or

$$\frac{\pi(\kappa+1)}{2\mu}T = \int_{-a}^{a} \frac{B_y(\xi)d\xi}{\xi - x}, \qquad -a < x < a \tag{2.182}$$

This integral actually is singular at $x = \xi$, but if we exclude this singular point, the integral does exist. This is called the principal value of the Cauchy-type integral, and (2.182) is called Cauchy-type singular integral of the first kind. The solution of (2.182) is (Muskhelishvili, 1953, 1975; Appendix 4 of Mura, 1987)

$$B_y(x) = -\frac{1}{\pi^2}\frac{1}{(a^2-x^2)^{1/2}}\{C + T\frac{\pi(\kappa+1)}{2\mu}\int_{-a}^{a}\frac{(a^2-\xi^2)^{1/2}d\xi}{\xi - x}\} \tag{2.183}$$

The constraint (2.181) requires that $C = 0$. We can apply the following formula

$$\int_{-a}^{a}\frac{(a^2-\xi^2)^{1/2}d\xi}{\xi - x} = -\pi x + (x^2-a^2)\frac{\pi \operatorname{sgn} x}{(x^2-a^2)^{1/2}}H(|x|-a) \tag{2.184}$$

However, the second term on the right of (2.184) vanishes for $-a < x < a$. Finally, we have

$$B_y(x) = T\frac{(\kappa+1)}{2\mu}\frac{x}{(a^2-x^2)^{1/2}}, \qquad -a < x < a \tag{2.185}$$

Back substitution of (2.185) into (2.180) gives

$$N(x) = -\frac{T}{\pi}\int_{-a}^{a}\frac{\xi d\xi}{(\xi - x)(a^2-\xi^2)^{/2}} \tag{2.186}$$

The integral can be evaluated by using the following integration formula:

$$\int_{-a}^{a}\frac{\xi d\xi}{(\xi - x)(a^2-\xi^2)^{/2}} = \pi - x\pi\frac{\operatorname{sgn} xH(|x|-a)}{(x^2-a^2)^{/2}} \tag{2.187}$$

Substitution of (2.187) into (2.186) gives

$$\begin{aligned} N(x) &= -T & |x| < a \\ &= -T + T\frac{|x|}{(x^2-a^2)^{1/2}} & |x| > a \end{aligned} \tag{2.188}$$

This solution can be used to get the Griffith crack by superposition as shown in Fig. 2.25.

The stress on the x-axis now becomes

$$N(x) = T\frac{|x|H(|x|-a)}{(x^2-a^2)^{1/2}} \qquad |x| < \infty \tag{2.189}$$

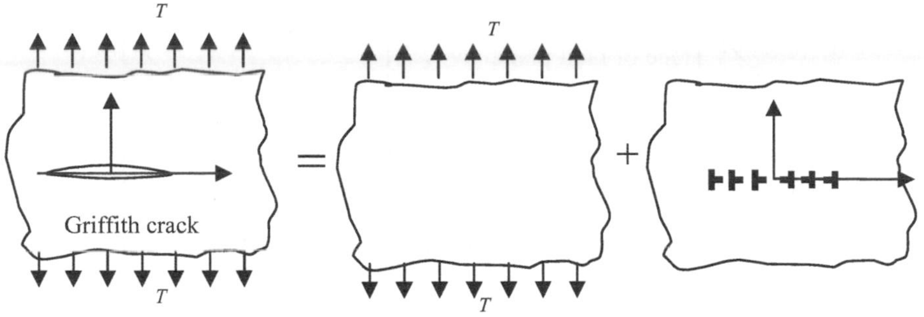

Figure 2.25 Griffith crack modeled by edge dislocation pile-up

Substitution of $x = a + r$ into (2.189) gives the normal stress ahead of the crack front:

$$\sigma_{yy}(x,0) = \frac{T\sqrt{\pi a}}{(2\pi r)^{1/2}} + O(1) \tag{2.190}$$

This inverse square root singularity in stress is actually a universal form at the crack tip, and it will be discussed in more detail in Chapter 6. For more in-depth discussions about the use of dislocation theory in fracture mechanics, the reader can refer to Weertman (1996), Hill et al. (1996), and Dundurs (2008). Wong (1990) applied dislocation pileup theory to model wedge crack nucleation in rocks similar to the Zerner–Stroh mechanism used in metals.

2.25 SCREW DISLOCATION AND FAULTING

As discussed by Weertmann and Weertmann (1964), dislocation can be used to model imperfection of both metals and nonmetallic crystalline solids. The theory of dislocation is closely related to plastic deformation in solids. The theory of infinitesimal dislocation distribution on a plane has been applied to crack problems as well as faulting problems in the Earth's crust (Chinnery 1961, 1963; Weertmann, 1964; Jeyakumaran et al., 1992).

Considering the geomechanics application of dislocation theory, we follow the discussion of Rudnicki (1988) on the slipping in strike-slip fault problems. Figure 2.26 shows an elastic half-space containing a vertical strike-slip fault. The fault slippage displacement is assumed uniform as b for $x < 0$ but zero for $x > 0$. The depth of the fault front to the free surface is H. This simple model closely resembles the situation at the San Andreas Fault in California. The lower part of the fault plane corresponds to the seismic creeping zone-driven tectonic stress, whereas the upper part of the fault plane is assumed locked.

Since the only nonzero displacement occurs along the z-axis, we have the so-called anti-plane problems in elasticity with displacement as (Milne-Thomson, 1962)

$$\boldsymbol{u} = u_z \boldsymbol{e}_z = w(x, y)\boldsymbol{e}_z \tag{2.191}$$

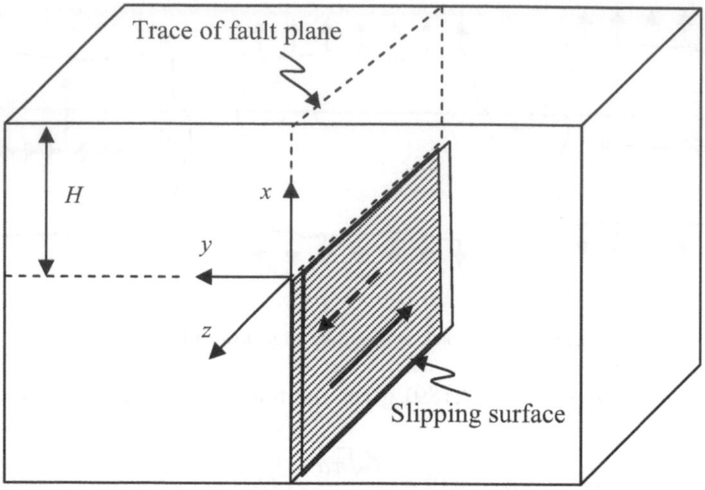

Figure 2.26 Screw dislocation model for strike-slip faulting

The term anti-plane was coined by Filon in 1937. The boundary condition on the fault plane is

$$\begin{aligned} w(x,0^+)-w(x,0^-) &= b \quad x<0 \\ &= 0 \quad x>0 \end{aligned} \tag{2.192}$$

Dimensional consideration for the linear elasticity suggests that the stress must be in the following form:

$$\frac{b}{r}\mu f(\theta) \tag{2.193}$$

where μ is the shear modulus and r is the radial distance from the origin as shown in Fig. 2.27. The only nonzero shear stress components of the problem are

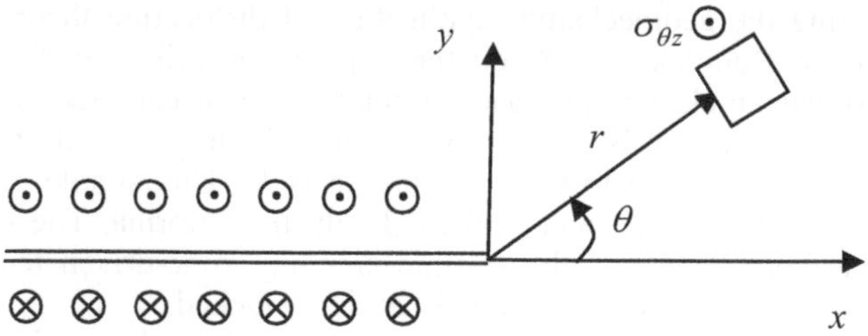

Figure 2.27 Definition of the polar coordinate around the fault tip

$$\varepsilon_{xz} = \frac{1}{2}\frac{\partial w}{\partial x} = \frac{\sigma_{xz}}{2\mu}, \quad \varepsilon_{yz} = \frac{1}{2}\frac{\partial w}{\partial y} = \frac{\sigma_{yz}}{2\mu} \tag{2.194}$$

The force equilibrium in the z-direction is

$$\frac{\partial \sigma_{xz}}{\partial x} + \frac{\partial \sigma_{yz}}{\partial y} = 0 \tag{2.195}$$

Substitution of (2.194) into (2.195) gives

$$\nabla^2 w = 0 \tag{2.196}$$

The compatibility equation in terms of stress is

$$\frac{\partial \varepsilon_{xz}}{\partial y} = \frac{\partial \varepsilon_{yz}}{\partial x} \quad \text{or} \quad \frac{\partial \sigma_{xz}}{\partial y} = \frac{\partial \sigma_{yz}}{\partial x} \tag{2.197}$$

Equations (2.195) and (2.197) can be considered the Cauchy–Riemann equations for real and imaginary parts of an analytical function (Silverman, 1974). Therefore, we can express the stresses as

$$\sigma_{xz} + i\sigma_{yz} = \frac{d\varphi}{dz} \tag{2.198}$$

for some analytical functions φ. Thus, the solution of w can be expressed as

$$\mu w = \text{Im}[\varphi(z)] \tag{2.199}$$

To satisfy the boundary condition (2.192), we can try the following solution form:

$$w = \frac{b}{2\pi}\theta = \frac{b}{2\pi}\tan^{-1}(\frac{y}{x}) \tag{2.200}$$

It is straightforward to show that (2.200) satisfies (2.192) identically. Substitution of (2.200) into (2.198) gives

$$\sigma_{rz} + i\sigma_{\theta z} = e^{-i\theta}(\sigma_{xz} + i\sigma_{yz}) = e^{-i\theta}\frac{\mu b}{2\pi r}(-\sin\theta + i\cos\theta) = \frac{i\mu b}{2\pi r} \tag{2.201}$$

Therefore, we have

$$\sigma_{\theta z} = \frac{\mu b}{2\pi r} \tag{2.202}$$

We can now shift the coordinate upward such that the origin is on the ground surface as:

$$w = \frac{b}{2\pi}\theta = \frac{b}{2\pi}\tan^{-1}(\frac{y}{x+H}) \tag{2.203}$$

$$\sigma_{xz} = -\frac{\mu b}{2\pi r}[\frac{y}{(x+H)^2 + y^2}], \quad \sigma_{yz} = \frac{\mu b}{2\pi r}[\frac{x+H}{(x+H)^2 + y^2}] \tag{2.204}$$

This is the solution for a screw dislocation in a full space. Thus, the surface traction on the ground surface is not zero. To remove this traction, we can apply the method of images (Weertman and Weertman, 1964). In particular, we can impose a negative Burgers vector b at $x = -H$ to get:

$$w = \frac{b}{2\pi}\left[\tan^{-1}(\frac{y}{x+H}) - \tan^{-1}(\frac{y}{x-H})\right] \tag{2.205}$$

$$\sigma_{xz} = -\frac{\mu b}{2\pi}\left[\frac{y}{(x+H)^2+y^2} - \frac{y}{(x-H)^2+y^2}\right], \tag{2.206}$$

$$\sigma_{yz} = \frac{\mu b}{2\pi}\left[\frac{x+H}{(x+H)^2+y^2} - \frac{x-H}{(x-H)^2+y^2}\right] \tag{2.207}$$

On the ground surface, the strain can be evaluated (by setting $x = 0$) as

$$\varepsilon_{yz} = \frac{b}{2\pi H}\left[\frac{1}{1+(y/H)^2}\right] \tag{2.208}$$

This strain prediction can be checked against field data to calibrate the value of b. Thus, the dislocation model provides a simple way to estimate the creeping strain underneath the ground surface. Subsequently, the average slip on the fault can be estimated.

This screw dislocation can also be applied to model external crack in a half-space, similar to the case of breaking of the locked zone during an earthquake. As shown in Fig. 2.28, the screw dislocation is applied to model external crack in a half-space. Now suppose the slipping displacement or Burgers vector b becomes a distributed function $\mu^*(t) = -(\partial b/\partial x)$ instead of a constant. The displacement can be expressed as

$$w = -\frac{1}{2\pi}\int_0^a \left[\tan^{-1}\left(\frac{y}{x-t}\right) - \tan^{-1}\left(\frac{y}{x+t}\right)\right]\mu^*(t)dt \tag{2.209}$$

The corresponding shear stress is

$$\sigma_{yz} = -\frac{\mu}{2\pi}\int_0^a \mu^*(t)\left[\frac{x-t}{(x-t)^2+y^2} - \frac{x+t}{(x+t)^2+y^2}\right]dt \tag{2.210}$$

$$\sigma_{xz} = \frac{\mu}{2\pi}\int_0^a \mu^*(t)\left[\frac{y}{(x-t)^2+y^2} - \frac{y}{(x+t)^2+y^2}\right]dt \tag{2.211}$$

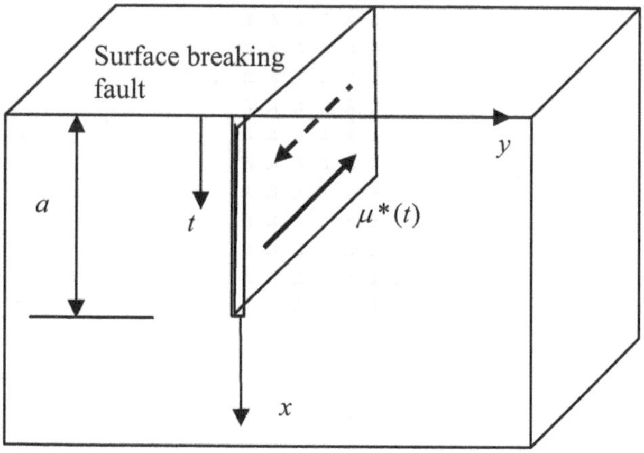

Figure 2.28 Surface breaking fault in half-space

On the fault surface $y = 0$, and the shear stress becomes

$$\sigma_{yz}(x,0) = -\frac{\mu}{2\pi}\int_0^a \mu^*(t)\left[\frac{x-t}{(x-t)^2} - \frac{x+t}{(x+t)^2}\right]dt = -\frac{\mu}{2\pi}\int_0^a \left[\frac{2t\mu^*(t)}{x^2-t^2}\right]dt \quad (2.212)$$

If the slipping is driven by the difference of the far field applied shear stress and the fault face shear resistance, we have the following integral equation for the distributed dislocation function $\mu^*(t)$

$$\tau_{yz}^{\infty} - \tau_f(x) = -\frac{\mu}{2\pi}\int_0^a \left[\frac{2t\mu^*(t)}{x^2-t^2}\right]dt \quad (2.213)$$

Recall that this solution for half-space is a result of the image method, and thus we have a subsidiary condition that

$$\mu^*(0) = 0 \quad (2.214)$$

and it is an even function of x. For the case of constant resistance (τ_f), the solution of this Cauchy-type singular integral can be expressed as (Muskhelishvili, 1953)

$$\mu^*(t) = \frac{2}{\mu}\frac{2t}{\sqrt{a^2-t^2}}(\tau_{yz}^{\infty} - \tau_f) \quad (2.215)$$

The corresponding stress components are

$$\sigma_{xz} + i\sigma_{yz} = \varphi'(z) = (\tau_{yz}^{\infty} - \tau_f)\frac{z}{(z^2-a^2)^{1/2}} \quad (2.216)$$

where $z = x + iy$. Integration of (2.216) gives

$$\varphi(z) = (\tau_{yz}^{\infty} - \tau_f)(z^2-a^2)^{1/2} \quad (2.217)$$

The displacement field can be found by using (2.199):

$$\mu w(x,0) = \mathrm{Im}[\varphi(z)] = (\tau_{yz}^{\infty} - \tau_f)(a^2-x^2)^{1/2} \quad (2.218)$$

on $y = 0$, whereas

$$\mu w(0,y) = (\tau_{yz}^{\infty} - \tau_f)(y^2+a^2)^{1/2} \quad (2.219)$$

on $x = 0$. We now examine the stress field around the fault tip at $x = a$ by defining $x = a + r$. The stress on $y = 0$ becomes

$$\begin{aligned}\sigma_{xz} + i\sigma_{yz} &= (\tau_{yz}^{\infty} - \tau_f)\frac{x}{(x^2-a^2)^{1/2}} = \frac{(\tau_{yz}^{\infty} - \tau_f)(a+r)}{(2ar+r^2)^{1/2}} \\ &\simeq \frac{(\tau_{yz}^{\infty} - \tau_f)\sqrt{\pi a}}{(2\pi r)^{1/2}} + O(1)\end{aligned} \quad (2.220)$$

The dominant stress field near the fault tip can be expressed as

$$\sigma_{xz} + i\sigma_{yz} \simeq \frac{(\tau_{yz}^{\infty} - \tau_f)\sqrt{\pi a}}{(2\pi r)^{1/2}} = \frac{K}{(2\pi r)^{1/2}} \quad (2.221)$$

In particular, an inverse square root singularity of the stress field with distance r is obtained, similar to the conclusion obtained in Section 2.24. In fact, we will show in Chapter 6 that this singular stress field is universal for all cracks, and the strike-slip fault tip considered here can be regarded as a mode III crack. The factor K, which is called the stress intensity factor, is only a function of loading and crack geometry.

2.26 MURA FORMULA FOR CURVED DISLOCATION

The dislocation solutions discussed so far have been restricted to isotropic solids and to straight dislocation. In this section, we will present the more general Mura's formula for calculating displacement and strain in anisotropic solids due to curved dislocation, as shown in Fig. 2.29. The following presentation is from Mura (1987). A slip plane S is inside the solid with a slip displacement of $\boldsymbol{b}$ from the upper surface S^+ relative to the lower surface S^-. The positive direction $\boldsymbol{\nu}$ of the curved dislocation line L is defined by using a right-handed screw with the Burgers circuit c, as shown in Fig. 2.29. The circuit c does not cross the boundary S from S^+ to S^-. The slip displacement vector $\boldsymbol{b}$ to close the Burgers circuit c is the Burgers vector proposed by Burgers in 1939 (Mura, 1987).

As shown by Mura (1987), the dislocation along dislocation line L induces a self-equilibrating elastic field, which can be simulated by an eigenstrain or the transformation strain of Eshelby (1957). The displacement gradient can be written in terms of elastic distortion $\boldsymbol{\beta}$ and plastic distortion $\boldsymbol{\beta}^*$ as

$$u_{i,j} = \beta_{ji} + \beta_{ji}^* \tag{2.222}$$

The plastic strain can be written in terms of the slip $\boldsymbol{b}$ and normal vector $\boldsymbol{n}$ from S^+ to S^- as (more or less like the slip line theory in plasticity):

$$\varepsilon_{ij}^*(\boldsymbol{x}) = -\frac{1}{2}(b_i n_j + b_j n_i)\delta(\boldsymbol{S} - \boldsymbol{x}) \tag{2.223}$$

and the one-dimensional Dirac delta function indicates the singular strain at the surface S and is zero elsewhere. The elastic field due to an eigenstrain can be formulated in terms of Green's function as (Mura, 1987)

$$u_i(\boldsymbol{x}) = -\int_{\boldsymbol{\Omega}} C_{ijmn}\varepsilon_{mn}^*(\boldsymbol{x}')G_{ij,i}(\boldsymbol{x} - \boldsymbol{x}')d\boldsymbol{x}' \tag{2.224}$$

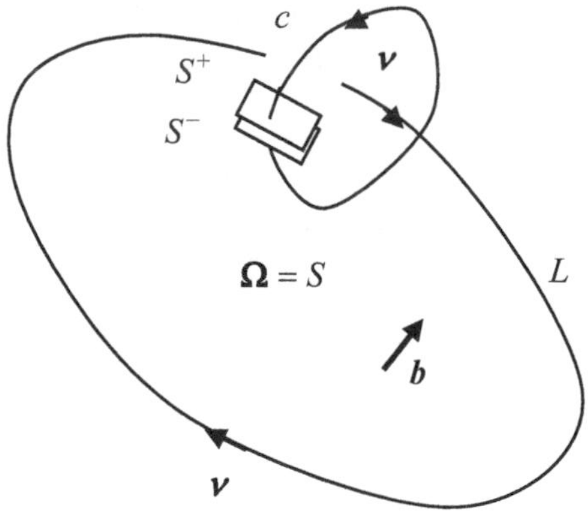

Figure 2.29 Dislocation L and Burgers circuit c (after Mura (1987) with permission from Springer, the Kluwer copyright controller)

where G_{ij} is the Green's tensor or the displacement field u_i at observation point $\boldsymbol{x}$ subject to a unit point force in the j-th direction at point $\boldsymbol{x}'$ in a full space or the so-called Kelvin's solution (Mura, 1987). Note that the Green's function method was proposed by George Green in 1828. Substitution of (2.223) into (2.224) yields

$$u_i(\boldsymbol{x}) = \int_S C_{ijmn} b_m G_{ij,i}(\boldsymbol{x}-\boldsymbol{x}') n_n dS(\boldsymbol{x}') \tag{2.225}$$

This is the Volterra formula obtained in 1907 by V. Volterra. Differentiating (2.225) and substituting the result into (2.222) gives

$$\beta_{ji}(\boldsymbol{x}) = \int_S C_{ijmn} b_m G_{ij,i}(\boldsymbol{x}-\boldsymbol{x}') n_n dS(\boldsymbol{x}') + b_i n_j \delta(\boldsymbol{S}-\boldsymbol{x}) \tag{2.226}$$

Mura (1963) found that this surface integral can be rewritten as line integral:

$$\beta_{ji}(\boldsymbol{x}) = \int_L \varepsilon_{jnh} C_{pqmn} G_{ip,q}(\boldsymbol{x}-\boldsymbol{x}') b_m \nu_h dl(\boldsymbol{x}') \tag{2.227}$$

where L and dl are the dislocation line and the dislocation line segment, respectively. This was called Mura formula by Willis (1970).

To show the equivalence of (2.227) and (2.226), we can use Stokes' theorem (see (1.60) of Chapter 1):

$$\int_L f \nu_h dl = \int_S \varepsilon_{klh} f_{,l} n_k dS \tag{2.228}$$

Applying (2.228) to (2.227) gives

$$\beta_{ji}(\boldsymbol{x}) = -\int_S \varepsilon_{klh} \varepsilon_{jnh} C_{pqmn} G_{ip,ql}(\boldsymbol{x}-\boldsymbol{x}') b_m n_k dS(\boldsymbol{x}') \tag{2.229}$$

The minus sign is from the differentiation with respect to $\boldsymbol{x}'$

$$\frac{\partial}{\partial x_l'} G_{ip,q} = -G_{ip,ql} \tag{2.230}$$

We now observe that the e-δ formula given in (1.18) of Chapter 1

$$\varepsilon_{klh} \varepsilon_{jnh} = \delta_{kj} \delta_{ln} - \delta_{kn} \delta_{lj} \tag{2.231}$$

can be applied to (2.229) as

$$\beta_{ji}(\boldsymbol{x}) = -\int_S (C_{pqmn} G_{ip,qn} b_m n_j - C_{pqmn} G_{ip,qj} b_m n_n) dS \tag{2.232}$$

Since the Green tensor satisfies the following equation:

$$C_{pqmn} G_{ip,qn}(\boldsymbol{x}-\boldsymbol{x}') = -\delta_{mi} \delta(\boldsymbol{x}-\boldsymbol{x}'), \tag{2.233}$$

the first term in (2.232) becomes

$$\int_S \delta(\boldsymbol{x}-\boldsymbol{x}') b_i n_j dS(x') = b_i n_j \delta(\boldsymbol{S}-\boldsymbol{x}) \tag{2.234}$$

Combining (2.334) and (2.232) gives (2.226). The proof of Mura formula is completed. Both Volterra and Mura formulas have been found useful for evaluating the strain by dislocation loop in anisotropic solids (Willis, 1970).

2.27 SUMMARY AND FURTHER READING

A number of elementary topics in linear elasticity, such as the concepts of stress and strain, constitutive relation, strain and complementary energies, hyperelasticity and hypoelasticity, strain compatibility, the Airy stress function, and dislocation theory are covered in this chapter. The method of solution using complex variable technique will be discussed in Chapter 3 and three-dimensional elasticity will be explained in Chapter 4.

A brief history of the theory of elasticity can be found in Westergaard (1952) and Timoshenko (1953). For general principles, Love (1944) and Malvern (1969) give a comprehensive coverage. Mal and Singh (1991) give an elementary introduction to mechanics of elastic solids. For the mechanics of anisotropic elastic solids, we recommend Hearmon (1961) and Lekhnitskii (1963). For more general applications, Timoshenko and Goodier (1982) provide excellent coverage. For transversely isotropic solids, the excellent paper by Hu (1954) is highly recommended. For anti-plane problems in elasticity, we recommend Milne-Thomson (1962).

For further general reading, we recommend the books by Sokolnikoff (1956), Fung (1965), Chou and Pagano (1967), Green and Zerna (1968), Karasudhi (1991), Davis and Selvadurai (1996), Little (1973), Barber (2002), Boresi et al. (2011), and Hetnarski and Ignacezak (2011). For elastic solutions available for foundation engineering applications, the book by Poulos and Davis (1974) offers a good start. Elastic solutions of particular relevance to rock mechanics can be found in Jaeger and Cook (1976). For dislocation theory, Weertman and Weertman (1964) give a good introduction, and the micromechanics book by Mura (1987) also provides a very elegant treatment of dislocation theory. Mura's (1987) book also provided a systematic treatment of micromechanics, and it is highly recommended for its mathematical elegance. Another comprehensive book on micromechanics is by Nemat-Nasser and Hori (1993). More advanced topics of dislocation can be found in Hirth and Lothe (1982). For the application of dislocation theory to earthquake rupture, one can refer to the review paper by Rice (1980) and Savage (1980).

2.28 PROBLEMS

Problem 2.1 Write the Cartesian components of first Piola–Kirchhoff stress given in (2.3) and the second Piola–Kirchhoff stress given in (2.6).

Problem 2.2 Write the Cartesian components of Eulerian strain tensor given in (2.24) (similar to that of Eq. (2.28)).

Problem 2.3 Write the Cartesian components of the compatibility given in (2.35).

Problem 2.4 Show the symmetry of the elasticity tensor $C_{ijkl} = C_{klij}$ given in (2.42).

Problem 2.5 Prove the following identity:

$$e_{ijk}x_i x_j = 0 \tag{2.235}$$

Problem 2.6 Show from (2.120) that in Cartesian coordinate the Airy stress function is defined as

$$\sigma_{11} = \frac{\partial^2 \varphi}{\partial x_2^2} + V, \quad \sigma_{22} = \frac{\partial^2 \varphi}{\partial x_1^2} + V, \quad \sigma_{12} = -\frac{\partial^2 \varphi}{\partial x_1 \partial x_2} \tag{2.236}$$

Problem 2.7 The Airy stress function has been found very useful in solving two-dimensional problems. Consider the following polynominal:

$$\varphi = A_0 x^5 + A_1 x^4 y + A_2 x^3 y^2 + A_3 x^2 y^3 + A_4 x^4 y + A_5 y^5 \tag{2.237}$$

Determine A_4 and A_5 in terms of A_0, A_1, A_2 and A_3 such that this polynomial satisfies the biharmonic equation.

Problem 2.8 Consider any two arbitrary harmonic functions ϕ and φ. That is,

$$\nabla^2 \phi = 0, \quad \nabla^2 \varphi = 0 \tag{2.238}$$

Show that the following function is a biharmonic function:

$$\psi = x_1 \phi + \varphi \tag{2.239}$$

Problem 2.9 Plot the principal stresses along r/R for the Jaeger's modified Brazilian shown in Fig. 2.21 for $\phi = 0$ for the concentrated line load and for $2\alpha = 5°$.

Problem 2.10 Sections 2.24 and 2.25 are restricted to mode I tensile crack and mode III tearing crack. Actually, mode II crack can also be formulated following a similar approach, as shown in Fig. 2.30. For the glide dislocation field (i.e., non-zero b_x), the following stress field is given by Weertman (1996) in Cartesian coordinates:

$$\sigma_{xy} = \frac{2Gb_x}{\pi(1+\kappa)} \frac{x(x^2 - y^2)}{(x^2 + y^2)^2} \tag{2.240}$$

$$\sigma_{yy} = \frac{Gb_x}{\pi(1+\kappa)} \frac{y(x^2 - y^2)}{(x^2 + y^2)^2} \tag{2.241}$$

$$\sigma_{xx} = -\frac{2Gb_x}{\pi(1+\kappa)} \frac{y(3x^2 + y^2)}{(x^2 + y^2)^2} \tag{2.242}$$

By using (2.240)–(2.242), show that the shear traction $T(x)$ on $y = 0$ can be formulated as

$$T(x) = -\frac{2\mu}{\pi(\kappa+1)} \int_{-a}^{a} \frac{B_x(\xi)d\xi}{\xi - x}, \quad -a < x < a \tag{2.243}$$

Problem 2.11 Solve the Cauchy integral equation (2.243) in Problem 2.10 and discuss the stress singularity near the crack tip for mode II shear crack (see Fig. 2.30).

Problem 2.12 Show the validity of Hooke's law given in (2.112) for plane problems.

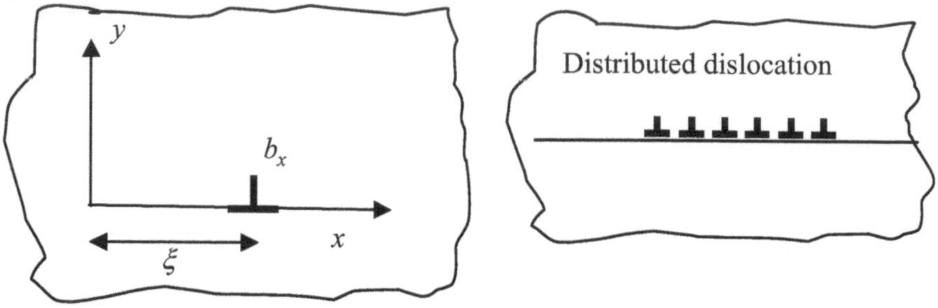

Figure 2.30 Dislocation pile-up for mode II shear crack

CHAPTER THREE

Complex Variable Method for 2-D Elasticity

3.1 INTRODUCTION

The idea of using the complex variable technique in solving elasticity problems was probably first proposed by E. Goursat in 1898 and by L.N.G. Filon in 1903 (Filon, 1903; Milne-Thomson, 1968; Meleshko, 2003). The formal treatment is usually credited to two Russians, Kolosov and Muskhelishvili, although similar results were obtained independently by others, including Stevenson, Green, and Milne-Thomson, much later. We summarize the essence of this approach in this chapter.

The main idea of this approach is that the Airy stress function introduced in Section 2.17 can, in general, be written in terms of two *analytic functions* of a complex variable $z = x_1+ix_2$, where i is the usual imaginary constant and equals $(-1)^{1/2}$. For a full account of the properties of analytic functions, we refer to Carrier et al. (1966) and Muskhelishvili (1975). In short, a function $\zeta(z)$ is said to be analytic in a region if it is finite single valued and has a definite derivative at all points of the region. Any analytic function $\zeta(z)$ can be expressed in the form

$$\zeta(z) = \mathrm{Re}(\zeta) + i\,\mathrm{Im}(\zeta) = \xi + i\eta \tag{3.1}$$

where Re and Im stand for the real and imaginary parts of the function ζ. The derivative of ζ with respect to z, analogous to that of real functions, is defined as:

$$\frac{d\zeta}{dz} = \zeta' = \lim_{\delta z \to 0} \frac{\delta\zeta}{\delta z} = \lim_{\delta z \to 0} \left(\frac{\delta\xi + i\delta\eta}{\delta x_1 + i\delta x_2}\right) \tag{3.2}$$

Applying the usual definition for a total derivative, we get

$$\begin{aligned} \frac{d\zeta}{dz} &= \lim_{\delta z \to 0} \left[\frac{(\frac{\partial\xi}{\partial x_1}\delta x_1 + \frac{\partial\xi}{\partial x_2}\delta x_2) + i(\frac{\partial\eta}{\partial x_1}\delta x_1 + \frac{\partial\eta}{\partial x_2}\delta x_2)}{\delta x_1 + i\delta x_2}\right] \\ &= \lim_{\delta z \to 0} \left[\frac{(\frac{\partial\xi}{\partial x_1} + i\frac{\partial\eta}{\partial x_1})\delta x_1 + i(\frac{\partial\eta}{\partial x_2} - i\frac{\partial\xi}{\partial x_2})\delta x_2}{\delta x_1 + i\delta x_2}\right] \end{aligned} \tag{3.3}$$

For analytic functions, the derivative in (3.3) should be independent of the path of determining limits. Therefore, no matter whether we set δx_1 to zero first, then δx_2 to zero or vice versa, the limit should be exactly the same. This implies

$$\frac{\partial\xi}{\partial x_1} + i\frac{\partial\eta}{\partial x_1} = \frac{\partial\eta}{\partial x_2} - i\frac{\partial\xi}{\partial x_2} \tag{3.4}$$

or equating the real and imaginary parts gives

$$\frac{\partial \xi}{\partial x_1} = \frac{\partial \eta}{\partial x_2}, \quad \frac{\partial \eta}{\partial x_1} = -\frac{\partial \xi}{\partial x_2} \tag{3.5}$$

This is the well-known Cauchy–Riemann relation (Carrier et al., 1966), and consequently the complex derivative is

$$\zeta'(z) = \frac{d\zeta}{dz} = \frac{\partial \xi}{\partial x_1} + i\frac{\partial \eta}{\partial x_1} = \frac{\partial \xi}{\partial x_1} - i\frac{\partial \xi}{\partial x_2} \tag{3.6}$$

Either ξ or η can be eliminated from either of the (3.5) equations, and the resulting equations are

$$\nabla \cdot \nabla\ \xi = \nabla^2 \xi = 0, \quad \nabla \cdot \nabla\ \eta = \nabla^2 \eta = 0 \tag{3.7}$$

where the Laplacian operator is denoted by ∇^2 and is defined as

$$\nabla^2 \equiv \frac{\partial^2}{\partial x_1^2} + \frac{\partial^2}{\partial x_2^2} \tag{3.8}$$

Therefore, it can be concluded that both the real and imaginary parts of an analytic function satisfy the Laplace equation. Note that η and ξ are called conjugate functions, and both are harmonic functions (since they satisfy the Laplace equation). We now consider the governing equation for the Airy stress function (2.124) if no body force exits:

$$\nabla^4 \varphi = \nabla^2 \nabla^2 \varphi = 0 \tag{3.9}$$

We now let P be the trace of the two-dimensional stress dyadic given in (2.123):

$$P = \mathrm{tr}(\boldsymbol{\sigma}) = \nabla \cdot \nabla \varphi = \nabla^2 \varphi \tag{3.10}$$

where P must be *harmonic* as shown by (3.9). We add a conjugate function of P to form an analytic function $f(z)$:

$$f(z) = P + iQ \tag{3.11}$$

and its integration, defined as follows, is also analytic:

$$\phi(z) = \frac{1}{4} \int f(z)\, dz = p + iq \tag{3.12}$$

Therefore, we have

$$\phi'(z) = \frac{\partial p}{\partial x_1} + i\frac{\partial q}{\partial x_1} = \frac{1}{4} f(z) = \frac{1}{4}(P + iQ) \tag{3.13}$$

Equating both real and imaginary parts in (3.13) gives

$$\frac{1}{4}P = \frac{\partial p}{\partial x_1} = \frac{\partial q}{\partial x_2}, \quad \frac{1}{4}Q = \frac{\partial q}{\partial x_1} = -\frac{\partial p}{\partial x_2} \tag{3.14}$$

Next we show that $\varphi - px_1 - qx_2$ is harmonic; by direct differentiation we have

$$\begin{aligned} \nabla^2(\varphi - px_1 - qx_2) &= \nabla^2 \varphi - \nabla^2(px_1) - \nabla^2(qx_2) \\ &= \nabla^2 \varphi - x_1 \nabla^2 p - 2\frac{\partial p}{\partial x_1} - x_2 \nabla^2 q - 2\frac{\partial q}{\partial x_2} = 0 \end{aligned} \tag{3.15}$$

The last of (3.15) is obtained in view of (3.10) and (3.14) and using the fact that both p and q must the satisfy the Laplace equation. Thus,

$$\varphi - px_1 - qx_2 = p_1 \tag{3.16}$$

where p_1 is a harmonic function, or it may be taken as the real part of some unknown analytic function $\chi(z)$. In addition, px_1+qx_2 is the real part of the following function:

$$px_1+qx_2=\mathrm{Re}[(x_1-ix_2)(p+iq)]=\mathrm{Re}[\bar{z}\phi(z)] \tag{3.17}$$

where the superimposed bar means complex conjugate. The Airy stress function φ can then be written as

$$\varphi=\mathrm{Re}[\bar{z}\phi(z)+\chi(z)]=\frac{1}{2}[\bar{z}\phi(z)+z\bar{\phi}(z)+\chi(z)+\bar{\chi}(z)] \tag{3.18}$$

Therefore, we have a general solution for the biharmonic equation in terms of two analytic functions $\phi(z)$ and $\chi(z)$. Differentiation of the solution form given in (3.18) gives

$$\frac{\partial\varphi}{\partial x_1}=\frac{1}{2}[\phi(z)+\bar{z}\phi'(z)+z\bar{\phi}'(z)+\bar{\phi}(z)+\chi'(z)+\bar{\chi}'(z)] \tag{3.19}$$

$$\frac{\partial\varphi}{\partial x_2}=\frac{1}{2}[-i\phi(z)+i\bar{z}\phi'(z)+i\bar{\phi}(z)-iz\bar{\phi}'(z)+i\chi'(z)-i\,\bar{\chi}'(z)] \tag{3.20}$$

$$\frac{\partial^2\varphi}{\partial x_1^2}=\frac{1}{2}[2\phi'(z)+\bar{z}\phi''(z)+2\overline{\phi}'(z)+z\overline{\phi}''(z)+\chi''(z)+\overline{\chi}''(z)] \tag{3.21}$$

$$\frac{\partial^2\varphi}{\partial x_2^2}=\frac{1}{2}[2\phi'(z)-\bar{z}\phi''(z)+2\overline{\phi}'(z)-z\overline{\phi}''(z)-\chi''(z)-\overline{\chi}''(z)] \tag{3.22}$$

For our later use, we combine (3.19) and (3.20) as

$$\frac{\partial\varphi}{\partial x_1}+i\frac{\partial\varphi}{\partial x_2}=\phi(z)+z\overline{\phi}'(z)+\overline{\chi}'(z) \tag{3.23}$$

In terms of $\phi(z)$ and $\chi(z)$, the stress components become

$$\begin{aligned}\sigma_{22}-\sigma_{11}+2i\sigma_{12}&=\frac{\partial^2\varphi}{\partial x_1^2}-\frac{\partial^2\varphi}{\partial x_2^2}-2i\frac{\partial^2\varphi}{\partial x_1\partial x_2}\\&=2[\bar{z}\phi''(z)+\chi''(z)]\end{aligned} \tag{3.24}$$

$$\sigma_{22}+\sigma_{11}=\frac{\partial^2\varphi}{\partial x_1^2}+\frac{\partial^2\varphi}{\partial x_2^2}=2[\phi'(z)+\overline{\phi}'(z)]=4\ \mathrm{Re}[\phi'(z)] \tag{3.25}$$

Using (2.112), we can write the displacement gradient as

$$\begin{aligned}2\mu u_{1,1}&=2\mu\varepsilon_{11}=\tfrac{1}{4}(1+\kappa)\sigma_{11}+\tfrac{1}{4}(\kappa-3)\sigma_{22}\\&=-\sigma_{22}+\tfrac{1}{4}(\kappa+1)tr(\boldsymbol{\sigma})\\&=-\varphi_{,11}+(1+\kappa)p_{,1}\end{aligned} \tag{3.26}$$

The last part of (3.26) results from (3.10) and (3.14). Following a similar procedure, we also have

$$2\mu u_{2,2}=-\varphi_{,22}+(1+\kappa)q_{,2} \tag{3.27}$$

Integration of (3.26) and (3.27) once gives

$$2\mu\ u_1=-\varphi_{,1}+(1+\kappa)p+f_1(x_2) \tag{3.28}$$

$$2\mu\ u_2=-\varphi_{,2}+(1+\kappa)q+f_2(x_1) \tag{3.29}$$

where $f_1(x_2)$ and $f_2(x_1)$ are unknown functions of x_2 and x_1. To find these functions, we substitute (3.28) and (3.29) into the following Hooke's law for shear deformation:

$$2\mu\,\varepsilon_{12} = \mu(u_{1,2} + u_{2,1}) = -\varphi_{,12} \tag{3.30}$$

This yields

$$f_1'(x_2) + f_2'(x_1) = 0 \tag{3.31}$$

Therefore, we have $f_1 = \omega x_2 + a$ and $f_2 = -\omega x_1 + b$, both of these displacements correspond to rigid body motion and can be neglected. Setting both f_1 and f_2 equal to zero, we can combine (3.28) and (3.29) to give:

$$\begin{aligned} 2\mu(u_1 + iu_2) &= -(\varphi_{,1} + i\varphi_{,2}) + (\kappa+1)(p+iq) \\ &= \kappa\phi(z) - z\bar{\phi}'(z) - \bar{\chi}'(z) \end{aligned} \tag{3.32}$$

Note that the analytic function $\chi(z)$ appears only in its derivative; therefore, these equations can be simplified by assuming:

$$\psi(z) = \chi'(z) \tag{3.33}$$

The stress and displacement components are given in terms of $\phi(z)$ and $\psi(z)$ as:

$$\sigma_{22} - \sigma_{11} + 2i\sigma_{12} = 2[\bar{z}\phi''(z) + \psi'(z)] \tag{3.34}$$

$$\sigma_{22} + \sigma_{11} = 2[\phi'(z) + \overline{\phi'}(z)] = 4\ \mathrm{Re}[\phi'(z)] \tag{3.35}$$

$$2\mu(u_1 + iu_2) = \kappa\phi(z) - z\bar{\phi}'(z) - \bar{\psi}(z) \tag{3.36}$$

We have just showed that the Airy stress function can be written in terms of two analytic functions $\phi(z)$ and $\psi(z)$. The solution for problems in 2-D elasticity reduces, therefore, to searching for the appropriate form of analytic functions. Many important problems can be solved by taking $\phi(z)$ and $\psi(z)$ to be polynomials or power series in z or z^{-1}. If the region in question includes the origin only the positive powers of z can be used for the functions to remain bounded at the origin, while for regions excluding the origin but including a far field region only negative powers can be used.

3.2 COORDINATE TRANSFORMATION IN COMPLEX VARIABLE THEORY

This section derives the following formula for coordinate transformation:

$$\sigma'_{22} - \sigma'_{11} + 2i\sigma'_{12} = (\sigma_{22} - \sigma_{11} + 2i\sigma_{12})e^{2i\theta}, \quad \sigma'_{11} + \sigma'_{22} = \sigma_{22} + \sigma_{11} \tag{3.37}$$

where θ is the angle between the 2-D coordinate x_1-x_2 and the rotated coordinate x'_1-x'_2. Note that the prime used in this section is for a rotated coordinate and it should not be confused with differentiation. This formula can readily be showed to be correct by recalling the following formulas for two-dimensional coordinate transformation:

$$\sigma'_{11} = \frac{1}{2}\sigma_{11}(1+\cos 2\theta) + \frac{1}{2}\sigma_{22}(1-\cos 2\theta) + \sigma_{12}\sin 2\theta \tag{3.38}$$

$$\sigma'_{22} = \frac{1}{2}\sigma_{22}(1+\cos 2\theta) + \frac{1}{2}\sigma_{11}(1-\cos 2\theta) - \sigma_{12}\sin 2\theta \tag{3.39}$$

$$\sigma'_{12} = \frac{1}{2}(\sigma_{22} - \sigma_{11})\sin 2\theta + \sigma_{12}\cos 2\theta \tag{3.40}$$

Combining (3.38) and (3.40), we have

$$\begin{aligned}\sigma'_{22}-\sigma'_{11}+2i\sigma'_{12}&=(\sigma_{22}-\sigma_{11})(\cos 2\theta+i\sin 2\theta)+2\sigma_{12}(i\cos 2\theta-\sin 2\theta)\\&=(\sigma_{22}-\sigma_{11}+2i\sigma_{12})(\cos 2\theta+i\sin 2\theta)\end{aligned} \tag{3.41}$$

Therefore, it is obvious that the first part of (3.37) is correct. The second part of (3.37) is obtained by simply summing (3.38) and (3.39). This completes the proof.

3.3 HOMOGENEOUS STRESSES IN TERMS ANALYTIC FUNCTIONS

We consider the following form of analytic functions:

$$\phi(z)=Cz\,,\quad \psi(z)=Dz \tag{3.42}$$

which C and D are constants which may be complex. Substitution of (3.42) into (3.34) to (3.36) gives

$$\sigma_{22}-\sigma_{11}+2i\sigma_{12}=2[\bar{z}\phi''(z)+\psi'(z)]=2D \tag{3.43}$$

$$\sigma_{22}+\sigma_{11}=2[\phi'(z)+\overline{\phi'}(z)]=4\ \mathrm{Re}[\phi'(z)]=2(C+\bar{C})=4E \tag{3.44}$$

$$2\mu(u_1+i\ u_2)=\kappa\phi(z)-z\overline{\phi'}(z)-\overline{\psi}(z)=(\kappa-1)Cz-D\bar{z} \tag{3.45}$$

Now let the principal stresses be σ_1 and σ_2 with σ_1 inclined at β to the x_1-axis. It follows from (3.37) that

$$\sigma_1-\sigma_2=2D\,e^{2i\beta}\,,\quad \sigma_1+\sigma_2=4E \tag{3.46}$$

The solution of (3.46) gives:

$$E=\tfrac{1}{4}(\sigma_1+\sigma_2)\,,\quad D=\tfrac{1}{2}(\sigma_2-\sigma_1)e^{-2i\beta} \tag{3.47}$$

Therefore we have hydrostatic compression if we set $E=-\sigma_1/2$ and $D=0$; uniaxial compression $-T$ along x_1 if $E=-T/4$ and $D=T/2$; and pure shear ($\sigma_2=-\sigma_1$ and $\beta=\pi/4$) if $E=0$ and $D=i\sigma_1$.

3.4 A BOREHOLE SUBJECT TO INTERNAL PRESSURE

Figure 3.1 shows a circular borehole with diameter R subject to an internal pressure p. We try the following analytic functions:

$$\phi(z)=0\,,\quad \psi(z)=\frac{D}{z} \tag{3.48}$$

where D is a real constant. In polar coordinates, we have

$$\sigma_{rr}+\sigma_{\theta\theta}=0 \tag{3.49}$$

$$\sigma_{\theta\theta}-\sigma_{rr}+2i\sigma_{r\theta}=2\overline{\psi}'(z)e^{2i\theta}=-\frac{2D}{r^2} \tag{3.50}$$

The first part of (3.50) follows immediately from the result of Section 3.2. Therefore, the stress dyadic is

$$\boldsymbol{\sigma}=\frac{D}{r^2}(\mathbf{e}_r\mathbf{e}_r-\mathbf{e}_\theta\mathbf{e}_\theta) \tag{3.51}$$

Note that this stress dyadic automatically satisfies the far field condition as $r\to\infty$. The displacement field in polar coordinates is

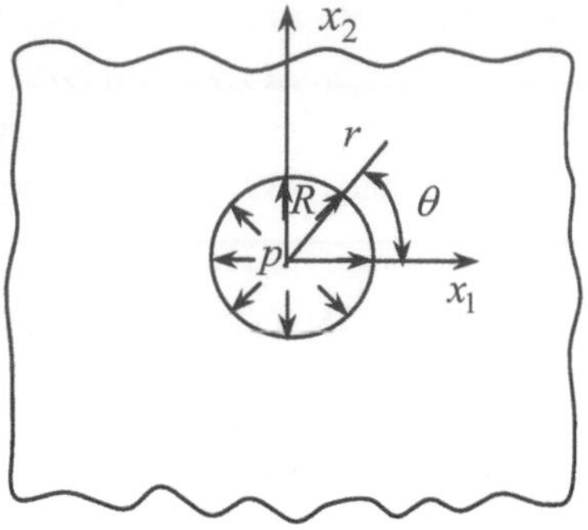

Figure 3.1 Borehole with internal pressure

$$2\mu(u_r + iu_\theta) = 2\mu(u_1 + iu_2)e^{-i\theta} = -\overline{\psi}(z)e^{-i\theta} = -\frac{D}{r} \tag{3.52}$$

The first part of (3.52) can be proved easily similar to the consideration given in Section 3.2 (see Problem 3.2). Now the boundary condition on $r = R$ is $\sigma_{rr} = -p$. Substitution of (3.51) into this condition gives $D = -pR^2$. Therefore, the solution becomes:

$$\boldsymbol{\sigma} = -\frac{pR^2}{r^2}(\boldsymbol{e}_r\boldsymbol{e}_r - \boldsymbol{e}_\theta\boldsymbol{e}_\theta), \quad \boldsymbol{u} = -\frac{pR^2}{r^2}\boldsymbol{e}_r \tag{3.53}$$

The solution for a hollow cylinder subject to both internal and external pressures will be considered in Problem 3.3.

3.5 KIRSCH SOLUTION BY COMPLEX VARIABLE METHOD

In Section 2.18, we solved the stress concentration near a circular hole in an infinite plane subject to far field uniaxial compression, which finds applications in both tunnel and borehole problems. We will re-derive this solution here using the complex variable technique. For the far field stress, we showed in Section 3.3 that it can be found by using $\phi(z) = -\frac{1}{4}Tz$ and $\psi(z) = \frac{1}{2}Tz$. For an infinite region with a hole, we must add terms to these analytic functions, such that the resulting stress decays at infinity, but cancels the traction on $r = R$ caused by the uniform stress field. We assume

$$\phi(z) = -\tfrac{1}{4}T(z + \frac{A}{z}), \quad \psi(z) = \tfrac{1}{2}T(z + \frac{B}{z} + \frac{C}{z^2}) \tag{3.54}$$

where A, B, and C are real constants. The choice is guided by the fact that all resulting stress components will be in terms of 2θ. Differentiating (3.54) yields

$$\phi'(z) = -\tfrac{1}{4}T(1 - \frac{A}{z^2}), \quad \psi'(z) = \tfrac{1}{2}T(1 - \frac{B}{z^2} - \frac{3C}{z^4}), \quad \phi''(z) = \frac{AT}{2z^3} \tag{3.55}$$

Therefore with $z = re^{i\theta}$ we have:

$$
\begin{aligned}
tr(\boldsymbol{\sigma}) &= \sigma_{rr}+\sigma_{\theta\theta} = 2[\phi'(z)+\bar{\phi}'(z)] \\
&= -\tfrac{1}{2}T\ [2-Ar^{-2}e^{-2i\theta}-Ar^{-2}e^{2i\theta}] \\
&= -T\ [1-Ar^{-2}\cos 2\theta]
\end{aligned} \tag{3.56}
$$

$$
\begin{aligned}
\sigma_{\theta\theta}-\sigma_{rr}+2i\sigma_{r\theta} &= (\sigma_{22}-\sigma_{11}+2i\sigma_{12})e^{2i\theta} \\
&= -T\ [re^{-i\theta}Ar^{-3}e^{-3i\theta}-1+Br^{-2}e^{-2i\theta}+3Cr^{-4}e^{-4i\theta}]e^{2i\theta} \\
&= -T\ [Br^{-2}-e^{2i\theta}+(Ar^{-2}+3Cr^{-4})e^{-2i\theta}]
\end{aligned} \tag{3.57}
$$

The real and imaginary parts of (3.57) give

$$\sigma_{\theta\theta}-\sigma_{rr} = -T\ [Br^{-2}-(1-Ar^{-2}-3Cr^{-4})\cos 2\theta] \tag{3.58}$$

$$\sigma_{12} = \tfrac{1}{2}T\ [1+Ar^{-2}+3Cr^{-4}]\sin 2\theta \tag{3.59}$$

Subtracting (3.58) from (3.56) yields:

$$2\sigma_{rr} = -T[1-Br^{-2}+(1-2Ar^{-2}-3Cr^{-4})\cos 2\theta] \tag{3.60}$$

Form the traction-free boundary condition on $r = R$:

$$\sigma_{rr} = \sigma_{r\theta} = 0 \tag{3.61}$$

we obtain

$$B = R^2\,,\quad A = 2R^2\,,\quad C = -R^4 \tag{3.62}$$

With back substitution of these constants into (3.56) to (3.57), we obtain (2.141) again, which was derived in Section 2.18 of Chapter 2 by using the Airy stress function.

3.6 DEFINITENESS AND UNIQUENESS OF THE ANALYTIC FUNCTION

We have shown in Section 3.1 that the Airy stress function can be expressed in terms of two arbitrary analytic functions. For an elastic body with known stress and displacement fields, one may ask whether we can find a unique pair of analytic functions to describe this solution. This question will be examined in this section.

We first assume that a pair of analytic functions $\phi_1(z)$ and $\psi_1(z)$ will give the known stress field [as given in (3.34)–(3.35)]:

$$\sigma_{22}-\sigma_{11}+2i\sigma_{12} = 2[\bar{z}\phi_1''(z)+\psi_1'(z)] \tag{3.63}$$

$$\sigma_{22}+\sigma_{11} = 4\,\mathrm{Re}[\phi_1'(z)] \tag{3.64}$$

We now suppose that another pair of analytic functions $\phi_2(z)$ and $\psi_2(z)$ will also lead to the same stress field. That is,

$$\sigma_{22}-\sigma_{11}+2i\sigma_{12} = 2[\bar{z}\phi_2''(z)+\psi_2'(z)] \tag{3.65}$$

$$\sigma_{22}+\sigma_{11} = 4\,\mathrm{Re}[\phi_2'(z)] \tag{3.66}$$

Let us now investigate the possible difference between these pairs of analytic functions.

Comparison of (3.64) and (3.66) indicates that $\phi'_2(z)$ and $\phi'_1(z)$ must have the same real part. Thus, we can write

$$\phi_2'(z) = \phi_1'(z)+iC \tag{3.67}$$

where C is an arbitrary real constant. Integration of (3.67) gives

$$\phi_2(z)=\phi_1(z)+iCz+\gamma \tag{3.68}$$

where $\gamma = A+ iB$ is an arbitrary complex constant. From (3.67), we must have

$$\phi_2''(z)=\phi_1''(z) \tag{3.69}$$

Thus, comparison of (3.63) and (3.65) gives:

$$\psi_2'(z)=\psi_1'(z) \tag{3.70}$$

or we have

$$\psi_2(z)=\psi_1(z)+\gamma' \tag{3.71}$$

where $\gamma' = A' + iB'$ is another arbitrary complex constant. Therefore, we find that $\phi_1(z)$ and $\psi_1(z)$ can be replaced by $\phi_1(z)+iCz+\gamma$ and $\psi_1(z)+\gamma'$, that is,

$$\phi_1(z)\leftarrow\phi_1(z)+iCz+\gamma, \quad \psi_1(z)\leftarrow\psi_1(z)+\gamma' \tag{3.72}$$

such that the stress field remains unchanged.

We now further restrict the arbitrary constants C, γ, and γ' by using the displacement expression:

$$2\mu(u_1+iu_2)=\kappa\phi_1(z)-z\overline{\phi_1'}(z)-\overline{\psi}_1(z) \tag{3.73}$$

In particular, we apply the transformation (3.72) into (3.73) to give

$$2\mu(u_1+iu_2)=\kappa\phi_1(z)-z\overline{\phi_1'}(z)-\overline{\psi}_1(z)+(1+\kappa)iCz+\kappa\gamma-\overline{\gamma}' \tag{3.74}$$

Therefore, we must have

$$C=0, \qquad \kappa\gamma-\overline{\gamma}'=0 \tag{3.75}$$

Finally, we conclude that $\phi_1(z)$ and $\psi_1(z)$ can be replaced by the following identification:

$$\phi_1(z)\leftarrow\phi_1(z)+\gamma, \quad \psi_1(z)\leftarrow\psi_1(z)+\kappa\overline{\gamma} \tag{3.76}$$

such that the stress field remains unchanged, where γ is an arbitrary complex constant.

3.7 BOUNDARY CONDITIONS FOR THE ANALYTIC FUNCTIONS

The boundary value problems solved in Sections 3.3–3.5 were solved by assuming appropriate analytic functions. But, in general, this cannot be done by pure inspection. In this section, we will formulate the traction and displacement boundary conditions for the analytic functions $\phi(z)$ and $\psi(z)$. Referring to Fig. 3.2, the surface traction vector $\boldsymbol{T} = \boldsymbol{n} \cdot \boldsymbol{\sigma}$ can be written in terms of the Airy stress function as:

$$n_1\left(\frac{\partial^2\varphi}{\partial x_2^2}\right)_S-n_2\left(\frac{\partial^2\varphi}{\partial x_1\partial x_2}\right)_S=T_1 \tag{3.77}$$

$$n_2\left(\frac{\partial^2\varphi}{\partial x_1^2}\right)_S-n_1\left(\frac{\partial^2\varphi}{\partial x_1\partial x_2}\right)_S=T_2 \tag{3.78}$$

where the subscript S means that the value in the bracket should be evaluated along the boundary S with normal $\boldsymbol{n}$ shown in Fig. 3.2. By pure geometric consideration, we have

$$n_1=\cos(\boldsymbol{n},x_1)=\cos\alpha=\frac{dx_2}{ds} \tag{3.79}$$

$$n_2 = \cos(\boldsymbol{n}, x_2) = \sin\alpha = -\frac{dx_1}{ds} \tag{3.80}$$

Substitution of (3.79) and (3.80) into (3.77) and (3.78) yields the following total differentials along *ds:*

$$\frac{d}{ds}\left(\frac{\partial \varphi}{\partial x_2}\right)_S = T_1, \qquad -\frac{d}{ds}\left(\frac{\partial \varphi}{\partial x_1}\right)_S = T_2 \tag{3.81}$$

We now define the complex boundary traction:

$$T_1 + iT_2 = \frac{d}{ds}\left(\frac{\partial \varphi}{\partial x_2}\right)_S - i\frac{d}{ds}\left(\frac{\partial \varphi}{\partial x_1}\right)_S = -i\frac{d}{ds}\left(\frac{\partial \varphi}{\partial x_1} + i\frac{\partial \varphi}{\partial x_2}\right)_S \tag{3.82}$$

Substitution of (3.23) into (3.82) and integration of the resulting equation with respect to *ds* from point *A* to point *B* (shown in Fig. 3.2) gives:

$$\left[\phi(z) + z\bar{\phi}'(z) + \bar{\psi}(z)\right]_A^B = i\int_A^B (T_1 + iT_2)ds \tag{3.83}$$

If the upper limit *B* is an arbitrary point along the boundary *s* and the lower limit evaluated at *A* is *k,* then we get

$$\left[\phi(z) + z\bar{\phi}'(z) + \bar{\psi}(z)\right]_s - k = i\int_A^B (T_1 + iT_2)ds \tag{3.84}$$

As shown in the previous section, $\phi(z)$ and $\psi(z)$ can arbitrarily be replaced by $\phi(z)+\gamma$ and $\psi(z)+\kappa\bar{\gamma}$, respectively. Therefore, we can always add a constant γ such that k is being cancelled. Thus, the traction boundary condition can finally be simplified to

$$\left[\phi(z) + z\bar{\phi}'(z) + \bar{\psi}(z)\right]_s = i\int (T_1 + iT_2)ds \tag{3.85}$$

Physically, the combination of the analytic functions on the boundary shown on the left of (3.85), which is evaluated at an arbitrary point *z*, must equal the total complex force on the surface times *i*.

By virtue of (3.36), the displacement boundary condition, say $u_1 = u_s$ and $u_2 = v_s$, is simply

$$2\mu(u_s + iv_s) = \left[\kappa\phi(z) - z\bar{\phi}'(z) - \bar{\psi}(z)\right]_s \tag{3.86}$$

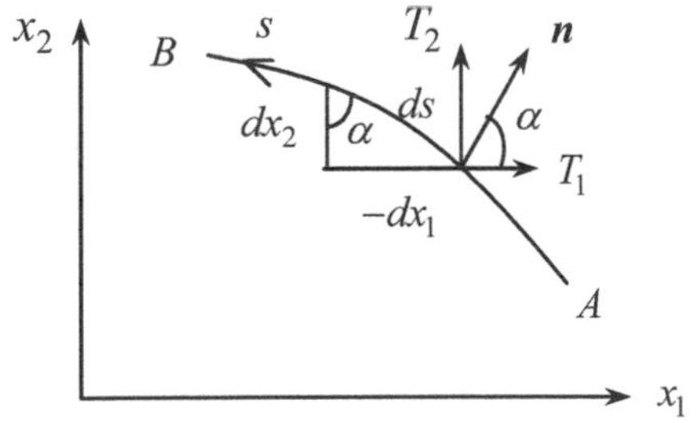

Figure 3.2 A segment of boundary with traction components T_1 and T_2

3.8 SINGLE-VALUED CONDITION FOR MULTI-CONNECTED BODIES

We are interested in obtaining solutions for bodies with holes. Such bodies are sometimes called multi-connected because there is more than one simply connected surface as boundary (e.g., see Fig. 3.3). For such bodies, the complex functions $\phi(z)$ and $\psi(z)$ may become multi-valued functions even though the same functions may be single-valued functions in a simply connected body. In this section, we investigate the necessary form of the analytic functions such that single-valued stress and displacement fields can be guaranteed. As shown in Fig. 3.3, let a body contains m inner boundaries $s_1,.., s_m$ and an outer boundary s_{m+1}. The stress field must be single valued, and as shown in (3.35) the sum of normal stresses is related to $\phi(z)$ as

$$\sigma_{22}+\sigma_{11}=4\,\mathrm{Re}[\phi'(z)] \tag{3.87}$$

Thus, the real part of $\phi'(z)$ must be single valued; however the imaginary part of $\phi'(z)$ may not be single-valued. For example, when we consider z to go along any inner boundary s_k, $\phi'(z)$ may increase by an imaginary quantity (i.e., an imaginary constant), say $2\pi i A_k$, where A_k is a real number.

To better understand this increment, let us consider Fig. 3.4 which shows the contour excluding the hole s_k (all other holes are ignored for the time being). The change in the value of $\phi'(z)$ for the variable z to undergo a complete clockwise loop Γ_k around the whole s_k is denoted by $[\phi'(z)]_{\Gamma k}$. This change in value can be written as an integral as

$$[\phi'(z)]_{\Gamma_k}=\int_{\Gamma_k}\frac{d\phi'(z)}{dz}dz \tag{3.88}$$

Then, the following closed contour integral cutting through the solid and excluding the hole can be shown to be zero by applying Cauchy's theorem as (Spiegel, 1964; Carrier et al., 1966)

$$\oint_{\Gamma=\Gamma_k+S_{km}+S_{m+1}-S_{km}}\frac{d\phi'(z)}{dz}dz=0 \tag{3.89}$$

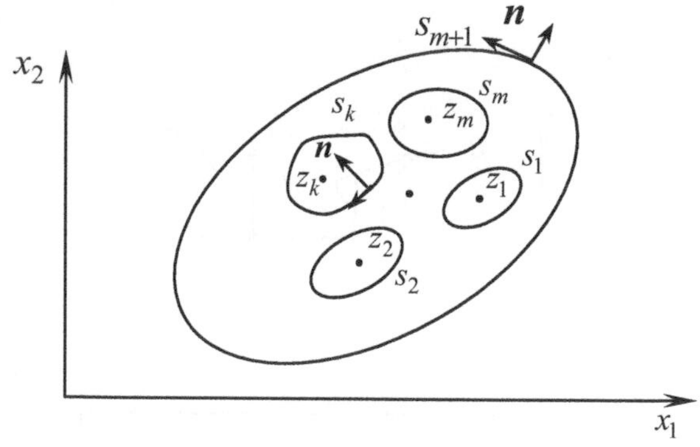

Figure 3.3 A multi-connected body with s_{m+1} boundaries

where, as shown in Fig. 3.4, Γ is a contour integral for a simply connected region. Thus, we have

$$\oint_{\Gamma_k + S_{m+1}} \frac{d\phi'(z)}{dz} dz = 0 \tag{3.90}$$

Therefore the integral for the loop Γ_k becomes

$$\oint_{\Gamma_k} \frac{d\phi'(z)}{dz} dz = -\oint_{S_{m+1}} \frac{d\phi'(z)}{dz} dz = \alpha_k = 2\pi i A_k \tag{3.91}$$

The last part of (3.91) is obtained by virtue of (3.87) that $\mathrm{Re}[\alpha_k] = 0$. In addition, since Cauchy's theorem applies to other contours as well as long as the hole is excluded, we can set $\Gamma_k = s_k$.

Now, we define another complex function:

$$\phi_*'(z) = \phi'(z) - \sum_{k=1}^{m} A_k \ln(z - z_k) \tag{3.92}$$

where z_k $(k = 1,...m)$ is an arbitrary point inside the inner boundary s_k (i.e., outside the body). Since $\ln(z-z_k)$ can be written as $\ln(r)+i\theta$ if $z-z_k = re^{i\theta}$ (the physical meaning of this $\ln(z-z_k)$ is shown in Fig. 3.5), when we go around s_k once, $A_k \ln(z-z_k)$ will increase by an amount $2i\pi A_k$ while all other terms in the summation remain unchanged. Consequently, $\phi'_*(z)$ must also remain unchanged. Therefore, we must have:

$$\phi'(z) = \phi_*'(z) + \sum_{k=1}^{m} A_k \ln(z - z_k) \tag{3.93}$$

where $\phi'_*(z)$ is a single-valued function for the multi-connected body. Integration of (3.93) gives

$$\phi(z) = \sum_{k=1}^{m} A_k [(z - z_k)\ln(z - z_k) - (z - z_k)] + \int_{z_0}^{z} \phi_*'(z) dz + \mathbf{const} \tag{3.94}$$

The integral on the right-hand side remains an analytic function of z. Thus, similarly, when we go around s_k once, it may increase by $2i\pi c_k$, where c_k is in general a complex number. However, we can follow the procedure that leads to (3.93) to yield

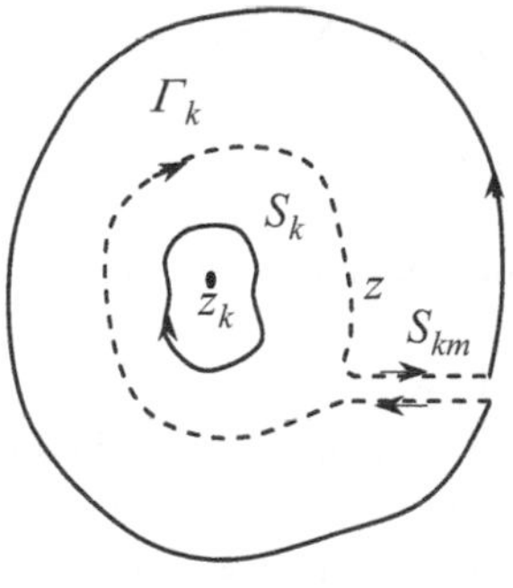

Figure 3.4 A contour integral excluding S_k

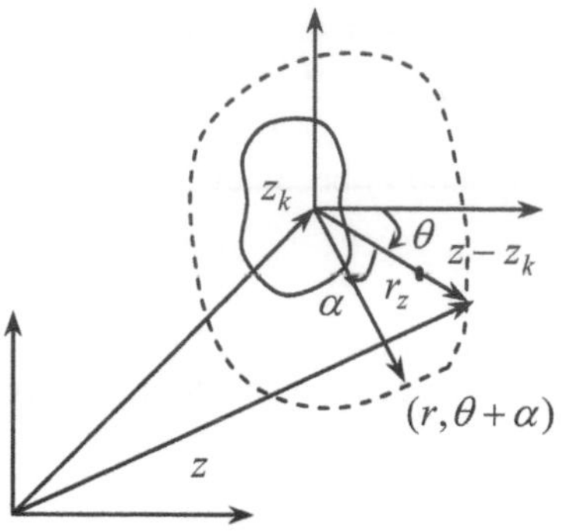

Fig. 3.5 The physical meaning of term ln($z-z_k$)

$$\int_{z_0}^{z} \phi'_*(z)dz = \sum_{k=1}^{m} c_k \ln(z-z_k) + \phi_i(z) \tag{3.95}$$

where $\phi_i(z)$ is the single-valued function for the integral in (3.94). Substitution of (3.95) into (3.94) gives

$$\phi(z) = z\sum_{k=1}^{m} A_k \ln(z-z_k) + \sum_{k=1}^{m} \gamma_k \ln(z-z_k) + \phi_1^*(z) \tag{3.96}$$

where ϕ_1^* is a single-valued function for the multi-connected body and γ_k is a complex constant.

We now recall (3.34), that is,

$$\sigma_{22} - \sigma_{11} + 2i\sigma_{12} = 2[\bar{z}\phi''(z) + \psi'(z)] \tag{3.97}$$

We see that $\phi''(z)$, in (3.97), must also be single-valued because the derivative of the single-valued function $\phi'_*(z)$ given in (3.93) is also single valued, and so is the derivative of $\ln(z-z_k)$, or $1/(z-z_k)$, since z_k is outside the body. Therefore, (3.97) indicates that $\psi'(z)$ must also be single-valued since the stresses on the left-hand side are single valued.

Consequently the argument leading to (3.93) can again be used to obtain

$$\psi(z) = \psi^*(z) + \sum_{k=1}^{m} \gamma'_k \ln(z-z_k) \tag{3.98}$$

where γ'_k is a complex constant and $\psi^*(z)$ is a single-valued function for the multi-connected body. Substituting (3.93), (3.96), and (3.98) into (3.36) and considering a complete path around the inner boundary s_k, the complex displacement will increase by:

$$2\pi i[(1+\kappa)A_k z + \kappa\gamma_k + \overline{\gamma'_k}] \tag{3.99}$$

Thus, the single-value condition of the displacement field gives

$$A_k = 0, \qquad \kappa\gamma_k + \overline{\gamma'_k} = 0 \tag{3.100}$$

The traction boundary condition on s_k provides another condition for γ_k and $\overline{\gamma'_k}$ to satisfy. In particular, (3.85) can be applied to the inner boundary s_k such that

$$[\phi(z)+z\bar{\phi}'(z)+\bar{\psi}(z)]_{S_k}=i(X_k+iY_k) \tag{3.101}$$

where X_k and Y_k are the components of the total resultant boundary force along the x_1- and x_2-axes, respectively. If we now take a complete cycle around s_k in clockwise direction as shown in Fig. 3.5 (i.e., the normal $\boldsymbol{n}$ should always point to our right-hand side), then (3.101) becomes

$$-2\pi i(\gamma_k-\bar{\gamma}_k')=i(X_k+iY_k) \tag{3.102}$$

Therefore, the second of (3.100) and (3.102) provides two equations for γ_k and $\bar{\gamma}_k'$ and the solution is

$$\gamma_k=-\frac{X_k+iY_k}{2\pi(1+\kappa)},\qquad \gamma_k'=\frac{\kappa(X_k-iY_k)}{2\pi(1+\kappa)} \tag{3.103}$$

Therefore, the analytic functions must have the following form:

$$\phi(z)=-\frac{1}{2\pi(1+\kappa)}\sum_{k=1}^{m}(X_k+iY_k)\ln(z-z_k)+\phi_*(z) \tag{3.104}$$

$$\psi(z)=\frac{\kappa}{2\pi(1+\kappa)}\sum_{k=1}^{m}(X_k-iY_k)\ln(z-z_k)+\psi_*(z) \tag{3.105}$$

where $\phi_*(z)$ and $\psi_*(z)$ are single-valued analytic functions for the multi-connected body.

3.9 MULTI-CONNECTED BODY OF INFINITE EXTENT

If we now let the exterior boundary s_{m+1} to be unbounded (i.e., the body becomes infinite in extent), (3.104) and (3.105) can be simplified. In particular, we can consider a circle s_R of radius R such that all inner boundaries are embraced. For any point z outside s_R we have $|z|\gg|z_k|$, and the expansion of $\ln(z-z_k)$ becomes

$$\ln(z-z_k)=\ln z+\ln(1-\frac{z_k}{z})=\ln z-\frac{z_k}{z}-\frac{1}{2}\left(\frac{z_k}{z}\right)^2+\cdots \tag{3.106}$$

by virtue of $\ln(1-x)=-x-x^2/2-\ldots$. If s_R is chosen to be large enough (i.e., $z\to\infty$), we have

$$\phi(z)=-\frac{(X+iY)}{2\pi(1+\kappa)}\ln z+\phi^{**}(z) \tag{3.107}$$

$$\psi(z)=\frac{\kappa(X-iY)}{2\pi(1+\kappa)}\ln z+\psi^{**}(z) \tag{3.108}$$

where X and Y are the sums of the horizontal and vertical force components over all inner boundaries, respectively, and $\phi^{**}(z)$ and $\psi^{**}(z)$ are the analytic functions outside s_R. It is normally proposed to expand both functions in the Laurent series (Muskhelishvili, 1953)

$$\phi^{**}(z)=\sum_{-\infty}^{\infty}a_nz^n,\qquad \psi^{**}(z)=\sum_{-\infty}^{\infty}b_nz^n \tag{3.109}$$

because the Laurent series converges uniform for all regions, except near infinity. Substituting (3.109) into (3.107) and (3.108), then into the stress expressions (3.34) and (3.35), we find that all a_n and b_n, where $n > 2$ must vanish if the stresses are bounded at infinity. Therefore, we have

$$\phi(z) = -\frac{(X+iY)}{2\pi(1+\kappa)}\ln z + Bz + \sum_{1}^{\infty}\frac{a_n}{z^n} \tag{3.110}$$

$$\psi(z) = \frac{\kappa(X-iY)}{2\pi(1+\kappa)}\ln z + (B'+iC')z + \sum_{1}^{\infty}\frac{b_n}{z^n} \tag{3.111}$$

In obtaining (3.110) and (3.111), we have removed the constant term (i.e., C, a_0 and b_0) since Section 3.6 has shown that arbitrary constants can be added to remove these terms without changing the resultant stresses. As will be shown in Section 3.11, B, B', and C' are related to far field applied tractions.

3.10 GENERAL TRANSFORMATION OF QUANTITIES

For later consideration of problems of an elastic body with holes, we consider the general coordinate transformation here:

$$z = \omega(\zeta) \tag{3.112}$$

which maps a body from the z-plane to a corresponding body in the ζ-plane. In the transformed plane, ζ can be expressed in polar form:

$$\zeta = \rho(\cos\theta + i\sin\theta) = \rho e^{i\theta} \tag{3.113}$$

The circles $\rho = C$ and the radial lines $\theta = C_2$ (where both C_1 and C_2 are constants) in the ζ-plane will appear as a curvilinear coordinate on the z-plane after mapping, as shown in Fig. 3.6.

As a consequence of conformal mapping, the coordinate lines $\rho = C_1$ and $\theta = C_2$ in the z-plane will remain orthogonal. Let A be some vector in the z-plane, the projections of which on the x_1- and x_2-directions are A_1 and A_2 , respectively. Applying (3.52) in Section 3.4, we have

$$A_\rho + iA_\theta = (A_1 + iA_2)e^{-i\lambda} \tag{3.114}$$

where A_ρ and A_θ are the projection of $\boldsymbol{A}$ along the ρ- and θ-directions and λ is the angle between the x_1- and ρ-directions (as shown in Fig. 3.6). If the point z is displaced by dz along the ρ-direction, the corresponding point ζ in the ζ-plane will move by $d\zeta$ along the radial direction. Hence,

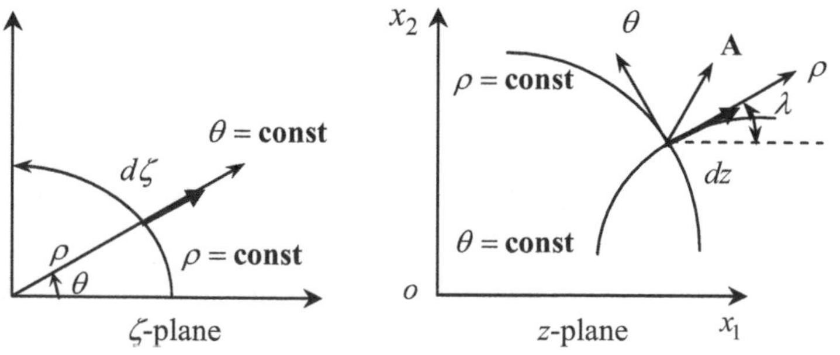

Fig. 3.6 Polar coordinate in the ζ-plane to curvilinear coordinate in the z-plane

$$dz = |dz|(\cos\lambda + i\sin\lambda) = e^{i\lambda}|dz| \tag{3.115}$$

$$d\zeta = |d\zeta|(\cos\theta + i\sin\theta) = e^{i\theta}|d\zeta| \tag{3.116}$$

Therefore, we have

$$e^{i\lambda} = \frac{dz}{|dz|} = \frac{\omega'(\zeta)d\zeta}{|\omega'(\zeta)||d\zeta|} = e^{i\theta}\frac{\omega'(\zeta)}{|\omega'(\zeta)|} = \frac{\zeta\omega'(\zeta)}{\rho|\omega'(\zeta)|} \tag{3.117}$$

Taking the conjugate for both sides gives

$$e^{-i\lambda} = \frac{\bar{\zeta}\bar{\omega}'(\bar{\zeta})}{\rho|\omega'(\zeta)|} \tag{3.118}$$

Substitution of (3.118) into (3.114) yields

$$A_\rho + iA_\theta = \frac{\bar{\zeta}\bar{\omega}'(\bar{\zeta})}{\rho|\omega'(\zeta)|}(A_1 + iA_2) \tag{3.119}$$

We now introduce $\phi_1(z)$ and $\psi_1(z)$ to denote the functions which were earlier written as $\phi(z)$ and $\psi(z)$. Then, we make the following definitions:

$$\phi(\zeta) = \phi_1(z) = \phi_1[\omega(\zeta)] \tag{3.120}$$

$$\psi(\zeta) = \psi_1(z) = \psi_1[\omega(\zeta)] \tag{3.121}$$

$$\Phi(\zeta) = \phi_1'(z) = \frac{\phi'(\zeta)}{\omega'(\zeta)} \tag{3.122}$$

$$\Psi(\zeta) = \psi_1'(z) = \frac{\psi'(\zeta)}{\omega'(\zeta)} \tag{3.123}$$

$$\Phi'(\zeta) = \phi_1''(z)\omega'(\zeta) \tag{3.124}$$

such that the stresses and displacements can be expressed in terms of the new variable ζ. Consequently, (3.36) becomes

$$2\mu(u_1 + iu_2) = \kappa\phi(\zeta) - \frac{\omega(\zeta)}{\bar{\omega}'(\bar{\zeta})}\bar{\phi}'(\bar{\zeta}) - \bar{\psi}(\bar{\zeta}) \tag{3.125}$$

The displacement in terms of the curvilinear coordinates u_ρ and u_θ is, according to (3.119),

$$u_\rho + iu_\theta = \frac{\bar{\zeta}\bar{\omega}'(\bar{\zeta})}{\rho|\omega'(\zeta)|}(u_1 + iu_2) \tag{3.126}$$

Consequently, we have

$$u_\rho + iu_\theta = \frac{1}{2\mu}\frac{\bar{\zeta}\bar{\omega}'(\bar{\zeta})}{\rho|\omega'(\zeta)|}[\kappa\phi(\zeta) - \frac{\omega(\zeta)}{\bar{\omega}'(\bar{\zeta})})\bar{\phi}'(\bar{\zeta}) - \bar{\psi}(\bar{\zeta})] \tag{3.127}$$

Recalling (3.37) we get

$$\sigma_{\theta\theta} + \sigma_{\rho\rho} = 2[\phi_1'(z) + \bar{\phi}_1'(z)] = 4\,\mathrm{Re}[\phi_1'(z)] \tag{3.128}$$

$$\sigma_{\theta\theta} - \sigma_{\rho\rho} + 2i\sigma_{\rho\theta} = 2[\bar{z}\phi_1''(z) + \psi_1'(z)]e^{2i\lambda} \tag{3.129}$$

where

$$e^{2i\lambda}=\frac{\zeta^2[\omega'(\zeta)]^2}{\rho^2\left|\omega'(\zeta)\right|^2}=\frac{\zeta^2\omega'(\zeta)}{\rho^2\bar{\omega}'(\bar{\zeta})} \tag{3.130}$$

Substitution of (3.130) into (3.129) and using (3.122) and (3.123), we get

$$\sigma_{\theta\theta}+\sigma_{\rho\rho}=2[\Phi(\zeta)+\bar{\Phi}(\bar{\zeta})]=4\,\mathrm{Re}[\Phi(\zeta)] \tag{3.131}$$

$$\sigma_{\theta\theta}-\sigma_{\rho\rho}+2i\sigma_{\rho\theta}=\frac{2\zeta^2}{\rho^2\bar{\omega}'(\zeta)}[\bar{\omega}(\zeta)\Phi'(z)+\omega'(\zeta)\Psi(\zeta)] \tag{3.132}$$

3.11 ELASTIC BODY WITH HOLES

Complex analysis has been found extremely useful in solving the stress concentration at a hole embedded in an infinite domain. This is because conformal mapping can be used to transform an opening of any shape in the z-plane into either the interior or exterior of a unit circle in the ζ-plane, as shown in Fig. 3.7. This section follows closely the presentation by Xu (1982). For mapping to the interior of a unit circle, $\omega(\zeta)$ can be expressed as

$$z=\omega(\zeta)=R(\frac{1}{\zeta}+\sum_{k=0}^{n}c_k\zeta^k) \tag{3.133}$$

where c_k is a complex constant and the sum of $|c_k|$ (k from 0 to n) should be smaller than unity. The number of terms n required to describe a particular shape of holes is normally small (i.e., only a few terms in the series are needed). For mapping to the exterior, we have $\omega_1(\zeta)$:

$$z=\omega_1(\zeta)=R(\zeta+\sum_{k=0}^{n}\frac{b_k}{\zeta^k}) \tag{3.134}$$

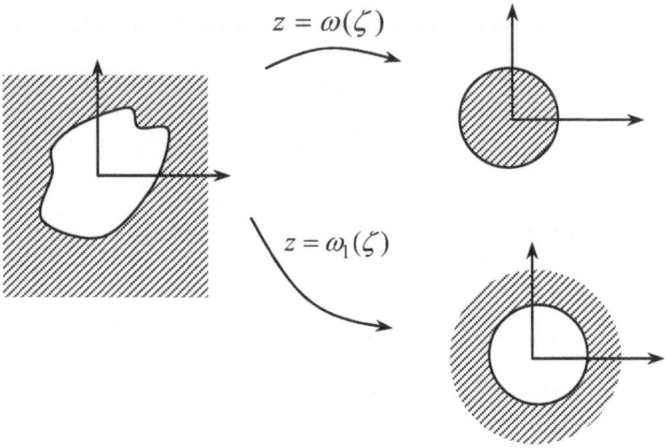

Figure 3.7 Mapping from the z-plane to either the interior or the exterior of a unit circle in the ζ-plane

where R is a real constant relating to the size of the hole. Although both interior and exterior transformations can be used, to date most results are obtained by the former technique.

To transform (3.110) and (3.111) in terms of ζ, we consider

$$\begin{aligned}\ln z &= \ln[\frac{R}{\zeta}(1+c_0\zeta+c_1\zeta^2+\cdots+c_n\zeta^{n+1})] \\ &= -\ln\zeta+\ln R+\ln[1+(c_0\zeta+c_1\zeta^2+\cdots+c_n\zeta^{n+1})]\end{aligned} \tag{3.135}$$

Since within the unit circle, we have $|\zeta|<1$, thus

$$\left|c_0\zeta+c_1\zeta^2+\cdots+c_n\zeta^{n+1}\right|<1 \tag{3.136}$$

Therefore, the logarithm can be expanded as:

$$\begin{aligned}&\ln[1+(c_0\zeta+c_1\zeta^2+\cdots+c_n\zeta^{n+1})] \\ &=(c_0\zeta+c_1\zeta^2+\cdots)-\frac{1}{2}(c_0\zeta+c_1\zeta^2+\cdots)^2+\cdots\end{aligned} \tag{3.137}$$

Thus, $\ln z$ equals $-\ln\zeta$ plus some analytic functions within the circle. In addition, the first term in the sums in (3.110) and (3.111) can also be expanded as:

$$\frac{a_1}{z}=\frac{a_1}{\frac{R}{\zeta}(1+c_0\zeta+c_1\zeta^2+\cdots)}=\frac{a_1\zeta}{R}(1-c_0\zeta+\cdots) \tag{3.138}$$

Similar consideration also applies to the higher-order terms in the series. Therefore, finally we write

$$\phi(\zeta)=\frac{(X+iY)}{2\pi(1+\kappa)}\ln\zeta+B\omega(\zeta)+\phi_0(\zeta) \tag{3.139}$$

$$\psi(\zeta)=-\frac{\kappa(X-iY)}{2\pi(1+\kappa)}\ln\zeta+(B'+iC')\omega(\zeta)+\psi_0(\zeta) \tag{3.140}$$

where

$$\phi_0(\zeta)=\sum_{k=1}^{\infty}\alpha_k\zeta^k \tag{3.141}$$

$$\psi_0(\zeta)=\sum_{k=1}^{\infty}\beta_k\zeta^k \tag{3.142}$$

Both of these series are analytic and continuous everywhere inside and on the circle. Note that all the constant terms are deleted since they do not affect the resulting stress (see Section 3.6).

The boundary force condition given in (3.85) can be expressed as

$$[\phi(\zeta)+\frac{\omega(\zeta)}{\overline{\omega}'(\zeta)}\overline{\phi}'(\zeta)+\overline{\psi}(\zeta)]_s=i\int(T_1+iT_2)ds \tag{3.143}$$

Now, on the circle's boundary we have $\rho=1$, that is $\zeta=e^{i\theta}=\sigma$ (which should not be mixed up with the stress term). Therefore, the boundary condition becomes

$$[\phi(\sigma)+\frac{\omega(\sigma)}{\bar{\omega}'(\sigma)}\bar{\phi}'(\sigma)+\bar{\psi}(\sigma)]_s = i\int(T_1+iT_2)ds \tag{3.144}$$

However, setting $\zeta = \sigma$ into (3.139) and (3.140) yields

$$\phi(\sigma)=\frac{(X+iY)}{2\pi(1+\kappa)}\ln\sigma + B\omega(\sigma)+\phi_0(\sigma) \tag{3.145}$$

$$\psi(\sigma)=-\frac{\kappa(X-iY)}{2\pi(1+\kappa)}\ln\sigma+(B'+iC')\omega(\sigma)+\psi_0(\sigma) \tag{3.146}$$

Substitution of (3.145) and (3.146) into (3.144) gives

$$\phi_0(\sigma)+\frac{\omega(\sigma)}{\bar{\omega}'(\sigma)}\bar{\phi}_0'(\sigma)+\bar{\psi}_0(\sigma)=f_0 \tag{3.147}$$

where f_0 is defined as

$$f_0 = i\int(T_1+iT_2)ds-\frac{(X+iY)}{2\pi}\ln\sigma-\frac{(X-iY)}{2\pi(1+\kappa)}\frac{\omega(\sigma)}{\bar{\omega}'(\sigma)}\sigma \\ -2B\omega(\sigma)-(B'-iC')\bar{\omega}(\sigma) \tag{3.148}$$

Another boundary condition is the conjugate of (3.147):

$$\bar{\phi}_0(\sigma)+\frac{\bar{\omega}(\sigma)}{\omega'(\sigma)}\phi_0'(\sigma)+\psi_0(\sigma)=\bar{f}_0 \tag{3.149}$$

When the hole boundary is free of traction, we have $T_1 = T_2 = 0$ and $X = Y = 0$. When the far field is considered (i.e., $z \to \infty$), we can demonstrate by substituting (3.110) and (3.111) into (3.34) and (3.35) that:

$$\sigma_{22}+\sigma_{11}=4B, \qquad \sigma_{22}-\sigma_{11}+2i\sigma_{12}=2(B'+iC') \tag{3.150}$$

The far field applied stress can be reflected by choosing B, B' and C'. Conversely, if only stress is applied on the hole's boundary, we have $B = B' = C' = 0$, and X and Y are the horizontal and vertical resultant force on the hole's surface. In any case, f_0 and $\bar{f}_0$ are prescribed, and we now have to decide on the forms for $\phi_0(\zeta)$ and $\psi_0(\zeta)$. In doing so, we apply the Cauchy integration formula to (3.147) and (3.149). In particular, we discuss here two types of Cauchy integration formulas:

CASE (1):
When a function $F(\zeta)$ is analytic within the unit circle, and is continuous inside and on the circle (its boundary denoted by σ), then for every point ζ inside the unit circle we have

$$\frac{1}{2\pi i}\int_\sigma \frac{F(\sigma)d\sigma}{\sigma-\zeta}=F(\zeta) \tag{3.151}$$

which is the Cauchy formula for finite regions

CASE (2):
When $F(\zeta)$ is analytic outside the unit circle, and is continuous everywhere outside and on the circle (its boundary denoted by σ), then we have for every point ζ inside the circle

$$\frac{1}{2\pi i}\int_\sigma \frac{F(\sigma)d\sigma}{\sigma-\zeta}=F(\infty) \tag{3.152}$$

which is the Cauchy formula for infinite regions. The proof of these formulas can be found in Muskhelishvili (1975).

Multiplying both sides of (3.147) by $[1/(2\pi i)][d\sigma/(\sigma - \zeta)]$ and integrating around the hole's boundary, we get

$$\begin{aligned}\frac{1}{2\pi i}\int_\sigma \frac{\phi_0(\sigma)}{\sigma-\zeta}d\sigma + \frac{1}{2\pi i}\int_\sigma \frac{\omega(\sigma)}{\overline{\omega}'(\sigma)}\frac{\overline{\phi}_0'(\sigma)}{\sigma-\zeta}d\sigma \\ +\frac{1}{2\pi i}\int_\sigma \frac{\overline{\psi}_0(\sigma)}{\sigma-\zeta}d\sigma = \frac{1}{2\pi i}\int_\sigma \frac{f_0(\sigma)}{\sigma-\zeta}d\sigma\end{aligned} \tag{3.153}$$

From the definition of $\phi_0(\zeta)$ given in (3.141), it is obvious that $\phi_0(\zeta)$ is analytic inside the unit circle and continuous everywhere inside and on the circle. Therefore, by Cauchy integral formula given in (3.151), we have

$$\frac{1}{2\pi i}\int_\sigma \frac{\phi_0(\sigma)}{\sigma-\zeta}d\sigma = \phi_0(\zeta) \tag{3.154}$$

On the other hand, we find from (3.142)

$$\overline{\psi}_0(\sigma) = \overline{\beta}_1\overline{\sigma} + \overline{\beta}_2\overline{\sigma}^2 + \cdots = \frac{\overline{\beta}_1}{\sigma} + \frac{\overline{\beta}_2}{\sigma^2} + \cdots \tag{3.155}$$

And, it is obvious that

$$\frac{\overline{\beta}_1}{\sigma} + \frac{\overline{\beta}_2}{\sigma^2} + \cdots \tag{3.156}$$

is analytic outside the circle and is continuous everywhere outside and on the unit circle; therefore, (3.152) gives

$$\frac{1}{2\pi i}\int_\sigma \frac{\overline{\psi}_0(\sigma)}{\sigma-\zeta}d\sigma = \overline{\psi}_0(\infty) = 0 \tag{3.157}$$

Substituting (3.154) and (3.157) into (3.153), we have

$$\phi_0(\zeta) + \frac{1}{2\pi i}\int_\sigma \frac{\omega(\sigma)}{\overline{\omega}'(\sigma)}\frac{\overline{\phi}_0'(\sigma)}{\sigma-\zeta}d\sigma = \frac{1}{2\pi i}\int_\sigma \frac{f_0(\sigma)}{\sigma-\zeta}d\sigma \tag{3.158}$$

Similar consideration to (3.149) yields

$$\psi_0(\zeta) + \frac{1}{2\pi i}\int_\sigma \frac{\overline{\omega}(\sigma)}{\omega'(\sigma)}\frac{\phi_0'(\sigma)}{\sigma-\zeta}d\sigma = \frac{1}{2\pi i}\int_\sigma \frac{\overline{f}_0(\sigma)}{\sigma-\zeta}d\sigma \tag{3.159}$$

Therefore, problems for stress concentration at holes reduce to finding an appropriate mapping function $\omega(\zeta)$ for the hole boundary and finding the appropriate form of $\phi_0(\zeta)$ and $\psi_0(\zeta)$. Fortunately, many useful functions for different shapes of holes have been tabulated in books on conformal mapping (e.g., Savin, 1961). In particular, substituting the appropriate $\omega(\zeta)$ and (3.141) into (3.158), the evaluation of the resulting equation using the Cauchy integration formula gives $\phi_0(\zeta)$ which can be further substituted into (3.159) yielding $\psi_0(\zeta)$. Then (3.139) and (3.140) can be used to find $\phi(\zeta)$ and $\psi(\zeta)$. Subsequently, (3.122) and (3.123) can be used to yield $\Phi(\zeta)$ and $\Psi(\zeta)$. Finally, the stress concentration can be found using (3.131) and (3.132).

3.12 STRESS CONCENTRATION AT A SQUARE HOLE

Using the Schwarz–Christoffel integral, one can show that the conformal mapping for the exterior of a square hole in the *z*-plane onto the interior of a unit circle in the ζ-plane is approximately (Savin, 1961)

$$z = \omega(\zeta) = R(\frac{1}{\zeta} - \frac{1}{6}\zeta^3 + \frac{1}{56}\zeta^7 - \frac{1}{176}\zeta^{11} + \frac{1}{384}\zeta^{15} - \frac{7}{4864}\zeta^{19} + \cdots) \quad (3.160)$$

where R is a real constant indicating the size of the square hole. It can be shown that the sharpness of the four corners depends on the number of terms used in (3.160). For example, if we can retain the first two terms, we have

$$z = \omega(\zeta) = R(\frac{1}{\zeta} - \frac{1}{6}\zeta^3) \quad (3.161)$$

On the circle boundary, $\zeta = \sigma = e^{i\theta}$ and

$$x + iy = R(e^{-i\theta} - \frac{1}{6}e^{3i\theta}) = R(\cos\theta - i\sin\theta - \frac{1}{6}\cos 3\theta - \frac{1}{6}i\sin 3\theta) \quad (3.162)$$

Equating the real and imaginary parts, we obtain the curve for the hole's boundary in the *z*-plane:

$$x = R(\cos\theta - \frac{1}{6}\cos 3\theta), \qquad y = -R(\sin\theta + \frac{1}{6}\sin 3\theta) \quad (3.163)$$

Setting θ to different values, we can sketch the hole's boundary as shown in Fig. 3.8(a). More specifically, when $\theta = 0$, $x = 5R/6$ and $y = 0$; when $\theta = \pi/2$, $x = 0$ and $y = -5R/6$; when $\theta = \pi/4$, $x = -y = 7R\sqrt{2}/12$. Therefore, at the center level, the width of the square is $a = 5R/3$ and the diagonal can be shown to be $d = 1.4a$. According to mathematics books on analytic geometry (e.g., Britton et al., 1966), the parametric form of the radius of curvature is

$$|r| = \left[\frac{[(\frac{dx}{d\theta})^2 + (\frac{dy}{d\theta})^2]^{3/2}}{\frac{dx}{d\theta}\frac{d^2y}{d\theta^2} - \frac{dy}{d\theta}\frac{d^2x}{d\theta^2}} \right]_{\theta=\pi/4} = \frac{R}{10} = 0.06a \quad (3.164)$$

Without showing the details, if we retain the first three terms in (3.160), we have $|r| = 0.025a$; and the corresponding hole is shown in Fig. 3.8(b). If the first four terms in (289) are obtained, $|r|$ becomes $0.014a$; the corresponding hole is shown in Fig. 3.8(c), which is essentially same as that shown in Fig. 3.8(b).

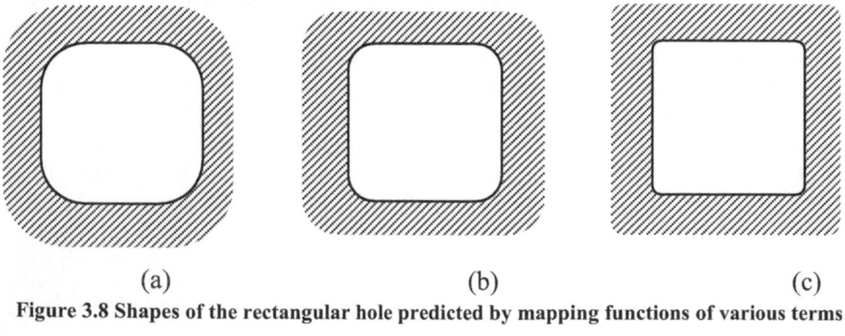

(a) (b) (c)

Figure 3.8 Shapes of the rectangular hole predicted by mapping functions of various terms

Using (3.110) and (3.111), it can be shown that the far field stresses σ_1 and σ_2 can be related to B, B', and C' as (see Problem 3.4)

$$\begin{aligned} \sigma_{22}+\sigma_{11} &= \sigma_1+\sigma_2 = 4B \\ \sigma_{22}-\sigma_{11}+2i\sigma_{12} &= -(\sigma_1-\sigma_2)e^{-2i\alpha} = 2(B'+iC') \end{aligned} \tag{3.165}$$

where σ_1 and σ_2 are the principal stresses as shown in Fig. 3.9.

For a square hole subject to a far field compression q along the orientation measuring α from the horizontal (as shown in Fig. 3.10), we have

$$B=-\frac{q}{4}, \qquad B'-iC'=\frac{q}{2}e^{2i\alpha} \tag{3.166}$$

Since the hole is traction free, we also get

$$X=Y=T_1=T_2=0 \tag{3.167}$$

Now, the mapping function (3.161) and its derivative and conjugate are

$$\omega(\sigma)=R(\frac{1}{\sigma}-\frac{1}{6}\sigma^3), \qquad \bar{\omega}(\sigma)=R(\sigma-\frac{1}{6\sigma^3}) \tag{3.168}$$

$$\omega'(\sigma)=-R(\frac{1}{\sigma^2}+\frac{1}{2}\sigma^2), \qquad \bar{\omega}'(\sigma)=-R(\sigma^2+\frac{1}{2\sigma^2}) \tag{3.169}$$

$$\frac{\omega(\sigma)}{\bar{\omega}'(\sigma)}=\frac{\sigma}{6}-\frac{13\sigma}{6(2\sigma^4+1)}, \qquad \frac{\bar{\omega}(\sigma)}{\omega'(\sigma)}=\frac{1}{6\sigma}-\frac{13\sigma^3}{6(2+\sigma^4)}) \tag{3.170}$$

Therefore, (3.148) becomes

$$f_0=\frac{qR}{2}[(\frac{1}{\sigma}-\frac{\sigma^3}{6})-(\sigma-\frac{1}{6\sigma^3})e^{2i\alpha}] \tag{3.171}$$

Substituting (3.171) into the right-hand side of (3.158), we get, in view of (3.151) and (3.152),

$$\frac{1}{2\pi i}\int_{\sigma}\frac{f_0 d\sigma}{\sigma-\zeta}=-\frac{qR}{12}(\zeta^2+6e^{2i\alpha})\zeta \tag{3.172}$$

Similarly, the second term on the left-hand side of (3.158) becomes

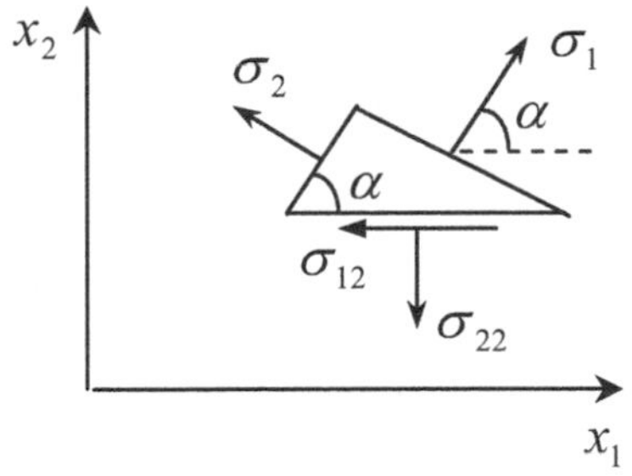

Fig. 3.9 The definitions for α, σ_1 and σ_2

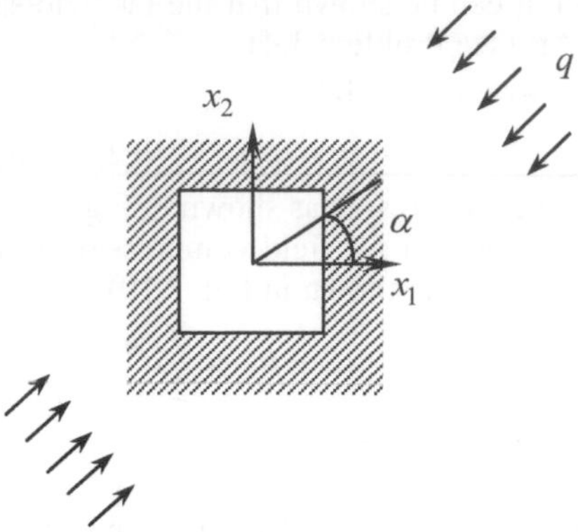

Fig. 3.10 A square hole subject to far field compression q inclined at α

$$\frac{1}{2\pi i}\int_\sigma \frac{\omega(\sigma)}{\overline{\omega}'(\sigma)}\frac{\overline{\phi}_0' d\sigma}{\sigma-\zeta} = \frac{1}{2\pi i}\int_\sigma \left[\frac{\sigma}{6}-\frac{13\sigma}{6(2\sigma^4+1)}\right]\left(\overline{\alpha}_1+\frac{2\overline{\alpha}_2}{\sigma}+\frac{3\overline{\alpha}_3}{\sigma^2}+\cdots\right)\frac{d\sigma}{\sigma-\zeta} \tag{3.173}$$
$$= \frac{1}{6}\overline{\alpha}_1\zeta+\frac{1}{3}\overline{\alpha}_2$$

Therefore, (3.158) becomes

$$\alpha_1\zeta+\alpha_2\zeta^2+\alpha_3\zeta^3+\ldots+\frac{1}{6}\overline{\alpha}_1\zeta+\frac{1}{3}\overline{\alpha}_2 = -\frac{qR}{12}(\zeta^2+6e^{2i\alpha})\zeta \tag{3.174}$$

Comparing the same order of ζ on both sides gives:

$$\alpha_1+\frac{1}{6}\overline{\alpha}_1 = -\frac{qR}{2}e^{2i\alpha}, \qquad \alpha_3 = -\frac{qR}{12}, \qquad \alpha_2=\alpha_4=\alpha_5=\cdots=0 \tag{3.175}$$

Hence, the α_1 and α_3 are

$$\alpha_1 = -qR(\frac{3}{7}\cos 2\alpha+i\frac{3}{5}\sin 2\alpha), \qquad \alpha_3 = -\frac{qR}{12} \tag{3.176}$$

Thus, $\phi_0(\zeta)$ becomes

$$\varphi_0(\zeta)=\alpha_1\zeta+\alpha_2\zeta^2+\alpha_3\zeta^3+\cdots = -qR[(\frac{3}{7}\cos 2\alpha+i\frac{3}{5}\sin 2\alpha)\zeta+\frac{1}{12}\zeta^3] \tag{3.177}$$

Substitution of (3.170) and (3.171) and (3.177) into (3.159) gives

$$\psi_0(\zeta)-\frac{1}{2\pi i}\int_\sigma \left[\frac{1}{6\sigma}-\frac{13\sigma^3}{6(2+\sigma^4)}\right]qR\left[(\frac{3}{7}\cos 2\alpha+i\frac{3}{5}\sin 2\alpha)+\frac{1}{4}\sigma^2\right]\frac{d\sigma}{\sigma-\zeta}$$
$$= \frac{1}{2\pi i}\int_\sigma \frac{qR}{2}\left[(\sigma-\frac{1}{6\sigma^3})-(\frac{1}{\sigma}-\frac{\sigma^3}{6})e^{-2i\alpha}\right]\frac{d\sigma}{\sigma-\zeta} \tag{3.178}$$

After integration, we get

$$\psi_0(\zeta)+\frac{13\zeta^3}{6(2+\zeta^4)}qR(\frac{3}{7}\cos 2\alpha+i\frac{3}{5}\sin 2\alpha)-\frac{qR}{4}\zeta^2\left[\frac{1}{6\zeta}-\frac{13\zeta^3}{6(2+\zeta^4)}\right] \quad (3.179)$$
$$=\frac{qR}{2}[\zeta+\frac{1}{6}\zeta^3 e^{-2i\alpha}]$$

Finally, substitution of (3.170)–(3.172), (3.177), and (3.179) into (3.139) and(3.140) yields the following analytic functions:

$$\phi(\zeta)=-qR[\frac{1}{4\zeta}+(\frac{3}{7}\cos 2\alpha+i\frac{3}{5}\sin 2\alpha)\zeta+\frac{1}{24}\zeta^3] \quad (3.180)$$

$$\psi(\zeta)=qR[\frac{e^{-2i\alpha}}{2\zeta}+\frac{13\zeta-26(\frac{3}{7}\cos 2\alpha+i\frac{3}{5}\sin 2\alpha)\zeta^3}{12(2+\zeta^4)}] \quad (3.181)$$

If uniaxial compression is applied along the x-axis, we have $\alpha = 0$ (i.e., $\sigma_{11} = -q$, $\sigma_{22} = \sigma_{12} = 0$) and the complex stress functions reduce to

$$\phi(\zeta)=-qR[\frac{1}{4\zeta}+\frac{1}{24}\zeta^3+\frac{3}{7}\zeta] \quad (3.182)$$

$$\psi(\zeta)=qR[\frac{1}{2\zeta}+\frac{91\zeta-78\zeta^3}{84(2+\zeta^4)}] \quad (3.183)$$

Substitution of (3.182) into (3.122) yields

$$\Phi(\zeta)=\left[\frac{24+7\zeta^2-\dfrac{14}{\zeta^2}}{28(\dfrac{2}{\zeta^2}+\zeta^2)}\right]q \quad (3.184)$$

Then, (3.184) can be combined with (3.113) and (3.131) to give

$$\sigma_{\theta\theta}+\sigma_{\rho\rho}=q\left[\frac{14\rho^8-56+(96\rho^2+48\rho^6)\cos 2\theta}{14(4+\rho^8+4\rho^4\cos 4\theta)}\right] \quad (3.185)$$

Since the far field in the z-plane is mapped to the origin in the ζ-plane, setting $\rho \to 0$ gives $\sigma_{\theta\theta}+\sigma_{\rho\rho} = -q$ as expected. On the hole's boundary (i.e., $\rho = 1$), $\sigma_{\rho\rho}$ is identically zero (traction-free hole boundary); thus the tangential stress on the hole boundary becomes

$$\sigma_{\theta\theta}=q\left[\frac{72\cos 2\theta-21}{7(5+4\cos 4\theta)}\right] \quad (3.186)$$

At point A shown in Fig. 3.11 (i.e., $\theta = 0$), the stress concentration is

$$[\sigma_{\theta\theta}]_{\rho=1,\theta=0}=\frac{51}{63}q=0.81q\,, \quad (3.187)$$

compared to $-q$ for a circular hole (see Section 2.18). At the corner, shown as point B in Fig. 3.11, we have $\theta = \pi/4$ and the tangential stress becomes

$$[\sigma_{\theta\theta}]_{\rho=1,\theta=\pi/4}=-3q \quad (3.188)$$

At the midpoint of the sides parallel to the compression (i.e., point C), we set $\theta = \pi/2$ and the tangential stress concentration becomes

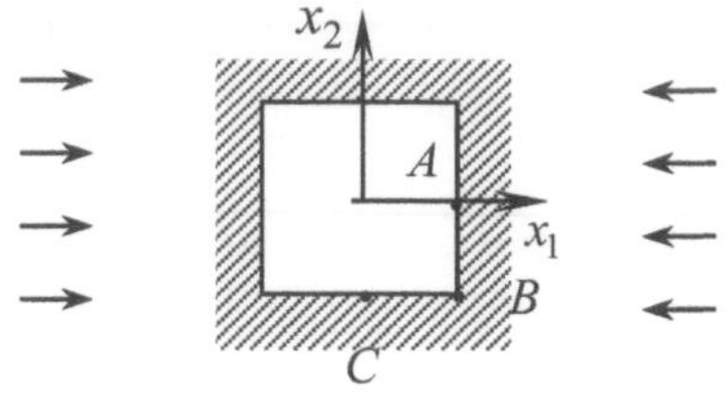

Figure 3.11 A square hole subject to far field uniaxial compression

$$[\sigma_{\theta\theta}]_{\rho=1,\theta=\pi/2} = -\frac{93}{63}q = -1.476q\,, \tag{3.189}$$

compared to $-3q$ for a circular hole (see Section 2.18). However, it can be shown that neither is the maximum compressive tangential stress at corner point B nor is the maximum tensile tangential at point A (see Problem 3.5). In particular, the maximum compression occurs at 47.087^0 measured from the x_1-axis (compared to 45^0 at the comer) and its magnitude is $-3.857q$; the maximum tensile occurs at 30.29^0 measured from the x_1-axis and its magnitude is $0.857q$.

Similar to the discussion given in Section 2.18 for a circular hole, this result can readily be extended to biaxial compression (see Fig. 3.12). For the stress concentration at the boundary, we add (3.186) with another solution for the vertical stress βq, which is obtained simply by replacing q and θ by βq and $\theta+\pi/2$ in (3.186). The final stress concentration is

$$\sigma_{\theta\theta} = q\left[\frac{72(1-\beta)\cos 2\theta - 21(1+\beta)}{7(5+4\cos 4\theta)}\right] \tag{3.190}$$

The stress concentrations at $\theta = 0$, $\pi/4$, and $\pi/2$ are

$$\sigma_{\theta\theta} = q\left(\frac{17-31\beta}{21}\right), \quad \sigma_{\theta\theta} = -3q(1+\beta), \quad \sigma_{\theta\theta} = -q\left(\frac{31-17\beta}{21}\right), \tag{3.191}$$

respectively. The maximum and minimum of the stress concentration can be shown to occur at

$$\theta = \frac{1}{2}\cos^{-1}\left\{\frac{7(1+\beta) \pm \sqrt{121(1+\beta^2) - 46\beta}}{24(1-\beta)}\right\} \tag{3.192}$$

We can also follow the argument by Terzaghi and Richart (1952) or (2.144) that the geostatic stress state is

$$\sigma_{22} = -\gamma z, \qquad \sigma_{11} = K\sigma_{22} = -\left(\frac{\nu}{1-\nu}\right)\gamma z \tag{3.193}$$

where γ and z are the unit weight of the overlying rock and the depth of the square tunnel, respectively. Thus, we can set $q = \nu\gamma z/(1-\nu)$ and $\beta = (1-\nu)/\nu$ into (3.190). For typical rocks, Poisson's ratio ν is 0.2, hence β becomes 4. For this case, the stress concentrations at $\theta = 0$, $\pi/4$, and $\pi/2$ are, respectively,

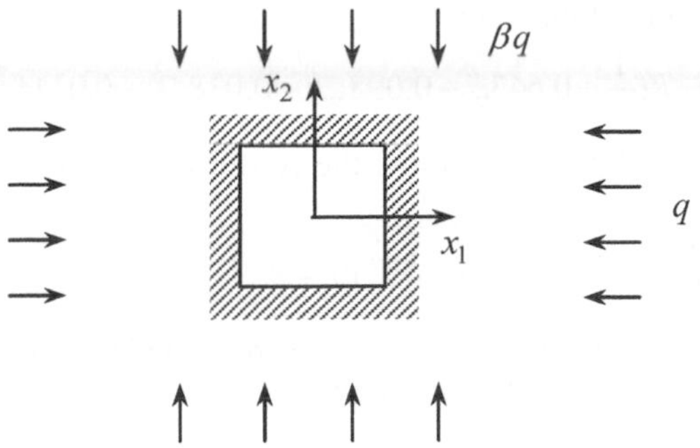

Fig. 3.12 A square hole subject to far field biaxial compression

$$\sigma_{\theta\theta} = -\frac{101}{21}q = -5.09q, \qquad \sigma_{\theta\theta} = -15q, \qquad \sigma_{\theta\theta} = \frac{111}{63}q = 1.762q \qquad (3.194)$$

In terms of γz, these concentrations are

$$\sigma_{\theta\theta} = -1.2725\gamma z, \qquad \sigma_{\theta\theta} = -3.75\gamma z, \qquad \sigma_{\theta\theta} = 0.44\gamma z \qquad (3.195)$$

The maximum compression occurs at 43.58^0 measured from the x_1-axis and the maximum compression is $-16.77q$. Comparing this result with those for the circular tunnel (see Section 2.18), it can be concluded the stress concentration (both tensile and compression) is stronger around a square tunnel than around a circular tunnel.

3.13 MAPPING FUNCTIONS FOR OTHER HOLES

To conclude this chapter, we cite here some useful conformal mapping functions. To map the exterior of an elliptical hole onto the interior of a unit circle, we can use

$$z = \omega(\zeta) = R(\frac{1}{\zeta} + m\zeta) \qquad (3.196)$$

where

$$R = \frac{a+b}{2}, \qquad m = \frac{a-b}{a+b} \qquad (3.197)$$

and $2a$ and $2b$ are the lengths of the major and minor axes of the ellipse, respectively. As a special case, to map the exterior of a 2-D crack onto the interior of a unit circle, we can use

$$z = \omega(\zeta) = \frac{a}{2}(\frac{1}{\zeta} + \zeta) \qquad (3.198)$$

where a is half the length of the crack. To map the exterior of a rectangular hole with side ratio 3.2:1 (the long side is parallel to the x_1-axis) onto the interior of a unit circle, we can use

$$z = \omega(\zeta) = R(\frac{1}{\zeta} + \frac{1}{2}\zeta - \frac{1}{8}\zeta^3 - \frac{3}{80}\zeta^5 - \frac{3}{896}\zeta^7 + \frac{5}{768}\zeta^9 + ...) \qquad (3.199)$$

where R indicates the size of the hole. When the side ratio of the rectangular hole becomes 5:1, the mapping function is

$$z=\omega(\zeta)=R(\frac{1}{\zeta}+0.643\zeta-0.098\zeta^3-0.038\zeta^5-0.011\zeta^7+...) \quad (3.200)$$

More generally, for an arbitrary side ratio, the conformal mapping can be expressed as

$$\omega(\zeta)=R\int_1^{\zeta}[(t-e^{k\pi i})(t-e^{(2-k)\pi i})(t-e^{(1+k)\pi i})(t-e^{(1-k)\pi i})]^{1/2}\frac{dt}{t^2} \quad (3.201)$$

where k characterizes the side ratio of the rectangular hole. Expanding the integrand and carrying out the integration gives (Savin, 1961)

$$\begin{aligned}\omega(\zeta)=R[&\frac{1}{\zeta}+\frac{a+\bar{a}}{2}\zeta+\frac{(a-\bar{a})^2}{24}\zeta^3+\frac{(a^2-\bar{a}^2)(a-\bar{a})}{80}\zeta^5\\&+\frac{5(a^4+\bar{a}^4)-4(a^2+\bar{a}^2)-2}{896}\zeta^7+...]\end{aligned} \quad (3.202)$$

where

$$a=e^{2k\pi i}, \qquad \bar{a}=e^{-2k\pi i} \quad (3.203)$$

For $k < 1/4$, the longer side of the rectangular hole will be parallel to x_1; for $k > 1/4$, the longer side will be parallel to x_2; thus, $k = 1/4$ corresponds to a square. For $k = 1/6$, we will obtain (3.199); for $k = 10/86$, we will obtain (3.200). Therefore, any side ratio can be considered by setting appropriate values of k.

3.14 SUMMARY AND FURTHER READING

The majority of this chapter discusses the stress concentrations at holes subject to far field stresses. In fact, complex variable techniques can be used to solve other 2-D problems in elasticity. For general discussions on complex variable technique, the reader should refer to the book by Muskhelishvili (1975) together with his book on singular integral equation (Muskhelishvili, 1953). Another Russian book by Kalandiya (1975) also discusses many useful solutions, including numerical technique for solving the resulting integral equations. However, these books are not intended for beginners. Readers are recommended to consult Silverman (1974), and Spiegel (1964) for general knowledge on complex variable analysis before reading Muskhelishvili (1953, 1975) more seriously.

Other books on this topic include Green and Zerna (1968), Milne-Thomson (1968), and England (1971). Many books also include an introduction to the complex variable method for elastic problems, such as Timoshenko and Goodier (1982), Little (1973), and Xu (1982). Broberg (1999) provides a short introduction of complex variable technique to crack and fracture problems. Savin (1961) applies the complex variable technique to stress concentration at holes in solids.

Rice and Cleary (1976) demonstrated that the complex variable technique can also be used to solve 2-D problems in poroelastic diffusive solids. However, such discussion is outside the scope of the present chapter.

3.15 PROBLEMS

Problem 3.1 The problem provides a different proof of the result that the Airy stress function can be written in terms of two analytic functions as shown in (3.18). In particular, we assume two new variables, $z = x_1+ix_2$ and $\bar{z} = x_1-ix_2$, instead of x_1 and x_2.

(a) Use this change of variables to show that

$$\frac{\partial\varphi}{\partial x_1} = (\frac{\partial}{\partial z} + \frac{\partial}{\partial \bar{z}})\varphi \tag{3.204}$$

$$\frac{\partial\varphi}{\partial x_2} = i(\frac{\partial}{\partial z} - \frac{\partial}{\partial \bar{z}})\varphi \tag{3.205}$$

(b) Use the result of (a) to further show that

$$\nabla^2\varphi = 4\frac{\partial^2\varphi}{\partial z\partial\bar{z}} \tag{3.206}$$

(c) Prove that the biharmonic equation for ϕ becomes

$$\frac{\partial^4\varphi}{\partial z^2\partial\bar{z}^2} = 0 \tag{3.207}$$

(d) Finally, by integrating (3.207), show the validity of (3.18).

Problem 3.2 Show the following identity between the displacements in Cartesian and cylindrical polar coordinates given in (3.52):

$$2\mu(u_r + iu_\theta) = 2\mu(u_1 + iu_2)e^{-i\theta} \tag{3.208}$$

Problem 3.3 Use the complex variable technique to solve the stress and displacement fields for a hollow cylinder subject to internal pressure p_i and external pressure p_e . *Hint:* Assuming $\varphi(z) = cz$ and $\psi(z) = d/z$.

Problem 3.4 Prove equation (3.165) in the text.

Problem 3.5 Prove that, for an infinite solid containing a square hole and subject to far field uniaxial compression, the maximum compression occurs at 47.087^0 measured from the x_1-axis (comparing to 45^0 at the corner) and its magnitude is $-3.857q$; the maximum tensile occurs at 30.29^0 measured from the x_1-axis and its magnitude is $0.857q$.

Problem 3.6 Plot the angular variation of the tangential stress on the boundary of a square hole subject to a far field uniaxial compression.

Problem 3.7 In the text, the Muskhelishvili (1975) method of complex variables was discussed for solving problems of two-dimensional elasticity. In particular, the technique relies on the fact that Airy stress function can be expressed in terms of two *analytic functions* (the real and imaginary parts of these complex functions satisfy the Cauchy–Riemann equations). But, the body force potential V can be incorporated with the Airy stress function to solve problems with nonzero body force. In this exercise, we want to extend the results of (3.34)–(3.36) to include the effect of body force. In particular, we want to prove the following equations in this problem:

$$\sigma_{11}+\sigma_{22}=4\,\mathrm{Re}[\varphi'(z)]+\frac{4}{1+\kappa}\frac{\partial W}{\partial z} \tag{3.209}$$

$$\sigma_{22}-\sigma_{11}+2i\sigma_{12}=2[\bar{z}\varphi''(z)+\psi'(z)]+2(\frac{1-\kappa}{1+\kappa})\frac{\partial \overline{W}}{\partial z} \tag{3.210}$$

$$2\mu(u_1+iu_2)=\kappa\varphi(z)-z\overline{\varphi}'(\bar{z})-\overline{\psi}(\bar{z})-(\frac{1-\kappa}{1+\kappa})W \tag{3.211}$$

where

$$\frac{\partial W}{\partial z}=V(z,\bar{z}) \tag{3.212}$$

Some useful equations are given below: the equilibrium equations along x_1- and x_2-directions are

$$\frac{\partial \sigma_{11}}{\partial x_1}+\frac{\partial \sigma_{12}}{\partial x_2}+F_1=0 \tag{3.213}$$

$$\frac{\partial \sigma_{12}}{\partial x_1}+\frac{\partial \sigma_{22}}{\partial x_2}+F_2=0 \tag{3.214}$$

Assume that the body force is conservative such that

$$F_2=-\frac{\partial V}{\partial x_2},\quad F_1=-\frac{\partial V}{\partial x_1} \tag{3.215}$$

The two-dimensional Hooke's law can be expressed as:

$$\varepsilon_{ij}=\frac{1}{2\mu}[\sigma_{ij}-\frac{3-\kappa}{4}(tr\boldsymbol{\sigma})\delta_{ij}] \tag{3.216}$$

where κ is same as that given in (2.112) of Chapter 2. Answer the following questions:

(i) We now introduce a change of variable:

$$z=x_1+ix_2,\quad \bar{z}=x_1-ix_2 \tag{3.217}$$

such that the new variables become z and $\bar{z}$. Show that for any function f:

$$\frac{\partial f}{\partial x_1}=\frac{\partial f}{\partial z}+\frac{\partial f}{\partial \bar{z}},\qquad \frac{\partial f}{\partial x_2}=i(\frac{\partial f}{\partial z}-\frac{\partial f}{\partial \bar{z}}) \tag{3.218}$$

(ii) Use the result in (i) to show

$$2\frac{\partial f}{\partial z}=\frac{\partial f}{\partial x_1}-i\frac{\partial f}{\partial x_2},\qquad 2\frac{\partial f}{\partial \bar{z}}=\frac{\partial f}{\partial x_1}+i\frac{\partial f}{\partial x_2} \tag{3.219}$$

(iii) Use the equilibrium equations (3.213) and (3.214) and the result in (i), to show that

$$\frac{\partial}{\partial z}(\sigma_{11}-\sigma_{22}+2i\sigma_{12})+\frac{\partial}{\partial \bar{z}}(\sigma_{11}+\sigma_{22}-2V)=0 \tag{3.220}$$

(iv) Let

$$\frac{\partial F}{\partial z}=\sigma_{11}+\sigma_{22}-2V \tag{3.221}$$

and show that

$$\frac{\partial F}{\partial \bar{z}} = -(\sigma_{11} - \sigma_{22} + 2i\sigma_{12}) \tag{3.222}$$

(v) We now define a complex displacement $u = u_1 + iu_2$. Use the result of (ii) to show that

$$2\frac{\partial u}{\partial \bar{z}} = \varepsilon_{11} - \varepsilon_{22} + 2i\varepsilon_{12} \tag{3.223}$$

(vi) Use Hooke's law given in (3.216) and the result of (v), to show that

$$4\mu u = -F(z, \bar{z}) + f(z) \tag{3.224}$$

(vii) Use the result in (ii) and Hooke's law to show

$$\frac{8\mu}{\kappa - 1}\frac{\partial u}{\partial z} = \sigma_{11} + \sigma_{22} + i\frac{4\mu}{\kappa - 1}\left(\frac{\partial u_2}{\partial x_1} - \frac{\partial u_1}{\partial x_2}\right) \tag{3.225}$$

(viii) Show that differentiating (3.224) with respect to z and combining the resulting expression with (3.225) to eliminate $\partial u/\partial z$ will yield

$$(\sigma_{11} + \sigma_{22}) - \frac{4}{\kappa - 1}V + i\frac{4\mu}{\kappa - 1}\left(\frac{\partial u_2}{\partial x_1} - \frac{\partial u_1}{\partial x_2}\right) = \frac{2f'(z)}{1+\kappa} \tag{3.226}$$

(ix) Let $f(z) = 2(1+\kappa)\varphi(z)$ and combine (3.226) with its conjugate to show the validity of (3.209).

(x) Use (3.209) and the definition of $F(z,\bar{z})$ given in (iv) to show that

$$F(z,\bar{z}) = 2[\phi(z) + z\bar{\phi}'(\bar{z})] + \frac{2(1-\kappa)}{1+\kappa}W + 2\bar{\psi}(\bar{z}) \tag{3.227}$$

where $\psi(z)$ is an arbitrary function of z.

(xi) Substitute (3.227) into (3.222) to show the validity of (3.210).

(xii) Substitute (3.227) into (3.224) to show the validity of (3.211).

Problem 3.8 Show that the following mapping can be used to map a triangular hole to a unit circle as shown in Fig. 3.13:

$$\omega(\zeta) = R\left[\frac{1}{\zeta} + \frac{1}{3}\zeta^2\right] \tag{3.228}$$

Problem 3.9 Show that the radius of curvature at the corners of the triangular hole given by (3.228) is

$$r = \frac{R}{21} \tag{3.229}$$

Problem 3.10 Show that the analytic functions for the triangular hole modeled by (3.228) subject to a far field inclined traction p shown in Fig. 3.14 are

$$\phi(\zeta) = -pR\left(\frac{1}{4\zeta} + \frac{e^{2i\alpha}}{2}\zeta - \frac{1}{12}\zeta^2\right) \tag{3.230}$$

$$\psi(\zeta) = pR\left(\frac{e^{-2i\alpha}}{2\zeta} + \frac{3e^{2i\alpha} + 9e^{2i\alpha}\zeta^3 - 11\zeta}{12\zeta^3 - 18}\right) \tag{3.231}$$

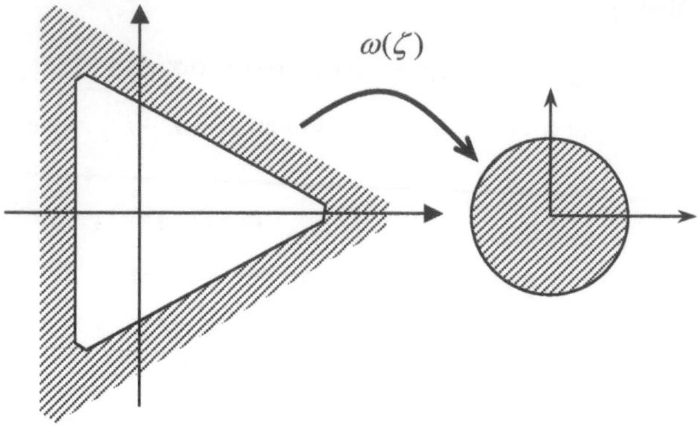

Figure 3.13 A triangular hole mapped to a unit circle

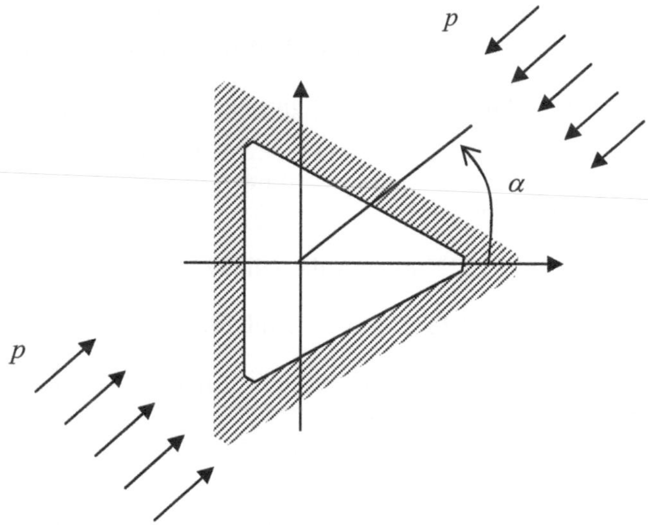

Figure 3.14 A triangular hole subject to far field compression *p*

CHAPTER FOUR

Method of Solution for 3-D Elasticity

4.1 INTRODUCTION

Real solids are always three dimensional (3-D) in nature; however, most textbooks on elasticity have been restricted to two-dimensional (2-D) situations (either plane strain or plane stress). The main reason is that the equations of motion or equilibrium (the so-called Navier equation in displacement formulation) or the compatibility equations (or the Beltrami–Michell equation in stress formulation) for 3-D solids are difficult to solve. This chapter introduces the method of solution for 3-D elasticity. More specifically, three coupled differential equations have to be solved for three unknown variables in the displacement formulation. Two major approaches have been adopted to uncouple these equations by either introducing displacement potentials or stress functions. Even though the resulting governing equations for displacement potentials or stress functions are uncoupled, the method of solutions for them is by no means straightforward even for simple practical problems.

As discussed in Chapter 2, plane stress or plane strain condition is normally assumed to idealize real situations. The problem of 2-D elasticity is much simpler than 3-D elasticity. Even when numerical methods (such as the finite element method) are used to solve real problems, 2-D idealization is usually adopted. In geomechanics, 3-D solutions are, however, essential in engineering applications. Examples include the Kelvin problem (point force in a full space), Boussinesq's problem (vertical surface point force applied on an elastic half-space), Cerruti's problem (horizontal surface point force applied on an elastic half-space), and Mindlin's problem (point force in an elastic half-space). These solutions have been used extensively to generate solutions for other practical problems. For example, the widely used Newmark influence charts and Fadum charts in soil mechanics were both obtained by superimposing (or integrating) the solution of Boussinesq's problem. These solutions also provide the fundamental solutions to the Green's method, the body force method, the boundary integral equation method, and the boundary element method.

Because of its mathematical complexity, the solution technique in solving these fundamental solutions is not covered in most textbooks in geomechanics or in elasticity. For example, the classical textbook on elasticity by Timoshenko and Goodier did not include a complete treatment on the method of solutions for 3-D elastic solid. Only some ad hoc 3-D elastic problems were considered. We believe, however, that the method of solutions for 3-D elasticity is of utmost importance and must be covered in a textbook on geomechanics or elasticity. This is the purpose of this chapter.

Readers familiar with the method of solutions for 3-D elasticity summarized in this chapter will find themselves better equipped to understand and tackle more complicated 3-D problems in thermoelasticity and poroelasticity. The former is important when one deals with geothermal energy extraction problems while the latter is important when one deals with problems of saturated soils or rocks (such as subsidence problems in clay or well water-level fluctuations before and during earthquakes).

In this chapter, the Galerkin vector potential and Papkovitch–Neuber displacement potentials will be covered under displacement formulation whereas the Maxwell stress functions and Morera stress functions will be covered under stress formulation. Some of these methods are applied to obtain the solutions of the Kelvin problem, Boussinesq's problem, Cerruti's problem, and Mindlin's problem. The general features of harmonic and biharmonic functions are discussed. The former finds applications in establishing solutions using the Papkovitch–Neuber potential, whilst the latter finds applications in establishing solutions using the Galerkin potential. The concept of indirect method is also introduced. Finally, the Muki (1960) vector potential for cylindrical coordinates is introduced together with the mathematical technique of the Hankel transform.

4.2 DISPLACEMENT FORMULATION

As shown in (2.72) in Chapter 2, the equations of equilibrium in the absence of body force can be expressed in terms of the displacement field $\boldsymbol{u}$ as

$$\mu\nabla^2\boldsymbol{u}+(\lambda+\mu)\boldsymbol{\nabla}\boldsymbol{\nabla}\bullet\boldsymbol{u}=0 \tag{4.1}$$

This equation is called Navier's equation. Recall that boldface indicates a vector or tensor. Various mathematical techniques will be discussed next in order to obtain the general solution of $\boldsymbol{u}$.

4.2.1 Helmholtz Decomposition

It is well known from the Helmholtz theorem that any vector which is finite, continuous, and vanishes at infinity can be decomposed in an irrotational part and a solenodial part (Chou and Pagano, 1967). Note that a field $\boldsymbol{A}$ is called irrotational if $\boldsymbol{\nabla}\times\boldsymbol{A}=\boldsymbol{0}$, and a field $\boldsymbol{B}$ is called solenodial if $\boldsymbol{\nabla}\cdot\boldsymbol{B}=0$. Therefore, a displacement field of a physically feasible solution can be decomposed into a scalar potential field $\phi(x_1, x_2, x_3)$ and a vector potential field $\boldsymbol{\psi}(x_1, x_2, x_3)$ as

$$\boldsymbol{u}=\boldsymbol{\nabla}\phi+\boldsymbol{\nabla}\times\boldsymbol{\psi} \tag{4.2}$$

The first term on the right of (4.2) is the irrotational part whereas the second term on the right of (4.2) is solenodial because of the following vector identities:

$$\boldsymbol{\nabla}\times\boldsymbol{\nabla}\phi=0 \qquad \boldsymbol{\nabla}\bullet\boldsymbol{\nabla}\times\boldsymbol{\psi}=0 \tag{4.3}$$

which are given in (1.46) and (1.47) in Chapter 1. Since there are four scalar functions ϕ, ψ_1, ψ_2, and ψ_3 in (4.2) but there are only three scalar functions for displacement $\boldsymbol{u}$ in (4.1); therefore, an additional constraint can be imposed on these potentials. Without loss of generality, the following condition is normally imposed on the vector potential:

$$\nabla \bullet \boldsymbol{\psi} = 0 \tag{4.4}$$

Substitute (4.2) into (4.1) and note that

$$\nabla^2(\boldsymbol{\nabla A}) = \boldsymbol{\nabla}(\nabla^2 \boldsymbol{A}) \tag{4.5}$$

in Cartesian coordinates yields the following governing equation for these potential functions:

$$(\lambda + 2\mu)\boldsymbol{\nabla}(\nabla^2\phi) + \mu\boldsymbol{\nabla} \times (\nabla^2\boldsymbol{\psi}) = 0 \tag{4.6}$$

This is the equilibrium equation in terms of ϕ and $\boldsymbol{\psi}$.

A particular solution of (4.6) is obviously:

$$\nabla^2\phi = c_1 \qquad \nabla^2\boldsymbol{\psi} = \boldsymbol{c}_2 \tag{4.7}$$

for any arbitrary scalar constant c_1 and vector constant $\boldsymbol{c}_2$. This is known as Poisson's equation. Therefore, any solution of (4.7) is also a solution for (4.6) but not vice versa. In other words, any solutions of Poisson's equation can be used to find a particular solution $\boldsymbol{u}$, but in general any solution $\boldsymbol{u}$ may not be expressible in terms of the solutions of Poisson's equation. In Cartesian coordinates, (4.7) provides a system of four uncoupled differential equations for four potential functions ϕ, ψ_1, ψ_2, and ψ_3. However, this is not true for curvilinear coordinates. That is, the components of $\boldsymbol{\psi}$ do not satisfy Poisson's equation. These components are in fact still coupled in the second equation of (4.7).

Another way of finding the general solution of ϕ and $\boldsymbol{\psi}$ is to take the divergence of (4.6):

$$(\lambda + 2\mu)\boldsymbol{\nabla} \bullet \boldsymbol{\nabla}(\nabla^2\phi) + \mu\boldsymbol{\nabla} \bullet (\boldsymbol{\nabla} \times \nabla^2\boldsymbol{\psi}) = 0 \tag{4.8}$$

The second term on the left is clearly identically zero by virtue of the second equation of (4.3). Thus, the scalar potential must satisfy the biharmonic equation:

$$\nabla^2(\nabla^2\phi) = \nabla^4\phi = 0 \tag{4.9}$$

Similarly, by taking the curl of (4.6), we have

$$(\lambda + 2\mu)\boldsymbol{\nabla} \times \boldsymbol{\nabla}(\nabla^2\phi) + \mu\boldsymbol{\nabla} \times (\boldsymbol{\nabla} \times \nabla^2\boldsymbol{\psi}) = 0 \tag{4.10}$$

By vector identity (1.46) or the first equation of (4.3), the first term of (4.10) must vanish and, in view of (1.49), the second term of (4.10) can be expressed as

$$\boldsymbol{\nabla} \times (\boldsymbol{\nabla} \times \nabla^2\boldsymbol{\psi}) = \boldsymbol{\nabla}(\nabla^2\boldsymbol{\nabla} \bullet \boldsymbol{\psi}) - \nabla^4\boldsymbol{\psi} = 0 \tag{4.11}$$

The first term of the second part of (4.11) vanishes by the constraint imposed in (4.4), and thus the vector potential also satisfies the biharmonic equation. Therefore, in summary we have

$$\nabla^4\phi = 0\,, \quad \nabla^4\boldsymbol{\psi} = 0 \tag{4.12}$$

Again all Cartesian components of $\boldsymbol{\psi}$ are part of a biharmonic function, but the corresponding components in curvilinear coordinates do not satisfy the biharmonic equation. In fact, (4.12) is still a coupled differential equation for the components of $\boldsymbol{\psi}$ in polar coordinates. Solving coupled differential equations is one of the most difficult problems in applied mathematics. Therefore, Helmholtz vector decomposition is only useful for solving problems in Cartesian coordinates.

Instead of solving the coupled differential equations (4.6), we now have a decoupled system of differential equations (4.12) in Cartesian coordinates to solve. There is of course a price to pay. In particular, we have to deal with fourth order of differentiation, comparing to third order of differentiation in (4.6).

The corresponding stress components in terms of Helmholtz decomposition can be evaluated by using the following formulas:

$$\nabla\boldsymbol{u}+[\nabla\boldsymbol{u}]^T = 2\nabla\nabla\phi+[\nabla(\nabla\times\boldsymbol{\psi})+(\nabla\times\boldsymbol{\psi})\nabla] \tag{4.13}$$

$$tr(\nabla\boldsymbol{u}) = \nabla\bullet\boldsymbol{u} = \nabla^2\phi+\nabla\bullet(\nabla\times\boldsymbol{\psi}) = \nabla^2\phi \tag{4.14}$$

For isotropic solids, Hooke's law can be expressed in terms of displacement as

$$\boldsymbol{\sigma} = \lambda tr(\nabla\boldsymbol{u})\boldsymbol{I} + \mu(\nabla\boldsymbol{u}+\boldsymbol{u}\nabla) \tag{4.15}$$

Substitution of (4.13) and (4.14) into (4.15) gives the dyadic form of the stress tensor as

$$\boldsymbol{\sigma} = \lambda\nabla^2\phi\boldsymbol{I} + \mu[2\nabla\nabla\phi+\nabla(\nabla\times\boldsymbol{\psi})+(\nabla\times\boldsymbol{\psi})\nabla] \tag{4.16}$$

The corresponding Cartesian component form of (4.16) can be expressed as

$$\sigma_{ij} = \lambda\nabla^2\phi\delta_{ij} + \mu[2\phi,_{ij} + (e_{klj}\psi_{l,ki} + e_{kli}\psi_{l,kj})] \tag{4.17}$$

More explicitly, we can write

$$\sigma_{11} = (\lambda\nabla^2 + 2\mu\frac{\partial^2}{\partial x_1^2})\phi + 2\mu\frac{\partial}{\partial x_1}(\frac{\partial\psi_3}{\partial x_2} - \frac{\partial\psi_2}{\partial x_3}) \tag{4.18}$$

$$\sigma_{22} = (\lambda\nabla^2 + 2\mu\frac{\partial^2}{\partial x_2^2})\phi + 2\mu\frac{\partial}{\partial x_2}(\frac{\partial\psi_1}{\partial x_3} - \frac{\partial\psi_3}{\partial x_1}) \tag{4.19}$$

$$\sigma_{33} = (\lambda\nabla^2 + 2\mu\frac{\partial^2}{\partial x_3^2})\phi + 2\mu\frac{\partial}{\partial x_3}(\frac{\partial\psi_2}{\partial x_1} - \frac{\partial\psi_1}{\partial x_2}) \tag{4.20}$$

$$\sigma_{12} = \mu[\frac{\partial^2\phi}{\partial x_1\partial x_2} + \frac{\partial}{\partial x_1}(\frac{\partial\psi_1}{\partial x_3} - \frac{\partial\psi_3}{\partial x_1}) + \frac{\partial}{\partial x_2}(\frac{\partial\psi_3}{\partial x_2} - \frac{\partial\psi_2}{\partial x_3})] \tag{4.21}$$

$$\sigma_{23} = \mu[\frac{\partial^2\phi}{\partial x_3\partial x_2} + \frac{\partial}{\partial x_2}(\frac{\partial\psi_2}{\partial x_1} - \frac{\partial\psi_1}{\partial x_2}) + \frac{\partial}{\partial x_3}(\frac{\partial\psi_1}{\partial x_3} - \frac{\partial\psi_3}{\partial x_1})] \tag{4.22}$$

$$\sigma_{13} = \mu[\frac{\partial^2\phi}{\partial x_3\partial x_1} + \frac{\partial}{\partial x_3}(\frac{\partial\psi_3}{\partial x_2} - \frac{\partial\psi_2}{\partial x_3}) + \frac{\partial}{\partial x_1}(\frac{\partial\psi_2}{\partial x_1} - \frac{\partial\psi_1}{\partial x_2})] \tag{4.23}$$

A special case of Helmholtz decomposition is Lamé's strain potential. If we set $\boldsymbol{\psi} = 0$, we have the following Lamé's strain potential (Malvern, 1969):

$$\nabla^2\phi = c_1 \qquad \boldsymbol{\psi} = 0 \tag{4.24}$$

4.2.2 Lamé's Strain Potential for Incompressible Solids

Another special case of Lamé's strain potential given in (4.7) that deserves special attention is the case of incompressible solids. In geomechanics, the short-term behavior of saturated clay can clearly be considered incompressible. For such a situation, Lamé's strain potential ϕ satisfies the Laplace equation instead of Poisson's equation:

$$u = \nabla\phi, \quad \nabla^2\phi = 0 \tag{4.25}$$

The volumetric strain e can be shown as

$$e = \mathrm{tr}(\boldsymbol{\varepsilon}) = \boldsymbol{\nabla} \bullet u = \nabla^2 \phi = 0 \tag{4.26}$$

Therefore, (4.25) clearly corresponds to incompressible solids. The stress components for this special case are extremely simple:

$$\begin{aligned}\frac{\boldsymbol{\sigma}}{2\mu} &= \boldsymbol{e}_x\boldsymbol{e}_x \frac{\partial^2 \phi}{\partial x^2} + \boldsymbol{e}_y\boldsymbol{e}_y \frac{\partial^2 \phi}{\partial y^2} + \boldsymbol{e}_z\boldsymbol{e}_z \frac{\partial^2 \phi}{\partial z^2} + (\boldsymbol{e}_x\boldsymbol{e}_y + \boldsymbol{e}_y\boldsymbol{e}_x)\frac{\partial^2 \phi}{\partial x \partial y} \\ &+ (\boldsymbol{e}_z\boldsymbol{e}_y + \boldsymbol{e}_y\boldsymbol{e}_z)\frac{\partial^2 \phi}{\partial z \partial y} + (\boldsymbol{e}_z\boldsymbol{e}_x + \boldsymbol{e}_x\boldsymbol{e}_z)\frac{\partial^2 \phi}{\partial x \partial z}\end{aligned} \tag{4.27}$$

Note that since the solid is incompressible or equivalently corresponding to $\nu = 1/2$, the stress given in (4.27) is independent of ν. As will be shown later, many special features of elastic solutions exist for incompressible solids.

In addition, as remarked by Fung (1965), Westergaard (1952) proposed a perturbation of elasticity solutions by a change of Poisson's ratio. Thus, the solutions obtained by solving (4.25)–(4.27) for incompressible solids can be easily extended to compressible elastic solids with arbitrary ν.

In particular, as shown by Fung (1965), the solutions in Cartesian coordinates for elastic solids with arbitrary ν can be obtained by adding to this solution for incompressible solids with another stress field resulting from a "twinned gradient" φ:

$$\sigma_{11} = \nabla^2 \varphi - \varphi_{,22}, \quad \sigma_{22} = \nabla^2 \varphi - \varphi_{,11}, \quad \sigma_{12} = \varphi_{,12} \tag{4.28}$$

where φ can be evaluated by using the incompressible solids solutions

$$\nabla^2 \varphi = 0, \quad \varphi_{,33} = \frac{(2\nu - 1)}{3}\sigma_{\alpha\alpha}^{*} \tag{4.29}$$

where $\sigma^*_{\alpha\alpha}$ is the sum of the normal stress of the incompressible solid. An example using this method can be found in Section 8.10 of Fung (1965) and Article 77 of Westergaard (1952).

4.2.3 Galerkin Vector

The Helmholtz decomposition discussed in Section 4.2.1 represents the displacement vector by a scalar potential and a vector potential. It is also possible to represent the displacement vector by a single vector potential called the Galerkin vector $\boldsymbol{G}$:

$$\boldsymbol{u} = 2(1-\nu)\nabla^2 \boldsymbol{G} - \boldsymbol{\nabla\nabla} \bullet \boldsymbol{G} \tag{4.30}$$

Note that

$$\nabla^2 \boldsymbol{u} = 2(1-\nu)\nabla^2\nabla^2 \boldsymbol{G} - \nabla^2(\boldsymbol{\nabla\nabla} \bullet \boldsymbol{G}) \tag{4.31}$$

$$\begin{aligned}\boldsymbol{\nabla\nabla} \bullet \boldsymbol{u} &= 2(1-\nu)\boldsymbol{\nabla\nabla} \bullet (\nabla^2 \boldsymbol{G}) - \boldsymbol{\nabla\nabla} \bullet (\boldsymbol{\nabla\nabla} \bullet \boldsymbol{G}) \\ &= 2(1-\nu)\nabla^2(\boldsymbol{\nabla\nabla} \bullet \boldsymbol{G}) - \nabla^2(\boldsymbol{\nabla\nabla} \bullet \boldsymbol{G}) \\ &= (1-2\nu)\nabla^2(\boldsymbol{\nabla\nabla} \bullet \boldsymbol{G})\end{aligned} \tag{4.32}$$

Substitution of (4.31) and (4.32) into (4.1) yields a biharmonic equation for $\boldsymbol{G}$:

$$\nabla^2\nabla^2 \boldsymbol{G} = \nabla^4 \boldsymbol{G} = 0 \tag{4.33}$$

Therefore, any biharmonic vector function may be used as the Galerkin vector. Comparison of (4.30) with (4.2) gives the following relation between the Helmholtz decomposition potential and the Galerkin vector:

$$\phi = -\nabla \bullet \boldsymbol{G}, \quad \boldsymbol{\nabla} \times \boldsymbol{\psi} = 2(1-2\nu)\nabla^2 \boldsymbol{G} \tag{4.34}$$

A subset of the general solution is, of course, that $\boldsymbol{G}$ is a harmonic function (i.e., $\nabla^2 \boldsymbol{G} = 0$). The second part of (4.34) implies that

$$\boldsymbol{\nabla} \times \boldsymbol{\psi} = 0 \tag{4.35}$$

Taking the Laplacian of the first part of (4.34) shows that ϕ is also harmonic. In summary,

$$u = \boldsymbol{\nabla} \phi, \quad \nabla^2 \phi = 0 \tag{4.36}$$

which is again the special case of Lamé's strain function given in Section 4.2.2.

Without going into the details, we record here the stress tensor in terms of Galerkin vector as

$$\boldsymbol{\sigma} = 2\mu[\nu\nabla^2(\boldsymbol{\nabla} \bullet \boldsymbol{G})\boldsymbol{I} + (1-\nu)\nabla^2(\boldsymbol{\nabla G} + \boldsymbol{G\nabla}) - \boldsymbol{\nabla\nabla}(\boldsymbol{\nabla} \bullet \boldsymbol{G})] \tag{4.37}$$

In Cartesian coordinate, stress components are

$$\sigma_{ij} = 2\mu[\nu G_{k,kmm}\delta_{ij} + (1-\nu)(G_{j,imm} + G_{i,jmm}) - G_{m,mij}] \tag{4.38}$$

$$\frac{\sigma_{\alpha\alpha}}{2\mu} = 2(1-\nu)\frac{\partial}{\partial x_\alpha}\nabla^2 G_\alpha + (\nu\nabla^2 - \frac{\partial^2}{\partial x_\alpha^2})(\frac{\partial G_1}{\partial x_1} + \frac{\partial G_2}{\partial x_2} + \frac{\partial G_3}{\partial x_3}) \tag{4.39}$$

$$\frac{\sigma_{\alpha\beta}}{2\mu} = (1-\nu)(\frac{\partial}{\partial x_\alpha}\nabla^2 G_\beta + \frac{\partial}{\partial x_\beta}\nabla^2 G_\alpha) - \frac{\partial^2}{\partial x_\alpha \partial x_\beta}(\frac{\partial G_1}{\partial x_1} + \frac{\partial G_2}{\partial x_2} + \frac{\partial G_3}{\partial x_3}) \tag{4.40}$$

where α, β = 1,2,3 (with no summation on α in (4.39)).

It is interesting to note that Galerkin started his academic career in prison. He also laid the mathematical foundation of today's finite element method (see his brief biography at the end of this book).

4.2.4 Love's Displacement Potential for Cylindrical Solids

A special case of the Galerkin vector is Love's displacement potential which corresponds to only the case of the nonzero axial component:

$$\boldsymbol{G} = G_z \boldsymbol{e}_z, \quad \nabla^4 G_z = 0 \tag{4.41}$$

In cylindrical coordinates, the corresponding displacement and stress tensors are

$$\boldsymbol{u} = \left[(1-2\nu)\nabla^2 - \frac{\partial^2}{\partial z^2}\right] G_z \boldsymbol{e}_z - \frac{\partial^2 G_z}{\partial r \partial z}\boldsymbol{e}_r - \frac{1}{r}\frac{\partial^2 G_z}{\partial \theta \partial z}\boldsymbol{e}_\theta, \tag{4.42}$$

$$\begin{aligned}\frac{\boldsymbol{\sigma}}{2\mu} &= (\nu\nabla^2 - \frac{\partial^2}{\partial r^2})\frac{\partial G_z}{\partial z}\boldsymbol{e}_r\boldsymbol{e}_r + (\nu\nabla^2 - \frac{1}{r}\frac{\partial}{\partial r} - \frac{1}{r^2}\frac{\partial^2}{\partial \theta^2})\frac{\partial G_z}{\partial z}\boldsymbol{e}_\theta\boldsymbol{e}_\theta \\ &+[(2-\nu)\nabla^2 - \frac{\partial^2}{\partial z^2}]\frac{\partial G_z}{\partial z}\boldsymbol{e}_z\boldsymbol{e}_z + \frac{1}{r}\frac{\partial}{\partial \theta}[(1-\nu)\nabla^2 - \frac{\partial^2}{\partial z^2}]G_z(\boldsymbol{e}_\theta\boldsymbol{e}_z + \boldsymbol{e}_z\boldsymbol{e}_\theta) \\ &+ \frac{\partial}{\partial r}[(1-\nu)\nabla^2 - \frac{\partial^2}{\partial z^2}]G_z(\boldsymbol{e}_r\boldsymbol{e}_z + \boldsymbol{e}_z\boldsymbol{e}_r) - \frac{\partial^3}{\partial r \partial \theta \partial z}(\frac{G_z}{r})(\boldsymbol{e}_\theta\boldsymbol{e}_r + \boldsymbol{e}_r\boldsymbol{e}_\theta)\end{aligned} \tag{4.43}$$

For the axisymmetric case or $G_z = G_z(r, z)$, (4.43) can further be simplified. A.E.H. Love was a major contributor to three-dimensional elasticity and his book *A Treatise on the Mathematical Theory of Elasticity* remains a standard reference book on elasticity (see biography section at the end of this book).

4.2.5 Papkovitch–Neuber Displacement Potential

The Helmholtz scalar and vector potentials satisfy a coupled third-order equation (4.6) or uncoupled fourth-order system of (4.12), whereas the governing equation for the Galerkin vector satisfies the fourth-order biharmonic equation. It is therefore desirable to find a solution system composed of second-order equations which are of the same order as the original Navier's equation given in (4.1). We have also remarked earlier that a harmonic function is also the solution of biharmonic equation (like Lamé's strain potential discussed earlier), but they are not general. For this reason, a new displacement potential was proposed independently by P.F. Papkovitch in 1932 and by H. Neuber in 1934 (Mindlin, 1936a). Goodman (1974) also cited a less-recognized Russian paper by Grodski from 1935 in which Grodski also independently proposed the same displacement potentials. We will, however, follow the more widely adopted term "Papkovitch–Neuber displacement potential" which can be derived from the Helmholtz decomposition given in (4.2). Taking the divergence of (4.2) gives

$$\boldsymbol{\nabla}\bullet u = \boldsymbol{\nabla}\bullet\boldsymbol{\nabla}\phi + \boldsymbol{\nabla}\bullet\boldsymbol{\nabla}\times\boldsymbol{\psi} = \nabla^2\phi \tag{4.44}$$

Substitution of (4.44) into (4.1) yields

$$\nabla^2[(\lambda+\mu)\boldsymbol{\nabla}\phi + \mu\boldsymbol{u}] = 0 \tag{4.45}$$

Integrating (4.45) gives

$$(\lambda+\mu)\boldsymbol{\nabla}\phi + \mu\boldsymbol{u} = \mu\boldsymbol{\Phi} \tag{4.46}$$

where $\boldsymbol{\Phi}$ is a harmonic vector function. Rearranging (4.46) gives

$$\boldsymbol{u} = \boldsymbol{\Phi} - (1+\lambda/\mu)\boldsymbol{\nabla}\phi \tag{4.47}$$

Taking the divergence of (4.47) and in view of (4.44), we obtain

$$\nabla^2\phi = \boldsymbol{\nabla}\bullet\boldsymbol{\Phi} - (1+\lambda/\mu)\nabla^2\phi \tag{4.48}$$

Solving for ϕ gives

$$(2+\frac{\lambda}{\mu})\nabla^2\phi = \boldsymbol{\nabla}\bullet\boldsymbol{\Phi} = \frac{1}{2}\nabla^2(\boldsymbol{r}\bullet\boldsymbol{\Phi}) \tag{4.49}$$

The last equation of (4.49) can be shown by using the following identity:

$$\begin{aligned}\nabla^2(\boldsymbol{r}\bullet\boldsymbol{\Phi}) &= (\Phi_i x_i)_{,kk} = (\Phi_{i,k}x_i + \Phi_i x_{i,k})_{,k} = (\Phi_{i,k}x_i + \Phi_i\delta_{ik})_{,k} \\ &= (\Phi_{i,kk}x_i + \Phi_{i,k}\delta_{ik} + \Phi_{i,k}\delta_{ik}) = (2\Phi_{k,k} + \Phi_{i,kk}x_i) \\ &= 2\boldsymbol{\nabla}\bullet\boldsymbol{\Phi} + \boldsymbol{r}\bullet\nabla^2\boldsymbol{\Phi}\end{aligned} \tag{4.50}$$

Recall from (4.46) that $\boldsymbol{\Phi}$ is a harmonic vector function; therefore, the last term in (4.50) must vanish. This completes the proof of the last equation in (4.49). Rearranging (4.49) gives

$$\nabla^2\phi = \frac{\mu}{2(\lambda+2\mu)}\nabla^2(\boldsymbol{r}\bullet\boldsymbol{\Phi}) \tag{4.51}$$

Integrating (4.51) twice finally gives

$$\phi = \frac{\mu}{2(\lambda+2\mu)}(\Phi_0 + \boldsymbol{r}\bullet\boldsymbol{\Phi}) = \frac{1-2\nu}{4(1-\nu)}(\Phi_0 + \boldsymbol{r}\bullet\boldsymbol{\Phi}) \tag{4.52}$$

where Φ_0 is another harmonic function. Note that integration constants have been ignored in (4.52). The last equation of (4.52) can be verified by using (2.50) and (2.51) given in Chapter 2. Back substitution of (4.52) into (4.47) gives the Papkovitch–Neuber displacement potential as

$$\boldsymbol{u} = \boldsymbol{\Phi} - \frac{1}{4(1-\nu)}\nabla(\Phi_0 + \boldsymbol{r}\bullet\boldsymbol{\Phi}), \quad \nabla^2\Phi_0 = 0, \quad \nabla^2\boldsymbol{\Phi} = 0 \tag{4.53}$$

Comparison of (4.53) with the Galerkin vector given in (4.30) gives the following relation between the Papkovitch–Neuber displacement potential and the Galerkin vector:

$$\boldsymbol{\Phi} = 2(1-\nu)\nabla^2\boldsymbol{G}, \quad \Phi_0 = 4(1-\nu)\nabla\bullet\boldsymbol{G} - \boldsymbol{r}\bullet\boldsymbol{\Phi} \tag{4.54}$$

This equivalence between the Papkovitch–Neuber displacement potential and the Galerkin vector was first noted by Mindlin (1936a).

Substitution of (4.53) into (4.15) gives

$$\boldsymbol{\sigma} = \lambda[\nabla\bullet\boldsymbol{\Phi} - \frac{1}{4(1-\nu)}\nabla^2(\boldsymbol{r}\bullet\boldsymbol{\Phi})]\boldsymbol{I} + \mu[(\nabla\boldsymbol{\Phi} + \boldsymbol{\Phi}\nabla) - \frac{1}{2(1-\nu)}\nabla\nabla(\Phi_0 + \boldsymbol{r}\bullet\boldsymbol{\Phi})] \tag{4.55}$$

$$\sigma_{ij} = \lambda\delta_{ij}[\Phi_{k,k} - \frac{1}{4(1-\nu)}(x_k\Phi_k),mm] + \mu[(\Phi_{i,j} + \Phi_{j,i}) - \frac{1}{2(1-\nu)}(\Phi_0 + x_k\Phi_k)_{,ij}] \tag{4.56}$$

$$\begin{aligned}\sigma_{\alpha\alpha} &= 2\mu\frac{\partial\Phi_\alpha}{\partial x_\alpha} + \lambda(\frac{\partial\Phi_1}{\partial x_1} + \frac{\partial\Phi_2}{\partial x_2} + \frac{\partial\Phi_3}{\partial x_3}) \\ &\quad - \frac{1}{4(1-\nu)}(\lambda\nabla^2 + 2\mu\frac{\partial^2}{\partial x_\alpha^2})(\Phi_0 + x_1\Phi_1 + x_2\Phi_2 + x_3\Phi_3)\end{aligned} \tag{4.57}$$

$$\sigma_{\alpha\beta} = \mu\left\{\frac{\partial\Phi_\alpha}{\partial x_\beta} + \frac{\partial\Phi_\beta}{\partial x_\alpha} - \frac{1}{2(1-\nu)}\frac{\partial^2}{\partial x_\alpha\partial x_\beta}(\Phi_0 + x_1\Phi_1 + x_2\Phi_2 + x_3\Phi_3)\right\} \tag{4.58}$$

where α, β = 1, 2, 3 (with no summation on α in (4.57)). A simplified form for σ_{ij} will be given later in (4.212).

There are, however, some disadvantages of using Papkovitch–Neuber displacement potentials. For example, there were originally three unknown displacements, but now we have to find four displacement potentials (one scalar potential plus three components of vector potentials). It was shown by Sternberg (1960) that for an arbitrary three-dimensional convex domain, the four potentials are reducible to three. In addition, it is straightforward to show that Papkovitch–Neuber displacement potentials are not invariant upon translation of the origin.

4.2.6 2-D Papkovitch–Neuber vs. Kolosov–Muskhelisvili Methods

The Papkovitch–Neuber solution method can be specialized to two dimensional and can be shown to be equivalent to the complex variable method of Kolosov-Muskhelisvili (Muskhelishvili, 1953) discussed in Chapter 3. First, we can rewrite (4.53) as

$$\boldsymbol{u} = \frac{3-4\nu}{4(1-\nu)}\boldsymbol{\Phi} - \frac{1}{4(1-\nu)}\nabla \Phi_0 - \frac{1}{4(1-\nu)}(\nabla \boldsymbol{\Phi}) \bullet \boldsymbol{r} \tag{4.59}$$

In Cartesian coordinates, we consider the special case that $u_3 = 0$ and $\Phi_3 = 0$ and $\boldsymbol{\Phi}$ is independent of x_3, and introduce the following harmonic functions:

$$\phi_\alpha(x_1, x_2) = \frac{\mu}{2(1-\nu)}\Phi_\alpha \quad (\alpha = 1,2) \tag{4.60}$$

$$\psi_1(x_1, x_2) = \frac{\mu}{2(1-\nu)}\frac{\partial \Phi_0}{\partial x_1}, \quad \psi_2(x_1, x_2) = -\frac{\mu}{2(1-\nu)}\frac{\partial \Phi_0}{\partial x_2} \tag{4.61}$$

Using these harmonic functions, the displacement components of (4.59) can be rewritten as

$$2\mu u_1(x_1, x_2) = (3-4\nu)\phi_1 - \psi_1 - x_1\frac{\partial \phi_1}{\partial x_1} - x_2\frac{\partial \phi_2}{\partial x_1} \tag{4.62}$$

$$2\mu u_2(x_1, x_2) = (3-4\nu)\phi_2 + \psi_2 - x_1\frac{\partial \phi_1}{\partial x_2} - x_2\frac{\partial \phi_2}{\partial x_2} \tag{4.63}$$

These two equations can be combined as

$$2\mu(u_2 + iu_2) = (3-4\nu)\phi(z) - \overline{\psi}(\overline{z}) - z\overline{\phi}'(\overline{z}) \tag{4.64}$$

where the superimposed bar means complex conjugate and the following definitions have been adopted:

$$\psi(z) = \psi_1 + i\psi_2, \quad \phi(z) = \phi_1 + i\phi_2, \tag{4.65}$$

$$\phi'(z) = \frac{d\phi(z)}{dz} = \frac{\partial \phi_1}{\partial x_1} + i\frac{\partial \phi_2}{\partial x_1} = \frac{\partial \phi_2}{\partial x_2} - i\frac{\partial \phi_1}{\partial x_2} \tag{4.66}$$

This is the Kolosov–Muskhelisvili complex variable method discussed in Chapter 3.

4.3 STRESS FORMULATION

4.3.1 Beltrami and Beltrami–Schaefer Stress Functions

When the body force is zero, the equilibrium equation is expressed as (see (2.67))

$$\nabla \bullet \boldsymbol{\sigma} = \boldsymbol{0} \tag{4.67}$$

and the Beltrami–Michell compatibility equation is (see (2.83))

$$\nabla^2 \boldsymbol{\sigma} + \frac{1}{1+\nu}\nabla\nabla \sigma_{kk} = 0 \tag{4.68}$$

The equilibrium equation can be satisfied as long as $\boldsymbol{\sigma}$ is expressed as a curl of some vector function, since the divergence of a curl vanishes identically. Since the stress tensor $\boldsymbol{\sigma}$ is symmetric, we can define a symmetric dyadic function

$$\boldsymbol{\sigma} = \nabla \times \boldsymbol{\Psi} \times \nabla \tag{4.69}$$

This form was first proposed by Beltrami in 1892 (Malvern, 1969, Wang, 2002). In Cartesian coordinates, the Beltrami stress function $\boldsymbol{\Psi}$ can be evaluated as

$$\nabla \times \boldsymbol{\Psi} = (\boldsymbol{e}_i \frac{\partial}{\partial x_i}) \times (\Psi_{jk} \boldsymbol{e}_j \boldsymbol{e}_k) = e_{ijl} \Psi_{jk,i} \boldsymbol{e}_l \boldsymbol{e}_k \tag{4.70}$$

$$(\nabla \times \boldsymbol{\Psi}) \times \nabla = e_{ijl} \Psi_{jk,im} \boldsymbol{e}_l (\boldsymbol{e}_k \times \boldsymbol{e}_m) = e_{kmn} e_{ijl} \Psi_{jk,im} \boldsymbol{e}_l \boldsymbol{e}_n \tag{4.71}$$

Take the divergence of (4.71) yields

$$\begin{aligned} \nabla \bullet (\nabla \times \boldsymbol{\Psi}) \times \nabla &= e_{kmn} e_{ijl} \Psi_{jk,iml} \boldsymbol{e}_n = e_{kmn} e_{lji} \Psi_{jk,lmi} \boldsymbol{e}_n \\ &= -e_{kmn} e_{ijl} \Psi_{jk,lmi} \boldsymbol{e}_n = 0 \end{aligned} \tag{4.72}$$

Since reversing the order of any two indices (i and l) of the permutation tensor will lead to a change of sign while the order of differentiation can be reversed arbitrarily, the vector given in (4.72) must be identically zero, as expected. Therefore, (4.67) is automatically satisfied by (4.69).

In view of the e-δ identity (see (1.18) of Chapter 1), the trace of the stress tensor becomes

$$\begin{aligned} \sigma_{kk} &= tr[\nabla \times \boldsymbol{\Psi} \times \nabla] = e_{kmn} e_{ijl} \Psi_{jk,im} \delta_{nl} = e_{kml} e_{ijl} \Psi_{jk,im} \\ &= \left(\delta_{ki} \delta_{mj} - \delta_{kj} \delta_{mi} \right) \Psi_{jk,im} = \Psi_{mi,im} - \Psi_{kk,mm} \\ &= \nabla \bullet (\nabla \bullet \boldsymbol{\Psi}) - \nabla^2 |\boldsymbol{\Psi}| \end{aligned} \tag{4.73}$$

By virtue of (4.69) and (4.73), the Beltrami–Michell compatibility equation becomes

$$\nabla^2 [\nabla \times \boldsymbol{\Psi} \times \nabla] + \frac{1}{1+\nu} \nabla\nabla [\nabla \bullet (\nabla \bullet \boldsymbol{\Psi}) - \nabla^2 |\boldsymbol{\Psi}|] = 0 \tag{4.74}$$

There are six components of $\boldsymbol{\Psi}$, but only three of them are independent which is what we discussed for the compatibility condition in Chapter 2. In most of the elasticity books only two common choices of selecting components of $\boldsymbol{\Psi}$ are discussed. If the diagonal terms are selected, the stress functions are called Maxwell stress functions. If the off-diagonal terms are selected, the stress functions are called Morera stress functions (Chou and Pagano, 1967; Fung, 1965).

As summarized by Wang (2002), Schaefer in 1953 extended the Beltrami stress function to the following form:

$$\boldsymbol{\sigma} = \nabla \times \boldsymbol{\Psi} \times \nabla + \boldsymbol{h} \nabla + \nabla \boldsymbol{h} - \boldsymbol{I} \nabla \bullet \boldsymbol{h} \tag{4.75}$$

where $\boldsymbol{h}$ is a harmonic vector function. It can be seen that the terms involving $\boldsymbol{h}$ satisfy (4.67) exactly by observing the following identities:

$$\nabla \bullet (\boldsymbol{h} \nabla) = (\nabla \bullet \boldsymbol{h}) \nabla \tag{4.76}$$

$$\nabla \bullet (\nabla \boldsymbol{h}) = \nabla^2 \boldsymbol{h} \tag{4.77}$$

$$\nabla \bullet (\boldsymbol{I} \nabla \bullet \boldsymbol{h}) = \nabla (\nabla \bullet \boldsymbol{h}) \tag{4.78}$$

With nonzero $\boldsymbol{h}$ in (4.75), the Beltrami–Michell compatibility equation can be revised as

$$\nabla^2[\nabla\times\boldsymbol{\Psi}\times\nabla]+\frac{1}{1+\nu}\nabla\nabla[\nabla\bullet(\nabla\bullet\boldsymbol{\Psi})-\nabla^2|\boldsymbol{\Psi}|-\nabla\bullet\boldsymbol{h}]=0 \tag{4.79}$$

This stress function is called the Beltrami–Schaefer stress function by Wang (2002). Wang (2002) further showed that the Beltrami–Schaefer stress function can be refined as

$$\boldsymbol{\sigma}=\nabla\times\boldsymbol{\Psi}^*\times\nabla-\frac{1}{2(1-\nu)}[\nabla\nabla(\boldsymbol{r}\bullet\boldsymbol{h})-\boldsymbol{I}\nabla^2(\boldsymbol{r}\bullet\boldsymbol{h})]+\boldsymbol{h}\nabla+\nabla\boldsymbol{h}-\boldsymbol{I}\nabla\bullet\boldsymbol{h} \tag{4.80}$$

$$\boldsymbol{\Psi}^*=\nabla^2\boldsymbol{B}^*+\frac{1}{1-\nu}\boldsymbol{I}(\nabla\bullet\boldsymbol{B}^*\bullet\nabla),\quad \nabla^4\boldsymbol{B}^*=\boldsymbol{0}\ ,\quad \nabla^2\boldsymbol{h}=\boldsymbol{0} \tag{4.81}$$

where $\boldsymbol{B}^*$ is a symmetric biharmonic second-order tensor while $\boldsymbol{h}$ is a harmonic vector. Details of its proof will not be given here.

4.3.2 Maxwell Stress Functions

To illustrate the nonzero terms used in the Maxwell stress function, we can use the matrix form to express the components of the second-order tensor $\boldsymbol{\Psi}$

$$[\boldsymbol{\Psi}]=\begin{pmatrix}\Psi_{11} & \Psi_{12} & \Psi_{13}\\ \Psi_{12} & \Psi_{22} & \Psi_{23}\\ \Psi_{13} & \Psi_{23} & \Psi_{33}\end{pmatrix} \tag{4.82}$$

Note that Maxwell is more widely known for his mathematical theory coupling electricity, magnetism, and light (see biography section of this book). Then, the nonzero terms of the Maxwell stress function components of the tensor $\boldsymbol{\Psi}$ can be expressed in matrix form as:

$$[\boldsymbol{\Psi}_1]=\begin{pmatrix}\Psi_{11} & 0 & 0\\ 0 & \Psi_{22} & 0\\ 0 & 0 & \Psi_{33}\end{pmatrix} \tag{4.83}$$

Note that there is only one way of choosing diagonal terms.

For the 2-D special case, we can express the Maxwell stress function as

$$[\boldsymbol{\Psi}_0]=\begin{pmatrix}0 & 0 & 0\\ 0 & 0 & 0\\ 0 & 0 & \Psi_{33}\end{pmatrix},\qquad [\boldsymbol{\Psi}_{1A}]=\begin{pmatrix}\nu\Psi_{33} & 0 & 0\\ 0 & \nu\Psi_{33} & 0\\ 0 & 0 & \Psi_{33}\end{pmatrix} \tag{4.84}$$

The stress functions $\boldsymbol{\Psi}_0$ and $\boldsymbol{\Psi}_{1A}$ correspond to plane stress and plane strain, respectively. In both cases, Ψ_{33} is the negative of the Airy stress function. When $\boldsymbol{\Psi}_0$ is chosen, the compatibility conditions along the x_1- and x_2-directions are not satisfied (recall from Chapter 2 that plane stress is an approximate solution). For the case of torsion, we have

$$[\boldsymbol{\Psi}_{1B}]=\begin{pmatrix}\Psi_{11} & 0 & 0\\ 0 & \Psi_{22} & 0\\ 0 & 0 & 0\end{pmatrix} \tag{4.85}$$

where Ψ_{11} and Ψ_{22} relate to Prandtl's stress function Ψ as

$$\Psi_{11,2}=x_3\Psi_{,1}, \quad \Psi_{22,1}=-x_3\Psi_{,2} \tag{4.86}$$

4.3.3 Morera Stress Function

When nondiagonal terms were chosen, we have the Morera stress function. The nonzero terms of the Morera stress function components of the tensor $\boldsymbol{\Psi}$ can be expressed in matrix form as

$$[\boldsymbol{\Psi}_2]=\begin{pmatrix} 0 & \Psi_{12} & \Psi_{13} \\ \Psi_{12} & 0 & \Psi_{23} \\ \Psi_{13} & \Psi_{23} & 0 \end{pmatrix} \tag{4.87}$$

Again there is only one way of choosing nondiagonal terms.

For the 2-D special case, we can express Morera stress function as

$$[\boldsymbol{\Psi}_{2A}]=\begin{pmatrix} 0 & 0 & x_3\Psi_{,1} \\ 0 & 0 & x_3\Psi_{,2} \\ x_3\Psi_{,1} & x_3\Psi_{,2} & 0 \end{pmatrix}, \quad [\boldsymbol{\Psi}_{2B}]=\begin{pmatrix} 0 & \Psi_{12} & x_3\Psi_{,1} \\ \Psi_{12} & 0 & x_3\Psi_{,2} \\ x_3\Psi_{,1} & x_3\Psi_{,2} & 0 \end{pmatrix} \tag{4.88}$$

where Ψ equals half of the Airy stress function and

$$\Psi_{12,12}=\nu\nabla^2\Psi \tag{4.89}$$

Plane stress and plane strain cases are represented by Morera stress functions $\boldsymbol{\Psi}_{2A}$ and $\boldsymbol{\Psi}_{2B}$, respectively. For the case of torsion, we have the following special case of Morera stress function:

$$[\boldsymbol{\Psi}_{2C}]=\begin{pmatrix} 0 & 0 & 0 \\ 0 & 0 & \Psi_{23} \\ 0 & \Psi_{23} & 0 \end{pmatrix} \tag{4.90}$$

where Ψ_{23} relates to Prandtl's stress function Ψ as

$$\Psi_{23,1}=\Psi \tag{4.91}$$

4.3.4 Other Beltrami Stress Functions

In addition to the Maxwell and Morera stress functions, Wang (2002) discussed the possibility of other choices. Explicit forms were not reported in Wang (2002), but they will be given here. In particular, four more types of stress functions can be selected:

Type 3:

$$[\boldsymbol{\Psi}_{3A}]=\begin{pmatrix} \Psi_{11} & \Psi_{12} & 0 \\ \Psi_{12} & \Psi_{22} & 0 \\ 0 & 0 & 0 \end{pmatrix}, \quad [\boldsymbol{\Psi}_{3B}]=\begin{pmatrix} 0 & 0 & 0 \\ 0 & \Psi_{22} & \Psi_{23} \\ 0 & \Psi_{23} & \Psi_{33} \end{pmatrix}, \quad [\boldsymbol{\Psi}_{3C}]=\begin{pmatrix} \Psi_{11} & 0 & \Psi_{13} \\ 0 & 0 & 0 \\ \Psi_{13} & 0 & \Psi_{33} \end{pmatrix} \tag{4.92}$$

In the Type 3 case, two diagonal terms plus the off-diagonal terms associated with both diagonal terms are selected. Three different combinations are shown in (4.92).

Type 4:

$$\left[\boldsymbol{\Psi}_{4A}\right]=\begin{pmatrix}\Psi_{11} & 0 & \Psi_{13}\\ 0 & 0 & \Psi_{23}\\ \Psi_{13} & \Psi_{23} & 0\end{pmatrix},\ \left[\boldsymbol{\Psi}_{4B}\right]=\begin{pmatrix}\Psi_{11} & \Psi_{12} & 0\\ \Psi_{12} & 0 & \Psi_{23}\\ 0 & \Psi_{23} & 0\end{pmatrix}$$

$$\left[\boldsymbol{\Psi}_{4C}\right]=\begin{pmatrix}0 & 0 & \Psi_{13}\\ 0 & \Psi_{22} & \Psi_{23}\\ \Psi_{13} & \Psi_{23} & 0\end{pmatrix}\ \left[\boldsymbol{\Psi}_{4D}\right]=\begin{pmatrix}0 & \Psi_{12} & \Psi_{13}\\ \Psi_{12} & \Psi_{22} & 0\\ \Psi_{13} & 0 & 0\end{pmatrix}$$

$$\left[\boldsymbol{\Psi}_{4E}\right]=\begin{pmatrix}0 & \Psi_{12} & 0\\ \Psi_{12} & 0 & \Psi_{23}\\ 0 & \Psi_{23} & \Psi_{33}\end{pmatrix},\left[\boldsymbol{\Psi}_{4F}\right]=\begin{pmatrix}0 & \Psi_{12} & \Psi_{13}\\ \Psi_{12} & 0 & 0\\ \Psi_{13} & 0 & \Psi_{33}\end{pmatrix} \qquad (4.93)$$

In the Type 4 case, one diagonal term plus two off-diagonal terms (one associated with the diagonal term, the other one not associated with the diagonal term). The six possible combinations are shown in (4.93).

Type 5:

$$\left[\boldsymbol{\Psi}_{5A}\right]=\begin{pmatrix}\Psi_{11} & 0 & 0\\ 0 & \Psi_{22} & \Psi_{23}\\ 0 & \Psi_{23} & 0\end{pmatrix},\ \left[\boldsymbol{\Psi}_{5B}\right]=\begin{pmatrix}\Psi_{11} & 0 & \Psi_{13}\\ 0 & \Psi_{22} & 0\\ \Psi_{13} & 0 & 0\end{pmatrix}$$

$$\left[\boldsymbol{\Psi}_{5C}\right]=\begin{pmatrix}0 & \Psi_{12} & 0\\ \Psi_{12} & \Psi_{22} & 0\\ 0 & 0 & \Psi_{33}\end{pmatrix},\ \left[\boldsymbol{\Psi}_{5D}\right]=\begin{pmatrix}0 & 0 & \Psi_{13}\\ 0 & \Psi_{22} & 0\\ \Psi_{13} & 0 & \Psi_{33}\end{pmatrix}$$

$$\left[\boldsymbol{\Psi}_{5E}\right]=\begin{pmatrix}\Psi_{11} & 0 & 0\\ 0 & 0 & \Psi_{23}\\ 0 & \Psi_{23} & \Psi_{33}\end{pmatrix},\ \left[\boldsymbol{\Psi}_{5F}\right]=\begin{pmatrix}\Psi_{11} & \Psi_{12} & 0\\ \Psi_{12} & 0 & 0\\ 0 & 0 & \Psi_{33}\end{pmatrix} \qquad (4.94)$$

In the Type 5 case, two diagonal terms plus one off-diagonal term associated with only one of the diagonal terms are selected. Six different combinations are shown in (4.94).

Type 6:

$$\left[\boldsymbol{\Psi}_{6A}\right]=\begin{pmatrix}0 & 0 & \Psi_{13}\\ 0 & 0 & \Psi_{23}\\ \Psi_{13} & \Psi_{23} & \Psi_{33}\end{pmatrix},\ \left[\boldsymbol{\Psi}_{6B}\right]=\begin{pmatrix}\Psi_{11} & \Psi_{12} & \Psi_{13}\\ \Psi_{12} & 0 & 0\\ \Psi_{13} & 0 & 0\end{pmatrix}$$

$$[\boldsymbol{\Psi}_{6C}]=\begin{pmatrix} 0 & \Psi_{12} & 0 \\ \Psi_{12} & \Psi_{22} & \Psi_{23} \\ 0 & \Psi_{23} & 0 \end{pmatrix} \tag{4.95}$$

In the Type 6 case, one diagonal term plus both off-diagonal terms associated with the diagonal term are selected. Three different combinations are shown in (4.95). There is a total of 20 combinations of choosing three components of the Beltrami stress functions. Types 1 and 2 are the Maxwell and Morera stress functions discussed in the last two sections. The completeness of the first 17 stress function matrices shown in (4.83), (4.87), and (4.92)–(4.94) of Type 1 to Type 5 can be demonstrated (Wang, 2002). The last three stress functions given in (4.95) are rather restrictive. For example, substitution of $\boldsymbol{\Psi}_{6A}$ into (4.69) reveals that there is no contribution for σ_{33} (or σ_{33} = 0). Therefore, any problem with nonzero σ_{33} cannot be solved by using $\boldsymbol{\Psi}_{6A}$. Similar conclusions can also be drawn for matrices $\boldsymbol{\Psi}_{6B}$ and $\boldsymbol{\Psi}_{6C}$.

Note that a special form of either $\boldsymbol{\Psi}_{4E}$ or $\boldsymbol{\Psi}_{4F}$ can also be specialized to recover the plane strain problem. In particular, we have (either setting $\Psi_{23} = 0$ in $\boldsymbol{\Psi}_{4E}$ or setting $\Psi_{13} = 0$ in $\boldsymbol{\Psi}_{4F}$):

$$[\boldsymbol{\Psi}_{4G}]=\begin{pmatrix} 0 & \Psi_{12} & 0 \\ \Psi_{12} & 0 & 0 \\ 0 & 0 & \Psi_{33} \end{pmatrix} \tag{4.96}$$

where

$$\Psi_{12,12} = -\frac{1}{2}\nu\nabla^2\Psi_{33} \tag{4.97}$$

4.4 SOME 3-D SOLUTIONS IN GEOMECHANICS

In the last two sections, both displacement and strain potential methods in the displacement formulation and the stress function method in the stress formulation have been discussed. To illustrate how to use this technique, some useful and well-known formulas for 3-D elasticity will be considered in this section.

4.4.1 Hollow Sphere Subject to Internal and External Pressures

To illustrate the use of this displacement potential, a hollow sphere subject to an internal pressure p_a at radius $r = a$ and an external pressure p_b at radius $r = b$ is considered, as shown in Fig. 4.1. For this case of spherical symmetry, Lamé's strain potential given (4.25) can be used.

In view of symmetry, the following Lamé's strain potential is assumed:

$$\phi(r) = \frac{A}{2}r - \frac{B}{r} \tag{4.98}$$

Applying the Laplacian operator on (4.98) and in view of (1.85), we have

$$\nabla^2\phi(r)=\frac{1}{r^2}\frac{\partial}{\partial r}(r^2\frac{\partial\phi}{\partial r})=3A \tag{4.99}$$

which is clearly Poisson's equation, and thus potentially a solution for displacement $\boldsymbol{u}$ of the problem. Applying the first part of (4.25) gives

$$u_r(r)=Ar+\frac{B}{r^2} \tag{4.100}$$

The corresponding strain tensor becomes

$$\boldsymbol{\varepsilon}(r)=\boldsymbol{e}_r\boldsymbol{e}_r\frac{\partial u_r}{\partial r}+(\boldsymbol{e}_\theta\boldsymbol{e}_\theta+\boldsymbol{e}_\phi\boldsymbol{e}_\phi)\frac{u_r}{r} \tag{4.101}$$

and the stress tensor is

$$\boldsymbol{\sigma}(r)=\left[(2\mu+\lambda)\frac{\partial u_r}{\partial r}+2\lambda\frac{u_r}{r}\right]\boldsymbol{e}_r\boldsymbol{e}_r+\left[(2\mu+\lambda)\frac{u_r}{r}+\lambda\frac{\partial u_r}{\partial r}\right](\boldsymbol{e}_\theta\boldsymbol{e}_\theta+\boldsymbol{e}_\phi\boldsymbol{e}_\phi) \tag{4.102}$$

Therefore, the stress components are

$$\sigma_{rr}=(\frac{E}{1-2\nu})A-(\frac{2E}{1+\nu})\frac{B}{r^3},\quad \sigma_{\theta\theta}=\sigma_{\phi\phi}=(\frac{E}{1-2\nu})A+(\frac{E}{1+\nu})\frac{B}{r^3} \tag{4.103}$$

As shown in Figure 4.1, the boundary conditions at $r=a$ and $r=b$ are

$$(\sigma_{rr})_{r=a}=-p_a,\quad (\sigma_{rr})_{r=b}=-p_b \tag{4.104}$$

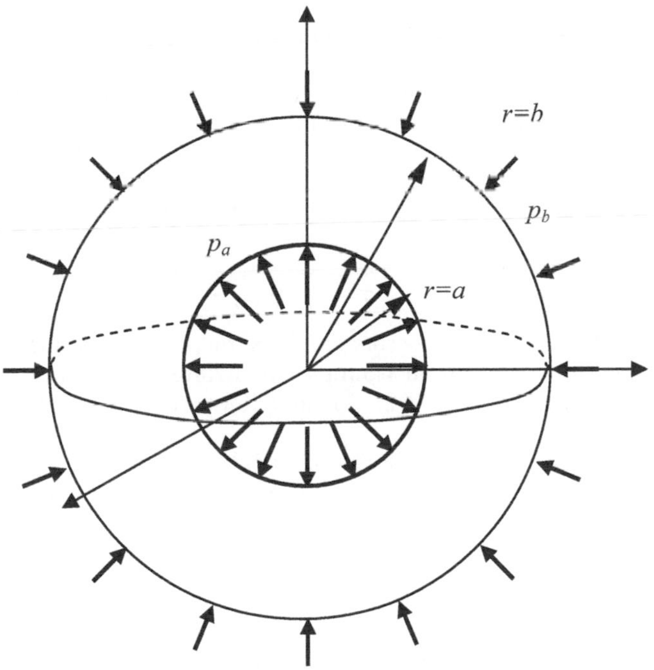

Figure 4.1 A hollow sphere subject to internal and external pressures

Substituting the first part of (4.102) into (4.104) gives two simultaneous equations for determining A and B:

$$\frac{E}{1-2\nu}A-\frac{2E}{1+\nu}\frac{B}{a^3}=-p_a, \quad \frac{E}{1-2\nu}A-\frac{2E}{1+\nu}\frac{B}{b^3}=-p_b \tag{4.105}$$

The solution of (4.105) gives

$$A=\frac{a^3p_a-b^3p_b}{E(b^3-a^3)}(1-2\nu), \quad B=\frac{a^3b^3(p_a-p_b)}{2E(b^3-a^3)}(1+\nu) \tag{4.106}$$

Substituting these constants into (4.100) and (4.103) gives

$$u_r=\frac{(1+\nu)r}{E(1-\alpha^3)}\left\{\left[\frac{1}{2\rho^3}+\alpha^3(\frac{1-2\nu}{1+\nu})\right]p_a-\left[\frac{1}{2\rho^3}+(\frac{1-2\nu}{1+\nu})\right]p_b\right\} \tag{4.107}$$

$$\sigma_{rr}=-\frac{1}{1-\alpha^3}\left\{(\frac{1}{\rho^3}-\alpha^3)p_a+(1-\frac{1}{\rho^3})p_b\right\} \tag{4.108}$$

$$\sigma_{\theta\theta}=\sigma_{\phi\phi}=\frac{1}{1-\alpha^3}\left\{(\frac{1}{2\rho^3}+\alpha^3)p_a-(1+\frac{1}{2\rho^3})p_b\right\} \tag{4.109}$$

where $\alpha = a/b$ and $\rho = r/a$. Note that 3-D elastic stress solutions are, in general, functions of Poisson's ratio ν whereas 2-D elastic stress solutions are independent of ν. Equations (4.108) and (4.109) show that the present 3-D solution is clearly independent of ν. To understand this observation, we can cut the sphere through any diametral plane in Fig. 4.1, and the resulting section is a section of axisymmetry. This effectively reduces the problem to 2-D mathematically. Thus, the stress solution is independent of ν for this special case.

For the special case of a solid sphere subject to external pressure, we can set $p_a = 0$ and $a \to 0$, and we find

$$\boldsymbol{u}=-p_b(\frac{1-2\nu}{E})r\boldsymbol{e}_r, \quad \boldsymbol{\sigma}=-p_b(\boldsymbol{e}_r\boldsymbol{e}_r+\boldsymbol{e}_\theta\boldsymbol{e}_\theta+\boldsymbol{e}_\phi\boldsymbol{e}_\phi) \tag{4.110}$$

That is, only radial deformation occurs and its shrinks to zero at the center of the sphere, and the stress state is in isotropic (or hydrostatic) compression. If the sphere is incompressible ($\nu = 1/2$), radial deformation is identically zero, as expected.

Another special case of interest is a spherical hole subject to internal pressure in an infinite medium. We can set $b \to \infty$ and $\alpha \to \infty$, and the resulting solution is

$$\boldsymbol{u}=\frac{ap_a}{2\rho^2}(\frac{1+\nu}{E})\boldsymbol{e}_r, \quad \boldsymbol{\sigma}=-\frac{p_a}{\rho^3}\boldsymbol{e}_r\boldsymbol{e}_r+\frac{p_a}{2\rho^3}(\boldsymbol{e}_\theta\boldsymbol{e}_\theta+\boldsymbol{e}_\phi\boldsymbol{e}_\phi) \tag{4.111}$$

where $\rho \geq 1$. On the surface of the spherical hole ($r = a$), the hoop stress equals $p_a/2$. This result can be contrasted with the result of Chapter 2 for a 2-D hole subject to far field compression stress $-T$ so that the maximum tensile hoop stress at a 2-D hole is T.

4.4.2 Kelvin's Fundamental Solution

One of the most commonly used elastic solutions is probably Kelvin's solution or point force solution in a full space (or infinite domain). Kelvin's problem is shown in Fig. 4.2. Its application is in the formulation of the boundary element method or in Green's function method. This solution was obtained by Lord Kelvin in 1848 by using the method of singularities in the theory of Newtonian potentials. Apparently, Lord Kelvin was aware of its usefulness in Green's function method, which was first recognized and publicised by himself shortly after the death of George Green in 1841 (see brief biography section at the end of the book). In this section, two different methods discussed earlier will be used to solve this problem, namely the Papkovitch–Neuber potential method and Love's strain potential method.

4.4.2.1 Papkovitch–Neuber Potential Method

In general, if a point force vector $\boldsymbol{P}$ is applied at the origin, the following Papkovitch–Neuber potential can be assumed:

$$\Phi_0 = 0\,, \qquad \boldsymbol{\Phi} = \frac{\boldsymbol{P}}{4\pi\mu R} \tag{4.112}$$

where R is the distance of the observation point from the origin (or the magnitude of position vector $\boldsymbol{r}$). It is straightforward to show that $1/R$ is harmonic:

$$\frac{\partial}{\partial x_j}\left(\frac{1}{R}\right) = -\frac{x_j}{R^3}\,, \quad \frac{\partial^2}{\partial x_j \partial x_i}\left(\frac{1}{R}\right) = -\frac{\delta_{ij}}{R^3} + \frac{3x_i x_j}{R^5}\,, \quad \frac{\partial^2}{\partial x_k \partial x_k}\left(\frac{1}{R}\right) = -\frac{\delta_{kk}}{R^3} + \frac{3x_k x_k}{R^5} = 0 \tag{4.113}$$

where

$$R = (x_1^2 + x_2^2 + x_3^2)^{1/2} \tag{4.114}$$

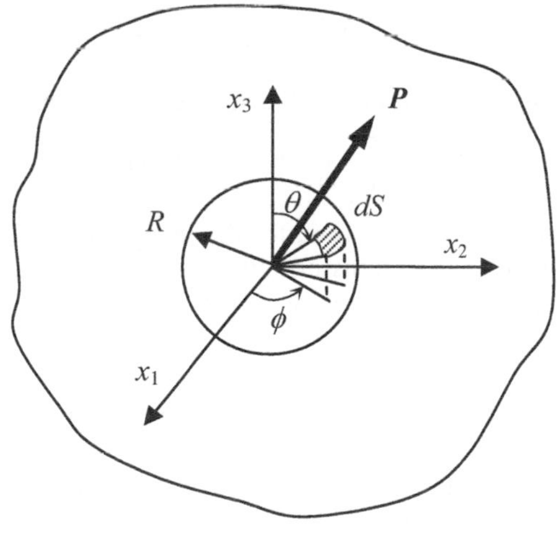

Figure 4.2 Kelvin's problem: A point force applied at the origin of a full space

Substitution of (4.112) into (4.53) yields

$$\boldsymbol{u} = \frac{1}{16\pi\mu(1-\nu)R}[(3-4\nu)\boldsymbol{P} + \frac{1}{R^2}\boldsymbol{r}(\boldsymbol{r} \bullet \boldsymbol{P})] \tag{4.115}$$

In component form, the displacement can be expressed as

$$u_i = \frac{P_j}{16\pi\mu(1-\nu)R}[(3-4\nu)\delta_{ij} + \frac{x_i x_j}{R^2}] = G_{ij}P_j \tag{4.116}$$

where G_{ij} can be interpreted as the displacement Green's tensor for infinite solid. The corresponding stress components are

$$\sigma_{ij} = -\frac{P_m}{8\pi(1-\nu)R^3}[\frac{3x_i x_j x_m}{R^2} + (1-2\nu)(-\delta_{ij}x_m + \delta_{im}x_j + \delta_{jm}x_i)] = G^*_{ijm}P_m \tag{4.117}$$

where G^*_{ijm} can be interpreted as the stress Green's tensor for an infinite solid. It is clear from (4.116) and (4.117) that both displacement and stress decay to zero at infinity, but are singular near the origin as $R \to 0$. The stress singularity is in the order of R^{-2}. For the case of vertical point force, we have $\boldsymbol{P} = (0, 0, P_3)$. The displacement components is simplified to

$$u_1 = A\frac{x_1 x_3}{R^3}, \quad u_2 = A\frac{x_2 x_3}{R^3}, \quad u_3 = A[(3-4\nu)\frac{1}{R} + \frac{x_3^2}{R^3}] \tag{4.118}$$

where

$$A = \frac{P_3}{16\pi\mu(1-\nu)} \tag{4.119}$$

We now consider a sphere of radius R around the origin as shown in Fig. 4.2. The traction on the surface of the sphere can be evaluated as

$$t_i = \sigma_{ij}n_j = \frac{\sigma_{ij}x_j}{R} \tag{4.120}$$

Substitution of (4.117) into (4.120) results in

$$t_1 = -6\mu A\frac{x_3 x_1}{R^4}, \quad t_2 = -6\mu A\frac{x_3 x_2}{R^4}, \quad t_3 = -2\mu A[\frac{1-2\nu}{R^2} + \frac{3x_3^2}{R^4}] \tag{4.121}$$

The resultant force acting on the surface of the sphere can be determined from

$$F_i = \int_S t_i dS = \int_0^{2\pi}\int_0^{\pi} t_i R^2 \sin\theta d\theta d\phi \tag{4.122}$$

where a spherical coordinate has been used for the integration: $0 < \phi < 2\pi$ and $0 < \theta < \pi$. Cartesian coordinates can be transformed to polar form by using

$$x_1 = R\sin\theta\cos\phi, \quad x_2 = R\sin\theta\sin\phi, \quad x_3 = R\cos\theta \tag{4.123}$$

The components of (4.122) become

$$F_1 = -6\mu A\int_0^{2\pi}\int_0^{\pi}\cos\theta\sin^2\theta\cos\phi d\theta d\phi = 0 \tag{4.124}$$

$$F_2 = -6\mu A\int_0^{2\pi}\int_0^{\pi}\cos\theta\sin^2\theta\sin\phi d\theta d\phi = 0 \tag{4.125}$$

$$F_3 = -2\mu A[(1-2\nu)\int_0^{2\pi}\int_0^{\pi} \sin\theta d\theta d\phi + 3\int_0^{2\pi}\int_0^{\pi} \cos^2\theta \sin\theta d\theta d\phi]$$
$$= -2\mu A[(1-2\nu)(4\pi) + 3(\frac{4\pi}{3})] \tag{4.126}$$
$$= -16\mu\pi(1-\nu)A = -P_3$$

Since the sphere must be in equilibrium, the resultant surface traction must balance the applied force at the origin. This verifies that the proposed solution given in (4.118) and (4.119) indeed provides the solution of Kelvin's problem. Note that in the derivation R has been cancelled. That is, this force equilibrium is true for spheres of any size with the center at the origin.

From the mathematical form of displacement given in (4.118), it is obvious that similar results can be established for horizontal point forces P_2 and P_3. In addition, if the point force is applied at a point $\boldsymbol{y}$ (with Cartesian components y_1, y_2, and y_3), the derivation is the same except is replaced $\boldsymbol{x}$ by $\boldsymbol{x}-\boldsymbol{y}$ and R by

$$R = [(x_i - y_i)(x_i - y_i)]^{1/2} \tag{4.127}$$

In summary, Kelvin's solution can be written as:

$$u_i = G_{ij}(\boldsymbol{x}, \boldsymbol{y})P_j \tag{4.128}$$

where

$$G_{ij}(\boldsymbol{x}, \boldsymbol{y}) = \frac{1}{16\pi\mu(1-\nu)R}[(3-4\nu)\delta_{ij} + \frac{(x_i - y_i)(x_j - y_j)}{R^2}] \tag{4.129}$$

This is the elastostatic Green's tensor or Green's function for isotropic infinite elastic solid. It is obvious that $\boldsymbol{G}$ is symmetric with respect to $\boldsymbol{x}$ and $\boldsymbol{y}$ as well as with respect to indices i and j.

4.4.2.2 Love's Displacement Potential Method

In this section, we will use Love's displacement potential given in Section 4.2.4 to reconsider Kelvin's problem presented in the last section. Recall that Love's displacement potential is a special case of the Galerkin vector, and it is especially useful in solving axisymmetric problems. Kelvin's problem is clearly one of those axisymmetric problems.

In particular, the following Love's displacement potential is assumed:

$$G_z = BR = B(z^2 + r^2)^{1/2} \tag{4.130}$$

Note from the last section that $1/R$ is a harmonic function in 3-D domain. Then the Almansi theorem states that $R = R^2(1/R)$ must be biharmonic (see Section 8.11 of Fung, 1965). Using Cartesian coordinates, R can be demonstrated as biharmonic:

$$\frac{\partial R}{\partial x_j \partial x_i} = \frac{\delta_{ij}}{R} - \frac{x_i x_j}{R^3}, \quad \nabla^2 R = \frac{\partial R}{\partial x_k \partial x_k} = \frac{3}{R} - \frac{x_k x_k}{R^3} = \frac{2}{R}, \quad \nabla^4 R = 2\nabla^2(\frac{1}{R}) = 0 \tag{4.131}$$

In polar coordinates, the derivation is somehow similar:

$$\frac{\partial G_z}{\partial r} = \frac{Br}{R}, \quad \frac{\partial^2 G_z}{\partial r^2} = \frac{Bz^2}{R^3}, \quad \frac{\partial G_z}{\partial z} = \frac{Bz}{R},$$

$$\frac{\partial^2 G_z}{\partial z^2} = \frac{Br^2}{R^3}, \quad \nabla^2 G_z = 2\frac{B}{R}, \quad \nabla^4 G_z = 2B\nabla^2(\frac{1}{R}) = 0 \tag{4.132}$$

Thus, G_z given in (4.130) automatically satisfies the biharmonic equation (4.41). In polar coordinates, the displacement becomes

$$u_r = B\frac{rz}{R^3}, \quad u_\theta = 0, \quad u_z = B(\frac{3-4\nu}{R} + \frac{z^2}{R^3}) \tag{4.133}$$

The corresponding stresses are

$$\sigma_{rr} = 2\mu B[\frac{(1-2\nu)z}{R^3} - \frac{3r^2 z}{R^5}], \qquad \sigma_{zz} = -2\mu B[\frac{(1-2\nu)z}{R^3} + \frac{3z^3}{R^5}]$$

$$\sigma_{rz} = -2\mu B[\frac{(1-2\nu)r}{R^3} + \frac{3rz^2}{R^5}], \qquad \sigma_{\theta\theta} = 2\mu B(1-2\nu)\frac{z}{R^3}$$

$$\sigma_{z\theta} = \sigma_{r\theta} = 0 \tag{4.134}$$

These stresses are singular at the origin in the order of R^{-2}, and vanish at infinity. This suggests a point force is applied at the origin. The axisymmetric nature of (4.133) suggests that the applied force is vertical.

To calculate the constant B, the applied point force at the origin is assumed to be enclosed in a circular cylinder of length $2a$ and radius r, as shown in Fig. 4.3. Since the cylinder must be in equilibrium, the resultant vertical surface traction on the surface of the cylinder must balance the applied vertical force at the origin. As $r \to \infty$, this equilibrium is

$$P = \int_0^\infty 2\pi r(-\sigma_{zz})_{z=a} dr + \int_0^\infty 2\pi r(\sigma_{zz})_{z=-a} dr + \lim_{r\to\infty} \int_{-a}^{a} 2\pi r \sigma_{rz} dz \tag{4.135}$$

The last term vanishes as $r \to \infty$ and the first two integrals are the same. Using the identity $R = (r^2+z^2)^{1/2}$ and $rdr = RdR$, (4.135) is simplified to

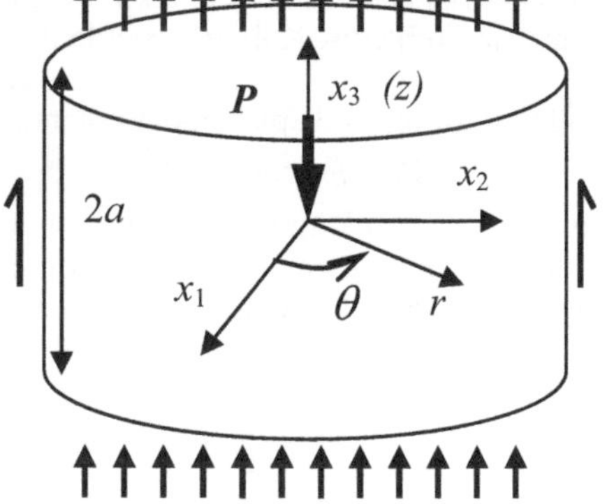

Figure 4.3 Kelvin's problem: A vertical point force applied at origin of a full space

$$P = 2\int_a^{\infty} 2\pi R(-\sigma_{zz})_{z=a} dR$$
$$= 8\pi\mu B[(1-2\nu)a\int_a^{\infty}\frac{1}{R^2}dR + 3a^3\int_a^{\infty}\frac{1}{R^4}dR] \tag{4.136}$$
$$= 16\pi\mu(1-\nu)B$$

Finally, we obtain

$$B = \frac{P}{16\pi\mu(1-\nu)} \tag{4.137}$$

The solutions given in (4.133), (4.134), and (4.137) in cylindrical coordinates can be shown to be equivalent to the solution derived in the last section in Cartesian coordinate.

4.4.3 Boussinesq's Fundamental Solution

Although Kelvin's solution is probably the most useful formula in solid mechanics, Boussinesq's problem considered in this section must be the most important solution in geomechanics. As shown in Fig. 4.4, Boussinesq's problem deals with a vertical point force P applied on the surface of a half-space. The problem was first solved by Boussinesq in 1878, 30 years after Lord Kelvin obtained the point solution in a full space. Any surface vertical loading applied on the ground surface can be determined by integrating Boussinesq's solution on a ground surface. Two particular results obtained from such integration, called Fadum chart and Newmark influence chart, are still used daily by geotechnical engineers in foundation engineering. Although Boussinesq's solution was considered by some the start of geomechanics, Boussinesq is more widely recognized for his contributions to fluid mechanics and turbulence (see biography section at the end of this book).

The boundary condition on $z = 0$ must be traction free except at the origin:

$$\sigma_{zz} = -P\delta(r), \quad \sigma_{zr} = 0 \tag{4.138}$$

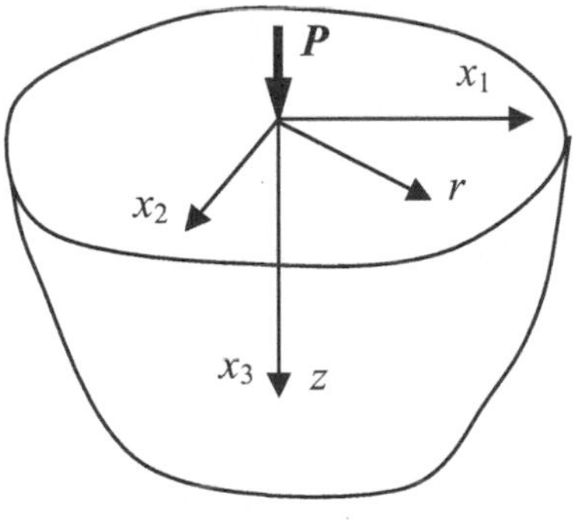

Figure 4.4 Boussinesq's problem: A vertical point force applied on a half-space

The total vertical force resulting from σ_{zz} at any depth z must balance with the vertical force P. That is,

$$\int_0^\infty 2\pi r(\sigma_{zz})dr + P = 0 \tag{4.139}$$

4.4.3.1 Love's and Lamé's Strain Potential Methods

We can start with the Love's potential used in Section 4.4.2.2 for Kelvin's problem since it satisfies the point force singularity at the origin:

$$G_z = \frac{A_1}{2\mu} R \tag{4.140}$$

The corresponding displacement and stress components are

$$u_r = \frac{A_1}{2\mu}(\frac{rz}{R^3}), \qquad u_z = \frac{A_1}{2\mu}(\frac{3-4\nu}{R} + \frac{z^2}{R^3}) \tag{4.141}$$

$$\sigma_{rr} = A_1[\frac{(1-2\nu)z}{R^3} - \frac{3r^2 z}{R^5}], \qquad \sigma_{zz} = -A_1[\frac{(1-2\nu)z}{R^3} + \frac{3z^3}{R^5}]$$

$$\sigma_{rz} = -A_1[\frac{(1-2\nu)r}{R^3} + \frac{3rz^2}{R^5}], \qquad \sigma_{\theta\theta} = A_1(1-2\nu)\frac{z}{R^3} \tag{4.142}$$

As discussed earlier, these stresses are singular at the origin in the order of R^{-2}, and vanish at infinity. The first part of (4.138) can be satisfied by this choice but the shear traction at $z = 0$ becomes

$$\sigma_{rz} = -A_1 \frac{(1-2\nu)}{r^2} \tag{4.143}$$

which clearly does not satisfy the second part of (4.138).

In order to cancel out this shear stress distribution on the surface, we have to superimpose another function that gives rise to zero σ_{zz} on $z = 0$ but a nonzero shear stress distribution that cancels the shear stress given in (4.143). By inspection, we can use Lamé's strain potential discussed in Section 4.2.2 in the following form:

$$\phi = \frac{A_2}{2\mu} \ln(R + z) \tag{4.144}$$

A special discussion on the logarithmic function given in (4.144) is needed. Physically, the argument of a logarithmic function cannot have any unit, but unfortunately most of the authors in elasticity do not mention this explicitly. This can be easily seen by referring to the following Taylor series expansion of the natural logarithmic function:

$$\ln(1+x) = x - \frac{x^2}{2} + \frac{x^3}{3} - \frac{x^4}{4} + \ldots \tag{4.145}$$

It is obvious that a number with a unit of length cannot be added to a number with a unit of area, or to a number with a unit of volume, etc. Therefore, the length scale must be normalized first with respect to an arbitrary length scale, such that R and z

can be interpreted as normalized length. The displacement and stress components in cylindrical coordinates can be shown to be

$$u_r = \frac{\partial \phi}{\partial r}, \qquad u_z = \frac{\partial \phi}{\partial z} \tag{4.146}$$

$$\sigma_{rr} = 2\mu \frac{\partial^2 \phi}{\partial r^2}, \quad \sigma_{\theta\theta} = 2\mu \frac{1}{r}\frac{\partial \phi}{\partial r}, \quad \sigma_{zz} = 2\mu \frac{\partial^2 \phi}{\partial z^2}, \quad \sigma_{rz} = 2\mu \frac{\partial^2 \phi}{\partial z \partial r} \tag{4.147}$$

Substitution of (4.144) into (4.146) and (4.147) results in

$$u_r = \frac{A_2 r}{2\mu R(R+z)}, \qquad u_z = \frac{A_2}{2\mu R} \tag{4.148}$$

$$\sigma_{rr} = A_2[\frac{z}{R^3} - \frac{1}{R(R+z)}], \quad \sigma_{\theta\theta} = \frac{A_2}{R(R+z)}, \quad \sigma_{zz} = -\frac{A_2 z}{R^3}, \quad \sigma_{rz} = -\frac{A_2 r}{R^3} \tag{4.149}$$

The stress components from these two solutions given in (4.143) and (4.149) are superimposed and the resulting stresses are enforced to satisfy boundary conditions (4.138) and (4.139). We have two equations for A_1 and A_2:

$$A_1(1-2\nu) + A_2 = 0, \qquad 4\pi(1-2\nu)A_1 + 2\pi A_2 = P \tag{4.150}$$

Note that the second part of (4.150) is independent of the depth of the vertical stress being evaluated. That is, it is true at any depth. The solution of (4.150) gives

$$A_1 = \frac{P}{2\pi}, \qquad A_2 = -\frac{(1-2\nu)P}{2\pi} \tag{4.151}$$

Finally, substituting (4.151) into the stress components and adding the two solutions, we have

$$u_r = \frac{P}{4\pi\mu R}[\frac{rz}{R^2} - \frac{(1-2\nu)r}{R+z}], \quad u_z = \frac{P}{4\pi\mu R}[2(1-\nu) + \frac{z^2}{R^2}] \tag{4.152}$$

$$\sigma_{rr} = \frac{P}{2\pi R^2}[\frac{(1-2\nu)R}{R+z} - \frac{3r^2 z}{R^3}], \quad \sigma_{\theta\theta} = \frac{(1-2\nu)P}{2\pi R^2}(\frac{z}{R} - \frac{R}{R+z})$$

$$\sigma_{zz} = -\frac{3Pz^3}{2\pi R^5}, \quad \sigma_{rz} = -\frac{3Pz^2 r}{2\pi R^5} \tag{4.153}$$

4.4.3.2 Papkovitch–Neuber Potential Method

In this section, we will follow another method to consider the Boussinesq problem. From Section 4.4.2.1 we can first use the following Papkovitch–Neuber potential for Kelvin's problem:

$$\Phi_0 = \Phi_1 = \Phi_2 = 0, \quad \Phi_3 = \frac{A_1}{R} \tag{4.154}$$

Similar to the results in Section 4.4.2.1, the resulting stress tensor is

$$\sigma_{3i} = -\frac{\mu A_1 x_i}{2(1-\nu)R^3}(1 - 2\nu + \frac{3x_3^2}{R^2}) \tag{4.155}$$

where $i = 1,2$. The shear stress on the boundary $x_3 = 0$ becomes

$$\sigma_{31} = -\frac{(1-2\nu)\mu A_1 x_1}{2(1-\nu)r^3}, \quad \sigma_{32} = -\frac{(1-2\nu)\mu A_1 x_2}{2(1-\nu)r^3} \tag{4.156}$$

Similar to the observation made in the previous section, we can consider another potential as

$$\Phi_0 = A_2 \ln(R + x_3), \quad \boldsymbol{\Phi} = 0 \tag{4.157}$$

The stress associated with this potential is

$$\sigma_{3i} = \frac{\mu A_2 x_i}{2(1-\nu)R^3} \tag{4.158}$$

Therefore, the stresses on the surface of the half-space become

$$\sigma_{31} = \frac{\mu A_2 x_1}{2(1-\nu)r^3}, \quad \sigma_{32} = \frac{\mu A_2 x_2}{2(1-\nu)r^3}, \quad \sigma_{33} = 0 \tag{4.159}$$

To ensure the traction-free condition, the sum of (4.156) and (4.159) must vanish, and this leads to

$$A_2 = (1-2\nu)A_1 \tag{4.160}$$

The sum of (4.155) and (4.158) gives

$$\begin{aligned} \sigma_{33} = \sigma_{33}^{(1)} + \sigma_{33}^{(2)} &= -\frac{\mu A_1 x_3}{2(1-\nu)R^3}(1 - 2\nu + \frac{3x_3^2}{R^2}) + \frac{\mu(1-2\nu)A_1 x_3}{2(1-\nu)R^3} \\ &= -\frac{3\mu A_1 x_3^3}{2(1-\nu)R^5} \end{aligned} \tag{4.161}$$

Force equilibrium at any depth given by (4.139) can be used again as

$$\int_0^\infty \int_0^{2\pi} \sigma_{33} r d\phi dr = -P \tag{4.162}$$

Substitution of (4.161) into (4.162) gives

$$\begin{aligned} P &= \frac{3\mu A_1 \pi}{(1-\nu)} \int_0^\infty \frac{r x_3^3 dr}{(r^2 + x_3^2)^{5/2}} = \frac{3\mu A_1 \pi}{(1-\nu)} \frac{x_3^3}{2} \int_0^\infty \frac{d(r^2 + x_3^2)}{(r^2 + x_3^2)^{5/2}} \\ &= \frac{3\mu A_1 \pi}{(1-\nu)} \frac{x_3^3}{2} \left[-\frac{2}{3(r^2 + x_3^2)^{3/2}} \right]_0^\infty = \frac{\mu A_1 \pi}{(1-\nu)} \end{aligned} \tag{4.163}$$

Therefore, we have

$$A_1 = (1-\nu)\frac{P}{\pi\mu} \tag{4.164}$$

Finally,

$$u_\alpha = \frac{P x_\alpha}{4\pi\mu R}(\frac{x_3}{R^2} - \frac{1-2\nu}{R+x_3}), \quad u_3 = \frac{P}{4\pi\mu R}[2(1-\nu) + \frac{x_3^2}{R^2}] \tag{4.165}$$

where α = 1, 2. We can rewrite these displacement components in cylindrical coordinates by recalling (3.52):

$$u_r = u_1 \cos\phi + u_2 \sin\phi = u_1 \frac{x_1}{r} + u_2 \frac{x_2}{r}$$
$$= \frac{Px_1^2}{4\pi\mu Rr}(\frac{x_3}{R^2} - \frac{1-2\nu}{R+x_3}) + \frac{Px_2^2}{4\pi\mu Rr}(\frac{x_3}{R^2} - \frac{1-2\nu}{R+x_3}) = \frac{Pr}{4\pi\mu R}(\frac{z}{R^2} - \frac{1-2\nu}{R+z}) \tag{4.166}$$

$$u_\phi = u_2 \cos\phi - u_1 \sin\phi = u_2 \frac{x_1}{r} - u_1 \frac{x_2}{r}$$
$$= \frac{Px_1x_2}{4\pi\mu Rr}(\frac{x_3}{R^2} - \frac{1-2\nu}{R+x_3}) - \frac{Px_1x_2}{4\pi\mu Rr}(\frac{x_3}{R^2} - \frac{1-2\nu}{R+x_3}) = 0 \tag{4.167}$$

$$u_z = \frac{P}{4\pi\mu R}[2(1-\nu) + \frac{z^2}{R^2}] \tag{4.168}$$

which are of course the same as those obtained by using Love's displacement potential and Lamé's strain potential given in (4.152). This completes the derivation.

As illustrated in Fig. 4.5, Boussinesq's solution given in (4.152) and (4.165) predicts a cone of expansion under the point load. To see this, we first define $z = R\cos\theta$. Then an expanding zone can be defined as $u_r > 0$ under the point load. For $u_r > 0$, from (4.152) we must have

$$\cos^2\theta + \cos\theta - (1-2\nu) > 0 \tag{4.169}$$

Thus, $u_r > 0$ (expanding zone) if $\theta < \theta_0$, where

$$\cos^2\theta_0 + \cos\theta_0 - (1-2\nu) = 0 \tag{4.170}$$

The size of the zone ranges from 51.8° to 90°, depending on Poisson's ratio:

$$\cos\theta_0 = \frac{-1 \pm \sqrt{1+4(1-2\nu)}}{2} \tag{4.171}$$

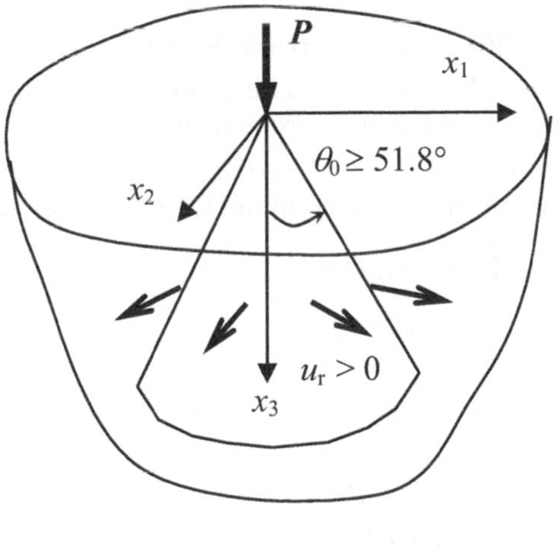

Figure 4.5 Expansion zone in Boussinesq's problem

For $\nu = 0$, we have $\theta_0 = 51.8°$. When $\nu = 1/4$, θ_0 increases to 68.5°. When $\nu = 1/3$, θ_0 further increases to 74.7° (note that this is a typical value assumed for rocks). For incompressible solids (i.e., $\nu = 1/2$), θ_0 becomes 90°. Therefore, the whole half-space is expanding for an incompressible solid, like saturated clay. In a sense, the solid is being squeezed sideways axismmetrically by the surface point load.

Similarly, we can also consider the zones of compression and tension in the half-space. For example, the hoop stress $\sigma_{\theta\theta}$ is found compressive within a cone under the point force:

$$\cos^2\theta + \cos\theta - 1 > 0 \tag{4.172}$$

This equals precisely (4.169) for the case of $\nu = 0$. Therefore, the tensile zone is defined by a cone with $\theta_0 = 51.8°$, independent of the value of Poisson's ratio. Instead, the value of tensile stress within the "fixed" cone varies with Poisson's ratio as $(1 - 2\nu)$. Similarly, there is also a compressive zone for radial stress within a cone under the point load. The size of the cone θ_0 can be found by

$$3\cos\theta_0 \sin^2\theta_0 (1 + \cos\theta_0) - (1 - 2\nu) = 0 \tag{4.173}$$

The analytical solution for θ_0 cannot be obtained for (4.173). Readers interested in the solution can find θ_0 numerically using standard software.

4.4.4 Cerruti's Fundamental Solution

Another fundamental solution of interest to geomechanics is a horizontal point force applied parallel to the surface of a half-space. The problem was obtained by Cerruti in 1882, 4 years after the Boussinesq problem was solved. The problem is illustrated in Fig. 4.6. There was an interesting encounter between V. Cerruti and C.A. Castigliano who was the originator of "Castigliano principle" in structural mechanics (see biography section).

To solve Cerruti's problem, we can propose a combined solution from the Galerkin vector and from Lamé's strain potential. In particular, the displacement vector is expressed as

$$\boldsymbol{u} = \nabla\phi + 2(1-\nu)\nabla^2\boldsymbol{G} - \nabla(\nabla \bullet \boldsymbol{G}) \tag{4.174}$$

and as before ϕ and $\boldsymbol{G}$ satisfy the Laplace and biharmonic equations, respectively.

The boundary conditions of Cerruti's problem are

$$\sigma_{33} = 0, \quad \sigma_{31} = 0, \quad \sigma_{32} = 0 \tag{4.175}$$

on $x_3 = 0$ (except for σ_{31} at the origin). In addition, force equilibrium requires

$$\int_{-\infty}^{\infty}\int_{-\infty}^{\infty} \sigma_{31} dx_1 dx_2 + P = 0 \tag{4.176}$$

The following Galerkin vector is assumed:

$$G_1 = A_1 R, \quad G_2 = 0, \quad G_3 = A_2 x_1 \ln(R + x_3) \tag{4.177}$$

In addition, the following Lamé's strain potential is also assumed:

$$\phi = \frac{A_3 x_1}{R + x_3} \tag{4.178}$$

The resulting stresses can be obtained by substitution of (4.177) into (4.39) and (4.40) and (4.178) into (4.27) and adding these stress components. The calculation is rather tedious, and we will try to include the steps as much as possible so that the reader can work out the details themselves.

To aid the derivation, the following formulas are found useful repeatedly:

$$R=(x_k x_k)^{1/2}, \quad \frac{\partial R}{\partial x_k}=\frac{x_k}{R}, \quad \nabla^2 R=\frac{\partial R}{\partial x_k \partial x_k}=\frac{2}{R}, \quad \frac{\partial}{\partial x_k}(\frac{1}{R})=-\frac{x_k}{R^3},$$

$$\frac{\partial}{\partial x_m}(\frac{x_k}{R})=(\frac{\delta_{km}}{R}-\frac{x_k x_m}{R^3}), \quad \frac{\partial}{\partial x_m}(\frac{1}{R+x_3})=-\frac{1}{(R+x_3)^2}(\frac{x_m}{R}+\delta_{3m}) \quad (4.179)$$

With these formulas, it is straightforward to show that

$$\nabla^2 G_1=\frac{2A_1}{R}, \quad \nabla^2 G_3=\frac{2x_1 A_2}{R(R+x_3)}, \quad \nabla \bullet \boldsymbol{G}=\frac{x_1}{R}(A_1+A_2) \quad (4.180)$$

$$\frac{\partial}{\partial x_k}\nabla \bullet \boldsymbol{G}=(\frac{\delta_{1k}}{R}-\frac{x_1 x_k}{R^3})(A_1+A_2) \quad (4.181)$$

$$\frac{\partial^2}{\partial x_k \partial x_m}\nabla \bullet \boldsymbol{G}=\frac{(A_1+A_2)}{R^5}[-R^2(\delta_{1m}x_k+\delta_{1k}x_m+\delta_{mk}x_1)+3x_1 x_m x_k] \quad (4.182)$$

$$\frac{\partial}{\partial x_m}(\nabla^2 G_3)=\frac{2A_2}{R+x_3}[\frac{\delta_{1m}}{R}-\frac{x_1 x_m}{R^3}-\frac{x_1}{R(R+x_3)}(\frac{x_m}{R}+\delta_{3m})] \quad (4.183)$$

$$\frac{\partial \phi}{\partial x_k}=\frac{A_3}{R+x_3}[\delta_{1k}-\frac{x_1}{(R+x_3)}(\frac{x_k}{R}+\delta_{3k})] \quad (4.184)$$

$$\frac{\partial^2 \phi}{\partial x_k \partial x_m}=\frac{A_3}{(R+x_3)^2}[-\delta_{1k}(\frac{x_m}{R}+\delta_{3m})-\delta_{1m}(\frac{x_k}{R}+\delta_{3k})-x_1(\frac{\delta_{km}}{R}-\frac{x_k x_m}{R^3})$$
$$+\frac{2x_1}{(R+x_3)}(\frac{x_k}{R}+\delta_{3k})(\frac{x_m}{R}+\delta_{3m})] \quad (4.185)$$

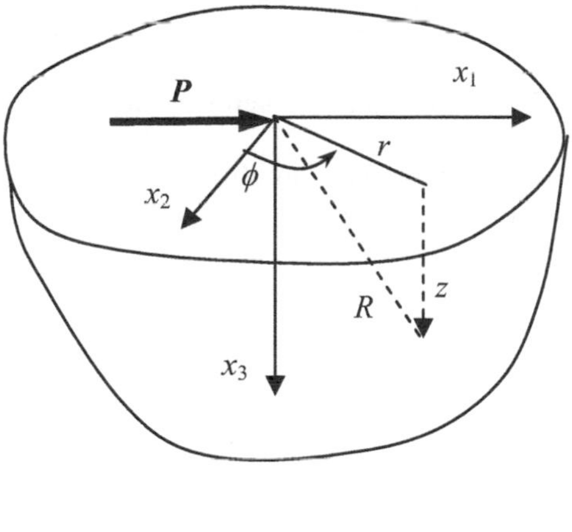

Figure 4.6 Cerruti's problem: A horizontal point force applied at the origin of a half-space

With the help of (4.180)–(4.185), the vertical normal stress σ_{33} resulting from (4.177) can be shown to be

$$\sigma_{33} = \frac{\partial^2 \phi}{\partial x_3^2} + 2(1-\nu)\frac{\partial}{\partial x_3}\nabla^2 G_3 + (\nu\nabla^2 - \frac{\partial^2}{\partial x_3^2})(\nabla \bullet \boldsymbol{G})$$
$$= \frac{A_3 x_1}{R^3} - 4(1-\nu)\frac{A_2 x_1}{R^3} - 2\nu\frac{x_1}{R^3}(A_2 + A_1) - \frac{(-R^2 + 3x_3^2)x_1}{R^5}(A_2 + A_1) \tag{4.186}$$

The first of (4.175) requires that σ_{33} vanishes at $x_3 = 0$ (except at the origin) and this leads to the following equation:

$$A_3 - 4(1-\nu)A_2 + (1-2\nu)(A_2 + A_1) = 0 \tag{4.187}$$

With the help of (4.180)–(4.185), σ_{31} can be determined as

$$\sigma_{31} = \frac{\partial^2 \phi}{\partial x_1 \partial x_3} + (1-\nu)(\frac{\partial}{\partial x_3}\nabla^2 G_1 + \frac{\partial}{\partial x_1}\nabla^2 G_3) - \frac{\partial^2}{\partial x_1 \partial x_3}(\nabla \bullet \boldsymbol{G})$$
$$= \frac{[A_3 - 2(1-\nu)A_2]}{R(R+x_3)}[-1 + \frac{x_1^2(2R+x_3)}{R^2(R+x_3)}] - (1-\nu)\frac{2A_1 x_3}{R^3} - \frac{(3x_1^2 - R^2)x_3}{R^5}(A_2 + A_1) \tag{4.188}$$

The second of (4.175) requires that σ_{31} vanishes at $x_3 = 0$ and this leads to

$$\frac{1}{r^2}[-A_3 + 2(1-\nu)A_2] + \frac{x_1^2}{r^4}[2A_3 - 4(1-\nu)A_2] = 0 \tag{4.189}$$

where r is defined in Fig. 4.6. Since both r and x_1 can be arbitrary, the following equation must be satisfied:

$$A_3 = 2(1-\nu)A_2 \tag{4.190}$$

With the help of (4.180)–(4.185), σ_{32} can be determined as

$$\sigma_{32} = \frac{\partial^2 \phi}{\partial x_2 \partial x_3} + (1-\nu)(\frac{\partial}{\partial x_2}\nabla^2 G_3) - \frac{\partial^2}{\partial x_2 \partial x_3}(\nabla \bullet \boldsymbol{G})$$
$$= \frac{[A_3 - 2(1-\nu)A_2](2R+x_3)}{R^3(R+x_3)^2} + \frac{3x_1 x_2 x_3}{R^5}(A_2 + A_1) \tag{4.191}$$

The third part of (4.175) requires that σ_{32} vanishes at $x_3 = 0$ and this leads exactly to (4.190) again.

Before we consider the force equilibrium required by (4.176), we first substitute (4.190) into (4.187) to eliminate A_3 and obtain the following relation between A_1 and A_2:

$$A_2 = (1-2\nu)A_1 \tag{4.192}$$

Then the expression for σ_{31} given by (4.188) can be greatly simplified in view of (4.190), and the result is

$$\sigma_{31} = -\frac{3x_1^2 x_3}{R^5}(A_2 + A_1) \tag{4.193}$$

Substitution of (4.193) into (4.176) gives

$$-3x_3(A_1 + A_2)\int_{-\infty}^{\infty}\int_{-\infty}^{\infty}\frac{x_1^2}{R^5}dx_1 dx_2 + P = 0 \tag{4.194}$$

We now apply the following change of variable:

$$x_1 = r\cos\phi, \quad x_2 = r\sin\phi, \quad dx_1 dx_2 = r d\phi dr \tag{4.195}$$

to the integration given in (4.194)

$$\begin{aligned}\int_{-\infty}^{\infty}\int_{-\infty}^{\infty}\frac{x_1^2}{R^5}dx_1 dx_2 &= \int_0^{\infty}\int_0^{2\pi}\frac{r^2\cos^2\phi}{R^5}r d\phi dr \\ &= \int_0^{\infty}\frac{r^3 dr}{(r^2+x_3^2)^{5/2}}\int_0^{2\pi}\cos^2\phi d\phi \\ &= \pi\int_0^{\infty}\frac{r^3 dr}{(r^2+x_3^2)^{5/2}}\end{aligned} \tag{4.196}$$

To evaluate the remaining integration, we can first apply partial fractions to rewrite the integrand as

$$\frac{r^3}{(r^2+x_3^2)^{5/2}} = \frac{r}{(r^2+x_3^2)^{3/2}} - \frac{rx_3^2}{(r^2+x_3^2)^{5/2}} \tag{4.197}$$

With (4.197), the integrand given in (4.196) can then be determined using the standard method, and the result is ($2/3x_3$). Substituting this result into (4.196) and then into (4.194), we find

$$(A_1 + A_2) = \frac{P}{2\pi} \tag{4.198}$$

Finally, combining equations (4.190), (4.192), and (4.198) give

$$A_1 = \frac{P}{4\pi(1-\nu)}, \quad A_2 = \frac{(1-2\nu)P}{4\pi(1-\nu)}, \quad A_3 = \frac{(1-2\nu)P}{2\pi} \tag{4.199}$$

Once the unknown constants are determined, all stress and displacements can be determined as

$$u_1 = \frac{P}{4\pi\mu R}\left\{1+\frac{x_1^2}{R^2}+(1-2\nu)\left[\frac{R}{R+x_3}-\frac{x_1^2}{(R+x_3)^2}\right]\right\} \tag{4.200}$$

$$u_2 = \frac{Px_1x_2}{4\pi\mu R}\left[\frac{1}{R^2}-\frac{1-2\nu}{(R+x_3)^2}\right] \tag{4.201}$$

$$u_3 = \frac{Px_1}{4\pi\mu R}\left[\frac{x_3}{R^2}+\frac{1-2\nu}{R+x_3}\right] \tag{4.202}$$

$$\sigma_{11} = \frac{Px_1}{2\pi R^3}\left[-\frac{3x_1^2}{R^2}+\frac{1-2\nu}{(R+x_3)^2}(R^2-x_2^2-\frac{2Rx_2^2}{R+x_3})\right] \tag{4.203}$$

$$\sigma_{22} = \frac{Px_1}{2\pi R^3}\left[-\frac{3x_2^2}{R^2}+\frac{1-2\nu}{(R+x_3)^2}(3R^2-x_1^2-\frac{2Rx_1^2}{R+x_3})\right] \tag{4.204}$$

$$\sigma_{33} = -\frac{3Px_1x_3^2}{2\pi R^5} \tag{4.205}$$

$$\sigma_{23} = -\frac{3Px_1x_2x_3}{2\pi R^5}, \quad \sigma_{31} = -\frac{3Px_1^2x_3}{2\pi R^5} \tag{4.206}$$

$$\sigma_{12} = \frac{Px_2}{2\pi R^3}\left[-\frac{3x_1^2}{R^2} + \frac{1-2\nu}{(R+x_3)^2}(-R^2 + x_1^2 + \frac{2Rx_1^2}{R+x_3}) \right] \tag{4.207}$$

4.4.5 Mindlin's Fundamental Solution in Half-Space

Another useful solution was obtained by Mindlin in 1936 for a point force applied in the interior of a half-space. Mindlin solved the problem first by the method of images, but the physical meaning of all those images is, however, not clear (Mindlin, 1936b). In 1953, Mindlin rederived the solution by using the Papkovitch–Neuber displacement potential (Mindlin, 1953). Mindlin's problem is illustrated in Fig. 4.7. This solution is very powerful generating other solutions for tunnels or cracks within the Earth. Of course, Boussinesq's and Cerruti's solutions can both be recovered as a special case for $c \to 0$. If we consider the limit that $c \to \infty$, Kelvin's solution can also be recovered as a special case. Therefore, this is the most general solution of all these fundamental point force solutions for homogeneous solids. The Mindlin solution was R.D. Mindlin's Ph.D. thesis, and more amazingly he obtained this solution without any guidance as a student at Columbia University (see biography section at the end of this book).

In this section, we adopt the Papkovitch–Neuber potential approach to solve Mindlin's problem. The presentation here somewhat follows that of Wang (2002). Consider a force vector $\boldsymbol{F}$ with components F_1, F_2, and F_3 applied at a point at the position $(0,0,c)$, as shown in Fig. 4.7. Similar to the discussion about solving Boussinesq's problem, we first consider Kelvin's solution and superimpose another solution from the Papkovitch–Neuber potential that would result in stresses that cancel all stresses from Kelvin's solution at the surface of the half-space.

The final Papkovitch–Neuber potentials consists of two parts:

$$\Phi_0 = \psi_0 + \varphi_0 , \quad \boldsymbol{\Phi} = \boldsymbol{\psi} + \boldsymbol{\varphi} \tag{4.208}$$

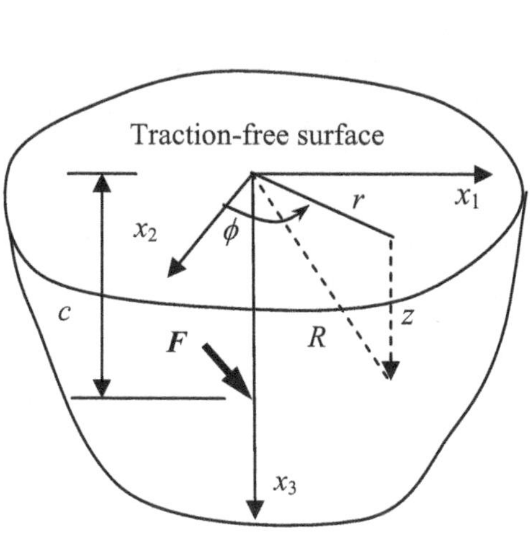

Figure 4.7 Mindlin's problem: A point force applied at the interior of a half-space

The first pair is given as

$$\psi_0 = -\frac{1}{4\pi\mu}\frac{cF_3}{R}, \quad \boldsymbol{\psi} = \frac{1}{4\pi\mu}\frac{\boldsymbol{F}}{R} \tag{4.209}$$

where

$$R = \sqrt{x_1^2 + x_2^2 + (x_3 - c)^2} \tag{4.210}$$

When $c = 0$, this Papkovitch–Neuber potential is exactly the same as the one given previously for Kelvin's problem when the point force was applied at the origin. The unknown harmonic functions φ_0 and $\boldsymbol{\varphi}$ have to be determined from the traction-free boundary condition given by (4.175).

First, the displacement components given in (4.53) can be rewritten explicitly as

$$u_i = \frac{1}{4(1-\nu)}[(3-4\nu)\Phi_i - \Phi_{0,i} - x_k\Phi_{k,i}] \tag{4.211}$$

By virtue of the fact that both Φ_0 and Φ_i are harmonic, (4.56) can be simplified to

$$\sigma_{ij} = \frac{\mu}{2(1-\nu)}\left\{2\nu\Phi_{k,k}\delta_{ij} + (1-2\nu)(\Phi_{i,j} + \Phi_{j,i}) - \Phi_{0,ij} - x_k\Phi_{k,ij}\right\} \tag{4.212}$$

Three particular components of the stress which would be used to satisfy the boundary conditions are

$$\sigma_{31} = \frac{\mu}{2(1-\nu)}\left\{(1-2\nu)(\Phi_{3,1} + \Phi_{1,3}) - \Phi_{0,13} - x_k\Phi_{k,13}\right\} \tag{4.213}$$

$$\sigma_{32} = \frac{\mu}{2(1-\nu)}\left\{(1-2\nu)(\Phi_{3,2} + \Phi_{2,3}) - \Phi_{0,32} - x_k\Phi_{k,32}\right\} \tag{4.214}$$

$$\sigma_{33} = \frac{\mu}{2(1-\nu)}\left\{2\nu(\Phi_{1,1} + \Phi_{2,2}) + 2(1-\nu)\Phi_{3,3} - \Phi_{0,33} - x_k\Phi_{k,33}\right\} \tag{4.215}$$

For the following derivations, we first define a harmonic function A which is defined in the upper half-space $x_3 \le 0$. Then, a mirror image of it is defined for $x_3 \ge 0$ as

$$\tilde{A}(x_1, x_2, x_3) = A(x_1, x_2, -x_3) \tag{4.216}$$

Then the image function must also be a harmonic function in the domain $x_3 \ge 0$:

$$\nabla^2\tilde{A} = 0 \quad (x_3 \ge 0) \tag{4.217}$$

At the plane $x_3 = 0$, these two functions are connected as

$$\tilde{A} = A, \quad \frac{\partial\tilde{A}}{\partial x_1} = \frac{\partial A}{\partial x_1}, \quad \frac{\partial\tilde{A}}{\partial x_2} = \frac{\partial A}{\partial x_2}, \quad \frac{\partial\tilde{A}}{\partial x_3} = -\frac{\partial A}{\partial x_3} \tag{4.218}$$

Note that $\boldsymbol{\psi}$ and ψ_0 have only singularity in the lower half-space and with no singularity in the upper half-space. Conversely, the mirror images of them will have no singularity in the lower half-space $x_3 \ge 0$.

Substitution of (4.213) into (4.215) and the result into (4.175) yields the following equations on $x_3 = 0$:

$$\begin{aligned}&-x_1(\psi_{1,13} + \varphi_{1,13}) - x_2(\psi_{2,13} + \varphi_{2,13}) - \psi_{0,13} - \varphi_{0,13}\\&+(1-2\nu)(\psi_{3,1} + \varphi_{3,1} + \psi_{1,3} + \varphi_{1,3}) = 0\end{aligned} \tag{4.219}$$

$$-x_1(\psi_{1,23}+\varphi_{1,23})-x_2(\psi_{2,23}+\varphi_{2,23})-\psi_{0,23}-\varphi_{0,23} \\ +(1-2\nu)(\psi_{3,2}+\varphi_{3,2}+\psi_{2,3}+\varphi_{2,3})=0 \tag{4.220}$$

$$2\nu(\psi_{1,1}+\varphi_{1,1}+\psi_{2,2}+\varphi_{2,2})+2(1-\nu)(\psi_{3,3}+\varphi_{3,3})-x_1(\psi_{1,33}+\varphi_{1,33}) \\ -x_2(\psi_{2,33}+\varphi_{2,33})-\psi_{0,33}-\varphi_{0,33}=0 \tag{4.221}$$

The remaining step is to find the appropriate φ_0 and $\boldsymbol{\varphi}$ that satisfy the boundary conditions given in (4.219)–(4.221). For the lower half-space of $x_3 \geq 0$, we first set

$$\varphi_1=\tilde{\psi}_1,\quad \varphi_2=\tilde{\psi}_2 \tag{4.222}$$

By doing so, we do not impose any stress singularity for $x_3 \geq 0$. In view of (4.222) and (4.218), the boundary conditions (4.219)–(4.221) can further be simplified as (on $x_3 = 0$)

$$\tilde{\psi}_{0,13}-\varphi_{0,13}+(1-2\nu)(\tilde{\psi}_{3,1}+\varphi_{3,1})=0 \tag{4.223}$$

$$\tilde{\psi}_{0,23}-\varphi_{0,23}+(1-2\nu)(\tilde{\psi}_{3,2}+\varphi_{3,2})=0 \tag{4.224}$$

$$4\nu(\tilde{\psi}_{1,1}+\tilde{\psi}_{2,2})+2(1-\nu)(-\tilde{\psi}_{3,3}+\varphi_{3,3})-2x_1\tilde{\psi}_{1,33}-2x_2\tilde{\psi}_{2,33}-\tilde{\psi}_{0,33}-\varphi_{0,33}=0 \tag{4.225}$$

By observation, we can set

$$\varphi_{0,3}=\tilde{\psi}_{0,3}+(1-2\nu)(\tilde{\psi}_3+\varphi_3) \tag{4.226}$$

for $x_3 \geq 0$, then both (4.223) and (4.224) are satisfied identically. Substituting of (4.226) into (4.225), we have

$$\varphi_{3,3}=-4\nu(\tilde{\psi}_{1,1}+\tilde{\psi}_{2,2})+(3-4\nu)\tilde{\psi}_{3,3}+2(x_1\tilde{\psi}_{1,33}+x_2\tilde{\psi}_{2,33})+2\tilde{\psi}_{0,33} \tag{4.227}$$

on $x_3 = 0$.

Note, however, that (4.227) is only valid on the boundary. We could not simply integrate (4.227) to get the final unknown function φ_3 for $x_3 \geq 0$, because the second-last term in (4.227) involving x_1 and x_2 does not satisfy the Laplace equation (or they are not harmonic). Evidently some terms which are functions of x_3 disappear when we set $x_3 = 0$ on the boundary in the original function φ_3. The most difficult step of solving (4.227) to get φ_3 for $x_3 \geq 0$ is to figure out what has been set to zero when the surface boundary on $x_3 = 0$ is approached.

To remedy this problem, we observe that

$$\nabla^2(x_1\tilde{\psi}_{1,3}-x_3\tilde{\psi}_{1,1})=0 \tag{4.228}$$

$$\nabla^2(x_2\tilde{\psi}_{2,3}-x_3\tilde{\psi}_{2,2})=0 \tag{4.229}$$

Therefore, these functions are harmonic, and more importantly they suggest what has been dropped when the boundary condition is assigned. We assert that

$$\varphi_3=f(x_1,x_2,x_3)+2(x_1\tilde{\psi}_{1,3}-x_3\tilde{\psi}_{1,1}+x_2\tilde{\psi}_{2,3}-x_3\tilde{\psi}_{2,2}) \tag{4.230}$$

for $x_3 \geq 0$ where f can be obtained by integrating other terms in (4.227). Taking the differentiation of (4.230) with respect to x_3 leads to

$$\varphi_{3,3}=f_{,3}+2(x_1\tilde{\psi}_{1,33}-\tilde{\psi}_{1,1}-x_3\tilde{\psi}_{1,13})+2(x_2\tilde{\psi}_{2,33}-\tilde{\psi}_{2,2}-x_3\tilde{\psi}_{2,23}) \tag{4.231}$$

for $x_3 \geq 0$. As we expected, two terms in (4.230) disappear when the boundary on $x_3 = 0$ is considered. However, one more complication is observed. The differentiation leads to two extra terms which are grouped in the last brackets in the following equation:

$$\varphi_{3,3}=f_{,3}+2(x_1\tilde{\psi}_{1,33}-x_3\tilde{\psi}_{1,13})+2(x_2\tilde{\psi}_{2,33}-x_3\tilde{\psi}_{2,23})-2(\tilde{\psi}_{1,1}+\tilde{\psi}_{2,2}) \tag{4.232}$$

for $x_3 \geq 0$. Therefore, in view of (4.227) and (4.232) we propose the following form:

$$\varphi_3 = C\int(\tilde{\psi}_{1,1}+\tilde{\psi}_{2,2})dx_3+(3-4\nu)\tilde{\psi}_3+2(x_1\tilde{\psi}_{1,3}-x_3\tilde{\psi}_{1,1})+2(x_2\tilde{\psi}_{2,3}-x_3\tilde{\psi}_{2,2})+2\tilde{\psi}_{0,3} \tag{4.233}$$

for $x_3 \geq 0$. Comparison of the coefficients of (4.227) and the differentiation of (4.233) with respect to x_3 suggests that C must satisfy

$$C-2=-4\nu \tag{4.234}$$

That is,

$$\begin{aligned}\varphi_3 &= 2(1-2\nu)\int(\tilde{\psi}_{1,1}+\tilde{\psi}_{2,2})dx_3+(3-4\nu)\tilde{\psi}_3+2(x_1\tilde{\psi}_{1,3}-x_3\tilde{\psi}_{1,1})\\ &+2(x_2\tilde{\psi}_{2,3}-x_3\tilde{\psi}_{2,2})+2\tilde{\psi}_{0,3}\end{aligned} \tag{4.235}$$

Functions on the right-hand side of (4.235) can be found easily using the following identities:

$$\tilde{\psi}_{i,j} = -\frac{F_i}{4\pi\mu}\left(\frac{R_j}{\tilde{R}^3}\right), \quad \tilde{\psi}_{0,3} = \frac{cF_3}{4\pi\mu}\left(\frac{x_3+c}{\tilde{R}^3}\right) \tag{4.236}$$

where

$$\begin{aligned}&\tilde{R} = [x_1^2+x_2^2+(x_3+c)^2]^{1/2}, \quad \tilde{R}_1 = R_1 = x_1, \quad \tilde{R}_2 = R_2 = x_2,\\ &\tilde{R}_3 = x_3+c, \quad R_3 = x_3-c\end{aligned} \tag{4.237}$$

$$\begin{aligned}\tilde{\psi}_{1,1}+\tilde{\psi}_{2,2} &= -\frac{F_1}{4\pi\mu}\left(\frac{x_1}{\tilde{R}^3}\right)-\frac{F_2}{4\pi\mu}\left(\frac{x_2}{\tilde{R}^3}\right)\\ &= \frac{F_1}{4\pi\mu}\frac{\partial}{\partial x_3}\left[\frac{x_1}{\tilde{R}(\tilde{R}+x_3+c)}\right]+\frac{F_2}{4\pi\mu}\frac{\partial}{\partial x_3}\left[\frac{x_2}{\tilde{R}(\tilde{R}+x_3+c)}\right]\end{aligned} \tag{4.238}$$

By virtue of (4.236)–(4.238), (4.235) can now be determined as

$$\begin{aligned}\varphi_3 &= \frac{F_1}{4\pi\mu}\left[2(1-2\nu)\frac{x_1}{\tilde{R}(\tilde{R}+x_3+c)}-\frac{2cx_1}{\tilde{R}^3}\right]+\frac{F_2}{4\pi\mu}\left[2(1-2\nu)\frac{x_2}{\tilde{R}(\tilde{R}+x_3+c)}-\frac{2cx_2}{\tilde{R}^3}\right]\\ &+\frac{F_3}{4\pi\mu}\left[(3-4\nu)\frac{1}{\tilde{R}}+\frac{2c(x_3+c)}{\tilde{R}^3}\right]\end{aligned} \tag{4.239}$$

This solution can now be back substituted into (4.226) to integrate for φ_0. By applying the following integration formulas

$$\begin{aligned}&\int\frac{dx_3}{\tilde{R}^3} = -\frac{1}{\tilde{R}(\tilde{R}+x_3+c)}, \quad \int\frac{dx_3}{\tilde{R}(\tilde{R}+x_3+c)} = -\frac{1}{\tilde{R}+x_3+c},\\ &\int\frac{(x_3+c)dx_3}{\tilde{R}^3} = -\frac{1}{\tilde{R}}, \quad \int\frac{dx_3}{\tilde{R}} = \ln(\tilde{R}+x_3+c)\end{aligned} \tag{4.240}$$

we finally obtain

$$\varphi_0 = \frac{F_1}{4\pi\mu}\left[-\frac{2(1-2\nu)^2 x_1}{\tilde{R}+x_3+c}+\frac{2(1-2\nu)cx_1}{\tilde{R}(\tilde{R}+x_3+c)}\right]+\frac{F_2}{4\pi\mu}\left[-\frac{2(1-2\nu)^2 x_2}{\tilde{R}+x_3+c}+\frac{2(1-2\nu)cx_2}{\tilde{R}(\tilde{R}+x_3+c)}\right]$$
$$+\frac{F_3}{4\pi\mu}\left[4(1-2\nu)(1-\nu)\ln(\tilde{R}+x_3+c)-(3-4\nu)\frac{c}{\tilde{R}}\right] \tag{4.241}$$

Finally, the Papkovitch–Neuber potential for Mindlin's problem can be obtained by substituting (4.209), (4.222), (4.239), and (4.241) into (4.208):

$$\Phi_1 = \frac{F_1}{4\pi\mu}\left[\frac{1}{R}+\frac{1}{\tilde{R}}\right], \quad \Phi_2 = \frac{F_2}{4\pi\mu}\left[\frac{1}{R}+\frac{1}{\tilde{R}}\right] \tag{4.242}$$

$$\Phi_3 = \frac{F_1}{4\pi\mu}\left[\frac{2(1-2\nu)x_1}{\tilde{R}(\tilde{R}+x_3+c)}-\frac{2cx_1}{\tilde{R}^3}\right]+\frac{F_2}{4\pi\mu}\left[\frac{2(1-2\nu)x_2}{\tilde{R}(\tilde{R}+x_3+c)}-\frac{2cx_2}{\tilde{R}^3}\right]$$
$$+\frac{F_3}{4\pi\mu}\left[(3-4\nu)\frac{1}{\tilde{R}}+\frac{2c(x_3+c)}{\tilde{R}^3}+\frac{1}{R}\right] \tag{4.243}$$

$$\Phi_0 = \frac{F_1}{4\pi\mu}\left[-\frac{2(1-2\nu)^2 x_1}{\tilde{R}+x_3+c}+\frac{2(1-2\nu)cx_1}{\tilde{R}(\tilde{R}+x_3+c)}\right]+\frac{F_2}{4\pi\mu}\left[-\frac{2(1-2\nu)^2 x_2}{\tilde{R}+x_3+c}+\frac{2(1-2\nu)cx_2}{\tilde{R}(\tilde{R}+x_3+c)}\right]$$
$$+\frac{F_3}{4\pi\mu}\left[4(1-2\nu)(1-\nu)\ln(\tilde{R}+x_3+c)-(3-4\nu)\frac{c}{\tilde{R}}\right]-\frac{cF_3}{4\pi\mu}\left(\frac{1}{R}\right) \tag{4.244}$$

The displacement can then be evaluated by substituting (4.242)–(4.244) into (4.211). The calculation is straightforward but rather tedious, and the resulting formula is rather lengthy. The best way is to group the displacement according to the force condition:

$$u_i = u_i^{(1)}(F_1)+u_i^{(2)}(F_2)+u_i^{(3)}(F_3) = \sum_{j=1}^{3} u_i^{(j)}(F_j) \tag{4.245}$$

for $i = 1,2,3$. Under applied force F_1, the three displacement components are:

$$u_1^{(1)} = \frac{F_1}{16\pi\mu(1-\nu)}\{(3-4\nu)(\frac{1}{R}+\frac{x_1^2}{\tilde{R}^3})+\frac{1}{\tilde{R}}+\frac{x_1^2}{R^3}+\frac{2cx_3}{\tilde{R}^3}-\frac{6cx_1^2x_3}{\tilde{R}^5}$$
$$+4(1-\nu)(1-2\nu)[\frac{1}{\tilde{R}+x_3+c}-\frac{x_1^2}{\tilde{R}(\tilde{R}+x_3+c)^2}]\} \tag{4.246}$$

$$u_2^{(1)} = \frac{F_1}{16\pi\mu(1-\nu)}\{\frac{x_1x_2}{R^3}+(3-4\nu)\frac{x_1x_2}{\tilde{R}^3}-\frac{6cx_1x_2x_3}{\tilde{R}^5}-4(1-\nu)(1-2\nu)\frac{x_1x_2}{\tilde{R}(\tilde{R}+x_3+c)^2}\} \tag{4.247}$$

$$u_3^{(1)} = \frac{F_1}{16\pi\mu(1-\nu)}\{\frac{x_1(x_3-c)}{R^3} + (3-4\nu)\frac{x_1(x_3-c)}{\tilde{R}^3} - \frac{6cx_1x_3(x_3+c)}{\tilde{R}^5}$$
$$+4(1-\nu)(1-2\nu)\frac{x_1}{\tilde{R}(\tilde{R}+x_3+c)}\} \tag{4.248}$$

Under the applied force F_2, the three displacement components are

$$u_1^{(2)} = \frac{F_2}{16\pi\mu(1-\nu)}\{\frac{x_1x_2}{R^3} + (3-4\nu)\frac{x_1x_2}{\tilde{R}^3} - \frac{6cx_1x_2x_3}{\tilde{R}^5} - 4(1-\nu)(1-2\nu)\frac{x_1x_2}{\tilde{R}(\tilde{R}+x_3+c)^2}\} \tag{4.249}$$

$$u_2^{(2)} = \frac{F_2}{16\pi\mu(1-\nu)}\{(3-4\nu)(\frac{1}{R} + \frac{x_2^2}{\tilde{R}^3}) + \frac{1}{\tilde{R}} + \frac{x_2^2}{R^3} + \frac{2cx_3}{\tilde{R}^3} - \frac{6cx_2^2x_3}{\tilde{R}^5}$$
$$+4(1-\nu)(1-2\nu)[\frac{1}{\tilde{R}+x_3+c} - \frac{x_2^2}{\tilde{R}(\tilde{R}+x_3+c)^2}]\} \tag{4.250}$$

$$u_3^{(2)} = \frac{F_2}{16\pi\mu(1-\nu)}\{\frac{x_2(x_3-c)}{R^3} + (3-4\nu)\frac{x_2(x_3-c)}{\tilde{R}^3} - \frac{6cx_2x_3(x_3+c)}{\tilde{R}^5}$$
$$+4(1-\nu)(1-2\nu)\frac{x_2}{\tilde{R}(\tilde{R}+x_3+c)}\} \tag{4.251}$$

Under the applied force F_3, the three displacement components are

$$u_1^{(3)} = \frac{F_3}{16\pi\mu(1-\nu)}\{\frac{x_1(x_3-c)}{R^3} + (3-4\nu)\frac{x_1(x_3-c)}{\tilde{R}^3} + \frac{6cx_1x_3(x_3+c)}{\tilde{R}^5}$$
$$-4(1-\nu)(1-2\nu)\frac{x_1}{\tilde{R}(\tilde{R}+x_3+c)}\} \tag{4.252}$$

$$u_2^{(3)} = \frac{F_3}{16\pi\mu(1-\nu)}\{\frac{x_2(x_3-c)}{R^3} + (3-4\nu)\frac{x_2(x_3-c)}{\tilde{R}^3} + \frac{6cx_2x_3(x_3+c)}{\tilde{R}^5}$$
$$-4(1-\nu)(1-2\nu)\frac{x_2}{\tilde{R}(\tilde{R}+x_3+c)}\} \tag{4.253}$$

$$u_3^{(3)} = \frac{F_3}{16\pi\mu(1-\nu)}\{(3-4\nu)[\frac{1}{R} + \frac{(x_3+c)^2}{\tilde{R}^3}] + [1+4(1-\nu)(1-2\nu)]\frac{1}{\tilde{R}} + \frac{(x_3-c)^2}{R^3}$$
$$-\frac{2cx_3}{\tilde{R}^3} + \frac{6cx_3(x_3+c)^2}{\tilde{R}^5}\} \tag{4.254}$$

There are similarities in these expressions and it is possible to rewrite in a unified form as

$$u_i^{(j)} = \frac{F_j}{16\pi\mu(1-\nu)}\{[(3-4\nu)\frac{1}{R}+\frac{1}{\tilde{R}}]\delta_{ij} + 4(1-\nu)(1-2\nu)\frac{1}{\tilde{R}}\delta_{3i}\delta_{3j} + \frac{R_i R_j}{R^3}$$
$$+(3-4\nu)\frac{R_i R_j(1-\delta_{ij}) + \tilde{R}_i\tilde{R}_j\delta_{ij}}{\tilde{R}^3} + \frac{4(1-\nu)(1-2\nu)}{\tilde{R}(\tilde{R}+\tilde{R}_3)}(-\tilde{R}_i\delta_{3j} + \tilde{R}_j\delta_{3i})(1-\delta_{ij})$$
$$+\frac{4(1-\nu)(1-2\nu)}{R+\tilde{R}_3}[\delta_{ij} - \frac{\tilde{R}_i\tilde{R}_j}{\tilde{R}(\tilde{R}+\tilde{R}_3)}](1-\delta_{3i})(1-\delta_{3j})$$
$$+\frac{2cx_3}{\tilde{R}^5}(\delta_{ij}\tilde{R}^2 - 3\tilde{R}_i\tilde{R}_j)(1-2\delta_{3j})\} \tag{4.255}$$

where $i, j = 1, 2, 3$ with no summation on double indices. The corresponding stress components are given in Mindlin (1936b, 1953) and Westergaard (1952), and they will not be given here.

4.4.6 Lorentz's Fundamental Solution

The problem that surface of the half-space is fixed (or zero displacement), instead of traction free, was considered by Lorentz in 1907 (see Fig. 4.8). The original reference can be found in Wang (2002). This problem is of less practical interest in geomechanics, but may find application in solving problems with soils under large rigid mat footing. The method of solution is similar to that used in the last section for Mindlin's problem. Therefore, only the key steps are reported briefly here.

The boundary condition for Lorentz's problem on $x_3 = 0$

$$\boldsymbol{u} = 0 \tag{4.256}$$

We again determine the Papkovitch–Neuber potentials in the following form:

$$\Phi_0 = \psi_0 + \varphi_0, \quad \boldsymbol{\Phi} = \boldsymbol{\psi} + \boldsymbol{\varphi} \tag{4.257}$$

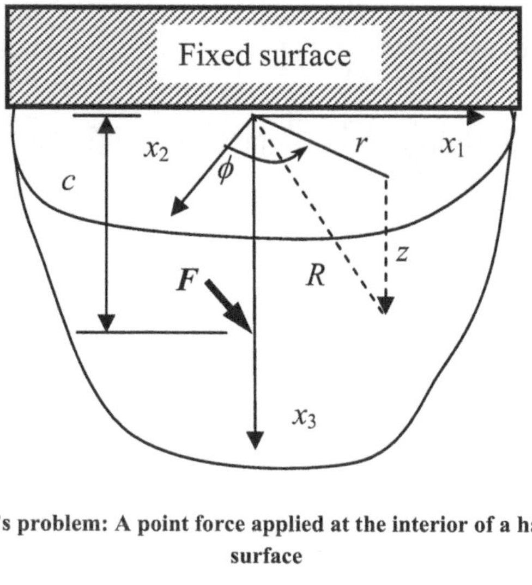

Figure 4.8 Lorentz's problem: A point force applied at the interior of a half-space with fixed surface

The unknown harmonic functions φ_0 and $\boldsymbol{\varphi}$ have to be determined from the fixed boundary condition given by (4.256).
The first pair is given as

$$\psi_0 = -\frac{1}{4\pi\mu}\frac{cF_3}{R}, \quad \boldsymbol{\psi} = \frac{1}{4\pi\mu}\frac{\boldsymbol{F}}{R} \tag{4.258}$$

The three displacement components are

$$u_1 = \frac{1}{4(1-\nu)}[(3-4\nu)\Phi_1 - \Phi_{0,1} - x_1\Phi_{1,1} - x_2\Phi_{2,1} - x_2\Phi_{3,1}] \tag{4.259}$$

$$u_2 = \frac{1}{4(1-\nu)}[(3-4\nu)\Phi_2 - \Phi_{0,2} - x_1\Phi_{1,2} - x_2\Phi_{2,2} - x_3\Phi_{3,2}] \tag{4.260}$$

$$u_3 = \frac{1}{4(1-\nu)}[(3-4\nu)\Phi_3 - \Phi_{0,3} - x_1\Phi_{1,3} - x_2\Phi_{2,3} - x_3\Phi_{3,3}] \tag{4.261}$$

Substitution of (4.257) and (4.259)–(4.261) into (4.256) yields

$$(3-4\nu)(\psi_1+\varphi_1)-(\psi_{0,1}+\varphi_{0,1})-x_1(\psi_{1,1}+\varphi_{1,1})-x_2(\psi_{2,1}+\varphi_{2,1})=0 \tag{4.262}$$

$$(3-4\nu)(\psi_2+\varphi_2)-(\psi_{0,2}+\varphi_{0,2})-x_1(\psi_{1,2}+\varphi_{1,2})-x_2(\psi_{2,2}+\varphi_{2,2})=0 \tag{4.263}$$

$$(3-4\nu)(\psi_3+\varphi_3)-(\psi_{0,3}+\varphi_{0,3})-x_1(\psi_{1,3}+\varphi_{1,3})-x_2(\psi_{2,3}+\varphi_{2,3})=0 \tag{4.264}$$

on $x_3 = 0$. For the lower half-space $x_3 \geq 0$, we first select the following harmonic functions:

$$\varphi_i = -\tilde{\psi}_i, \quad (i = 0,1,2) \tag{4.265}$$

With this selection, it is obvious, that both (4.262) and (4.263) are satisfied. Substitution of (4.265) into (4.264) gives

$$(3-4\nu)(\psi_3+\varphi_3)+2\tilde{\psi}_{0,3}+2x_1\tilde{\psi}_{1,3}+2x_2\tilde{\psi}_{2,3}=0 \tag{4.266}$$

on $x_3 = 0$. As in the previous section, the second-last two terms in (4.266) involving x_1 and x_2 do not satisfy the Laplace equation (or they are not harmonic functions). To add back the missing terms, we must have

$$\varphi_3 = -\psi_3 - \frac{2}{(3-4\nu)}\{\tilde{\psi}_{0,3} + x_1\tilde{\psi}_{1,3} - x_3\tilde{\psi}_{1,1} + x_2\tilde{\psi}_{2,3} - x_3\tilde{\psi}_{2,2}\} \tag{4.267}$$

Note that the added extra terms will vanish on $x_3 = 0$, and thus (4.266) is recovered on the boundary.

With this result, the final Papkovitch–Neuber potentials are

$$\Phi_0 = -\frac{cF_3}{4\pi\mu}\left[\frac{1}{R}-\frac{1}{\tilde{R}}\right], \quad \Phi_1 = \frac{F_1}{4\pi\mu}\left[\frac{1}{R}-\frac{1}{\tilde{R}}\right], \quad \Phi_2 = \frac{F_2}{4\pi\mu}\left[\frac{1}{R}-\frac{1}{\tilde{R}}\right] \tag{4.268}$$

$$\Phi_3 = \frac{F_3}{4\pi\mu}\left(\frac{1}{R}-\frac{1}{\tilde{R}}\right) + \frac{2}{(3-4\nu)}\frac{1}{4\pi\mu}\left[F_1\frac{cx_1}{\tilde{R}^3} + F_2\frac{cx_2}{\tilde{R}^3} - F_3\frac{c(x_3+c)}{\tilde{R}^3}\right] \tag{4.269}$$

The associated displacements can also be given in a compact form:

$$u_i^{(j)} = \frac{F_j}{16\pi\mu(1-\nu)}\{(3-4\nu)(\frac{1}{R}-\frac{1}{\tilde{R}})\delta_{ij} + \frac{R_iR_j}{R^3} - \frac{R_iR_j(1-\delta_{ij})+\tilde{R}_i\tilde{R}_j\delta_{ij}}{\tilde{R}^3}$$
$$+\frac{2}{3-4\nu}\frac{cx_3}{\tilde{R}^3}(2\delta_{3i}\delta_{3j}-\delta_{ij}) + \frac{1}{3-4\nu}\frac{6cx_3\tilde{R}_i}{\tilde{R}^5}(\tilde{R}_j - 2\delta_{3j}\tilde{R}_3)\} \tag{4.270}$$

where $i, j = 1, 2, 3$ with no summation on double indices.

4.4.7 Melan's Fundamental Solution

In 1932, Melan derived a 2-D line load $\boldsymbol{F}$ solution within a half-plane, as shown in Fig. 4.9. In this case, $u_2 = 0$ and all functions do not depend on x_2. The 2-D traction-free boundary conditions are

$$\sigma_{13} = 0, \qquad \sigma_{33} = 0 \tag{4.271}$$

on $x_3 = 0$. The 2-D Papkovitch–Neuber potentials consist of two parts,

$$\Phi_0 = \psi_0 + \varphi_0, \quad \boldsymbol{\Phi} = \boldsymbol{\psi} + \boldsymbol{\varphi} \tag{4.272}$$

where

$$\psi_0 = \frac{cF_3}{2\pi\mu}\ln r, \quad \psi_i = -\frac{F_i}{2\pi\mu}\ln r \tag{4.273}$$

where $i = 1,2$ and

$$r = \sqrt{x_1^2 + (x_3 - c)^2} \tag{4.274}$$

The stress components now become

$$\sigma_{31} = \frac{\mu}{2(1-\nu)}\left\{(1-2\nu)(\Phi_{3,1} + \Phi_{1,3}) - \Phi_{0,13} - x_k\Phi_{k,13}\right\} \tag{4.275}$$

$$\sigma_{33} = \frac{\mu}{2(1-\nu)}\left\{2\nu\Phi_{1,1} + 2(1-\nu)\Phi_{3,3} - \Phi_{0,33} - x_k\Phi_{k,33}\right\} \tag{4.276}$$

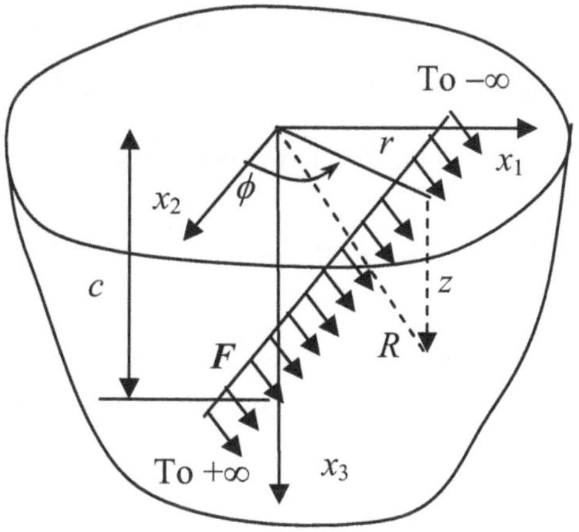

Figure 4.9 Melan problem: A 2-D line load F applied parallel to the x_2-axis and at a depth c

where $k=1, 2$. The boundary conditions on $x_3=0$ for the image function become

$$\tilde{A}=A\,,\qquad \frac{\partial \tilde{A}}{\partial x_1}=\frac{\partial A}{\partial x_1}\,,\qquad \frac{\partial \tilde{A}}{\partial x_3}=-\frac{\partial A}{\partial x_3} \tag{4.277}$$

Substitution of (4.275) and (4.276) into (4.271) gives two equations for $x_3=0$:

$$-x_1(\psi_{1,13}+\varphi_{1,13})-(\psi_{0,13}+\varphi_{0,13})+(1-2\nu)(\psi_{3,1}+\varphi_{3,1}+\psi_{1,3}+\varphi_{1,3})=0 \tag{4.278}$$

$$2\nu(\psi_{1,1}+\varphi_{1,1})+2(1-\nu)(\psi_{3,3}+\varphi_{3,3})-x_1(\psi_{1,33}+\varphi_{1,33})-(\psi_{0,33}+\varphi_{0,33})=0 \tag{4.279}$$

We can first set

$$\varphi_1=\tilde{\psi}_1 \tag{4.280}$$

With this setting, (4.278) and (4.279) become

$$\frac{1}{(1-2\nu)}(-\tilde{\psi}_{0,13}+\varphi_{0,13})-(\tilde{\psi}_{3,1}+\varphi_{3,1})=0 \tag{4.281}$$

$$4\nu(\tilde{\psi}_{1,1})+2(1-\nu)(-\tilde{\psi}_{3,3}+\varphi_{3,3})-2x_1\tilde{\psi}_{133}-(\tilde{\psi}_{0,33}+\varphi_{0,33})=0 \tag{4.282}$$

We assume φ_3 equals

$$\varphi_3=-\tilde{\psi}_3+\frac{1}{1-2\nu}(-\tilde{\psi}_{0,3}+\varphi_{0,3}) \tag{4.283}$$

for $x_3\geq 0$. We can see that (4.283) satisfies (4.281) exactly, and (4.282) becomes

$$\varphi_{0,33}=(3-4\nu)\tilde{\psi}_{0,33}+2(1-2\nu)x_1\tilde{\psi}_{1,33}-4\nu(1-2\nu)\tilde{\psi}_{1,1}+4(1-2\nu)(1-\nu)\tilde{\psi}_{3,3} \tag{4.284}$$

Similar to the previous arguments for seeking harmonic function and matching coefficients, we have

$$\begin{aligned}\varphi_{0,3}&=(3-4\nu)\tilde{\psi}_{0,3}+2(1-2\nu)(x_1\tilde{\psi}_{1,3}-x_3\tilde{\psi}_{1,1})\\&\quad-2(1-2\nu)^2\int_{x_3}^{\infty}\tilde{\psi}_{1,1}dx_3+4(1-2\nu)(1-\nu)\tilde{\psi}_3\end{aligned} \tag{4.285}$$

Substitution of (4.273) into (4.285) gives

$$\begin{aligned}\varphi_{0,3}&=\frac{F_1}{2\pi\mu}[2(1-2\nu)^2\tan^{-1}(\frac{x_1}{x_3+c})-2(1-2\nu)\frac{cx_1}{\tilde{r}^2}]\\&\quad+\frac{F_2}{2\pi\mu}[(3-4\nu)\frac{c(x_3+c)}{\tilde{r}^2}-4(1-\nu)(1-2\nu)\ln\tilde{r}]\end{aligned} \tag{4.286}$$

Integrating this equation results in

$$\begin{aligned}\varphi_0&=\frac{F_1}{2\pi\mu}\{2(1-2\nu)^2[x_1\ln\tilde{r}+(x_3+c)\tan^{-1}(\frac{x_1}{x_3+c})]+2(1-2\nu)c\tan^{-1}(\frac{x_1}{x_3+c})\}\\&\quad+\frac{F_2}{2\pi\mu}\{-4(1-\nu)(1-2\nu)[(x_3+c)\ln\tilde{r}-(x_3+c)-x_1\tan^{-1}(\frac{x_1}{x_3+c})]\\&\quad+(3-4\nu)c\ln\tilde{r}\}\end{aligned} \tag{4.287}$$

The final Papkovitch–Neuber potentials are

$$\Phi_1=-\frac{F_1}{2\pi\mu}\ln(r\tilde{r})\,, \tag{4.288}$$

$$\Phi_2 = -\frac{F_2}{2\pi\mu}\ln(\frac{r}{\tilde{r}}) + \frac{F_1}{2\pi\mu}[2(1-2\nu)\tan^{-1}(\frac{x_1}{x_3+c}) - \frac{2cx_1}{\tilde{r}^2}]$$
$$+\frac{F_2}{2\pi\mu}[-4(1-\nu)\ln\tilde{r} + \frac{2c(x_3+c)}{\tilde{r}^2}] \quad (4.289)$$

$$\Phi_0 = -\frac{cF_2}{2\pi\mu}\ln(r\tilde{r}) + \frac{F_1}{2\pi\mu}\{2(1-2\nu)^2[x_1\ln\tilde{r} + (x_3+c)\tan^{-1}(\frac{x_1}{x_3+c})]$$
$$+2(1-2\nu)c\tan^{-1}(\frac{x_1}{x_3+c})\} + \frac{F_2}{2\pi\mu}\{-4(1-\nu)(1-2\nu)[(x_3+c)\ln\tilde{r} \quad (4.290)$$
$$-(x_3+c) - x_1\tan^{-1}(\frac{x_1}{x_3+c})] + (3-4\nu)c\ln\tilde{r}\}$$

The associated displacements can also be given in a compact form as

$$u_\alpha^{(\beta)} = \frac{F_\beta}{8\pi\mu(1-\nu)}\{-[(3-4\nu)\ln r + \ln\tilde{r}]\delta_{\alpha\beta} + \frac{r_\alpha r_\beta}{r^2} + (3-4\nu)\frac{r_\alpha r_\beta(1-\delta_{\alpha\beta}) + \tilde{r}_\alpha\tilde{r}_\beta\delta_{\alpha\beta}}{\tilde{r}^2}$$
$$+\frac{2cx_3}{\tilde{r}^4}(\delta_{\alpha\beta}\tilde{r}^2 - 2\tilde{r}_\alpha\tilde{r}_\beta)(1-2\delta_{\alpha\beta}) + 4(1-\nu)(1-2\nu)\tan^{-1}(\frac{x_1}{x_3+c})$$
$$\times(\delta_{1\beta}\delta_{3\alpha} - \delta_{1\alpha}\delta_{3\beta}) - 4(1-\nu)(1-2\nu)(\ln\tilde{r})\delta_{\alpha\beta}\} \quad (4.291)$$

where α, β = 1, 3 with no summation on double indices.

4.5 HARMONIC FUNCTIONS AND INDIRECT METHOD

We have seen that the Papkovitch–Neuber displacement potentials have been useful in obtaining 3-D solutions and they are harmonic functions. The Lamé strain potential is also a harmonic function. It is instructive to give an overall introduction to harmonic functions. The following presentation mainly follows the Elasticity II class notes of John Dundurs of Northwestern University delivered in 1988 (Dundurs, 1988).

It was shown in Section 4.4.2 that the appropriate Papkovitch–Neuber displacement potential for Kelvin's fundamental point force solutions in full-space is $1/R$, where

$$R = (x_1^2 + x_2^2 + x_3^2)^{1/2} \quad (4.292)$$

Recall from (4.113) that it is straightforward to show that $1/R$ is harmonic. Actually, $1/R$ can be considered the granddaddy of many harmonic functions. For example, the partial derivative with respect to x_j and the Laplacian operator is commutative:

$$\frac{\partial}{\partial x_j}\nabla^2 f = \nabla^2(\frac{\partial f}{\partial x_j}) \quad (4.293)$$

Thus, it is obvious that

$$\frac{\partial}{\partial x_j}\nabla^2(\frac{1}{R}) = \nabla^2[\frac{\partial}{\partial x_j}(\frac{1}{R})] = \nabla^2(-\frac{x_j}{R^3}) = 0 \tag{4.294}$$

Therefore, we obtain a new harmonic function x_j/R^3. Following the same logic, we can take more derivatives of $1/R$ as:

$$\frac{\partial^2}{\partial x_j \partial x_i}\nabla^2(\frac{1}{R}) = \nabla^2[\frac{\partial^2}{\partial x_j \partial x_i}(\frac{1}{R})] = \nabla^2(-\frac{\delta_{ij}}{R^3} - \frac{3x_i x_j}{R^5}) = 0 \tag{4.295}$$

Thus, we obtain another set of harmonic functions. We can continue this process of differentiation and obtain an infinite set of harmonic functions. All these harmonic functions can be considered as point singularity of higher order since all these functions are singular at $R = 0$.

The next question that we ask is whether integration of $1/R$ leads to another family of harmonic functions. First, we integrate $1/R$ with respect to x_1:

$$\int \frac{dx_1}{R} = \ln[R + x_1] \tag{4.296}$$

Differentiation of the right-hand side of (4.296) gives

$$\frac{\partial}{\partial x_j}\ln[R + x_1] = \frac{1}{R + x_1}(\frac{x_j}{R} + \delta_{1j}) \tag{4.297}$$

Applying the second differentiation to (4.297) gives

$$\frac{\partial^2}{\partial x_j \partial x_i}\ln[R + x_1] = \frac{1}{(R + x_1)^2}[-(\frac{x_j}{R} + \delta_{1j})(\frac{x_i}{R} + \delta_{1i}) + (R + x_1)(\frac{\delta_{kj}}{R} - \frac{x_k x_j}{R^3})] \tag{4.298}$$

Thus, the Laplacian of $\ln(R+x_1)$ becomes

$$\nabla^2 \ln[R + x_1] = \frac{\partial^2}{\partial x_k \partial x_k}\ln[R + x_1] = \frac{1}{(R + x_1)^2}(-2 + 3 - 1) = 0 \tag{4.299}$$

Indeed, $\ln(R + x_1)$ is a harmonic function. Similarly, we can obtain other harmonic functions by integrating $1/R$ with respect to x_2 and x_3. Therefore, we have another series of harmonic functions:

$$\ln[R + x_1], \quad \ln[R + x_2], \quad \ln[R + x_3] \tag{4.300}$$

Similarly, we have the following series of

$$\ln[R - x_1], \quad \ln[R - x_2], \quad \ln[R - x_3] \tag{4.301}$$

Since integration can be interpreted as superposition, the Boussinesq potentials can be considered a superposition of $1/R$ along certain axes. If we add the line of singularities of the positive axis to the negative axis, we have

$$\ln(R + x_1) + \ln(R - x_1) = \ln(R^2 - x_1^2) = 2\ln(\sqrt{x_2^2 + x_3^2}) = 2\ln r \tag{4.302}$$

The last of (4.302) is independent of x_1 and is actually the line of singularity for the 2-D plane of x_2-x_3 space (i.e., for 2-D problems).

We can continue this integration (or superposition) process as

$$\int \ln[R + x_1]dx_1 = x_1\ln(R + x_1) - R \tag{4.303}$$

$$\int \ln[R - x_1]dx_1 = x_1\ln(R - x_1) + R \tag{4.304}$$

Thus, the new series of harmonic functions are

$$x_1 \ln(R+x_1)-R, \quad x_2 \ln(R+x_2)-R, \quad x_3 \ln(R+x_3)-R \tag{4.305}$$

and

$$x_1 \ln(R-x_1)+R, \quad x_2 \ln(R-x_2)+R, \quad x_3 \ln(R-x_3)+R \tag{4.306}$$

We can continue this integration process. For example, integrate (4.302) one more time gives

$$\int [x_1 \ln(R+x_1)-R]dx_1 = \frac{1}{4}\{(3x_1^2-R^2)\ln(R+x_1)-x_1^2-3x_1R+R^2\} \tag{4.307}$$

Following a similar procedure, we can simply replace x_1 by x_2 and x_3 in (4.306) for another series of harmonic functions:

$$(3x_1^2-R^2)\ln(R+x_1)-x_1^2-3x_1R+R^2 \tag{4.308}$$

$$(3x_2^2-R^2)\ln(R+x_2)-x_2^2-3x_2R+R^2 \tag{4.309}$$

$$(3x_3^2-R^2)\ln(R+x_3)-x_3^2-3x_3R+R^2 \tag{4.310}$$

We note that $\ln(R + x_1)$ is singular along the negative x_1-axis, and it has been used in (4.157) in solving Boussinesq's problem in elastic half-space. Similarly, $\ln(R - x_1)$ is singular along the positive x_1 axis. Therefore, it is a line of singularity as illustrated in Fig. 4.10. All these logarithmic functions can thus be considered as lines of singularities.

Since the integration and differentiation processes are interchangeable, the function on the right of (4.307) is again harmonic. We can also differentiate the right of (4.296) to get another series of harmonic function

$$\frac{\partial}{\partial x_2}\ln[R+x_1] = \frac{1}{R+x_1}\left(\frac{x_2}{R}\right) \tag{4.311}$$

Therefore, we have a new series of harmonic functions:

$$\frac{1}{R+x_1}\left(\frac{x_2}{R}\right), \quad \frac{1}{R+x_1}\left(\frac{x_3}{R}\right) \tag{4.312}$$

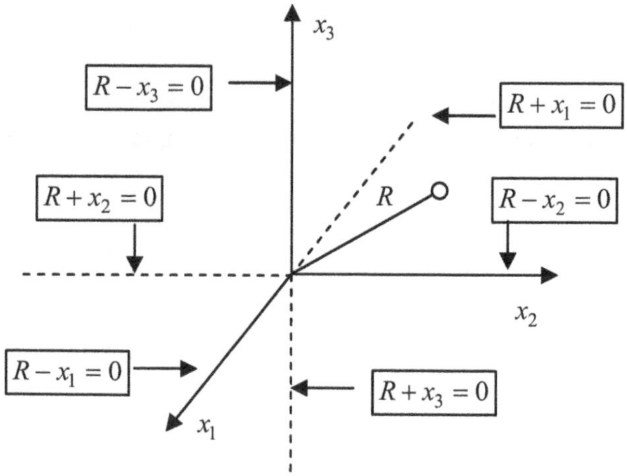

Figure 4.10 Lines of singularities

Further differentiation of (4.311) with respect to x_2 gives us a new series of harmonic functions:

$$\frac{\partial^2}{\partial x_2^2}\ln[R+x_1]=\frac{1}{R(R+x_1)}\{1-\frac{x_2^2}{R}[\frac{1}{R+x_1}+\frac{1}{R}]\} \tag{4.313}$$

Thus, other series of harmonic functions are:

$$\frac{1}{R(R+x_1)}\{1-\frac{x_2^2}{R}[\frac{1}{R+x_1}+\frac{1}{R}]\},\ \frac{1}{R(R+x_1)}\{1-\frac{x_3^2}{R}[\frac{1}{R+x_1}+\frac{1}{R}]\} \tag{4.314}$$

$$\frac{1}{R(R+x_2)}\{1-\frac{x_3^2}{R}[\frac{1}{R+x_2}+\frac{1}{R}]\},\ \frac{1}{R(R+x_2)}\{1-\frac{x_1^2}{R}[\frac{1}{R+x_2}+\frac{1}{R}]\} \tag{4.315}$$

$$\frac{1}{R(R+x_3)}\{1-\frac{x_1^2}{R}[\frac{1}{R+x_3}+\frac{1}{R}]\},\ \frac{1}{R(R+x_3)}\{1-\frac{x_2^2}{R}[\frac{1}{R+x_3}+\frac{1}{R}]\} \tag{4.316}$$

If we differentiate (4.311) with respect to x_3, we have another series of harmonic functions:

$$\frac{\partial^2}{\partial x_2 \partial x_3}\ln[R+x_1]=-\frac{x_2x_3}{R^2(R+x_1)}\{\frac{1}{R+x_1}+\frac{1}{R}\} \tag{4.317}$$

Thus, we have

$$\frac{x_2x_3}{R^2(R+x_1)}\{\frac{1}{R+x_1}+\frac{1}{R}\},\ \frac{x_1x_3}{R^2(R+x_2)}\{\frac{1}{R+x_2}+\frac{1}{R}\},\ \frac{x_2x_1}{R^2(R+x_3)}\{\frac{1}{R+x_3}+\frac{1}{R}\} \tag{4.318}$$

This process of integrating and differentiating harmonic functions provides a systematic way to generate infinite series of harmonic functions, as shown in Fig. 4.11.

Another family of harmonic functions is polynomials in x_1, x_2, and x_3:

$$\begin{aligned}&1,\\ &x_1, x_2, x_3,\\ &x_1x_2, x_3x_2, x_1x_3\\ &R^2-3x_1^2, R^2-3x_2^2, R^2-3x_3^2,\\ &\text{etc.}\end{aligned} \tag{4.319}$$

Of course, the method of generating harmonic functions discussed in this section does not allow us to solve a particular boundary value problem in 3-D elasticity. However, a table of 3-D displacement and stress fields resulting from these series of harmonic functions can be used to inspect the form of Papkovitch–Neuber displacement potential that could be used to cancel out stresses at certain boundary (say the surface of a half-space, etc.). This approach can be called the *indirect method.* Apparently, the original derivation of the Mindlin solution (Mindlin, 1936b) followed this indirect method. A lot of image functions (line of singularities, etc.) have been used to cancel out the stress on the surface of the half-space induced by the point force. Mindlin ingeniously recognized from his tables of biharmonic functions what kind of singular functions can be used for superposition. Apparently, J. Dundurs of Northwestern University also employed such an indirect method in obtaining many of his solution in 3-D elasticity. One example is the point force solution for two jointed half-spaces (Dundurs and Hetenyi, 1965). Such a method is also discussed in the books by Karasudhi (1991) and Barber (2002).

4.6 HARMONIC FUNCTIONS IN SPHERICAL COORDINATES

As shown by Little (1973) and Barber (2002), the family of harmonic functions in spherical coordinates can be done more systematically. In particular, the Laplacian in polar coordinate is

$$\nabla^2\phi = \frac{1}{R^2}\frac{\partial}{\partial R}(R^2\frac{\partial\phi}{\partial R}) + \frac{1}{R^2\sin\theta}\frac{\partial}{\partial\theta}(\sin\theta\frac{\partial\phi}{\partial\theta}) + \frac{1}{R^2\sin^2\theta}\frac{\partial^2\phi}{\partial\varphi^2} = 0 \quad (4.320)$$

Assuming a Fourier sine or cosine expansion of sin(mθ) and cos(mθ), we have

$$\phi = \sum_{m=0}^{\infty} f_m(R,\theta)\begin{cases}\cos(m\varphi)\\ \sin(m\varphi)\end{cases} \quad (4.321)$$

Substitution of (4.321) into (4.320) gives

$$\frac{\partial^2 f_m}{\partial R^2} + \frac{2}{R}\frac{\partial f_m}{\partial R} - \frac{m^2 f_m}{R^2\sin^2\theta} + \frac{1}{R^2}\frac{\partial^2 f_m}{\partial\theta^2} + \frac{\cot\theta}{R^2}\frac{\partial f_m}{\partial\theta} = 0 \quad (4.322)$$

Adopting the following change of variable,

$$x = \cos\theta \quad (4.323)$$

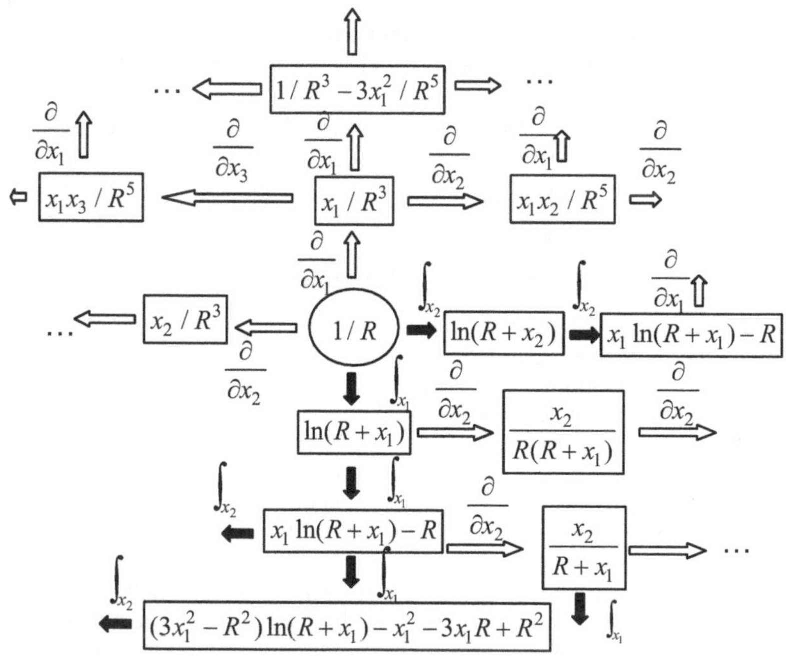

Figure 4.11 Illustration of generating a series of harmonic functions from 1/R

we have

$$\frac{\partial^2 f_m}{\partial R^2}+\frac{2}{R}\frac{\partial f_m}{\partial R}-\frac{m^2 f_m}{R^2(1-x^2)}-\frac{2x}{R^2}\frac{\partial f_m}{\partial x}+\frac{(1-x^2)}{R^2}\frac{\partial^2 f_m}{\partial x^2}=0 \tag{4.324}$$

We can further expand f_m in power series of R:

$$f_m(R,x)=\sum_{n=-\infty}^{\infty} R^n g_{mn}(x) \tag{4.325}$$

Substitution of (4.325) into (4.324) leads to the following Legendre equation (Abramowitz and Stegun, 1964):

$$(1-x^2)\frac{d^2 g_{mn}}{dx^2}-2x\frac{dg_{mn}}{dx}+[n(n+1)-\frac{m^2}{(1-x^2)}]g_{mn}=0 \tag{4.326}$$

The solution of (4.326) is the Legendre function. The Laplace equation and function has its origin in celestial mechanics (see the biography of A.M. Legendre at the end of this book). Note also that if we define

$$n=-p-1 \tag{4.327}$$

we find that (4.326) becomes

$$(1-x^2)\frac{d^2 g_{mn}}{dx^2}-2x\frac{dg_{mn}}{dx}+[p(p+1)-\frac{m^2}{(1-x^2)}]g_{mn}=0 \tag{4.328}$$

Thus, we have another series of harmonic functions. Finally, the harmonic functions or spherical harmonics for polar coordinates are (Hobson, 1955)

$$R^n P_n^m(\cos\theta)\begin{Bmatrix}\cos(m\varphi)\\ \sin(m\varphi)\end{Bmatrix},\quad R^{-n-1}P_n^m(\cos\theta)\begin{Bmatrix}\cos(m\varphi)\\ \sin(m\varphi)\end{Bmatrix} \tag{4.329}$$

The first few harmonics are

$$\phi_{11}=-R\sin\theta\begin{Bmatrix}\cos\varphi\\ \sin\varphi\end{Bmatrix},\quad \phi_{-11}=-\frac{1}{R^2}\sin\theta\begin{Bmatrix}\cos\varphi\\ \sin\varphi\end{Bmatrix} \tag{4.330}$$

$$\phi_{12}=-\frac{3}{2}R^2\sin(2\theta)\begin{Bmatrix}\cos\varphi\\ \sin\varphi\end{Bmatrix},\quad \phi_{-12}=-\frac{3}{2}\frac{1}{R^3}\sin(2\theta)\begin{Bmatrix}\cos\varphi\\ \sin\varphi\end{Bmatrix} \tag{4.331}$$

$$\phi_{22}=\frac{3}{2}R^2[1-\cos(2\theta)]\begin{Bmatrix}\cos 2\varphi\\ \sin 2\varphi\end{Bmatrix},\quad \phi_{-22}=\frac{3}{2}\frac{1}{R^3}[1-\cos(2\theta)]\begin{Bmatrix}\cos 2\varphi\\ \sin 2\varphi\end{Bmatrix} \tag{4.332}$$

Therefore, an infinite series of harmonic functions can be generated automatically.

4.7 HARMONIC FUNCTIONS IN CYLINDRICAL COORDINATES

Similar to spherical coordinates, harmonic functions in cylindrical coordinate can also be generated systematically (Little, 1973). In particular, the Laplace equation in cylindrical coordinates is

$$\nabla^2\phi=\frac{\partial^2\phi}{\partial r^2}+\frac{1}{r}\frac{\partial\phi}{\partial r}+\frac{1}{r^2}\frac{\partial^2\phi}{\partial\theta^2}+\frac{\partial^2\phi}{\partial z^2}=0 \tag{4.333}$$

Assuming a Fourier sine or cosine expansion of $\sin(m\theta)$ and $\cos(m\theta)$, we have

$$\phi = \sum_{m=0}^{\infty} \phi_m = \sum_{m=0}^{\infty} f_m(r,z) \begin{cases} \cos(m\theta) \\ \sin(m\theta) \end{cases} \tag{4.334}$$

Substitution of (4.334) into (4.333) gives

$$\frac{\partial^2 f_m}{\partial r^2} + \frac{1}{r}\frac{\partial f_m}{\partial r} - \frac{m^2 f_m}{r^2} + \frac{\partial^2 f_m}{\partial z^2} = 0 \tag{4.335}$$

A simple recursive formula can be derived as follows. Assume a harmonic function ϕ_m, and define ϕ_{m+1} as

$$\phi_{m+1} = f_{m+1}(r,z)\cos[(m+1)\theta] \tag{4.336}$$

We can show that it is also harmonic if

$$f_{m+1}(r,z) = \frac{\partial f_m}{\partial r} - \frac{m f_m}{r} \tag{4.337}$$

This equation provides a simple way to generate infinite series of harmonic functions. We can start with

$$f_0(r,z) = \frac{1}{8}(8z^2 - 24z^2r^2 + 3r^4) \tag{4.338}$$

Thus, the first three harmonic functions are

$$\phi_0 = \frac{1}{8}(8z^2 - 24z^2r^2 + 3r^4) \tag{4.339}$$

$$\phi_1 = \frac{3}{2}(-4z^2r + r^3)\begin{cases} \cos\theta \\ \sin\theta \end{cases} \tag{4.340}$$

$$\phi_2 = 3r^2 \begin{cases} \cos 2\theta \\ \sin 2\theta \end{cases} \tag{4.341}$$

In this way, infinite series of harmonic functions can be generated.

4.8 BIHARMONIC FUNCTIONS

In the last few sections, we have learned that there are infinite series of harmonic functions and their generations can be made systematically in Cartesian coordinates, in cylindrical coordinates and in spherical coordinates. Thus, Papkovitch–Neuber displacement potentials can be generated systematically although an indirect method has to be used to solve any practical problem. Sometimes, it may be more preferable to solve 3-D elasticity problems by Love's strain potential (such as Kelvin's problem) and the Galerkin vector. Since both Love's strain potential and the Galerkin vector satisfy biharmonic equations, it is informative to discuss the generation of biharmonic functions.

Earlier in this chapter, we mentioned one particular form of the Almansi theorems.

Theorem 1: If ϕ_1 and ϕ_2 are harmonic functions,

$$\varphi = x_1\phi_1 + \phi_2 \tag{4.342}$$

is biharmonic. The proof was given by Fung (1965).

Theorem 2: If ϕ_1 and ϕ_2 are harmonic functions,

$$\varphi = (R^2 - R_0^2)\phi_1 + \phi_2 \tag{4.343}$$

is biharmonic, and R_0 is an arbitrary constant. The proof was given by Fung (1965).

For cylindrical coordinates, by applying the Almansi theorem, the following functions are biharmonic (Fung, 1965):

$$\varphi = z\psi_n, \quad \varphi = (z^2 + r^2)\psi_n \tag{4.344}$$

where $n = 2, 3, 4, \ldots$ and

$$\psi_2 = z^2 - \frac{1}{2}r^2, \tag{4.345}$$

$$\psi_3 = z^3 - \frac{3}{2}zr^2, \tag{4.346}$$

$$\psi_4 = z^4 - 3z^2r^2 + \frac{3}{8}r^4, \tag{4.347}$$

$$\psi_5 = z^5 - 5z^3r^2 + \frac{15}{8}zr^4, \tag{4.348}$$

$$\psi_6 = z^6 - \frac{15}{2}z^4r^2 + \frac{45}{8}z^2r^4 - \frac{5}{16}r^6 \tag{4.349}$$

For example, for the special case that $R_0 = 0$ and $\phi_2 = 0$, and $\phi_1 = 1/R$, we have

$$\varphi = R^2\frac{1}{R} = R \tag{4.350}$$

This is the Love's strain potential for Kelvin's solution discussed in Section 4.4.2.2. For the infinite series of harmonic functions given in the last section, the corresponding infinite series of biharmonic functions can also be generated.

Other applications of biharmonic equations in engineering can be found in Selvadurai (2000), and the most general review article on biharmonic functions used in elasticity is given by Meleshko (2003).

4.9 MUKI'S FORMULATION IN CYLINDRICAL COORDINATES

Three-dimensional problems in cylindrical coordinates were formulated by Muki (1960), and his formalism will be introduced in this section. As discussed earlier in Section 4.2.1, the Galerkin vector in terms of cylindrical coordinates does not lead to uncoupling governing equations of biharmonic type. That is, the cylindrical components of the Galerkin vector do not satisfy the biharmonic equation (see also Problem 4.6). Muki (1960) proposed a combination of the Galerkin vector plus the solenodial part of the Helmholtz vector to represent the displacement vector and was able to uncouple the equilibrium equations in displacements into two differential equations. In particular, Muki (1960) used the z-component of the Galerkin vector and z-component of the irrotational vector component of Helmholtz. These vector components satisfy the biharmonic equation and the

Laplace equation, respectively. Then, Hankel transform is introduced to solve boundary value problems in cylindrical coordinates. As an example, the Boussinesq problem will be reconsidered using Muki's potentials and the Hankel transform.

4.9.1 Muki's Vector Potentials

The equilibrium equations in cylindrical coordinates have been derived and given in (1.99) to (1.101) in Chapter 1 as problems for the reader. For the case of zero body force, we have

$$\nabla^2 u_r + \frac{1}{1-2\nu}\frac{\partial e}{\partial r} - \frac{u_r}{r^2} - \frac{2}{r^2}\frac{\partial u_\phi}{\partial \phi} = 0 \tag{4.351}$$

$$\nabla^2 u_\phi + \frac{1}{1-2\nu}\frac{1}{r}\frac{\partial e}{\partial \phi} - \frac{u_\phi}{r^2} + \frac{2}{r^2}\frac{\partial u_r}{\partial \phi} = 0 \tag{4.352}$$

$$\nabla^2 u_z + \frac{1}{1-2\nu}\frac{\partial e}{\partial r} = 0 \tag{4.353}$$

$$e = \boldsymbol{\nabla} \bullet \boldsymbol{u} = \frac{\partial u_r}{\partial r} + \frac{1}{r}u_r + \frac{1}{r}\frac{\partial u_\phi}{\partial \phi} + \frac{\partial u_z}{\partial z} \tag{4.354}$$

where ν is Poisson's ratio and the Laplacian operator has been defined in (1.72).

In order to uncouple the equilibrium equations in cylindrical coordinates, we can propose the following form of vector potentials:

$$\boldsymbol{u} = 2(1-2\nu)\nabla^2 \boldsymbol{G} - \boldsymbol{\nabla}(\boldsymbol{\nabla} \bullet \boldsymbol{G}) + \boldsymbol{\nabla} \times \boldsymbol{A} \tag{4.355}$$

Note that $\boldsymbol{G}$ is the Galerkin vector given in (4.30) and $\boldsymbol{A}$ is the irrotational part of the Helmholtz decomposition given in (4.2). However, Muki (1960) showed that only the z-components of these vectors were needed (i.e., $G_z = \Phi$ and $A_z = 2\psi$). That is, Muki (1960) proposed the following vector potentials:

$$u_r = -\frac{\partial^2 \Phi}{\partial z \partial r} + \frac{2}{r}\frac{\partial \psi}{\partial \phi} \tag{4.356}$$

$$u_\phi = -\frac{1}{r}\frac{\partial^2 \Phi}{\partial z \partial \phi} - 2\frac{\partial \psi}{\partial r} \tag{4.357}$$

$$u_z = 2(1-\nu)\nabla^2 \Phi - \frac{\partial^2 \Phi}{\partial z^2} \tag{4.358}$$

These displacement potentials will be called Muki's (1960) displacement potentials. The governing equations for these vector potentials are

$$\nabla^4 \Phi = 0, \quad \nabla^2 \psi = 0 \tag{4.359}$$

The validity of (4.356) to (4.358) can be demonstrated by direct substitution of them into (4.351) to (4.353). The corresponding stress field is:

$$\sigma_{rr} = 2\mu\{\frac{\partial}{\partial z}(\nu\nabla^2\Phi - \frac{\partial^2 \Phi}{\partial r^2}) + \frac{2}{r}\frac{\partial^2 \psi}{\partial \phi \partial r} - \frac{2}{r^2}\frac{\partial \psi}{\partial \phi}\} \tag{4.360}$$

$$\sigma_{\phi\phi} = 2\mu\{\frac{\partial}{\partial z}(\nu\nabla^2\Phi - \frac{1}{r}\frac{\partial \Phi}{\partial r} - \frac{1}{r^2}\frac{\partial^2 \Phi}{\partial \phi^2}) - \frac{2}{r}\frac{\partial \psi}{\partial \phi \partial r} + \frac{2}{r^2}\frac{\partial \psi}{\partial \phi}\} \tag{4.361}$$

$$\sigma_{zz} = 2\mu \frac{\partial}{\partial z}[(2-\nu)\nabla^2 \Phi - \frac{\partial^2 \Phi}{\partial z^2}] \tag{4.362}$$

$$\sigma_{\phi z} = 2\mu\{\frac{1}{r}\frac{\partial}{\partial \phi}[(1-\nu)\nabla^2 \Phi - \frac{\partial^2 \Phi}{\partial z^2}] - \frac{\partial^2 \psi}{\partial r \partial z}\} \tag{4.363}$$

$$\sigma_{zr} = 2\mu\{\frac{\partial}{\partial r}[(1-\nu)\nabla^2 \Phi - \frac{\partial^2 \Phi}{\partial z^2}] + \frac{\partial^2 \psi}{r \partial \phi \partial z}\} \tag{4.364}$$

$$\sigma_{r\phi} = 2\mu\{\frac{\partial^2}{r\partial\phi\partial z}(\frac{\Phi}{r} - \frac{\partial \Phi}{\partial r}) - 2\frac{\partial^2 \psi}{\partial r^2} - \frac{\partial^2 \psi}{\partial z^2}\} \tag{4.365}$$

4.9.2 Method of Solution by the Hankel Transform

For general 3-D solutions, we can expand the tangential dependence of Muki's (1960) displacement potentials in cosine and sine series as

$$\Phi(r,\phi,z) = \sum_{m=0}^{\infty}[\Phi_m(r,z)\cos m\phi - \bar{\Phi}_m(r,z)\sin m\phi] \tag{4.366}$$

$$\psi(r,\phi,z) = \sum_{m=0}^{\infty}[\psi_m(r,z)\sin m\phi + \bar{\psi}_m(r,z)\cos m\phi] \tag{4.367}$$

It is clear from (4.366) and (4.367) that $m = 0$ and $\psi = 0$ corresponds to the case axisymmetric problems (Love's displacement potential is recovered), such as Boussinesq's problem considered in Section 4.4.3, and $m = 1$ corresponds to antisymmetric problems, such as Cerruti's problem considered earlier in Section 4.4.4.

Without loss of generality, we now only consider the first term in the series expansions in (4.366) and (4.367). In particular, substitution of (4.366) and (4.367) gives

$$\nabla_m^4 \Phi_m = (\frac{\partial^2}{\partial r^2} + \frac{1}{r}\frac{\partial}{\partial r} - \frac{m^2}{r^2} + \frac{\partial^2}{\partial z^2})^2 \Phi_m = 0 \tag{4.368}$$

$$\nabla_m^2 \psi_m = (\frac{\partial^2}{\partial r^2} + \frac{1}{r}\frac{\partial}{\partial r} - \frac{m^2}{r^2} + \frac{\partial^2}{\partial z^2})\psi_m = 0 \tag{4.369}$$

We can now apply the Hankel transform to these equations as

$$\int_0^\infty r\nabla_m^4 \Phi_m J_m(\xi r)dr = (\frac{d^2}{dz^2} - \xi^2)^2 \int_0^\infty r\Phi_m J_m(\xi r)dr = 0 \tag{4.370}$$

$$\int_0^\infty r\nabla_m^2 \psi_m J_m(\xi r)dr = (\frac{d^2}{dz^2} - \xi^2) \int_0^\infty r\psi_m J_m(\xi r)dr = 0 \tag{4.371}$$

In obtaining (4.370) and (4.371), we have used the following Bessel equation (Abramowitz and Stegun, 1964):

$$\frac{d^2}{dr^2}J_m(\xi r) + \frac{1}{r}\frac{d}{dr}J_m(\xi r) + (\xi^2 - \frac{m^2}{r})J_m(\xi r) = 0 \tag{4.372}$$

The Bessel equation has its origin in planetary and stellar motions (see biography section), but it also appears naturally as the solution of the Laplace equation in cylindrical coordinates. We can define the Hankel transform of these displacement potentials as (Sneddon, 1951):

$$G_m(\xi, z) = \int_0^\infty r\Phi_m J_m(\xi r)dr \tag{4.373}$$

$$H_m(\xi, z) = \int_0^\infty r\psi_m J_m(\xi r)dr \tag{4.374}$$

With these Hankel transforms, the differentiation with respect to r in (4.368) and (4.369) becomes an algebraic operation in parameter ξ in the transform space. Thus, the following ordinary differential equations result:

$$(\frac{d^2}{dz^2} - \xi^2)^2 G_m(\xi, z) = 0 \tag{4.375}$$

$$(\frac{d^2}{dz^2} - \xi^2) H_m(\xi, z) = 0 \tag{4.376}$$

The general solutions of (4.375) and (4.376) are

$$G_m(\xi, z) = (A_m + B_m z)e^{\xi z} + (C_m + D_m z)e^{-\xi z} \tag{4.377}$$

$$H_m(\xi, z) = E_m e^{\xi z} + F_m e^{-\xi z} \tag{4.378}$$

where these unknown constants have to be determined by boundary conditions. Once they are obtained, the inverse Hankel transform can be applied as (Sneddon, 1951)

$$\Phi_m(r, z) = \int_0^\infty \xi G_m(\xi, z) J_m(\xi r)d\xi \tag{4.379}$$

$$\psi_m(r, z) = \int_0^\infty \xi H_m(\xi, z) J_m(\xi r)d\xi \tag{4.380}$$

Next, we can express all displacements and stresses in terms of G_m and H_m, and evaluate their inverse either analytically or numerically. For example, the displacement in terms of the m term in the series solution is

$$u_z = [2(1-\nu)\nabla_m^2 \Phi_m - \frac{\partial^2 \Phi_m}{\partial z^2}]\cos m\phi \tag{4.381}$$

We can multiply both sides of (4.358) by $rJ_m(\xi r)$ and integrate from 0 to ∞ to get

$$\int_0^\infty r u_z J_m(\xi r)dr = [(1-2\nu)\frac{d^2 G_m}{dz^2} - 2(1-\nu)\xi^2 G_m]\cos m\phi \tag{4.382}$$

Inversion of this Hankel transform gives

$$u_z = \int_0^\infty [(1-2\nu)\frac{d^2 G_m}{dz^2} - 2(1-\nu)\xi^2 G_m]\xi \cos m\phi J_m(\xi r)d\xi \tag{4.383}$$

Similarly, all other displacement and stress components can be expressed in inverse Hankel transforms as (Muki, 1960)

$$\frac{u_r}{\cos m\phi} + \frac{u_\phi}{\sin m\phi} = \int_0^\infty [\frac{dG_m}{dz} + 2H_m]\xi^2 J_{m+1}(\xi r)d\xi \tag{4.384}$$

$$\frac{u_r}{\cos m\phi} - \frac{u_\phi}{\sin m\phi} = -\int_0^\infty [\frac{dG_m}{dz} - 2H_m]\xi^2 J_{m-1}(\xi r)d\xi \tag{4.385}$$

$$\frac{\sigma_{rr}}{\cos m\phi} + \frac{\sigma_{\phi\phi}}{\cos m\phi} = 2\mu\int_0^\infty [2\nu\frac{d^2G_m}{dz^2} + (1-2\nu)\xi^2\frac{dG_m}{dz}]\xi J_m(\xi r)d\xi \tag{4.386}$$

$$\frac{\sigma_{\phi z}}{\sin m\phi} + \frac{\sigma_{zr}}{\cos m\phi} = 2\mu\int_0^\infty [\nu\frac{d^2G_m}{dz^2} + (1-\nu)\xi^2 G_m + \frac{dH_m}{dz}]\xi^2 J_{m+1}(\xi r)d\xi \tag{4.387}$$

$$\frac{\sigma_{\phi z}}{\sin m\phi} - \frac{\sigma_{zr}}{\cos m\phi} = 2\mu\int_0^\infty [\nu\frac{d^2G_m}{dz^2} + (1-\nu)\xi^2 G_m - \frac{dH_m}{dz}]\xi^2 J_{m-1}(\xi r)d\xi \tag{4.388}$$

$$\frac{\sigma_{rr}}{\cos m\phi} + \frac{2\mu u_r}{r\cos m\phi} + \frac{2\mu u_\phi}{r\sin m\phi} = 2\mu\int_0^\infty [\nu\frac{d^3G_m}{dz^3} + (1-\nu)\xi^2 G_m]\xi J_m(\xi r)d\xi \tag{4.389}$$

$$\frac{\sigma_{r\phi}}{\sin m\phi} + \frac{2\mu m u_r}{r\cos m\phi} + \frac{2\mu m u_\phi}{r\sin m\phi} = 2\mu\int_0^\infty H_m\xi^3 J_m(\xi r)d\xi \tag{4.390}$$

In deriving these expressions, we have used the following identities (Watson, 1952):

$$\int_0^\infty r[\frac{1}{r}\frac{d}{dr}(r\psi_m)]J_m(\xi r)dr = \xi\int_0^\infty r\psi_m J_{m+1}(\xi r)dr \tag{4.391}$$

$$\int_0^\infty r[r^{m-1}\frac{d}{dr}(r^{1-m}\psi_m)]J_m(\xi r)dr = -\xi\int_0^\infty r\psi_m J_{m-1}(\xi r)dr \tag{4.392}$$

$$\frac{d}{dr}\{r^m J_m(\xi r)\} = \xi r^m J_{m-1}(\xi r) \tag{4.393}$$

$$\frac{d}{dr}\{r^{-m} J_m(\xi r)\} = -\xi r^{-m} J_{m+1}(\xi r) \tag{4.394}$$

Once we know the contribution from each m, the displacement components can now be summed from 0 to ∞ as

$$u_r = \frac{1}{2}\sum_{m=0}^\infty [U_{m+1}(r,z) - V_{m-1}(r,z)]\cos m\phi \tag{4.395}$$

$$u_\phi = \frac{1}{2}\sum_{m=0}^\infty [U_{m+1}(r,z) + V_{m-1}(r,z)]\sin m\phi \tag{4.396}$$

$$u_z = \sum_{m=0}^\infty [\int_0^\infty \{(1-2\nu)\frac{d^2G_m}{dz^2} - 2(1-\nu)\xi^2 G_m\}\xi J_m(\xi r)d\xi]\cos m\phi \tag{4.397}$$

where

$$U_{m+1}(r,z) = \int_0^\infty [\frac{dG_m}{dz} + 2H_m]\xi^2 J_{m+1}(\xi r)d\xi \tag{4.398}$$

$$V_{m-1}(r,z) = \int_0^\infty [\frac{dG_m}{dz} - 2H_m]\xi^2 J_{m-1}(\xi r)d\xi \tag{4.399}$$

Similarly, the stress components are

$$\frac{\sigma_{rr}}{2\mu}=\sum_{m=0}^{\infty}[\int_0^{\infty}\{\nu\frac{d^3G_m}{dz^3}+(1-\nu)\xi^2\frac{dG_m}{dz}\}\xi J_m(\xi r)d\xi-\frac{(m+1)}{2r}U_{m+1} \\ -\frac{(m-1)}{2r}V_{m-1}]\cos m\phi \tag{4.400}$$

$$\frac{\sigma_{\phi\phi}}{2\mu}=\sum_{m=0}^{\infty}[\int_0^{\infty}\{\frac{d^3G_m}{dz^3}-\xi^2\frac{dG_m}{dz}\}\xi J_m(\xi r)d\xi+\frac{(m+1)}{2r}U_{m+1}+\frac{(m-1)}{2r}V_{m-1}]\cos m\phi \tag{4.401}$$

$$\frac{\sigma_{zz}}{2\mu}=\sum_{m=0}^{\infty}[\int_0^{\infty}\{(1-\nu)\frac{d^3G_m}{dz^3}-(2-\nu)\xi^2\frac{dG_m}{dz}\}\xi J_m(\xi r)d\xi]\cos m\phi \tag{4.402}$$

$$\frac{\sigma_{\phi z}}{2\mu}=\frac{1}{2}\sum_{m=0}^{\infty}[\int_0^{\infty}\{\nu\frac{d^2G_m}{dz^2}+(1-\nu)\xi^2G_m+\frac{dH_m}{dz}\}\xi^2J_{m+1}(\xi r)d\xi \\ +\int_0^{\infty}\{\nu\frac{d^2G_m}{dz^2}+(1-\nu)\xi^2G_m-\frac{dH_m}{dz}\}\xi^2J_{m-1}(\xi r)d\xi]\sin m\phi \tag{4.403}$$

$$\frac{\sigma_{zr}}{2\mu}=\frac{1}{2}\sum_{m=0}^{\infty}[\int_0^{\infty}\{\nu\frac{d^2G_m}{dz^2}+(1-\nu)\xi^2G_m+\frac{dH_m}{dz}\}\xi^2J_{m+1}(\xi r)d\xi \\ -\int_0^{\infty}\{\nu\frac{d^2G_m}{dz^2}+(1-\nu)\xi^2G_m-\frac{dH_m}{dz}\}\xi^2J_{m-1}(\xi r)d\xi]\cos m\phi \tag{4.404}$$

$$\frac{\sigma_{r\phi}}{2\mu}=\sum_{m=0}^{\infty}[\int_0^{\infty}H_m\xi^3J_m(\xi r)d\xi-\frac{(m+1)}{2r}U_{m+1}+\frac{(m-1)}{2r}V_{m-1}]\sin m\phi \tag{4.405}$$

For axisymmetric cases, we have both $H_m = 0$ and $m = 0$, and obviously, the Love displacement discussed in Section 4.2.4 is recovered. Therefore, it can be concluded that the Hankel transform technique can be applied equally well to the Love potential. For problems formulated in Cartesian coordinate, we can apply Fourier transform in a similar manner. The details are referred to in Sneddon (1951).

4.9.3 Boussinesq Solution by Hankel Transform

In this section, Boussinesq's problem will be reconsidered by using the Hankel transform. For this axisymmetric case, $H_m = 0$ and $m = 0$, and only the Love displacement potential is needed.

Consider the case that surface tractions are applied on the surface of a half-space $z = 0$:

$$\sigma_{zz}=f(r),\quad \sigma_{rz}=g(r) \tag{4.406}$$

In view of (4.402), the normal traction boundary condition in the Hankel transform space is

$$2\mu[(1-\nu)\frac{d^3G_0}{dz^3}-(2-\nu)\xi^2\frac{dG_0}{dz}]_{z=0}=\bar{f}_0(\xi)=\int_0^{\infty}rf(r)J_0(\xi r)d\xi \tag{4.407}$$

Similarly, the shear traction boundary condition in the Hankel transform space is

$$2\mu[\xi^3(1-\nu)G_0+\nu\xi\frac{d^2G_0}{dz^2}]_{z=0}=\bar{g}_1(\xi)=\int_0^\infty rg(r)J_1(\xi r)d\xi \tag{4.408}$$

Substitution of (4.377) into (4.407) gives

$$\xi A+(1-2\nu)B=\frac{1}{2\mu\xi^2}\bar{f}_0(\xi) \tag{4.409}$$

Substitution of (4.377) into (4.408) leads to

$$\xi A-2\nu B=\frac{1}{2\mu\xi^2}\bar{g}_1(\xi) \tag{4.410}$$

The solutions of (4.409) and (4.410) are

$$B=\frac{1}{2\mu\xi^2}(\bar{f}_0+\bar{g}_1) \tag{4.411}$$

$$A=\frac{1}{2\mu\xi^3}[2\nu\bar{f}_0+(1-2\nu)\bar{g}_1] \tag{4.412}$$

Back substitution of A and B into (4.377) gives

$$G_0(\xi,z)=\frac{1}{2\mu\xi^3}[2\nu\bar{f}_0+(1-2\nu)\bar{g}_1+z\xi(\bar{f}_0-\bar{g}_1)]e^{-\xi z} \tag{4.413}$$

As shown in Fig. 4.12, the Boussinesq problem can be modeled by a uniform $f(r)$ within a circular patch with the radius a approaching zero as

$$\sigma_{zz}=-\lim_{a\to 0}\frac{P}{\pi a^2},\qquad \sigma_{rz}=0 \tag{4.414}$$

Thus, the Hankel transform of (4.414) gives

$$\frac{1}{2\mu}\bar{f}_0=\lim_{a\to 0}\int_0^a\frac{P}{\pi a^2}rJ_0(\xi r)dr=\lim_{a\to 0}\frac{P}{\pi a^2}[\frac{a}{\xi}J_1(\xi a)]=-\frac{P}{2\pi} \tag{4.415}$$

$$\bar{g}_1=0 \tag{4.416}$$

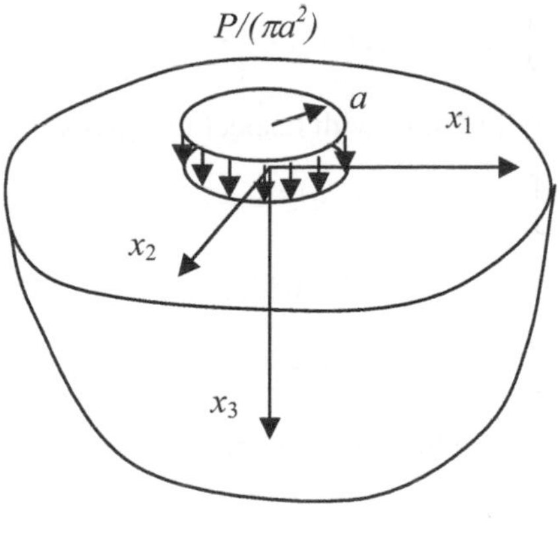

Figure 4.12 Boussinesq problem modeled as uniform circular vertical loads

Now all stress components can be expressed in terms of the inverse Hankel transform as

$$\sigma_{zz} = -\int_0^\infty \frac{P\xi}{2\pi}(z\xi+1)e^{-z\xi}J_0(\xi r)d\xi \tag{4.417}$$

$$\sigma_{zr} = -\int_0^\infty \frac{P\xi}{2\pi}z\xi e^{-z\xi}J_1(\xi r)d\xi \tag{4.418}$$

$$\sigma_{\phi\phi} = -\int_0^\infty \frac{P\xi}{2\pi}e^{-z\xi}J_0(\xi r)d\xi - \int_0^\infty \frac{P}{2\pi}[(1-2\nu)-z\xi]e^{-z\xi}\frac{J_1(\xi r)}{r}d\xi \tag{4.419}$$

$$\sigma_{rr} = -\int_0^\infty \frac{P\xi}{2\pi}(1-z\xi)e^{-\xi z}J_0(\xi r)d\xi + \int_0^\infty \frac{P}{2\pi}[(1-2\nu)-z\xi]e^{-\xi z}\frac{J_1(\xi r)}{r}d\xi \tag{4.420}$$

$$u_r = -\int_0^\infty \frac{P}{2\pi}[(1-2\nu)-z\xi]e^{-\xi z}J_1(\xi r)d\xi \tag{4.421}$$

$$u_z = \int_0^\infty \frac{P}{2\pi}[2(1-\nu)+z\xi]e^{-\xi z}J_0(\xi r)d\xi \tag{4.422}$$

$$u_\phi = 0 \tag{4.423}$$

To evaluate this inverse, we note the following formula (6.621 of Gradshteyn and Ryzhik, 1980):

$$\begin{aligned}\int_0^\infty x^n e^{-\lambda z}J_m(\alpha x)dx &= \frac{(n-m)!}{(\lambda^2+\alpha^2)^{1/[2(n+1)]}}P_n^m\left\{\frac{\lambda}{(\lambda^2+\alpha^2)^{1/2}}\right\} \quad m \le n \\ &= \frac{(n+m)!}{(\lambda^2+\alpha^2)^{1/[2(n+1)]}}P_n^m\left\{\frac{\lambda}{(\lambda^2+\alpha^2)^{1/2}}\right\} \quad m > n\end{aligned} \tag{4.424}$$

where the associated Legendre function becomes Legendre polynomials for $m = 0$. For $n = 1$, we further have (Spiegel, 1968)

$$P_1^0(x) = P_1(x) = x \tag{4.425}$$

Applying (4.424) and (4.425), we have

$$\int_0^\infty \xi e^{-\xi z}J_0(\xi r)d\xi = \frac{z}{(r^2+z^2)^{3/2}} \tag{4.426}$$

Taking the differentiation of (4.426) with respect to z gives

$$\int_0^\infty \xi^2 e^{-\xi z}J_0(\xi r)d\xi = \frac{2z^2-r^2}{(r^2+z^2)^{5/2}} \tag{4.427}$$

Considering the differentiation of (4.426) with respect to r, we obtain

$$\int_0^\infty \xi^2 e^{-\xi z}J_1(\xi r)d\xi = \frac{3zr}{(r^2+z^2)^{5/2}} \tag{4.428}$$

Substitution of (4.426)–(4.428) into (4.417) and (4.418) gives

$$\sigma_{zz} = -\frac{P}{2\pi}\int_0^\infty (z\xi^2+\xi)e^{-z\xi}J_0(\xi r)d\xi$$
$$= -\frac{P}{2\pi}\left[\frac{(2z^2-r^2)z}{(r^2+z^2)^{5/2}}+\frac{z}{(r^2+z^2)^{3/2}}\right] \tag{4.429}$$
$$= -\frac{3P}{2\pi}\left(\frac{z^3}{R^5}\right)$$

$$\sigma_{zr} = \int_0^\infty \frac{P\xi}{2\pi} z\xi e^{-z\xi}J_1(\xi r)d\xi = \frac{Pz}{2\pi}\frac{3zr}{(r^2+z^2)^{5/2}} = -\frac{3P}{2\pi}\left(\frac{rz^2}{R^5}\right) \tag{4.430}$$

where

$$R = (r^2+z^2)^{1/2} \tag{4.431}$$

These equations are of course equal to those given in (4.153). Other stress and displacement components of Boussinesq's problem given in (4.152) and (4.153) can be obtained by carrying out the integration in (4.419) to (4.422) (see Problem 4.17). This completes the evaluation of the Boussinesq solution by the Hankel transform. According to Little (1973), the Hankel transform calculations discussed in this section were first given by Lamb (1902). Apparently, Lamb (1902) was the first to introduce the use of the Hankel transform in 3-D elasticity problems (Goodman, 1974).

4.10 SUMMARY AND FURTHER READING

4.10.1 Summary

In this chapter, we introduce 3-D elasticity through both displacement and stress formulations. For the displacement approach, we present the Helmholtz decomposition, the Galerkin vector, and the Papkovitch–Neuber displacement potential. The special cases of Lamé's strain potential for incompressible solids and Love's displacement potential for cylindrical solids are also included in the discussion. For stress approach, we summarize the Beltrami and Beltrami–Schaefer stress functions, the Maxwell stress functions, the Morera stress function, and other combinations of Beltrami stress functions which are not covered in most elasticity textbooks. Various 3-D elasticity problems are presented, including hollow sphere subject to internal and external pressures, Kelvin's problem, Boussinesq's fundamental solution, Cerruti's fundamental solution, Mindlin's fundamental solution, Lorentz's fundamental solution, and Melan's fundamental solution. A systematic method of generating harmonic and biharmonic functions is introduced and its use as an "indirect method" is discussed.

4.10.2 Further reading

4.10.2.1 General Method of Solutions for 3-D Elasticity

Chapter 6 of Westergaard (1952) is devoted entirely to the Galerkin vector approach for 3-D elasticity. Chapter 13 of Chou and Pagano gives a comprehensive introduction to both displacement and stress approaches to solve 3-D elasticity problems. Although a systematic approach to solving 3-D problems is not presented, many results of 3-D elasticity are included in Love (1944). The book by Little (1973) also provides a concise introduction to 3-D elasticity. The nearly 500-pages book of Luré (1964) is devoted entirely to three-dimensional elasticity problems. Although the book introduced both displacement and stress approaches, the majority of the problems considered in the book are solved by using the Papkovitch–Neuber displacement potential. It should be mentioned that the Sadowsky and Sternberg (1949) stress components in curvilinear coordinate resulting from the Papkovitch–Neuber displacement potential are also included. These expressions may be used for very special problems. The book compiled some advanced topics in 3-D elasticity. Wang (2002) and Wang et al. (2008) discuss general solution of elasticity with applications. Some 3-D problems can be found in Kupradze (1979). The short review article by Goodman (1974) also discussed the important development of potential theory in 3-D elasticity, including the complex potential methods.

4.10.2.2 Integral Transform in Solving 3-D Problems

According to Goodman (1974), integral transform methods were introduced to 3-D elasticity by Lamb (1902). Only the Hankel transform is introduced in the present chapter. The mixed boundary value problems in 3-D elasticity can be solved using the systematic approach by taking the Hankel transform (e.g., Muki, 1960; Sneddon, 1951; Chan et al., 1974), leading to a pair of dual integral equations for inner and outer regions (Keer, 1967; Gladwell, 1980). For example, penny-shaped crack problems can be considered by using this approach (Westmann, 1965a,b). The Hankel and Fourier approaches can also be used to solve half-space problems of poroelasticity (e.g., Senjuntichai, 1994a,b; Rajapakse and Senjuntichai, 1993). Due to space limitations, such approach will not be discussed here but will be covered briefly in Chapter 8 when we deal with poroelasticity.

4.10.2.3 General Method of Solutions for Circular Cylinders

Because of its application of the uniaxial compression test for rocks and the triaxial test for both soils and rocks, solutions for solid circular cylinders have been found useful. Early papers include those of Filon (1902), Lure (1964), Roberts and Keer (1987a,b), and Watanabe (1996), to name a few. The most comprehensive approach to solving finite circular cylinder problems is given by Chau and Wei (2000). Diffuse mode bifurcations (such as buckling, barrelling, and surface instabilities) of circular cylinders under compressions are considered in a series of papers by Chau (1992, 1993, 1995a), by Bardet and Iai (2002), and by Sulem and Vardoulakis (1990). The effect of end constraint on the non-uniform stress in solid cylinders is given by Chau (1997) and Wei and Chau (2009). The stress analysis for the double punch test on solid cylinders is considered by Wei and Chau (2000). For axial point load tests, the stress analysis is given by Wei et al. (1999) and Wei and Chau (2002). For diametral point load tests, the stress analysis is given by Chau (1998c) and Chau and Wei (2001a).

4.10.2.4 General Method of Solutions for Spheres

The general method of solution for isotropic spheres subject to arbitrary traction and prescribed displacement is given by Lure (1964). The application to the diametral point load test on spheres is given by Chau and Wei (1999). Diametral impacts on spheres was considered by Chau et al. (2000), Wu et al. (2004), and Wu and Chau (2006). Another problem of elastic spheres relates to the vibrations of the Earth. It was recorded that after the 1952 Kamchatka earthquake and the 1960 Chile earthquake the natural period of oscillations of the Earth was measured at about 58 minutes. Vibrations of spheres can be classified into toroidal and spheroidal modes (Chau, 1998b). For a more detailed analysis of the vibrations of the Earth see Ben-Menahem and Singh (2000).

4.11 PROBLEMS

Problem 4.1. Show that a uniform tensile field T applied along the x_3-direction in a solid can be represented by the following Papkovitch–Neuber potentials:

$$\Phi_0 = \frac{T\nu}{2(1+\nu)}(R^2 - 3x_3^2), \quad \Phi_1 = \Phi_2 = 0, \quad \Phi_3 = \frac{T}{2(1+\nu)}x_3 \tag{4.432}$$

Problem 4.2. Consider the Southwell problem shown in Fig. 4.13 (Southwell and Gough, 1926). Show that a uniform tensile field T applied along the x_3-direction in a solid containing a spherical cavity with radius a (shown in Fig. 4.13) can be modeled by the following Papkovitch–Neuber potentials:

$$\Phi_0 = \frac{T}{2}\{\frac{\nu}{(1+\nu)}(R^2 - 3x_3^2) - \frac{a^3}{7-5\nu}[(6-5\nu)\frac{1}{R} - a^2(1 - \frac{3x_3^2}{R^2})]\},$$

$$\Phi_1 = \Phi_2 = 0, \quad \Phi_3 = \frac{T}{2}\{\frac{x_3}{(1+\nu)} + \frac{5a^3}{7-5\nu}\frac{x_3}{R^3}\} \tag{4.433}$$

Problem 4.3. Solve Cerruti's problem shown in Fig. 4.6 by using the following Papkovitch–Neuber potentials:

$$\Phi_0 = \frac{P(1-2\nu)^2}{4\pi(1-\nu)}\frac{x_1}{R+x_3}, \quad \Phi_1 = \frac{P}{4\pi(1-\nu)}\frac{1}{R}, \quad \Phi_2 = 0, \quad \Phi_3 = \frac{P(1-2\nu)}{4\pi(1-\nu)}\frac{x_1}{R(R+x_3)} \tag{4.434}$$

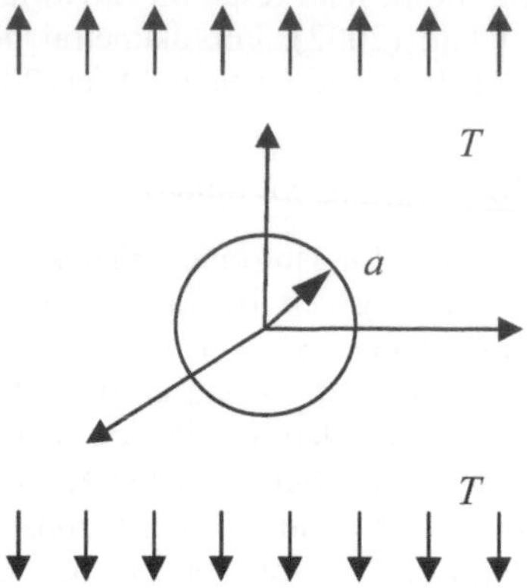

Figure 4.13 Spherical cavity under far field uniform tension

Problem 4.4. Solve Mindlin's problem of a vertical point force P_3 in a half-space by using the following Papkovitch–Neuber potentials:

$$\Phi_0 = \frac{P_3}{8\pi(1-\nu)}\{-\frac{c}{R_1} - (3-4\nu)\frac{c}{R_2} + \frac{1}{2}[(3-4\nu)^2 - 1]\ln(R_2 + x_3 + c)\} \tag{4.435}$$

$$\Phi_3 = \frac{P_3}{8\pi(1-\nu)}\{\frac{1}{R_1} + (3-4\nu)\frac{1}{R_2} + 2c\frac{x_3 + c}{R_2^3}\} \tag{4.436}$$

$$\Phi_1 = \Phi_2 = 0 \tag{4.437}$$

where

$$R_1^2 = (x_3 - c)^2 + x_1^2, \quad R_2^2 = (x_3 + c)^2 + x_1^2 \tag{4.438}$$

Problem 4.5. Solve Mindlin's problem of a horizontal point force P_1 in a half-space by using the following Papkovitch–Neuber potentials:

$$\Phi_0 = \frac{P_1(1-2\nu)}{4\pi(1-\nu)}\{-(1-2\nu)\frac{x_1}{R_2 + x_3 + c} + \frac{cx_1}{R_2(R_2 + x_3 + c)}\} \tag{4.439}$$

$$\Phi_1 = \frac{P_1}{8\pi(1-\nu)}\{\frac{1}{R_2} + \frac{1}{R_1}\}, \quad \Phi_2 = 0 \tag{4.440}$$

$$\Phi_3 = \frac{P_1}{8\pi(1-\nu)}\{2(1-2\nu)\frac{x_1}{R_2(R_2 + x_3 + c)} - \frac{2cx_1}{R_2^3}\} \tag{4.441}$$

Problem 4.6. Show that the component form of (4.12) in cylindrical coordinates can be expressed as

$$\nabla^4\psi_z = 0, \quad \nabla^4\psi_r + L_r(\psi_r, \psi_\theta) = 0, \quad \nabla^4\psi_\theta + L_\theta(\psi_r, \psi_\theta) = 0 \tag{E.4.1}$$

where L_r and L_θ are nonzero functions.

Problem 4.7. As shown in Fig. 4.14, Mindlin's problem is reconsidered for a mixed boundary condition imposed on the surface. Physically, this problem can be related to problems of a soil half-space subject with an incompressible but flexible membrane glued to the surface, that is, the problem of point force $\boldsymbol{F}$ applied at a point $(0,0,c)$ within a half-space subject to the following mixed boundary conditions on the surface $x_3 = 0$:

$$u_1 = 0, \quad u_2 = 0, \quad \sigma_{33} = 0 \tag{4.442}$$

(i) Show that the appropriate Papkovitch–Neuber potentials in solving this problem are

$$\Phi_0 = -\frac{cF_3}{4\pi\mu}\left[\frac{1}{R} - \frac{1}{\tilde{R}}\right], \quad \Phi_1 = \frac{F_1}{4\pi\mu}\left[\frac{1}{R} - \frac{1}{\tilde{R}}\right],$$

$$\Phi_2 = \frac{F_2}{4\pi\mu}\left[\frac{1}{R} - \frac{1}{\tilde{R}}\right], \quad \Phi_3 = \frac{F_3}{4\pi\mu}\left[\frac{1}{R} + \frac{1}{\tilde{R}}\right] \tag{4.443}$$

(ii) Show that the associated displacement can be written as

$$u_i^{(j)} = \frac{F_j}{16\pi\mu(1-\nu)}\{(3-4\nu)[\frac{1}{R} - (1-2\delta_{3j})\frac{1}{\tilde{R}}]\delta_{ij} + \frac{R_i R_j}{R^3} - \frac{\tilde{R}_i \tilde{R}_j}{\tilde{R}^3}(1-2\delta_{3j})\} \tag{4.444}$$

(no sum on i, j)

Problem 4.8. As shown in Fig. 4.15, Mindlin's problem is reconsidered for a mixed boundary condition imposed on the surface. Physically, this problem can be related to problems of a soil half-space subject to a large smooth rigid footing, that is, the problem of point force $\boldsymbol{F}$ applied at a point (0,0,c) within a half-space subject to the following mixed boundary conditions on the surface $x_3 = 0$:

$$\sigma_{31} = 0, \quad \sigma_{32} = 0, \quad u_3 = 0 \tag{4.445}$$

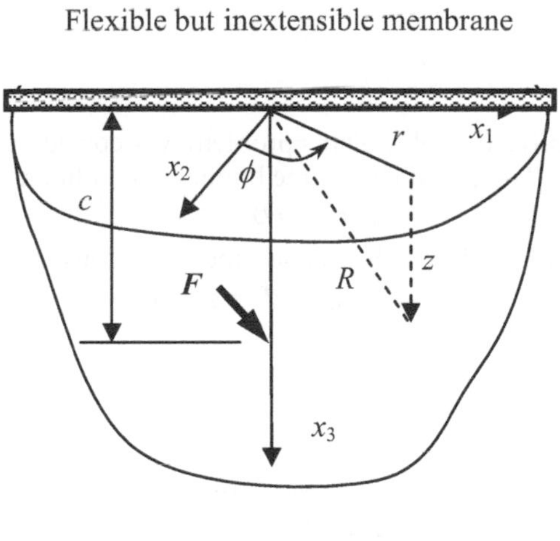

Figure 4.14 Mixed boundary problem: Mindlin's problem with a flexible incompressible membrane on the surface

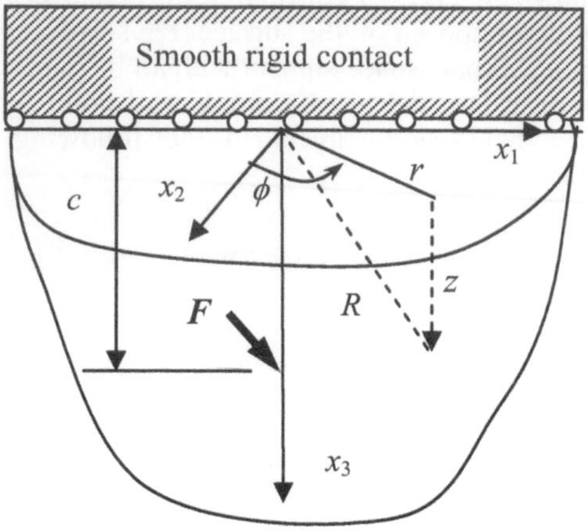

Figure 4.15 Mixed boundary problem: Mindlin's problem with a smooth rigid contact on the surface

(i) Show that the appropriate Papkovitch–Neuber potentials in solving this problem are

$$\Phi_0 = -\frac{cF_3}{4\pi\mu}\left[\frac{1}{R}+\frac{1}{\tilde{R}}\right], \quad \Phi_1 = \frac{F_1}{4\pi\mu}\left[\frac{1}{R}+\frac{1}{\tilde{R}}\right],$$

$$\Phi_2 = \frac{F_2}{4\pi\mu}\left[\frac{1}{R}+\frac{1}{\tilde{R}}\right], \quad \Phi_3 = \frac{F_3}{4\pi\mu}\left[\frac{1}{R}-\frac{1}{\tilde{R}}\right] \tag{4.446}$$

(ii) Show that the associated displacement can be written as

$$u_i^{(j)} = \frac{F_j}{16\pi\mu(1-\nu)}\{(3-4\nu)[\frac{1}{R}+(1-2\delta_{3j})\frac{1}{\tilde{R}}]\delta_{ij}+\frac{R_iR_j}{R^3}+\frac{\tilde{R}_i\tilde{R}_j}{\tilde{R}^3}(1-2\delta_{3j})\} \tag{4.447}$$

(no sum on i, j)

Problem 4.9. A modification of Melan's problem was considered in this problem, as shown in Fig. 4.16. The 2-D traction-free boundary conditions are

$$u_1 = u_3 = 0 \tag{4.448}$$

on $x_3 = 0$. Again the 2-D Papkovitch–Neuber potentials consist of two parts:

$$\Phi_0 = \psi_0 + \varphi_0, \quad \boldsymbol{\Phi} = \boldsymbol{\psi} + \boldsymbol{\varphi} \tag{4.449}$$

where

$$\psi_0 = \frac{cF_3}{2\pi\mu}\ln r, \quad \psi_i = -\frac{F_i}{2\pi\mu}\ln r \tag{4.450}$$

where $i = 1,3$ and

$$r = \sqrt{x_1^2+(x_3-c)^2} \tag{4.451}$$

(i) Show that the unknown harmonic functions φ_0 and $\boldsymbol{\varphi}$ satisfying (4.448) are

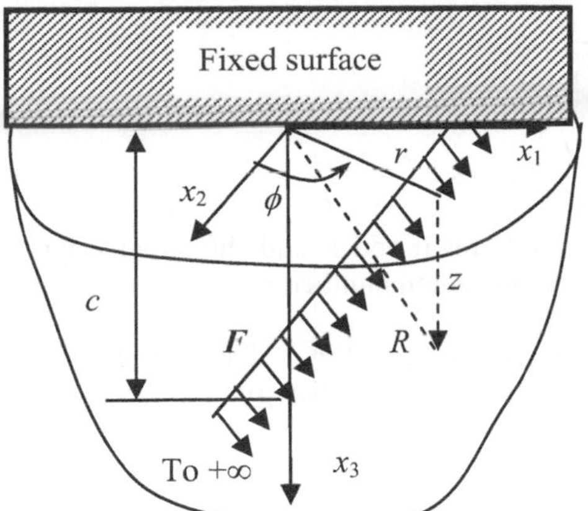

Figure 4.16 Modified Melan's problem: A 2-D line load F applied parallel to the x_2-axis on a fixed boundary

$$\Phi_0 = \frac{cF_3}{2\pi\mu}\ln(\frac{r}{\tilde{r}}), \quad \Phi_1 = -\frac{F_1}{2\pi\mu}\ln(\frac{r}{\tilde{r}}) \tag{4.452}$$

$$\Phi_3 = -\frac{F_3}{2\pi\mu}\ln(\frac{r}{\tilde{r}}) + \frac{2}{3-4\nu}\left[\frac{F_1}{2\pi\mu}\frac{cx_1}{\tilde{r}^2} - \frac{F_3}{2\pi\mu}\frac{c(x_3+c)}{\tilde{r}^2}\right] \tag{4.453}$$

(ii) Show that the associated displacements are

$$u_\alpha^{(\beta)} = \frac{F_\beta}{8\pi\mu(1-\nu)}\{-(3-4\nu)\ln(\frac{r}{\tilde{r}})\delta_{\alpha\beta} + \frac{r_\alpha r_\beta}{r^2} - \frac{r_\alpha r_\beta(1-\delta_{\alpha\beta}) + \tilde{r}_\alpha\tilde{r}_\beta\delta_{\alpha\beta}}{\tilde{r}^2}$$
$$+\frac{2}{3-4\nu}\frac{cx_3}{\tilde{r}^2}(2\delta_{3\alpha}\delta_{3\beta} - \delta_{\alpha\beta}) + (\frac{4}{3-4\nu})\frac{cx_3\tilde{r}_\alpha}{\tilde{r}^4}(\tilde{r}_\beta - 2\delta_{3\beta}\tilde{r}_3)\} \tag{4.454}$$

where $\alpha, \beta = 1, 3$ with no summation on double indices.

Problem 4.10. Referring to Fig. 4.17, show that the following Papkovitch–Neuber function satisfies Rongved's (1955) problem, an interior perpendicular force P_z applied to two perfectly bonded half-spaces:

Region 1:

$$\Phi_0^1 = \frac{P_z}{2\pi(\kappa_1+1)}\left\{-c\frac{1}{R_1} - A\kappa_1 c\frac{1}{R_2} + \frac{1}{2}(A\kappa_1^2 - B)\ln(R_2+z+c)\right\} \tag{4.455}$$

$$\Phi_z^1 = \frac{P_z}{2\pi(\kappa_1+1)}\left\{\frac{1}{R_1} + A\kappa_1\frac{1}{R_2} + 2Ac(\frac{z+c}{R_2^3})\right\} \tag{4.456}$$

Region 2:

$$\Phi_0^2 = \frac{P_z}{2\pi(\kappa_1+1)}\left\{-(1-A)c\frac{1}{R_1} + \frac{1}{2}[(1-A)\kappa_1 - (1-B)\kappa_2]\ln(R_1-z+c)\right\} \tag{4.457}$$

$$\Phi_z^2 = \frac{(1-B)P_z}{2\pi(\kappa_1+1)}\frac{1}{R_1} \tag{4.458}$$

where

$$A = \frac{1-\Gamma}{1+\kappa_1\Gamma}, \quad B = \frac{\kappa_2-\kappa_1\Gamma}{\kappa_2+\Gamma}, \quad \Gamma = \frac{\mu_2}{\mu_1}, \quad \kappa_i = 3-4\nu_i \tag{4.459}$$

for i = 1, 2. The subscripts for μ and κ and the superscripts for the Papkovitch–Neuber function denote the region number.

Problem 4.11. Referring to Fig. 4.17, show that the following Papkovitch–Neuber function satisfies Rongved's (1955) problem, an interior parallel force P_x applied to two perfectly bonded half-spaces:
Region 1:

$$\Phi_0^1 = \frac{P_x}{2\pi(\kappa_1+1)}\left\{-\frac{1}{2}(A\kappa_1^2+B-2S\kappa_1)\frac{x}{R_2+z+c}+\frac{x(A\kappa_1-S)c}{R_2(R_2+z+c)}\right\} \tag{4.460}$$

$$\Phi_z^1 = \frac{P_x}{2\pi(\kappa_1+1)}\left\{(A\kappa_1-S)\frac{x}{R_2(R_2+z+c)}-2Ac\frac{x}{R_2^3}\right\} \tag{4.461}$$

$$\Phi_x^1 = \frac{P_x}{2\pi(\kappa_1+1)}\left\{\frac{1}{R_1}+S\frac{1}{R_2}\right\} \tag{4.462}$$

Region 2:

$$\Phi_0^2 = \frac{P_x}{2\pi(\kappa_1+1)}\left\{-\frac{x[(1-A)\kappa_1+(1-B-2T)\kappa_2]}{2(R_1-z+c)}+\frac{x(1-A-T)c}{R_1(R_1-z+c)}\right\} \tag{4.463}$$

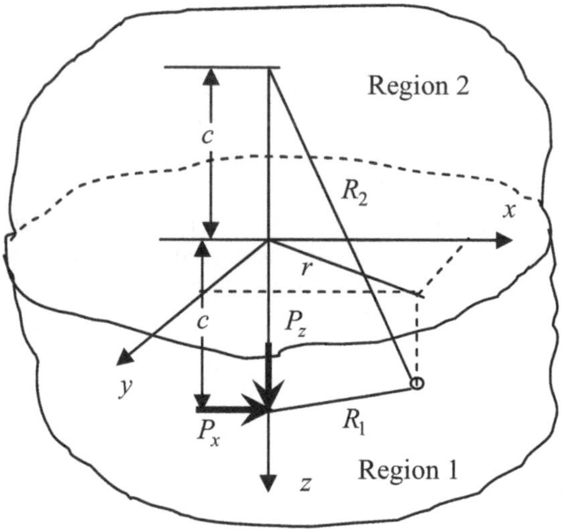

Figure 4.17 Rongved and Dundurs-Hetenyi problem: Force transmission between two half-spaces with smooth interface

$$\Phi_z^2 = -\frac{(1-B-T)P_x}{2\pi(\kappa_1+1)}\frac{x}{R_1(R_1-z+c)} \tag{4.464}$$

$$\Phi_x^2 = \frac{TP_x}{2\pi(\kappa_1+1)}\frac{1}{R_1} \tag{4.465}$$

where

$$S = \frac{1-\Gamma}{1+\Gamma}, \quad T = \frac{2(\kappa_1+1)\Gamma}{(\kappa_1+1)(1+\Gamma)} = (1-S)\frac{\kappa_1+1}{\kappa_2+1}, \quad \Gamma = \frac{\mu_2}{\mu_1}, \quad \kappa_i = 3-4\nu_i \tag{4.466}$$

with $i = 1,2$.

Problem 4.12. Referring to Fig. 4.17, show that the following Papkovitch–Neuber function satisfies Dundurs–Hetenyi's (1965) problem, an interior perpendicular force P_z applied to two half-spaces with smooth contact (zero shear tractions between the two half-spaces):

Region 1:

$$\Phi_0^1 = \frac{P_z}{2\pi(\kappa_1+1)}\left\{-\frac{c}{R_1} - [\kappa_1 - D(\kappa_1+1)]\frac{c}{R_2} + \frac{(1-D)(\kappa_1^2-1)}{2}\ln(R_2+z+c)\right\} \tag{4.467}$$

$$\Phi_z^1 = \frac{P_z}{2\pi(\kappa_1+1)}\left\{\frac{1}{R_1} + [\kappa_1 - D(\kappa_1+1)]\frac{1}{R_2} + 2(1-D)c(\frac{z+c}{R_2^3})\right\} \tag{4.468}$$

Region 2:

$$\Phi_0^2 = \frac{D(\kappa_2-1)P_z}{2\pi(\kappa_1+1)}\left\{c\frac{1}{R_1} - \frac{1}{2}(\kappa_1+1)\ln(R_1-z+c)\right\} \tag{4.469}$$

$$\Phi_z^2 = \frac{DP_z}{2\pi(\kappa_1+1)}\left\{(\kappa_1+1)\frac{1}{R_1} - 2c\frac{z-c}{R_1^3}\right\} \tag{4.470}$$

where

$$D = \frac{(1+\kappa_1)\Gamma}{(1+\kappa_1)\Gamma+1+\kappa_2}, \quad \Gamma = \frac{\mu_2}{\mu_1}, \quad \kappa_i = 3-4\nu_i \tag{4.471}$$

with $i = 1,2$.

Problem 4.13. Referring to Fig. 4.17, show that the following Papkovitch–Neuber function satisfies Dundurs–Hetenyi's (1965) problem, an interior parallel force P_x applied to two half-spaces with smooth contact (zero shear tractions between the two half-spaces):

Region 1:

$$\Phi_0^1 = \frac{(1-D)(\kappa_1-1)P_x}{2\pi(\kappa_1+1)}\left\{-\frac{1}{2}(\kappa_1-1)\frac{x}{R_2+z+c} + \frac{xc}{R_2(R_2+z+c)}\right\} \tag{4.472}$$

$$\Phi_z^1 = \frac{(1-D)P_x}{2\pi(\kappa_1+1)}\left\{(\kappa_1-1)\frac{x}{R_2(R_2+z+c)} - 2c\frac{x}{R_2^3}\right\} \tag{4.473}$$

$$\Phi_x^1 = \frac{P_x}{2\pi(\kappa_1+1)}\left\{\frac{1}{R_1}+\frac{1}{R_2}\right\} \tag{4.474}$$

Region 2:

$$\Phi_0^2 = \frac{D(\kappa_2-1)P_x}{2\pi(\kappa_1+1)}\left\{\frac{x(\kappa_1-1)}{2(R_1-z+c)} - \frac{xc}{R_1(R_1-z+c)}\right\} \tag{4.475}$$

$$\Phi_z^2 = \frac{DP_x}{2\pi(\kappa_1+1)}\left\{\frac{(\kappa_1-1)x}{R_1(R_1-z+c)} - 2c\frac{x}{R_1^3}\right\} \tag{4.476}$$

$$\Phi_x^2 = 0 \tag{4.477}$$

where

$$D = \frac{(1+\kappa_1)\Gamma}{(1+\kappa_1)\Gamma+1+\kappa_2}, \quad \Gamma = \frac{\mu_2}{\mu_1}, \quad \kappa_i = 3-4\nu_i \tag{4.478}$$

with $i = 1, 2$.

Problem 4.14. Show that (4.344) satisfies cylindrical biharmonic equations.

Problem 4.15. Prove (4.59) from (4.53).

Problem 4.16. Prove (4.79).

Problem 4.17. Carry out the integration in (4.418) to (4.421) to obtain all stress and displacement components of Boussinesq's problem given in (4.152) and (4.153).

Hints:

$$\int_0^\infty \xi e^{-\xi z} J_1(\xi r)d\xi = \frac{r}{(r^2+z^2)^{3/2}} \tag{4.479}$$

$$\int_0^\infty e^{-\xi z} J_1(\xi r)d\xi = \frac{(r^2+z^2)^{1/2}-z}{r(r^2+z^2)^{1/2}} \tag{4.480}$$

Problem 4.18. Solve Cerruti's Problem discussed in Section 4.4.4 again by using the Hankel transform formalism of Muki (1960) given in Section 4.9. Note that it is an antisymmetric problem, and as shown in Fig. 4.18 we can set the following tractions on $z = 0$ (with limit $a \to 0$) and $m = 1$ in (4.394) to (4.404):

$$\sigma_{zz} = 0, \quad \sigma_{rz} = \frac{P}{\pi a^2}\cos\phi, \quad \sigma_{rz} = -\frac{P}{\pi a^2}\sin\phi \tag{4.481}$$

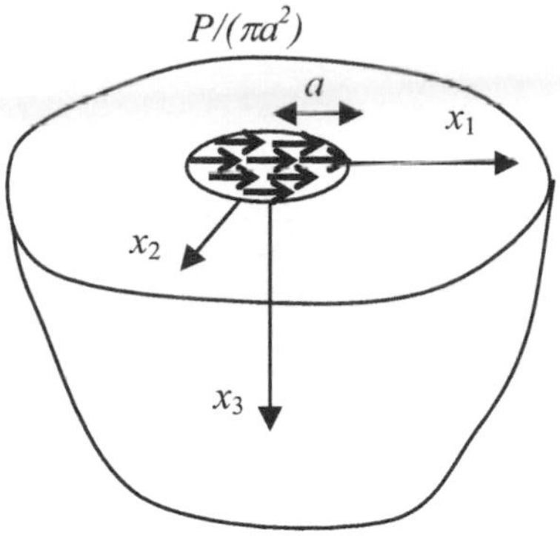

Figure 4.18 Cerruti problem modeled by uniform circular patch load

Problem 4.19. Show that Papkovitch–Neuber displacement potentials are not invariant upon translation of the origin.

Problem 4.20. Show the equivalence of (4.116) and (4.133) with (4.137) in view of (4.166).

CHAPTER FIVE

Plasticity and Its Applications

5.1 INTRODUCTION

Plasticity was originally developed for modeling the behavior of a wide variety of metals, which are pressure-insensitive and incompressible. Figure 5.1 shows some idealizations of these material responses, including both elastic and plastic materials: (a) nonlinearly elastic; (b) linearly elastic; (c) inelastic or plastic; (d) rigid, perfectly plastic; (e) elastic, perfectly plastic; (f) rigid, work-hardening; and (g) elastic, work-hardening. The traditional plastic materials were assumed to satisfy *associated flow rule* or *normality* law during plastic deformation. This, however, is not true for geomaterials, such as soil and rocks. For geomaterials, terms like *non-associated flow rule* or *non-normality rule* emerge naturally due to their constitutive properties, including pressure-sensitivity (i.e., frictional effect), compressibility, and plastic dilatancy. The application of plasticity to soil mechanics starts probably with Drucker and Prager (1952), who extended the classical form of Coulomb criteria to three-dimensional cases (Pietruszczak, 2010), and its application to rock-like materials is more recent (Rudnicki and Rice, 1975). In this chapter, we will present a generalized constitutive form which is applicable to both soils and rocks.

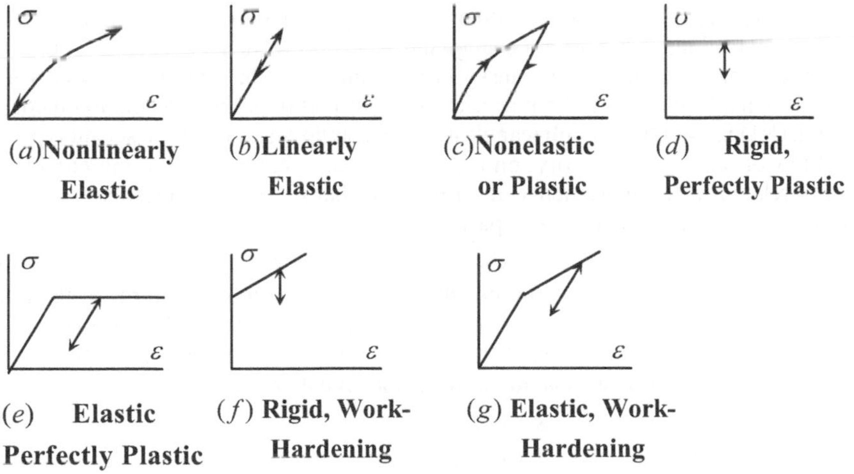

Figure 5.1 Some idealizations of stress-strain responses

There are some features of soil plasticity that are quite different metal plasticity. Under external applied load, soil particles may undergo crushing, rearrangement, interlocking, and shear-induced dilatancy. Many of these processes are irreversible. Soil may also yield under isotropic compression, and such phenomenon is not observed in metal and other materials. Soil which is initially densely packed may behave differently from loosely packed soils. Dense soil normally dilates before failure while loose soil compresses until failure. A key feature is that regardless of whether the soil is initially densely packed or loosely packed, the final volume (or void ratio) at failure is normally a constant value. This constant volume state is called critical state in soil mechanics. Cap models have been proposed to model such phenomena. We will start this chapter with some basic concepts of plasticity theory before we discuss cap models and Cam-clay models. Internal variables formulation based on thermodynamics is also introduced. The chapter concludes with an introduction to viscoplasticity.

5.2 FLOW THEORY AND DEFORMATION THEORY

There are two main types of theories for plasticity: *flow theory or incremental theory* and *deformation theory.*

Deformation theory assumes that the total plastic strain is related to the final stress, and that the total strain of a solid can be subdivided into two parts, the elastic part and plastic part:

$$\boldsymbol{\varepsilon} = \boldsymbol{\varepsilon}^e + \boldsymbol{\varepsilon}^p = \boldsymbol{\varepsilon}^e + \phi \frac{\partial F(\boldsymbol{\sigma})}{\partial \boldsymbol{\sigma}} \tag{5.1}$$

where the superscripts e and p denote the elastic and plastic parts of the total strain, respectively and ϕ is a scalar function. The second part of (5.1) implicitly assumes that the total plastic strain relates to the final stress state by a scalar function of the stress state $F(\boldsymbol{\sigma})$. Strictly speaking, deformation theory applies only to *proportional loading case* (i.e., the loading path does not change direction during the whole loading process in the stress space), although Budiansky (1959) illustrated that slight deviation from the proportional loading should not restrict the applicability of the deformation theory. Equation (5.1) is essentially nonlinear elastic stress-strain relations of the secant type.

Flow or incremental theory, on the other hand, assumes that the increment of strain is related to the increments of stress, and that the strain increments can be decomposed into elastic and plastic parts:

$$d\boldsymbol{\varepsilon} = d\boldsymbol{\varepsilon}^e + d\boldsymbol{\varepsilon}^p, \quad or \quad \dot{\boldsymbol{\varepsilon}} = \dot{\boldsymbol{\varepsilon}}^e + \dot{\boldsymbol{\varepsilon}}^p \tag{5.2}$$

The second part of (5.2) simply rewrites the increments of strains in terms of the rate of strain; both forms are, however, essentially the same if small deformation is considered. In flow and incremental theory, it is normally assumed that both elastic and plastic strains are proportional to the stress increment or stress rate:

$$\dot{\boldsymbol{\varepsilon}} = (\boldsymbol{M}^e + \boldsymbol{M}^p) : \dot{\boldsymbol{\sigma}} \tag{5.3}$$

where $\boldsymbol{M}^e$ and $\boldsymbol{M}^p$ are the elastic and plastic tangential compliance tensors, respectively. Inversion of this equation gives:

$$\dot{\boldsymbol{\sigma}} = (\boldsymbol{C}^e + \boldsymbol{C}^p) : \dot{\boldsymbol{\varepsilon}} \tag{5.4}$$

where $\boldsymbol{C}^{e}$ and $\boldsymbol{C}^{p}$ are the elastic and plastic tangential stiffness tensors, respectively. For finite deformation, the strain rate is normally replaced by the rate of deformation and Jaumann's rate of Kirchhoff stress, in view of the frame-indifference requirement:

$$\boldsymbol{D} = \boldsymbol{M} : \boldsymbol{\tau}^{\nabla} = (\boldsymbol{M}^{e} + \boldsymbol{M}^{p}) : \boldsymbol{\tau}^{\nabla} \tag{5.5}$$

where $\boldsymbol{\tau}^{\nabla}$ is defined in (2.106) of Chapter 2.

5.3 YIELD FUNCTION AND PLASTIC POTENTIAL

To determine $\boldsymbol{M}^{p}$, we must introduce two scalar functions of stress: the yield function $f(\boldsymbol{\sigma})$ and the plastic potential $g(\boldsymbol{\sigma})$. When the yield function is satisfied, that is,

$$f(\boldsymbol{\sigma}) = 0 \tag{5.6}$$

then the material starts to yield. However, for strain-hardening models the yield surface on the stress space can evolve (normally expand for strain-hardening models) as the stress increases such that the stress state must always be on or within the current yield surface. Mathematically, this requirement yields the following *consistency requirement*:

$$df = \frac{\partial f}{\partial \boldsymbol{\sigma}} : d\boldsymbol{\sigma} = 0, \quad or \quad df = \frac{\partial f}{\partial \sigma_{ij}} d\sigma_{ij} = Q_{ij} d\sigma_{ij} > 0 \tag{5.7}$$

where Q_{ij} are the normal to the yield surface in the stress space. When $Q_{ij}d\sigma_{ij} = 0$, the stress increment is tangential to the yield surface, and only elastic response is possible. Such a case is normally referred to as neutral loading. On the other hand, the plastic potential governs the direction of the plastic strain increments:

$$D_{ij}^{p} = \frac{\partial g}{\partial \sigma_{ij}} d\lambda = P_{ij} \; d\lambda \quad or \quad \boldsymbol{D}^{p} = \boldsymbol{P} d\lambda \tag{5.8}$$

where $d\lambda$ is a scalar parameter. Equation (5.8) is normally referred to as the *flow rule*. When $P_{ij} = Q_{ij}$, the flow rule is called the *associative flow rule*. In addition, for this case the normal to yield surface will parallel the normal to the plastic potential; therefore, this is also called the *normality rule*. However, for geomaterials we must have $P_{ij} \neq Q_{ij}$ if the yielding behavior is to be described properly. There is plenty of experimental evidence to support this assertion.

We now examine the possible form of $d\lambda$. Since, according to the consistency requirement (5.7), plastic deformation only takes place when the $Q_{ij}d\sigma_{ij} > 0$, and $Q_{ij}d\sigma_{ij}$ is precisely the component of stress tensor pointing outward of the yield surface $f = 0$. Therefore, it is natural to postulate that the plastic strain (5.8) will only be proportional to the part of stress tensor which is normal to the yield surface, but not the components of stress tensor tangential to the yield surface. In particular, we have

$$\begin{aligned} d\lambda &= \frac{1}{h} \boldsymbol{Q} : d\boldsymbol{\sigma} \qquad && \boldsymbol{Q} : d\boldsymbol{\sigma} > 0 \\ &= 0 && \boldsymbol{Q} : d\boldsymbol{\sigma} < 0 \end{aligned} \tag{5.9}$$

where h is a kind of hardening modulus. It is also called the Kuhn-Tucker condition (Lubliner, 1990).

So far, we have assumed that yield surface is *smooth*, or no *vertex* is formed at the yield surface. Rudnicki and Rice (1975) argued, based on the mechanism of sliding of fissures in rock, that yield vertex may be formed in the yield surface if the loading is applied beyond the initial yield surface. If this is the case, (5.9) will no longer be valid and modification is needed. Rudnicki and Rice (1975) also proposed a simple model to account for such a yield-vertex effect, and the details will be given in later sections.

5.4 ELASTO-PLASTIC CONSTITUTIVE MODEL

Combining (5.5), (5.8), and (5.9), we can write the compliance tensor as

$$D_{ij} = (M^e_{ijkl} + \frac{1}{h} P_{ij} Q_{kl}) \tau^{\nabla}_{kl} \quad or \quad \boldsymbol{D} = (\boldsymbol{M}^e + \frac{1}{h} \boldsymbol{P}\,\boldsymbol{Q}) : \boldsymbol{\tau}^{\nabla} \tag{5.10}$$

where h is the hardening modulus. This form can inverted to give

$$\boldsymbol{\tau}^{\nabla} = \left[\boldsymbol{C}^e - \frac{(\boldsymbol{C}^e : \boldsymbol{P})(\boldsymbol{Q} : \boldsymbol{C}^e)}{h + \boldsymbol{Q} : \boldsymbol{C}^e : \boldsymbol{P}} \right] : \boldsymbol{D} \tag{5.11}$$

The proof of (5.11) can be done by applying $\boldsymbol{C}^e$ to both side of (5.10) such that

$$\boldsymbol{C}^e : \boldsymbol{D} = [\boldsymbol{C}^e : \boldsymbol{M}^e + \frac{1}{h} (\boldsymbol{C}^e : \boldsymbol{P}) \boldsymbol{Q}] : \boldsymbol{\tau}^{\nabla} = [\boldsymbol{I} + \frac{1}{h} (\boldsymbol{C}^e : \boldsymbol{P}) \boldsymbol{Q}] : \boldsymbol{\tau}^{\nabla} \tag{5.12}$$

since $\boldsymbol{C}^e = (\boldsymbol{M}^e)^{-1}$ and $\boldsymbol{I}$ is the fourth-order unit tensor. Applying the dot product of $\boldsymbol{Q}$ to both sides of (5.12) gives

$$\boldsymbol{Q} : \boldsymbol{C}^e : \boldsymbol{D} = [1 + \frac{1}{h} (\boldsymbol{Q} : \boldsymbol{C}^e : \boldsymbol{P})]\; \boldsymbol{Q} : \boldsymbol{\tau}^{\nabla} \tag{5.13}$$

Note that (5.13) is a scalar equation and normal algebraic analysis applies. Therefore, we have

$$\boldsymbol{Q} : \boldsymbol{\tau}^{\nabla} = \frac{h\; \boldsymbol{Q} : \boldsymbol{C}^e : \boldsymbol{D}}{h + \boldsymbol{Q} : \boldsymbol{C}^e : \boldsymbol{P}} \tag{5.14}$$

Rearranging the second part of (5.12) gives

$$\boldsymbol{\tau}^{\nabla} = \boldsymbol{C}^e : \boldsymbol{D} - \frac{1}{h} \boldsymbol{C}^e : \boldsymbol{P}\; \boldsymbol{Q} : \boldsymbol{\tau}^{\nabla} \tag{5.15}$$

Finally, substitution of (5.14) into (5.15) gives (5.11). This completes the proof.

Alternatively, (5.11) can be obtained using the well-known *Sherman–Morrison formula* in linear algebra (e.g., Noble and Daniel, 1988; Campbell, 1980), which states

$$[\boldsymbol{I} + \boldsymbol{a}\, \boldsymbol{b}]^{-1} = [\boldsymbol{I} - \beta\, \boldsymbol{a}\, \boldsymbol{b}] \tag{5.16}$$

where $\beta = 1 + \boldsymbol{a}:\boldsymbol{b}$. In particular, the second part of (5.12) can be recognized as:

$$\boldsymbol{C}^e : \boldsymbol{D} = [\boldsymbol{I} + \frac{1}{h} \boldsymbol{R}\, \boldsymbol{Q}] : \boldsymbol{\tau}^{\nabla} \tag{5.17}$$

where $\boldsymbol{R} = \boldsymbol{C}^e:\boldsymbol{P}$, and then (5.16) can be applied to show the validity of (5.11). The validity of the Sherman–Morrison formula will be left as a problem for the readers (Problem 5.1).

5.5 RUDNICKI–RICE (1975) MODEL

Rudnicki and Rice (1975) derived the condition of localization of deformation in rock-like solids, and in doing so they proposed an attractive model incorporating the effects of frictional and plastic dilatancy effects. As mentioned in Rudnicki (1982), the Rudnicki and Rice (1975) model can be interpreted as the deformation theory of plasticity. In particular, as shown by Senseny et al. (1983), the Rudnicki–Rice model can be obtained by setting

$$f=\tau_e+\mu\sigma-k=0,\quad g=\tau_e+\beta\sigma-k'=0 \tag{5.18}$$

where $\tau_e=(J_2')^{1/2}=[\frac{1}{2}\boldsymbol{s}:\boldsymbol{s}]^{1/2}$, and $\sigma = 1/3\,\mathrm{tr}(\boldsymbol{\sigma})$, then

$$\boldsymbol{Q}=\frac{\boldsymbol{s}}{2\tau_e}+\frac{1}{3}\mu\boldsymbol{I},\quad \boldsymbol{P}=\frac{\boldsymbol{s}}{2\tau_e}+\frac{1}{3}\beta\boldsymbol{I} \tag{5.19}$$

Note that the deviatoric stress tensor is defined as

$$\boldsymbol{s}=\boldsymbol{\sigma}-\tfrac{1}{3}\mathrm{tr}(\boldsymbol{\sigma})\boldsymbol{I} \tag{5.20}$$

Professor J.R. Rice is one of the most versatile researchers in solid mechanics and a recipient of the Timoshenko medal. Professor J.W. Rudnicki made seminal contributions to localization analysis in geomaterials (see biography section at the end of this book).

Substituting (5.19) into (5.11), the famous model by Rudnicki and Rice (1975) is recovered:

$$\overset{\nabla}{\tau}_{kl}=\{G(\delta_{mk}\delta_{nl}+\delta_{ml}\delta_{kn})+(K-\tfrac{2}{3}G)\delta_{kl}\delta_{mn} - \frac{(\frac{G}{\tau_e}s_{kl}+\beta K\delta_{kl})(\frac{G}{\tau_e}s_{mn}+\mu K\delta_{mn})}{h+G+\mu K\beta}\}D_{mn} \tag{5.21}$$

where β and μ are the coefficient of plastic dilatancy and the frictional coefficient, G and K are the elastic shear and bulk moduli, respectively, and h is the plastic hardening modulus. Experimental calibration of the Rudnicki–Rice (1975) model was done by Wawersik et al. (1990) and Holcomb and Rudnicki (2001). A similar constitutive model proposed by Nemat-Nasser and Shokoon (1980) was found appropriate for granular materials as well (see also the discussion by Rudnicki, 1982). Note that when $\beta=\mu$, we have $P_{ij}=Q_{ij}$ or the normality flow rule; the Drucker–Prager (1952) model is recovered as a special case. When $\beta=\mu=0$, the Prandtl–Ruess elastic-plastic model or the *J_2-flow theory* for metals is recovered (Hill, 1950). Their paper also presents the classic solution for the condition of localization of deformation. This pioneering work and that of Rice (1976) have triggered intense interest on the localization of deformation and bifurcation analyses in geomechanics (e.g., Chau and Rudnicki, 1990; Chau, 1992, 1993, 1994a, 1995a,b).

5.6 DRUCKER'S POSTULATE, PMPR, AND IL'IUSHIN'S POSTULATE

In this section, we will summarize a number of postulates about plastic strain increment in plasticity. They are related to uniqueness, normality, strain hardening, and the direction of plastic strain. Their main consequences are in metal plasticity satisfying the associated flow rule, which is clearly inappropriate for geomaterials. Their importance to hardening and yield vertex effect deserves coverage in this chapter.

One of the most important inequalities in plasticity is called Drucker's postulate (Lubliner, 1990):

$$\dot{\sigma}_{ij}\dot{\varepsilon}_{ij}^{p} \geq 0 \tag{5.22}$$

The postulate is best illustrated in the one-dimensional case shown in Fig. 5.2. Professor D.C. Drucker was the first Ph.D. student of R. Mindlin, and together with Prager he had made fundamental contributions to soil plasticity (see biography section). It is clear from Fig. 5.2 that the larger than zero sign in (5.22) applies in the hardening regime, the equal sign applies in the perfectly plastic regime, and (5.22) would be violated in the softening regime. Thus, the first obvious consequence of Drucker's postulate is the existence of work-hardening or softening implies instability. It also implies that the work done during incremental loading is positive and the work done during a loading-unloading cycle is non-negative. The left-hand side of (5.22) is related to work increment, and it is often referred to as the thermodynamic requirement although it is independent of basic laws of thermodynamics. The strain rate and stress rate must also be of the same direction. Actually, the stress increment can start from an internal stress point, and thus the stress rate can be written as

$$(\sigma_{ij} - \sigma_{ij}^{*})\dot{\varepsilon}_{ij}^{p} \geq 0 \tag{5.23}$$

This equation is actually called the principle of maximum plastic resistance (PMPR), and it was proposed independently by R. von Mises, G.I. Taylor, and R. Hill (see Lubliner, 1990). It is also sometimes referred to as the principle of maximum plastic work. This is a special case of Drucker's postulate. A direct consequence can be shown in Fig. 5.3. Two main consequences of PMPR are the normality of flow rule (which may be violated in geomaterials) and the convexity of the yield surface, as shown in Fig. 5.4.

Another related postulate is called Il'iushin's postulate (Lubliner, 1990):

$$\oint \sigma_{ij} d\varepsilon_{ij} \geq 0 \tag{5.24}$$

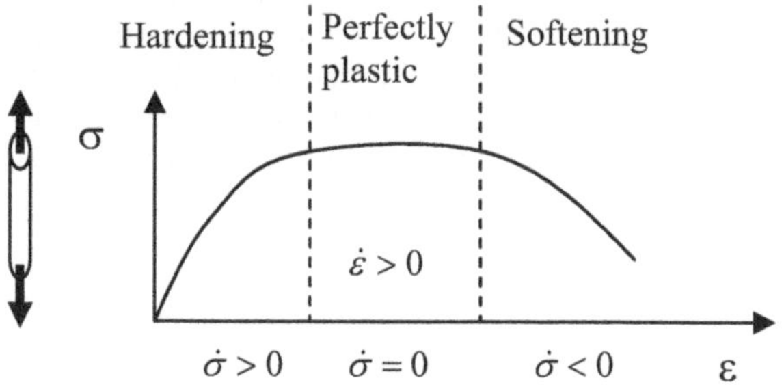

Figure 5.2 Drucker's postulate (1-D illustration)

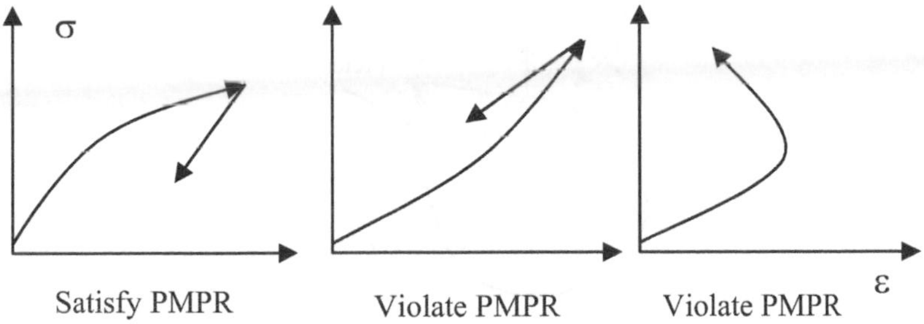

Figure 5.3 Stress-strain behavior violating PMPR

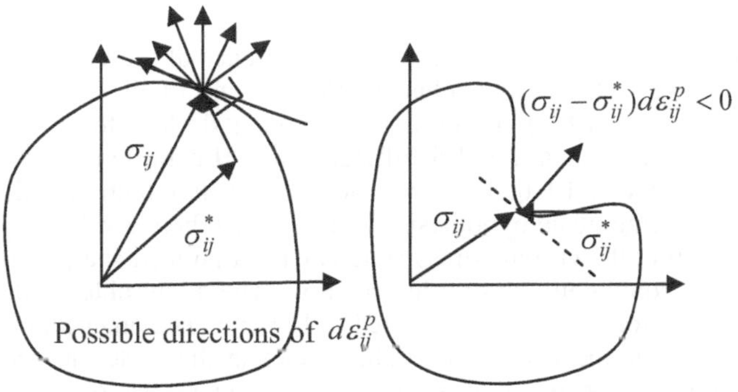

Figure 5.4 Normality requirement and convexity requirement

It requires that the work done is positive in any strain cycle. Il'iushin's postulate can be shown implying the PMPR (Lubliner, 1990), but it only works for isotropic hardening. Thus, in a sense it is a stronger (less general) hypothesis than the PMPR.

5.7 YIELD VERTEX

Based on the arguments of multiple slip plane theory in metal plasticity (Budiansky, 1959) or based on the multiple fissure plane argument in rock mass (Rudnicki and Rice, 1975), it is speculated that a vertex will be formed under continuous proportional loading. The loading induced yield vertex is illustrated in Fig. 5.5, together with the yield surface of the isotropic hardening model and the kinematic hardening model. Isotropic hardening is showed by a uniform expansion of the yield stress in all directions even through proportional loading is only applied along a particular loading path. Kinematic hardening is showed as a lateral translation of the yield surface, and it is the consequence of the Bauschinger effect (Sanders, 1954).

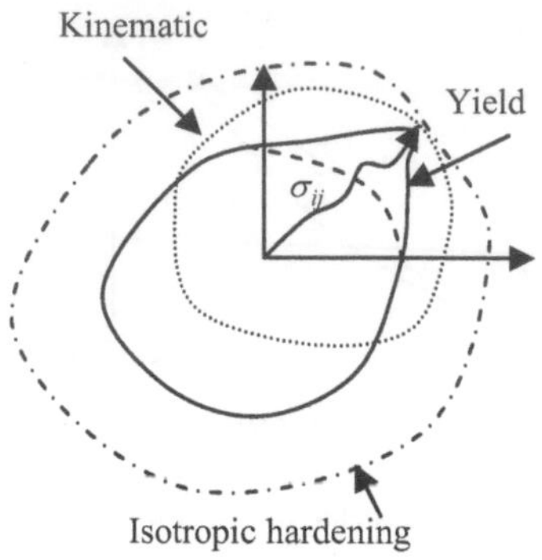

Figure 5.5 Yield vertex formation vs. isotropic hardening and kinematic hardening

There is experimental evidence of the existence of the yield vertex effect on the yield surface, but it is inconclusive (Bertsch and Findley, 1962; Phillips and Gray, 1961; Sewell, 1974). Theoretical arguments do lead to the prediction of a vertex in the yield surface (Christoffersen and Hutchinson, 1979; Hill, 1967; Cleary and Rudnicki, 1976; Pan and Rice, 1983). For both slip theory and the independent loading plane hypothesis (no interaction between loading planes), the point representing the current stress state carries a pointed sharp vertex or corner with it during continuous plastic deformation. The admissible loading cones for further plastic deformation are illustrated in Fig. 5.6. The outer admissible zone is bounded by the normals of the two slip planes at the vertex as required by the PMPR or Drucker's postulate, discussed in the previous section. The inner zone is that given by Sanders, formed by extending the slip planes at the sharp corner. In the context of deformation theory, the intermediate zone is that given by Budiansky (1959), with α given by

$$\alpha \le \bar{\beta}, \quad \text{and } \alpha \le \tan^{-1}[\frac{\tan^2 \bar{\beta}}{\tan \beta}] \tag{5.25}$$

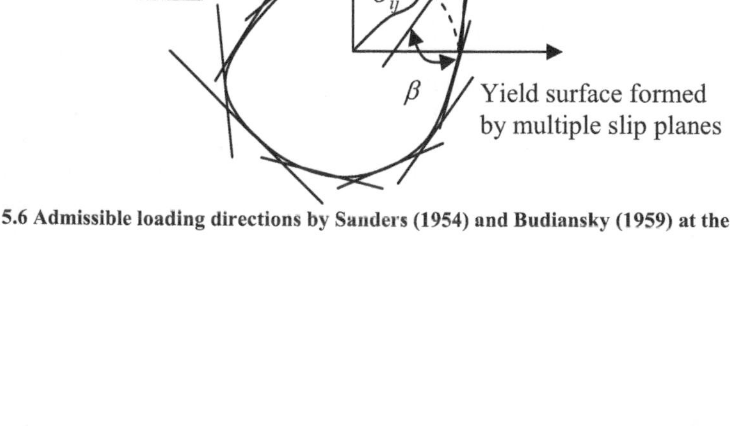

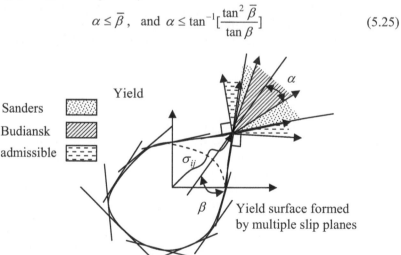

Figure 5.6 Admissible loading directions by Sanders (1954) and Budiansky (1959) at the vertex

where β is defined in Fig. 5.6. The following definitions have been used:

$$\bar{\beta} = \cos^{-1}[(\frac{1}{1+N})^{1/2}], \quad N = (\frac{1/E_t - 1/E}{1/E_s - 1/E}) \tag{5.26}$$

where E, E_t, and E_s are defined as the elastic, tangent, and second modulus, respectively (Budiansky, 1959). These moduli are defined as:

$$\varepsilon_{ij}^p = \frac{3}{2}(\frac{1}{E_s} - \frac{1}{E})s_{ij} \tag{5.27}$$

$$\dot{\varepsilon}_{ij}^p = \frac{3}{2}(\frac{1}{E_s} - \frac{1}{E})\dot{s}_{ij} + (\frac{1}{E_t} - \frac{1}{E_s})\frac{3}{2}(\frac{\dot{\tau}_e}{\tau_e})s_{ij} \tag{5.28}$$

where τ_e is the equivalent shear stress defined after (5.18). For uniaxial compression, these moduli are depicted in Fig. 5.7. In (5.25), we assumed that the angle β at the yield vertex is known as shown in Fig. 5.6. It is obvious that even from the deformation theory, we have seen that the loading cone should be smaller than the admissible zone suggested by Drucker's postulate.

In the context of damage in rocks, based on the microcrack model of Costin (1983, 1985) Holcomb and Costin (1986) demonstrated how the damage surface in the stress space (equivalent to the yield surface in the stress space discussed here) evolves with loading and how corners developed under continuous loading. The details can be found in Holcomb and Costin (1986).

For the case of rock mass containing an isotropic infinite set of frictional sliding fissures, the yield vertex expected is similar to that shown in Fig. 5.6. Rudnicki and Rice (1975) proposed that the tangential stress increment on the idealized isotropic hardening surface will also induce plastic deformation at the yield vertex. The main concept of the plastic strain component is demonstrated in Fig. 5.8 (after Rudnicki, 1984). In the smooth yield stress, stress increment tangent to the yield surface leads to elastic response (Fig. 5.8(a)), whereas in the yield vertex model shown in Fig. 5.8(b) the plastic response is controlled by modulus h_1, which is much smaller than the elastic modulus G. In general, we expect:

$$h \ll h_1 < G \tag{5.29}$$

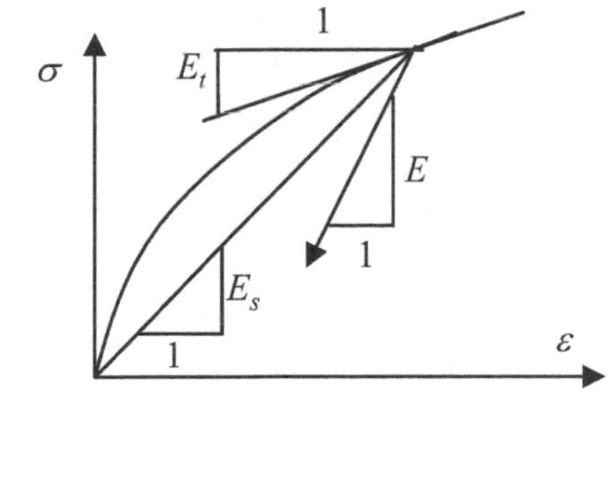

Figure 5.7 Definitions of elastic, secant, and tangent moduli in Budiansky's (1959) theory

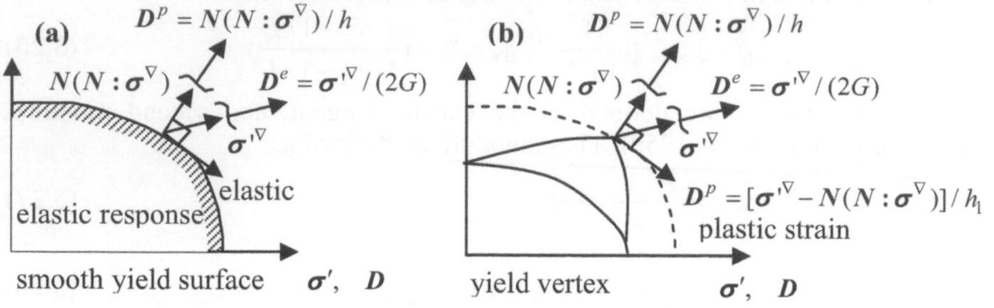

Figure 5.8 Smooth and cornered yield surface (after Rudnicki, 1984; with permission from The Society of Rheology)

It can be shown that subject to the tangent stress increment the smooth yield surface response is stiffer than the yield vertex response.

By adding this additional term and for the case of $\beta = \mu = 0$, the deviatoric part of the rate of the deformation tensor becomes

$$D' = \frac{\sigma'^\nabla}{2G} + \frac{1}{h} N(N:\sigma'^\nabla) + \frac{1}{h_1}[\sigma'^\nabla - N(N:\sigma^\nabla)] \tag{5.30}$$

where

$$N = \frac{\sigma'}{\sqrt{2}\bar{\tau}}, \text{ or } N_{ij} = \frac{\sigma'_{ij}}{\sqrt{2}\bar{\tau}}, \text{ and } \bar{\tau} = (\frac{1}{2}\sigma'_{ij}\sigma'_{ij})^{1/2} \tag{5.31}$$

where σ' is the deviatoric stress. Note that the second term and the third term in (5.30) are perpendicular, that is

$$\begin{aligned} N\cdot\cdot[\sigma'^\nabla - N(N\cdot\cdot\sigma^\nabla)] &= N:\sigma'^\nabla - N:N(N:\sigma^\nabla) \\ &= \frac{\sigma'_{kl}\sigma'^\nabla_{kl}}{\sqrt{2}\bar{\tau}} - \frac{\sigma'_{kl}\sigma'_{kl}}{2\bar{\tau}^2}(\frac{\sigma'_{mn}\sigma'^\nabla_{mn}}{\sqrt{2}\bar{\tau}}) \\ &= \frac{\sigma'_{kl}\sigma'^\nabla_{kl}}{\sqrt{2}\bar{\tau}}(\sigma'_{mn}\sigma'^\nabla_{mn} - \sigma'_{mn}\sigma'^\nabla_{mn}) = 0 \end{aligned} \tag{5.32}$$

Apparently, the yield vertex theory of Rudnicki and Rice (1975) can easily be incorporated into numerical codes and is better accepted than other yield vertex theories (Yu, 2006).

5.8 MOHR–COULOMB MODEL

The following form of the Mohr–Coulomb model is given by Senseny et al. (1983). In particular, f and g are given as:

$$\begin{aligned} f &= \tau_e(\cos\theta - \frac{\mu^*}{\sqrt{3}}\sin\theta) + \mu^*\sigma - k^* = 0, \\ g &= \tau_e(\cos\theta - \frac{\mu^*}{\sqrt{3}}\sin\theta) + \beta^*\sigma - k^{*\prime} = 0 \end{aligned} \tag{5.33}$$

where $\sigma = \sigma_{kk}/3$ and θ is the Lode angle measured in the π-plane or *octahedral* plane shown in Fig. 5.9. The Lode angle was proposed by German engineer W. Lode in 1926. The concept of Lode angle θ and the π-plane will be discussed in Sections 5.9 and 5.10. The tensors $\boldsymbol{P}$ and $\boldsymbol{Q}$ are

$$\boldsymbol{Q} = \Theta_1 \frac{\boldsymbol{\sigma}'}{2\tau_e} + \Theta_2 (\frac{\boldsymbol{\sigma}' \bullet \boldsymbol{\sigma}'}{2\tau_e^2} - \frac{1}{3}\boldsymbol{I}) + \frac{1}{3}\mu^* \boldsymbol{I} \tag{5.34}$$

$$\boldsymbol{P} = \Theta_1 \frac{\boldsymbol{\sigma}'}{2\tau_e} + \Theta_2 (\frac{\boldsymbol{\sigma}' \bullet \boldsymbol{\sigma}'}{2\tau_e^2} - \frac{1}{3}\boldsymbol{I}) + \frac{1}{3}\beta^* \boldsymbol{I} \tag{5.35}$$

where

$$\Theta_1 = \cos\theta[(1 + \tan\theta \tan 3\theta) + \frac{\mu^*}{\sqrt{3}}(\tan 3\theta - \tan\theta)] \tag{5.36}$$

$$\Theta_2 = \frac{\sqrt{3}\sin\theta + \mu^* \cos\theta}{\cos 3\theta} \tag{5.37}$$

The parameters μ^* and k^* can be related to the parameters for the usual Mohr-Coulomb law

$$\tau_e = c' + \sigma \tan\phi \tag{5.38}$$

as:

$$\mu^* = \sin\phi, \quad k^* = 2c'\cos\phi \tag{5.39}$$

To derive (5.33), we have to first consider the *Lode angle* θ. The details are given in the next section.

5.9 LODE ANGLE OR PARAMETER

The origin of the Lode angle relates to the finding of the principal deviatoric stress. In particular, the principal deviatoric stress s must satisfy the following eigenvalue problem:

$$\det|s_{ij} - s\delta_{ij}| = 0 \tag{5.40}$$

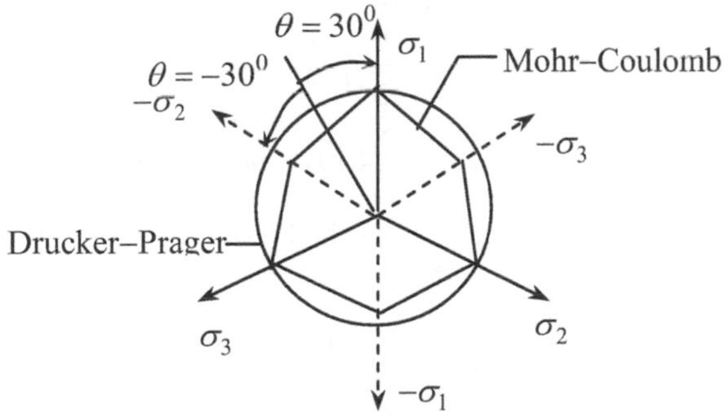

Figure 5.9 Drucker–Prager and Mohr–Coulomb yield surface in the π-plane

Expanding this determinant results in the following eigenvalue equation for s:

$$s^3 - J_1 s^2 - J_2 s - J_3 = 0\,, \tag{5.41}$$

where

$$\begin{aligned} J_1 &= s_{ii} = s_1 + s_2 + s_3 = 0, \\ J_2 &= \frac{1}{2} s_{ij} s_{ij} = \frac{1}{2}(s_1^2 + s_2^2 + s_3^2) = -(s_1 s_2 + s_2 s_3 + s_1 s_3), \\ J_3 &= \det|\mathbf{s}| = \frac{1}{3} tr(\mathbf{s} \bullet \mathbf{s} \bullet \mathbf{s}) = s_1 s_2 s_3 \end{aligned} \tag{5.42}$$

where s_1, s_2, and s_2 are the principal values of s. The key factor that leads to the introduction of the Lode angle is the similarity between (5.41) and the following trigonometry identity:

$$\sin 3\theta = 3\sin\theta - 4\sin^3\theta\,, \quad or \quad \sin^3\theta - \tfrac{3}{4}\sin\theta + \tfrac{1}{4}\sin 3\theta = 0 \tag{5.43}$$

If we let $s = r\sin\theta$, (5.41) becomes

$$r^3 \sin^3\theta - J_2 r \sin\theta - J_3 = 0 \quad or \quad \sin^3\theta - \frac{J_2}{r^2}\sin\theta - \frac{J_3}{r^3} = 0 \tag{5.44}$$

Comparing (5.43) and (5.44), we immediately identify that

$$\frac{J_2}{r^2} = \frac{3}{4}, \quad \frac{J_3}{r^3} = -\frac{1}{4}\sin 3\theta \tag{5.45}$$

Solving (5.44) gives

$$r = \pm\frac{2\sqrt{J_2}}{\sqrt{3}}, \quad \sin 3\theta = \frac{3\sqrt{3}}{2}\left(\frac{J_3}{\sqrt{J_2^3}}\right) \tag{5.46}$$

where θ is called the Lode angle. It provides an attractive alternative to the J_3 invariant and is also a quantitative indicator of the relative magnitude of the intermediate principal stress to the maximum and minimum principal stresses, σ_2/σ_1 and σ_3/σ_1. Physically, r and θ correspond to the polar coordinates in the π-plane, which will be discussed in the next section. Since $\sin 3\theta$ is periodic such that $\sin 3\theta = \sin(3\theta + 2\pi)$; Nayak and Zienkiewicz (1972) proposed the following solutions for s:

$$s_1 = -\frac{2}{\sqrt{3}}\sqrt{J_2}\sin(\theta + \frac{4\pi}{3}) \tag{5.47}$$

$$s_2 = -\frac{2}{\sqrt{3}}\sqrt{J_2}\sin\theta \tag{5.48}$$

$$s_3 = -\frac{2}{\sqrt{3}}\sqrt{J_2}\sin(\theta + \frac{2\pi}{3}) \tag{5.49}$$

The validity of (5.47)–(5.49) can be checked easily by substituting them into (5.41). Then, the principal stresses become

$$\sigma_1 = -\frac{2}{\sqrt{3}}\sqrt{J_2}\sin(\theta + \frac{4\pi}{3}) + \frac{1}{3}\sigma_{kk} \tag{5.50}$$

$$\sigma_2 = -\frac{2}{\sqrt{3}}\sqrt{J_2}\sin\theta + \frac{1}{3}\sigma_{kk} \tag{5.51}$$

$$\sigma_3 = -\frac{2}{\sqrt{3}}\sqrt{J_2}\sin(\theta + \frac{2\pi}{3}) + \frac{1}{3}\sigma_{kk} \tag{5.52}$$

Equation (5.51) can be rearranged to give a simple definition for the Lode angle θ:

$$\sin\theta = \frac{\sigma_1 + \sigma_3 - 2\sigma_2}{2\sqrt{3J_2}} \tag{5.53}$$

Using the Mohr circle, the traditional Mohr–Coulomb failure criteria can be expressed as

$$\sin\phi = \frac{(\sigma_1 - \sigma_3)/2}{c'\cot\varphi + (\sigma_1 + \sigma_3)/2} \tag{5.54}$$

Rearranging (5.54) gives

$$\sigma_1 - \sigma_3 - (\sigma_1 + \sigma_3)\sin\phi = 2c'\cos\phi \tag{5.55}$$

We now note that

$$\begin{aligned}\sigma_1 + \sigma_3 &= \sigma_1 + \sigma_3 - \frac{2}{3}(\sigma_1 + \sigma_2 + \sigma_3) + \frac{2}{3}\sigma_{kk} \\ &= \frac{1}{3}(\sigma_1 + \sigma_3 - 2\sigma_2) + 2\sigma = \frac{2}{3}\sqrt{3J_2}\sin\theta + 2\sigma\end{aligned} \tag{5.56}$$

The last part of (5.56) is the result of utilizing (5.53). Substitution of (5.56) into (5.55) leads to

$$\sigma_1 - \sigma_3 - \frac{2\sqrt{3}}{3}\tau_e \sin\theta\ \sin\phi = 2c'\cos\varphi + 2\sigma\ \sin\phi \tag{5.57}$$

Finally substitution of (5.50) and (5.52) into (5.57) yields

$$\tau_e(\cos\theta - \frac{\sin\phi}{\sqrt{3}}\sin\theta) - \sin\phi\sigma - c'\cos\phi = 0\,. \tag{5.58}$$

If we now replace σ by σ (i.e., σ is the mean compression), comparison of (5.58) and (5.33) gives the required results given in (5.39).

5.10 YIELD CRITERIA ON THE π-PLANE

One of the most popular ways to visualize the yield function in plasticity is the use of the π–plane (see Fig. 5.9). Note that the Mohr–Coulomb yield criterion is an irregular hexagonal pyramid. Compression has been taken as positive in Fig. 5.9. The construction of Fig. 5.9 will be considered as problems at the end of the chapter. In particular, any stress field $\boldsymbol{\sigma}$ can be expressed in terms of its principal stresses, say σ_1, σ_2, and σ_3. Thus, the yield stress may be depicted by a three-dimensional surface in the principal stress-space, which is also called the *Haigh-Westergaard* space (Hill, 1950). Since all yield conditions are more or less governed by the magnitude of deviatoric stress or equivalently J_2, it is advantageous to see the yield surface through the hydrostatic axis, which is defined as the line

$$\sigma_1 = \sigma_2 = \sigma_3 \tag{5.59}$$

We then can define a cylindrical coordinate (r, θ, z) with the hydrostatic axis as the z-axis (see Fig. 5.10). The plane perpendicular to the hydrostatic axis is called the *deviatoric* or *octahedral* plane, and is given by

$$\sigma_1 + \sigma_2 + \sigma_3 = C \tag{5.60}$$

where C is a constant. When $C = 0$, the octahedral plane passes through the origin and is called the π-plane. The shape of the yield surface can be projected onto the π-plane for different levels of mean stress. The polar coordinates (r, θ) locates any stress point on the deviatoric plane.

By proper transformation any σ can be expressed in polar coordinates (r, θ, z). The angle between the hydrostatic axis and the σ_1-, σ_2-, and σ_3-axis is 54.74° [≈ $\cos^{-1}(1/\sqrt{3})$] (see Fig. 5.10). As shown in Fig. 5.11, any stress state (σ_1, σ_2, σ_3) can be represented by a vector OA. The unit normal along the hydrostatic axis ξ is

$$\boldsymbol{n} = \frac{1}{\sqrt{3}}(\boldsymbol{e}_1 + \boldsymbol{e}_2 + \boldsymbol{e}_3) \tag{5.61}$$

The projection of OA onto the ξ-axis is OB and its magnitude is

$$|OB| = \sigma_i n_i = \frac{1}{\sqrt{3}}(\sigma_1 + \sigma_2 + \sigma_3) = \frac{1}{\sqrt{3}} I_1 = \sqrt{3} p \tag{5.62}$$

As shown in Fig. 5.11, OB = (p, p, p) and OA = (σ_1, σ_2, σ_3). Therefore, BA = OA – OB = ($\sigma_1 - p$, $\sigma_2 - p$, $\sigma_3 - p$) = (s_1, s_2, s_3). Then, the length of vector BA is

$$\rho = \sqrt{s_1^2 + s_2^2 + s_3^2} = \sqrt{2J_2} = \sqrt{3} r / \sqrt{2} \tag{5.63}$$

where r is the Lode parameter discussed earlier in Section 5.9. Thus, physically r relates to the distance of the stress state from the origin of the π-plane. We now refer to the π-plane given in Fig. 5.12, in which the axes σ_1', σ_2', and σ_3' are the projection of σ_1, σ_2 and σ_3 on the π-plane. Let the unit vector along the σ_1'-axis be $\boldsymbol{n}^{(1)}$, which is perpendicular to $\boldsymbol{n}$ and therefore must satisfy:

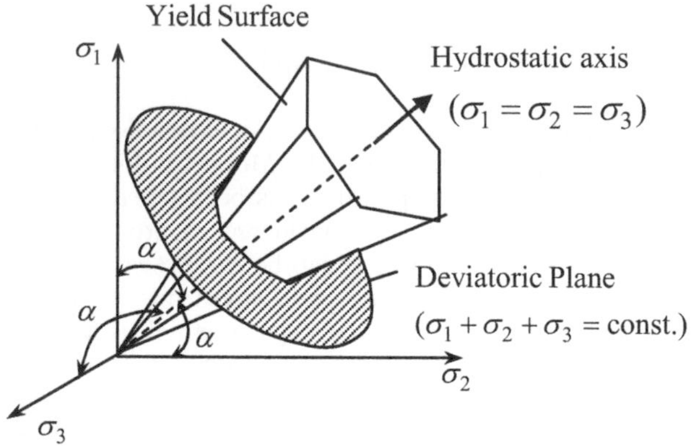

Figure 5.10 Yield surface representation in Haigh–Westergaard space and the π-plane

$$n^{(1)} \bullet n = (n_1 e_1 - n_2 e_2 - n_2 e_3) \bullet \frac{1}{\sqrt{3}}(e_1 + e_2 + e_3) = \frac{1}{\sqrt{3}}(n_1 - 2n_2) = 0 \quad (5.64)$$

where e_i (i = 1, 2, 3) is the base vector in the Haigh–Westergaard stress space. However, we also have $|n^{(1)}| = 1$ or

$$n_1^2 + 2n_2^2 = 1 \quad (5.65)$$

Solving (5.64) and (5.65), we obtain the following unit vector along σ_1'-axis:

$$n^{(1)} = \frac{1}{\sqrt{6}}(2e_1 - e_2 - e_3) \quad (5.66)$$

Note that BA = B′A′, thus B′C′ equals

$$\begin{aligned} \rho \cos\theta = \mathbf{B'A'} \times n^{(1)} &= (s_1 e_1 + s_2 e_2 + s_3 e_3) \times \frac{1}{\sqrt{6}}(2e_1 - e_2 - e_3) \\ &= \frac{1}{\sqrt{6}}(2s_1 - s_2 - s_3) \end{aligned} \quad (5.67)$$

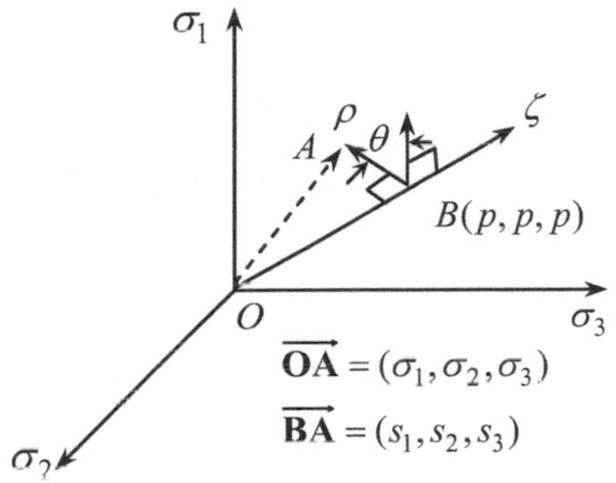

Figure 5.11 Stress state in principal stress space (after Chen and Mizuno (1990) with permission from Elsevier)

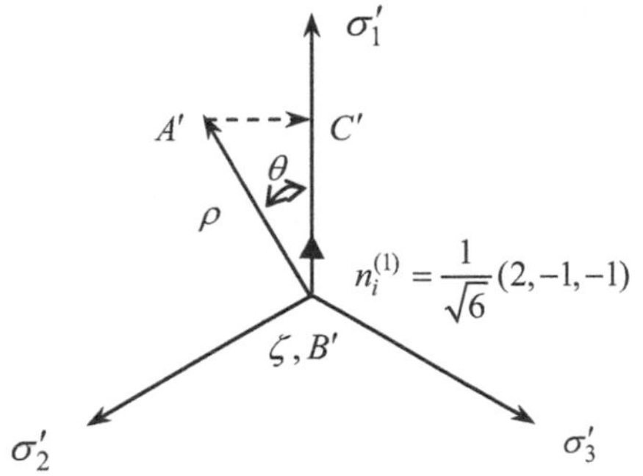

Figure 5.12 Stress state on the π-plane with $\sigma_1 > \sigma_2 > \sigma_3$ (after Chen and Mizuno (1990) with permission from Elsevier)

Since $J_1 = s_1+s_2+s_3 = 0$, therefore we have

$$\cos\theta = \frac{3}{\sqrt{6}}\frac{s_1}{\rho} = \frac{\sqrt{3}}{2}\frac{s_1}{\sqrt{J_2}} \tag{5.68}$$

The second part of (5.68) is obtained by using the result in (5.63). Using the trigonometric identity $\cos 3\theta = 4\cos^3\theta - 3\cos\theta$, we have

$$\cos 3\theta = \frac{3\sqrt{3}}{2\,J_2^{3/2}}(s_1^3 - s_1 J_2) \tag{5.69}$$

Noting that $J_2 = -(s_1s_2+s_2s_3+s_3s_1)$, $s_1+s_2+s_3 = 0$ and $J_3 = s_1s_2s_3$, we get

$$\theta = \frac{1}{3}\cos^{-1}[3\frac{\sqrt{3}}{2}\frac{J_3}{J_2^{3/2}}] \quad \text{where} \quad 0 \le \theta \le \pi/3 \tag{5.70}$$

for $\sigma_1 \ge \sigma_2 \ge \sigma_3$. This is another form of the Lode angle discussed in Section 5.9.

5.11 OTHER SOIL YIELD MODELS

The main disadvantage of the Mohr–Coulomb failure model is that there are sharp corners on the yield surface, as shown in Fig. 5.9. When numerical methods are used to solve plasticity problems, special care is needed to handle yield stress at the corner. Therefore, various models have been proposed to smooth out the yield corners. For example, Lade and Duncan (1975) proposed the following form of yield function:

$$f = I_1^3 / I_3 - k_1 = 0 \tag{5.71}$$

This yield function can be rewritten as (Chen and Saleeb, 1988)

$$f = \frac{2}{3\sqrt{3}}J_2^{3/2}\cos 3\theta - \frac{1}{3}I_1 J_2 + (\frac{1}{27} - \frac{1}{k_1})I_1^3 = 0 \tag{5.72}$$

where

$$k_1 = \frac{[\alpha(1+b)+(2-b)]^3}{b\alpha^2+(1-b)\alpha}, \quad \alpha = \frac{1+\sin\phi}{1-\sin\phi}, \quad b = \frac{\sigma_2-\sigma_3}{\sigma_1-\sigma_3} \tag{5.73}$$

Matsuoka and Nakai (1974) proposed

$$f = \frac{I_1 I_2}{I_3} - (9+8\tan^2\phi) = 0 \tag{5.74}$$

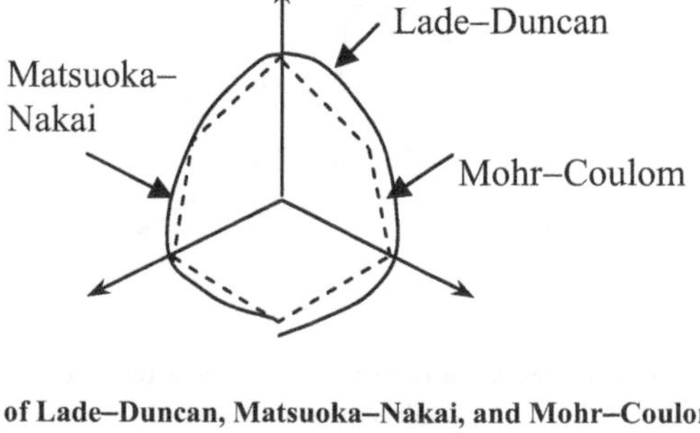

Figure 5.13 Comparison of Lade–Duncan, Matsuoka–Nakai, and Mohr–Coulomb yield functions (after Davis and Selvadurai (2002) with permission from Cambridge University Press)

In both of these models, the sharp corners on the yield surface have been smoothed out, while the shape of the Coulomb yield surface remains, as shown in Fig. 5.13.

5.12 CAP MODELS

For both granular and cohesive soils, one main restriction of the previous models is that the so-called *critical state concept* is not incorporated. The critical state idea can be illustrated using Figs. 5.14 and 5.15 (Schofield and Wroth, 1968; Atkinson and Bransby, 1978). In particular, we consider a *normally consolidated soil* 2 and an *overconsolidated soil* 1 having the same void ratio, and both are loaded by the conventional triaxial test to failure. Note that the triaxial test was originally proposed by von Karman, the father of rocket science, in 1910 for testing Carrara marble and Mutenberg sandstone (von Karman and Edison, 1967; Van and Vasarhelyi, 2010). It was subsequently adopted for soil testing. When an *undrained* compression is applied, both specimens fail at point A (following the solid lines in Fig. 5.14). The normally consolidated soil fails with a positive pore pressure, and the overconsolidated soil fails with a negative pore pressure. However, for *drained compression test* specimens 1 and 2 behave quite differently. Specimen 2 contracts under drained compression and fails at point B in Figs. 5.14 and 5.15, while specimen 1 initially contracts before reaching its peak strength at point C, then dilates as it reaches the residual strength point D (following the dotted lines in Fig. 5.14). The important fact is that all specimens whether normally consolidated or overconsolidated will fail along the *critical state line* in the *p*-*q* plane as shown in Fig. 5.14, provided that all soil specimens have the same void ratio.

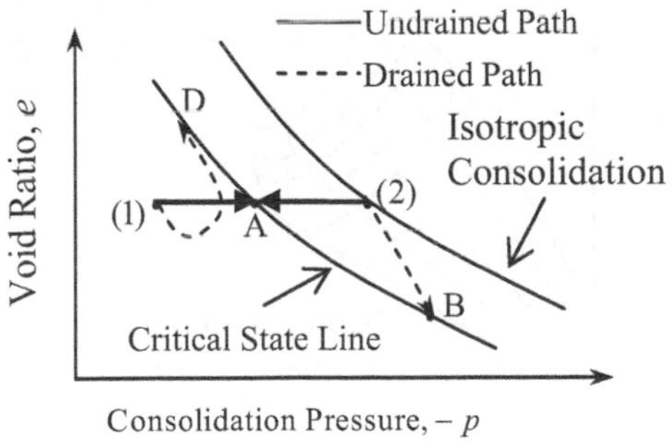

Figure 5.14 The loading response in the *e*-*p* plane (after Chen and Mizuno (1990) with permission from Elsevier)

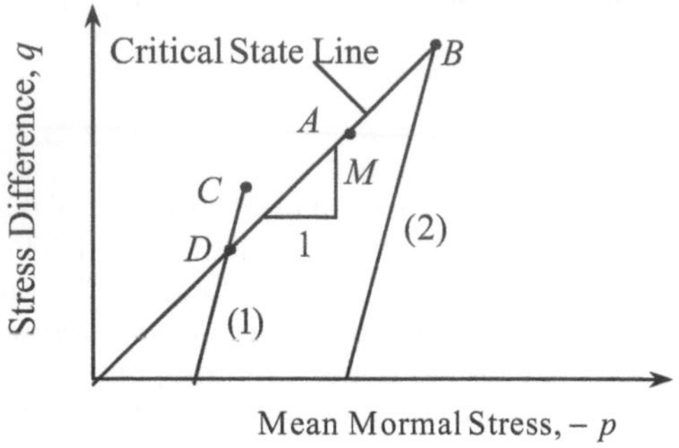

Figure 5.15 Drain response in the *p-q* plane (after Chen and Mizuno (1990) with permission from Elsevier)

To capture such a void ratio dependency, a cap model was proposed by Drucker et al. (1957). As shown in Fig. 5.16, a spherical cap is introduced to restrict the plastic dilatancy in the model. Note that the level of the spherical cap depends on the current density or *void ratio* of the soils. Now, the yield surface consists of two surfaces—the usual failure surface f_f and a strain-hardening cap f_c. Mathematically, they are

$$f_f(I_1, J_2)=0, \quad f_c(I_1, J_2, \zeta)=0 \tag{5.75}$$

where ζ is a strain-hardening parameter and depends on the plastic volumetric strain as

$$\zeta = \zeta[tr(\boldsymbol{\varepsilon}^p)] \tag{5.76}$$

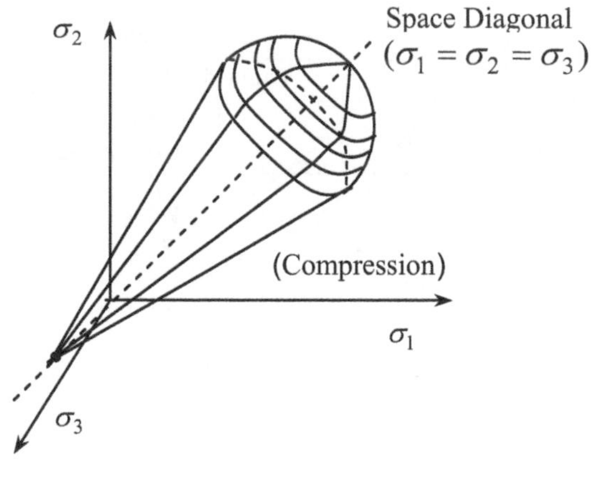

Figure 5.16 Drucker–Prager type of strain-hardening cap model (after Chen and Mizuno (1990) with permission from Elsevier)

5.13 PHYSICAL MEANING OF CAM-CLAY MODEL

With all the cap-type models, the *Cam-clay* model developed by Roscoe and co-workers at Cambridge (U.K.) is probably the most popular model. The Cam-clay model can capture the experimentally observed critical state. Following the presentation by Davis and Selvadurai (2002), we illustrate in this section a physical meaning of the Cam-clay model before we discuss the modified Cam-clay model in the next section.

Consider the shear experiment shown in Fig. 5.17 where shear stress τ is applied with normal stress σ. There is no lateral normal strain in the experiment but dilatancy is allowed. The yield condition is only a function of σ and τ, or $f(\sigma, \tau)$. The rate of plastic work for the specimen shown in Fig. 5.17 can be written as:

$$\dot{W}_p = \sigma\dot{\varepsilon}_p + \tau\dot{\gamma}_p \tag{5.77}$$

At a constant stress σ, the rate of dissipation D can be assumed proportional to the shear strain rate as well as the stress level as

$$\dot{D} = k\sigma\dot{\gamma}_p \tag{5.78}$$

With this equation, we can equate (5.77) and (5.78) to give

$$\frac{\dot{\varepsilon}_p}{\dot{\gamma}_p} = k - \frac{\tau}{\sigma} \tag{5.79}$$

where k is a material constant. According to Drucker's postulate discussed in Section 5.6, a small perturbation of the stress state from an equilibrium state must result in a positive rate of plastic work:

$$\delta\dot{W}_p = \delta\sigma\dot{\varepsilon}_p + \delta\tau\dot{\gamma}_p \geq 0 \tag{5.80}$$

For the case of normality (taking the equality sign in (5.80)), we have the following differential equation by replacing the incremental form by the differential form.

$$\frac{d\tau}{d\sigma} + \frac{\dot{\varepsilon}_p}{\dot{\gamma}_p} = \frac{d\tau}{d\sigma} + k - \frac{\tau}{\sigma} = 0 \tag{5.81}$$

The second part of (5.81) is a consequence of (5.79). Integrating the equation gives

$$\tau = \sigma(C_1 - k\ln\sigma) \tag{5.82}$$

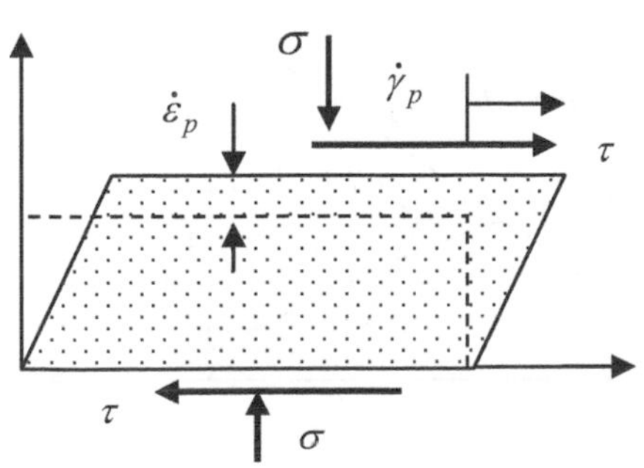

Figure 5.17 Response of idealized soil to hydrostatic pressure

To find the constant C_1, we can observe from (5.79) that $\sigma > \tau/k$ implies compression, and $\sigma < \tau/k$ implies dilation. Clearly, when σ equals τ/k, there is no volume change or it is at its so-called critical state. Using this condition, we find that

$$C_1 = k(1 + k\ln\sigma_c) \tag{5.83}$$

where $\sigma_c = \tau/k$. Substitution of (5.83) into (5.82) leads to

$$\tau + k\sigma\left[\ln(\frac{\sigma}{\sigma_c}) - 1\right] = 0 \tag{5.84}$$

If we extend this idea of yielding for simple shear test shown in Fig. 5.17 to the 3-D situations, we can replace σ and τ by p and q and the yield function given in (5.84) becomes

$$q + Mp\left[\ln(\frac{p}{p_c}) - 1\right] = 0 \tag{5.85}$$

where M is a material parameter and p and q are defined as

$$q = \sigma_1 - \sigma_3, \quad q = \frac{1}{3}(\sigma_1 + \sigma_2 + \sigma_3) \tag{5.86}$$

The yield function (5.85) of the Cam-clay model can be visualized in Fig. 5.18. A modified version of Cam-clay model has been proposed by dropping the natural logarithm as

$$q + Mp\left[(p/p_c) - 1\right] = 0 \tag{5.87}$$

which is also plotted in Fig. 5.18.

Equations (5.85) and (5.87) are the yield functions for the Cam-clay and modified Cam-clay models.

5.14 MODIFIED CAM-CLAY MODEL

We present in this section the *modified Cam-clay model* by Roscoe et al. (1958) and Roscoe and Burland (1968). In particular, the equation for the virgin consolidation line shown in Fig. 5.19 is:

$$e = e_1 - \Lambda\ln(-p) \tag{5.88}$$

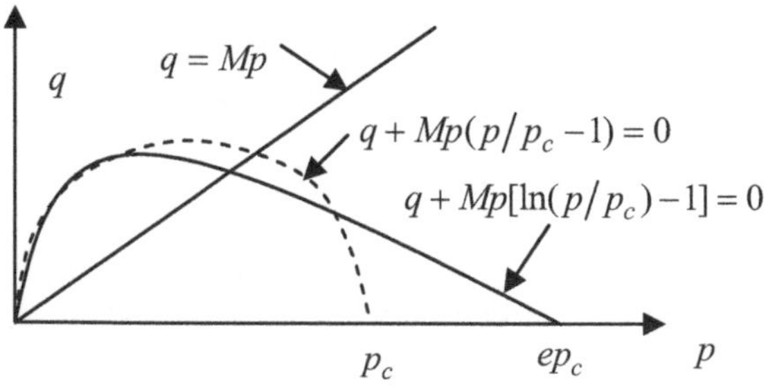

Figure 5.18 Yield functions for the Cam-clay model and modified Cam-clay model

where e is the void ratio and e_1 and Λ are material constants. Note that positive p is considered tension in (5.88). The rebound-reloading curve is similarly defined by

$$e = e_2 - \eta \ln(-p) \tag{5.89}$$

where e_2 and η are material constants.

The void ratio changes associated with the increase and the decrease in the hydrostatic pressure are, respectively,

$$de = -\Lambda \frac{dp}{p}, \quad de = -\eta \frac{dp}{p} \tag{5.90}$$

Since the volumetric strain can be related to the void ratio, we have

$$d\varepsilon_{kk} = \frac{de}{1+e} = -\frac{\Lambda dp}{(1+e)p} \tag{5.91}$$

The recoverable or elastic volumetric change is

$$d\varepsilon_{kk}^{e} = -\frac{\eta dp}{(1+e)p} \tag{5.92}$$

Combining (5.91) and (5.92) gives the irrecoverable or plastic volumetric strain:

$$d\varepsilon_{kk}^{p} = d\varepsilon_{kk} - d\varepsilon_{kk}^{e} = -\frac{(\Lambda - \eta)dp}{(1+e)p} \tag{5.93}$$

By (5.92) and the definition of the tangential elastic bulk modulus, we have

$$K = \frac{dp}{d\varepsilon_{kk}^{e}} = -\frac{(1+e)p}{\eta} \tag{5.94}$$

In this model, the elastic distortion is assumed negligible, i.e., $G \gg K$.

The modified Cam-clay yield surface and the critical state line are shown in Fig. 5.20. The yield curve is assumed to be elliptic shaped and is given as

$$f = p^2 - p_0 p + \frac{J_2}{M^2} = 0 \tag{5.95}$$

where M is a material constant and p_0 is a strain-hardening parameter. As shown in (5.42), the physical meaning of J_2 relates closely to the magnitude of deviatoric stress, and in turn to shear stress. Thus, (5.95) is equivalent to (5.87) and its origin was illustrated in Section 5.13. Alternatively, (5.95) can be expressed in terms of I_1 as

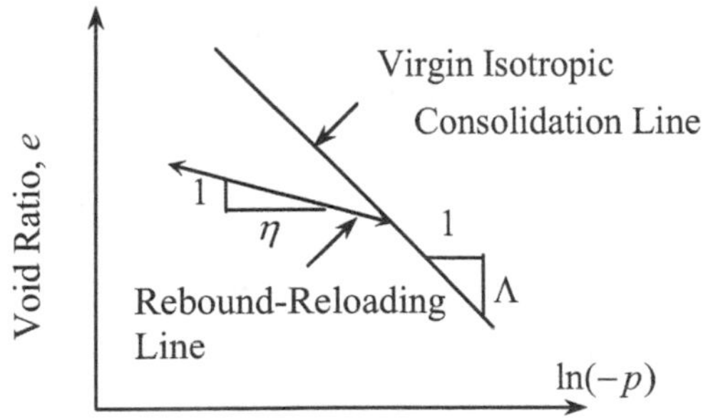

Figure 5.19 Response of idealized soil to hydrostatic pressure (after Chen and Mizuno (1990) with permission from Elsevier)

$$f = I_1^2 - I_1^0 I_1 + \frac{9J_2}{M^2} = 0 \tag{5.96}$$

where I_1^o is the value of I_1 at the yield cap, as shown in Fig. 5.20. In addition, since $I_1^o = 3\,p_o$, we have, using (5.93)

$$dI_1^0 = -\frac{(1+e)I_1^0}{\Lambda - \eta} d\varepsilon_{kk}^p \tag{5.97}$$

On the other hand, the critical state line, which controls the failure of the material, can be expressed as

$$\sqrt{J_2} = -Mp \tag{5.98}$$

Then, the stress state of soil must be within the region bounded by the elliptic curve and the critical state line, as shown in Fig. 5.21. In particular, Fig. 5.21 shows the loading path for a slightly overconsolidated clay at *A* to failure point at *D*, where the critical state line is reached (the lower curve passing point *D* in Fig. 5.21(a) and the inclined line passing point *D* in Fig. 5.21(b). The initial yielding starts at point *B* on the initial cap surface. Then, isotropic hardening occurs with the enlarged cap surface at *C* and *D*.

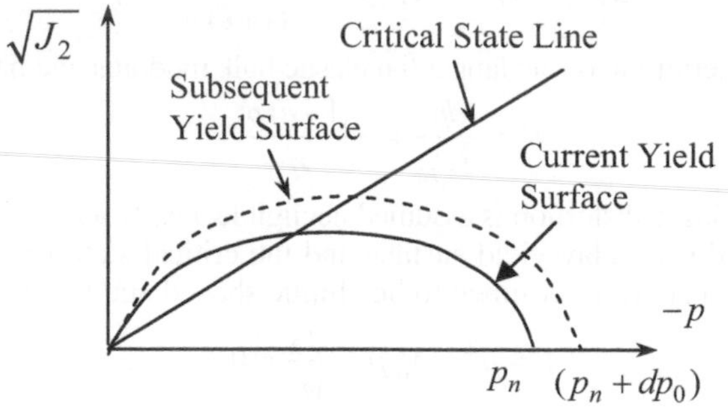

Figure 5.20 Modified Cam-clay yield surface in *p*-√*J*₂ space (after Chen and Mizuno (1990) with permission from Elsevier)

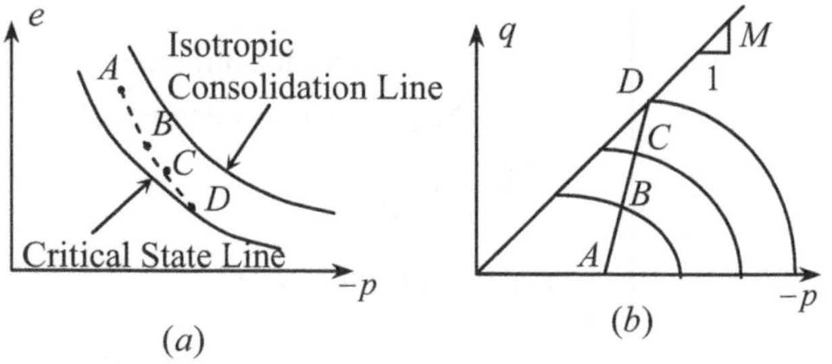

Figure 5.21 Modified Cam-clay in the *e*-*p* plane and *q*-*p* plane (after Chen and Mizuno (1990) with permission from Elsevier)

5.15 A CAM-CLAY MODEL FOR FINITE STRAIN

In this section, we present a generalized form of the Cam-clay model for finite strain that agrees with the form given in (5.11). This model is proposed by Yatomi et al. (1989). For the sake of simplicity, compression is treated as positive in this section. For fully saturated clay, the *effective Cauchy stress tensor* $\boldsymbol{\sigma}'$ is defined as

$$\boldsymbol{\sigma}' = \boldsymbol{\sigma} - u\boldsymbol{I} \tag{5.99}$$

where u is the pore water pressure in the clay, $\boldsymbol{\sigma}$ is the total stress, and $\boldsymbol{I}$ is again the second-order unit tensor.

The effective mean normal stress p' and the generalized stress deviator q are defined as

$$p' = \frac{1}{3} tr(\boldsymbol{\sigma}'),\ q = \sqrt{\frac{3}{2}}(s_{ij}s_{ij})^{1/2} = \sqrt{3}\tau_e \tag{5.100}$$

where τ_e is the stress as defined in (5.18).

We now recall the deformation tensor defined in (2.32) of Chapter 2 (i.e., $d\boldsymbol{x} = \boldsymbol{F} \cdot d\boldsymbol{X}$), and consider its time derivative as

$$d\boldsymbol{v} = \dot{\boldsymbol{F}} \bullet d\boldsymbol{X} = \dot{\boldsymbol{F}} \bullet \boldsymbol{F}^{-1} \bullet d\boldsymbol{x} = \boldsymbol{L} \bullet d\boldsymbol{x} = (\nabla \boldsymbol{v})^{\mathrm{T}} \bullet d\boldsymbol{x} \tag{5.101}$$

where $\boldsymbol{L}$ is the velocity gradient tensor defined in (2.31). The deformation gradient $\boldsymbol{F}$ is assumed to be smooth and the determinant is strictly positive, i.e.,

$$J = \det \boldsymbol{F} > 0 \tag{5.102}$$

Thus, the stretching tensor and spin tensors become

$$\boldsymbol{D} = \frac{1}{2}(\boldsymbol{L} + \boldsymbol{L}^T),\ \boldsymbol{\Omega} = \frac{1}{2}(\boldsymbol{L} - \boldsymbol{L}^T) \tag{5.103}$$

Again the decomposition of $\boldsymbol{D}$ into the elastic and plastic parts is presumed to be

$$\boldsymbol{D} = \boldsymbol{D}^e + \boldsymbol{D}^p \tag{5.104}$$

Equation (5.90) for the void ratio change is first rewritten as

$$\dot{e} = -\lambda \frac{\dot{p}'}{p'} \tag{5.105}$$

Similarly, the total and elastic volumetric strains given in (5.91 and 5.92) can be expressed as

$$\dot{v} = \frac{\lambda}{1+e}\frac{\dot{p}'}{p'} = tr(\boldsymbol{D}) = \frac{\dot{e}}{1+e},\quad \dot{v}^e = \frac{\kappa}{1+e}\frac{\dot{p}'}{p'} \tag{5.106}$$

where the slopes Λ and η in Fig. 5.19 are rewritten as λ and κ in (5.106). Then, the bulk and shear moduli are given as

$$\tilde{K} = \frac{1+e}{\kappa} p',\ \tilde{G} = \frac{3(1-2\nu)}{2(1+\nu)}\tilde{K} \tag{5.107}$$

The inelastic volumetric strain given in (5.93) becomes

$$\dot{v}^p = \frac{\lambda - \kappa}{1+e}\frac{\dot{p}'}{p'} \tag{5.108}$$

which, however, only takes into account the inelastic term due to the difference between the loading and unloading terms. Yatomi et al. (1989) proposed one more plastic volumetric strain term due to dilatancy (i.e., volumetric change due to shear) as

$$\dot{v}^p = D\frac{d}{dt}(\frac{q}{p'}) = \frac{\lambda-\kappa}{M(1+e)}\frac{d}{dt}(\frac{q}{p'}) \tag{5.109}$$

where D is the coefficient of dilatancy. The last term of (5.109) results from the definition of M given in Fig. 5.15 (i.e., $q = Mp'$). Therefore, the total volumetric plastic strain becomes

$$\dot{v}^p = \frac{\lambda-\kappa}{1+e}\frac{\dot{p}'}{p'} + D\frac{d}{dt}(\frac{q}{p'}) \tag{5.110}$$

In general, for finite strain e is not equal to e_0. Let us consider the change in volume of a small element of the original volume $dV_0 = dX_1dX_2dX_3$ due to the displacement $\boldsymbol{r}$ (e.g., Spiegel, 1963):

$$\begin{aligned} dV &= \left|\frac{\partial \boldsymbol{r}}{\partial X_1}\bullet(\frac{\partial \boldsymbol{r}}{\partial X_2}\times\frac{\partial \boldsymbol{r}}{\partial X_3})\right| dX_1 dX_2 dX_3 = e_{mij}\frac{\partial x_m}{\partial X_1}\frac{\partial x_i}{\partial X_2}\frac{\partial x_j}{\partial X_3}dV_0 \\ &= e_{mij}F_{m1}F_{i2}F_{j3}dV_0 = \det(\boldsymbol{F})dV_0 = JdV_0 \end{aligned} \tag{5.111}$$

where J is defined in (5.102) and also in (2.10) of Chapter 2, and (1.17) of Chapter 1 is used in obtaining (5.111). Consequently, we have equivalently: $1 + e = J(1 + e_0)$. Then, the time derivative of (5.111) yields

$$d\dot{V} = \dot{J}\,dV_0 = \frac{\dot{J}}{J}dV \tag{5.112}$$

Therefore, we have

$$\frac{\dot{J}}{J} = tr(\boldsymbol{D}) \tag{5.113}$$

Then, integration of (5.113) leads to

$$J = \exp[\textstyle\int_0^t \mathrm{tr}(\boldsymbol{D})dt] \tag{5.114}$$

Then the current volume or void ratio e can be calculated using (5.114). Similarly to the definition for yield function for small strain, Yatomi et al. (1989) proposed the following yield function for finite strain theory

$$f = \int_0^t (\frac{\lambda-\kappa}{1+e}\frac{\dot{p}'}{p'} + D\dot{\eta})dt - v^p \tag{5.115}$$

where $v^p = 0$ at $t = 0$, and $\eta = q/p'$. Differentiating (5.115) with respect to time gives

$$\dot{f} = P_{ij}\frac{\partial \sigma'_{ij}}{\partial t} - \dot{v}^p \tag{5.116}$$

where P_{ij} can be found to be (see Problem 5.3):

$$P_{ij} = D(\frac{M}{p'}\frac{\partial p'}{\partial \sigma'_{ij}} + \frac{\partial \eta}{\partial \sigma'_{ij}}) = \frac{D}{p'}(\frac{3}{2\eta}\frac{s_{ij}}{p'} + \frac{1}{3}\beta\delta_{ij}) \tag{5.117}$$

and $\beta = M - \eta$.

The elastic part of $\boldsymbol{D}$ can be defined as (see (5.5))

$$\boldsymbol{\sigma}^{\nabla} = \boldsymbol{C}^e : \boldsymbol{D} \tag{5.118}$$

while the plastic part of $\boldsymbol{D}$ is defined as

$$\begin{aligned} D_{ij}^p &= \Lambda P_{ij} \quad \text{if} \quad f = 0 \quad \text{and} \quad P_{ij}\dot{\sigma}'_{ij} > 0 \\ &= 0 \quad \text{if} \quad f < 0 \quad \text{or} \quad \text{if} \quad f = 0 \quad \text{and} \quad P_{ij}\dot{\sigma}'_{ij} \le 0 \end{aligned} \tag{5.119}$$

depending on whether continuous loading or unloading occurs. By using (see Problem 5.4)

$$P_{ij}\dot{\sigma}'_{ij} = P_{ij}\overset{\nabla}{\sigma}_{ij} \tag{5.120}$$

and (5.104), (5.116) with $f = 0$, (5.118), and (5.119), we finally have (see Problem 5.5)

$$\Lambda = \frac{\boldsymbol{P} : \boldsymbol{C}^e : \boldsymbol{D}}{\boldsymbol{P} : \boldsymbol{C}^e : \boldsymbol{P} + \mathrm{tr}(\boldsymbol{P})} \tag{5.121}$$

Finally, we have the following generalized form of constitutive law similar to the Rudnicki–Rice (1975) model given in (5.21):

$$\overset{\nabla}{\sigma}_{kl} = \{\tilde{G}(\delta_{mk}\delta_{nl} + \delta_{ml}\delta_{kn}) + (\tilde{K} - \tfrac{2}{3}\tilde{G})\delta_{kl}\delta_{mn} - \frac{(\frac{\tilde{G}}{\tau_e}s_{kl} + \beta\tilde{K}\delta_{kl})(\frac{\tilde{G}}{\tau_e}s_{mn} + \beta\tilde{K}\delta_{mn})}{h + \tilde{G} + \tilde{K}\bar{\beta}^2}\}D_{mn} \tag{5.122}$$

where $\bar{\beta} = \beta/\sqrt{3}$ and h is the hardening modulus given by

$$h = \frac{p'\bar{\beta}}{\sqrt{3}D} \tag{5.123}$$

When comparing to (5.21), it is clear that this model is associative. A non-coaxial model (i.e., the inelastic strain increment is not directly parallel to the stress increment in the strain and stress spaces, respectively) similar to that of Rudnicki and Rice (1975) was also given by Yatomi et al. (1989), but the details will not be discussed here. More discussion on finite deformation plasticity is given by Nemat-Nasser (1983, 2004).

5.16 PLASTICITY BY INTERNAL VARIABLES

Instead of assuming the existence of plastic potential and yield function (i.e., g and f introduced in Section 5.3), a more recent approach for estimating the plastic or inelastic deformations in nonlinear solids is to formulate the macroscopic deformation as a function of the internal variable change (such as the change in damage or slippage at the microscopic level). Mechanisms of the deformation at the microscopic level, say involving changes in microcracks, micropores, or microdefects, and their relationship to the macroscopic deformation form a new branch of mechanics called *micromechanics*. We refer the reader to the comprehensive book by Mura (1987).

In this section, we will summarize briefly the essential form being used for the approach (e.g., Rice, 1971, 1975; Hill and Rice, 1973; Nemat-Nasser, 1983). The formulation starts with the Helmholtz free energy, which is assumed to be a function of a strain measure $\boldsymbol{E}$ (a second-order tensor), the temperature θ, and a set of internal variables $\boldsymbol{\xi}$ (a first-order tensor):

$$\phi = \phi(\boldsymbol{E}, \theta, \boldsymbol{\xi}) \tag{5.124}$$

Then, the stress $\boldsymbol{S}$ conjugate to the strain measure (i.e., the double dot product between $d\boldsymbol{E}$ and $\boldsymbol{S}$ equals the external work), entropy η, and the thermodynamic force $\boldsymbol{\Lambda}$ acting on $\boldsymbol{\xi}$ are

$$S = \frac{\partial \phi}{\partial \boldsymbol{E}}, \quad \eta = -\frac{\partial \phi}{\partial \theta}, \quad \boldsymbol{\Lambda} = -\frac{\partial \phi}{\partial \boldsymbol{\xi}} = \Lambda(\boldsymbol{E}, \theta, \boldsymbol{\xi}) \tag{5.125}$$

The total change in ϕ can be written as

$$d\phi = d^e\phi + d^{in}\phi = [\boldsymbol{S} : d\boldsymbol{E} - \eta d\theta] - \boldsymbol{\Lambda} \cdot d\xi \tag{5.126}$$

The elastic deformation only leads to $d^e\phi$ (the change in ϕ due to elastic deformation), while inelastic deformation at constant $\boldsymbol{E}$ and ϕ contributes only to $d^{in}\phi$ (the change in ϕ due to inelastic deformation). The time derivative of $\boldsymbol{S}$ can be expressed as

$$\dot{\boldsymbol{S}} = \dot{\boldsymbol{S}}^{\mathbf{e}} + \dot{\boldsymbol{S}}^{\mathbf{in}} = \left\{ \boldsymbol{L} : \dot{\boldsymbol{E}} + \frac{\partial \boldsymbol{S}}{\partial \theta} \dot{\theta} \right\} + \frac{\partial \boldsymbol{S}}{\partial \xi_i} \dot{\xi}_i \tag{5.127}$$

where

$$\boldsymbol{L} = \frac{\partial^2 \phi}{\partial \boldsymbol{E} \partial \boldsymbol{E}} \tag{5.128}$$

is the tangential elastic modulus tensor, which in general changes with the deformation process. As shown by Hill and Rice (1973), the inelastic stress rate can also be written as

$$\dot{\boldsymbol{S}}^{\mathrm{in}} = \frac{\partial^2 \phi}{\partial \xi_i \partial \boldsymbol{E}} \dot{\xi}_i = \frac{\partial}{\partial \boldsymbol{E}} \dot{\phi}^{in} \tag{5.129}$$

Conversely, by applying the following Legendre transformation for the complementary potential (see Appendix C)

$$\psi = \boldsymbol{S} : \boldsymbol{E} - \phi = \psi(\boldsymbol{S}, \theta, \xi) \tag{5.130}$$

we have

$$\dot{\mathbf{E}} = \dot{\mathbf{E}}^{\mathbf{e}} + \dot{\mathbf{E}}^{\mathbf{in}} = \left\{ \boldsymbol{M} : \dot{\mathbf{S}} + \frac{\partial \mathbf{E}}{\partial \theta} \dot{\theta} \right\} + \frac{\partial \mathbf{E}}{\partial \xi_i} \dot{\xi}_i \tag{5.131}$$

where

$$\boldsymbol{M} = \frac{\partial^2 \psi}{\partial \mathbf{S} \partial \mathbf{S}}, \quad \mathbf{E} = \frac{\partial \psi}{\partial \mathbf{S}}, \quad \eta = \frac{\partial \psi}{\partial \theta}, \quad \boldsymbol{\Lambda} = \frac{\partial \psi}{\partial \boldsymbol{\xi}} \tag{5.132}$$

This approach looks very promising in estimating the constitutive response of rocks with respect to the damage evolution. As shown by Chau and Wong (1997), the effective moduli of a solid containing microcracks can be found by following a similar idea. See also Section 6.16 for the formulation of continuum damage mechanics by following a similar approach.

5.17 VISCOPLASTICITY

5.17.1 One-Dimensional Model

The simplest form of dynamic constitutive law of a viscoplastic solid is the Bingham fluid model (Lubliner, 1990). For a more general form, the constitutive form can be written as

$$\boldsymbol{\sigma} = \boldsymbol{\varphi}(\boldsymbol{\varepsilon}^p, \dot{\boldsymbol{\varepsilon}}^{vp}) \tag{5.133}$$

That is, the stress is a function of both plastic strain and viscoplastic strain rate. For the case of a one-dimensional problem, the simplest viscoplastic model can be modeled by the dashpot-spring model with a yield stress Y, as shown in Fig. 5.22. The elastic deformation is modeled by the spring with Young's modulus of E when the applied stress σ is less than the yield stress Y. Once Y is exceeded, viscoplastic yielding occurs and the additional deformation is modeled by the dashpot. Except for the yield stress lock, the model is exactly same as that for the Maxwell viscoelastic model to be discussed in Chapter 7. The strain rate of the model can be formulated as

$$\dot{\boldsymbol{\varepsilon}} = \dot{\boldsymbol{\varepsilon}}^e + \dot{\boldsymbol{\varepsilon}}^{vp} = \frac{\dot{\boldsymbol{\sigma}}}{E} + \frac{\langle \Phi \rangle}{\eta}\,\mathrm{sgn}(\sigma) \tag{5.134}$$

where the viscosity is given by η and Φ is the overstress function. The second term in (5.134) leads to an inhomogeneous equation between the rate of stress and the rate of strain. It is this term that leads to creeping. The Macauley bracket ⟨ ⟩ has the following definition:

$$\begin{aligned} \langle \Phi \rangle &= \Phi \qquad \Phi > 0 \\ &= 0 \qquad \Phi \le 0 \end{aligned} \tag{5.135}$$

The simplest form of overstress function is given as

$$\Phi = |\sigma| - Y \tag{5.136}$$

This is equivalent to the Bingham flow model. Malvern (1951) gave the following form of viscoplastic strain based on experimental results:

$$\dot{\varepsilon}^{vp} = \frac{1}{\eta}[\exp(\frac{\sigma - f(\varepsilon)}{a}) - 1] \tag{5.137}$$

where a and η are material parameters. The static stress-strain function is given by $f(\varepsilon)$. To examine the time effect, we can consider a sudden imposed strain on the Bingham model as

$$\varepsilon = \varepsilon_0 H(t) \tag{5.138}$$

The associated stress is then

$$\sigma_0 = E\varepsilon_0 \tag{5.139}$$

This is a relaxation problem and it can be shown by taking the Laplace transform of (5.134) that the solution is (see Chapter 7 and Appendix B for details of the Laplace transform)

$$\sigma = Y + (\sigma_0 - Y)e^{-Et/\eta} \tag{5.140}$$

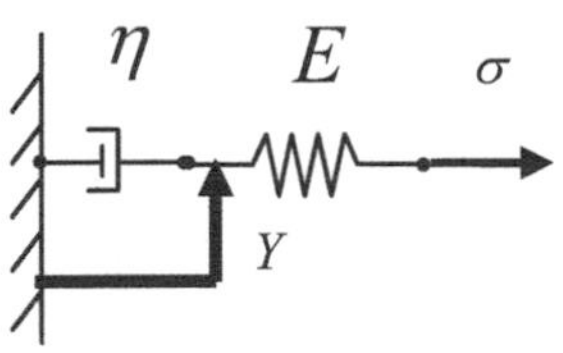

Figure 5.22 One-dimensional viscoplastic model

It is obvious that for $t \to 0$ we have $\sigma = \sigma_0$ and when $t \to \infty$ we have $\sigma \to Y$. One main feature of viscoplasticity is that the stress can (actually must) go outside the yield surface in order to have viscoplastic response, whereas for plasticity the applied stress can only be on the yield surface and cannot go beyond the yield surface.

5.17.2 Three-Dimensional Models

We can now extend (5.134) to the three-dimensional case:

$$\dot{\varepsilon}_{ij} = \dot{\varepsilon}_{ij}^{e} + \dot{\varepsilon}_{ij}^{vp} = \dot{\varepsilon}_{ij}^{e} + \frac{\langle \Phi \rangle}{\eta}\frac{\partial \Phi}{\partial \sigma_{ij}} \tag{5.141}$$

where Φ is the viscoplastic potential (Lubliner, 1990). The simplest model is given by the overstress function of (compare (5.136))

$$\Phi = \bar{\sigma} - Y \tag{5.142}$$

where Y is the yield stress and

$$\bar{\sigma} = (\frac{3}{2} s_{ij} s_{ij})^{1/2} = (3J_2)^{1/2} \tag{5.143}$$

When Φ is given by

$$\Phi = \sqrt{J_2} / k - 1 \tag{5.144}$$

where k is the yield shear stress, we recover the Hohenemser–Prager (1932) model. When the power law is used,

$$\Phi = Y_0 (\frac{\bar{\sigma}}{Y} - 1)^m \tag{5.145}$$

where m is the rate sensitivity exponent, we recover the Bodner and Symonds (1960) model. Therefore, with (5.141) we have

$$\dot{\sigma}_{ij} = C_{ijkl}^{e} (\dot{\varepsilon}_{kl} - \frac{\langle \Phi \rangle}{\eta}\frac{\partial \Phi}{\partial \sigma_{ij}}) \tag{5.146}$$

Another popular viscoplastic model is given by Perzyna (1963), and various versions of this model have also been used for clay and other geomaterials (e.g., Adachi and Oka, 1982; Kimoto et al., 2004). In particular, for isotropic solids, the Perzyna (1963) model is in the following form:

$$\dot{\varepsilon}_{ij} = \frac{1}{2\mu}\dot{s}_{ij} + \frac{1-2\nu}{3E}\dot{\sigma}_{kk}\delta_{ij} + \gamma \langle \Phi(F) \rangle \frac{\partial f}{\partial \sigma_{ij}} \tag{5.147}$$

where γ is the viscosity of soil and F is the static yield function and is defined as

$$F = \frac{1}{\kappa} f(\sigma_{ij}, \varepsilon_{ij}^{vp}) - 1 \tag{5.148}$$

The function $\Phi(F)$ should be determined by experiments but is expected as a function of stress difference (or the so-called overstress function as defined in (5.144) or (5.145)). The material constant κ is a function of the hardening parameter. When there is no viscoplastic effect, we have the yield function as

$$\varphi(\sigma_{ij}, \varepsilon_{ij}^{vp}, \kappa) = \kappa F = f(\sigma_{ij}, \varepsilon_{ij}^{vp}) - \kappa = 0 \tag{5.149}$$

Therefore, when $F = 0$, we have the static yielding, whereas when $F > 0$, we have a dynamic effect with viscoplastic behavior. We can take the self-product of the plastic strain from (5.147) and obtain

$$\sqrt{\dot{I}_2} = \sqrt{\frac{1}{2}\dot{\varepsilon}_{ij}^{vp}\dot{\varepsilon}_{ij}^{vp}} = \gamma\Phi(F)\sqrt{\frac{1}{2}\frac{\partial f}{\partial \sigma_{ij}}\frac{\partial f}{\partial \sigma_{ij}}} \tag{5.150}$$

If the inversion of Φ exists, we can write f as

$$f = \kappa\{1 + \Phi^{-1}[\frac{1}{\gamma}\sqrt{2\dot{I}_2 / (\frac{\partial f}{\partial \sigma_{ij}}\frac{\partial f}{\partial \sigma_{ij}})}]\} \tag{5.151}$$

since for static yielding

$$\kappa F = f(\sigma_{ij}, \dot{\varepsilon}_{ij}^{p}) - \kappa(\alpha) = 0 \tag{5.152}$$

where

$$\alpha = \int \sigma_{ij} d\varepsilon_{ij}^{p} \tag{5.153}$$

The static and dynamic yielding overlaps for the case of zero viscoplastic strain. For dynamic yielding, we can assume

$$\varphi(\sigma_{ij}, \varepsilon_{ij}^{vp}, \kappa, \dot{\varepsilon}_{ij}^{vp}) = F(\sigma_{ij}, \varepsilon_{ij}^{vp}, \kappa) - H(\sqrt{\dot{I}_2}) = 0 \tag{5.154}$$

It is clear from (5.154) and (5.151) that

$$H(\sqrt{\dot{I}_2}) = \Phi^{-1}[\frac{1}{\gamma}\sqrt{2\dot{I}_2 / (\frac{\partial f}{\partial \sigma_{ij}}\frac{\partial f}{\partial \sigma_{ij}})}] \tag{5.155}$$

The yield surface $\varphi = 0$ expands uniformly with the loading. The viscoplastic strain rate of the Perzyna (1963) can be expressed as

$$\dot{\varepsilon}_{ij}^{vp} = [\sqrt{2\dot{I}_2 / (\frac{\partial f}{\partial \sigma_{mn}}\frac{\partial f}{\partial \sigma_{mn}})}]\frac{\partial f}{\partial \sigma_{ij}} \tag{5.156}$$

The behavior of the Perzyna (1963) model is somewhat similar to the Maxwell model shown in Fig. 5.22. It is also clear from (5.154) and (5.151) that the normality rule has been assumed in the Perzyna (1963) model. We can also consider the special case that $F = 0$ (static yielding), $\Phi(F) \to 0$, and $\gamma \to \infty$ such that $\Phi(F)\,\gamma = d\lambda$. It is clear that the mathematical form for the plastic strain given in (5.8) is recovered.

For geomaterials, the following power law has been adopted for $\Phi(f)$ as (Desai and Zhang, 1987)

$$\Phi(f) = (\frac{f}{Y})^m \tag{5.157}$$

where $m \geq 1$ and Y is the initial yield stress.

5.17.3 Consistency Condition for Perzyna Model

Instead of assuming the yield function and plastic potential are the same, we can also modify Perzyna's formulation slightly to the non-normality rule. In particular, we can assume

$$\dot{\varepsilon}_{ij}^{vp} = \gamma \langle \Phi(f) \rangle \frac{\partial g}{\partial \sigma_{ij}} \tag{5.158}$$

To consider the consistency condition for the Perzyna model, we now use the normality flow rule for the viscoplastic strain as

$$\dot{\varepsilon}_{ij}^{vp} = \dot{\lambda} \frac{\partial g}{\partial \sigma_{ij}} \tag{5.159}$$

The consistency parameter can be related to Perzyna's formulation given in (5.158) for the case of loading as

$$\dot{\lambda} = \gamma \Phi(f) \tag{5.160}$$

Taking the derivative of (5.160) one more time gives

$$\ddot{\lambda} = \gamma \frac{d\Phi}{df} \dot{f} \tag{5.161}$$

The function f in Perzyna's formulation is only a function of stress $\boldsymbol{\sigma}$ and λ, and thus

$$\dot{f} = \frac{\partial f}{\partial \boldsymbol{\sigma}} : \dot{\boldsymbol{\sigma}} - h\dot{\lambda} \tag{5.162}$$

Substitution of (5.162) into (5.161) gives

$$\frac{\partial f}{\partial \boldsymbol{\sigma}} : \dot{\boldsymbol{\sigma}} - h\dot{\lambda} - (\gamma \frac{d\Phi}{df})^{-1} \ddot{\lambda} = 0 \tag{5.163}$$

5.17.4 Consistency Model of Wang et al. (1997)

On the other hand, a rate-dependent yield function has been proposed by Wang et al. (1997) as:

$$f_{rd}(\sigma_{ij}, \kappa, \dot{\kappa}) = 0 \tag{5.164}$$

where

$$\dot{\kappa} = \sqrt{\frac{2}{3} \dot{\boldsymbol{\varepsilon}}^{vp} : \dot{\boldsymbol{\varepsilon}}^{vp}} \tag{5.165}$$

Substitution of (5.159) into (5.165) gives

$$\dot{\kappa} = \dot{\lambda} \sqrt{\frac{2}{3} \frac{\partial g}{\boldsymbol{\sigma}} : \frac{\partial g}{\boldsymbol{\sigma}}} \tag{5.166}$$

Then (5.164) can be written as

$$f_{rd}(\sigma_{ij}, \lambda, \dot{\lambda}) = 0 \tag{5.167}$$

The consistency condition requires

$$\dot{f}_{rd} = \frac{\partial f_{rd}}{\partial \boldsymbol{\sigma}} : \dot{\boldsymbol{\sigma}} - h\dot{\lambda} - \zeta \ddot{\lambda} = 0 \tag{5.168}$$

where

$$h = -\frac{\partial f_{rd}}{\partial \lambda}, \quad \zeta = -\frac{\partial f_{rd}}{\partial \dot{\lambda}} \tag{5.169}$$

Comparing (5.163) with (5.168), we have

$$\frac{\partial f}{\partial \boldsymbol{\sigma}}=\frac{\partial f_{rd}}{\partial \boldsymbol{\sigma}},\quad \zeta=(\gamma\frac{d\Phi}{df})^{-1} \tag{5.170}$$

Thus, although Perzyna's formulation given in (5.158) appears quite different from the consistency model of viscoplasticity, they are actually the same. This proof was first noted by Heeres et al. (2002). However, for the unloading behavior, these models are not the same.

If the yield function is independent of the strain-rate (i.e., $\zeta = 0$), we can recover the rate-independent plasticity model as

$$\dot{\varepsilon}_{ij}^{p}=\frac{1}{h}\frac{\partial f}{\partial \sigma_{kl}}\dot{\sigma}_{kl}\frac{\partial g}{\partial \sigma_{ij}} \tag{5.171}$$

Similar to our discussion for the Rudnicki–Rice model, we can take the inverse constitutive model as

$$\dot{\boldsymbol{\sigma}}=\left[\boldsymbol{C}^{e}-\frac{(\boldsymbol{C}^{e}:\boldsymbol{P})(\boldsymbol{Q}:\boldsymbol{C}^{e})}{h+\boldsymbol{Q}:\boldsymbol{C}^{e}:\boldsymbol{P}}\right]:\dot{\boldsymbol{\varepsilon}} \tag{5.172}$$

where

$$P_{ij}=\frac{\partial g}{\partial \sigma_{ij}},\quad Q_{ij}=\frac{\partial f}{\partial \sigma_{ij}} \tag{5.173}$$

5.17.5 Adachi-Oka (1982) Model

Based on experimental data on clay under constant strain rate, the following form of Perzyna's model is given by Adachi and Oka (1982) as:

$$\gamma\langle\Phi(f)\rangle=c\exp\{m'f\} \tag{5.174}$$

where c and m' are material constants and f is the yield function of the Cam-clay model. The following form of f is given by Kimoto et al. (2004):

$$f=\bar{\eta}_{(0)}^{*}+\tilde{M}^{*}\ln(\frac{\sigma_m'}{\sigma_{my}'}) \tag{5.175}$$

The mean stress is denoted by σ'_m and σ'_{my} is the mean effective stress in the static equilibrium state. The deviatoric stress to mean effective stress ratio is defined as

$$\eta^{*}=\sqrt{\eta_{mn}^{*}\eta_{mn}^{*}},\quad \eta_{ij}^{*}=\frac{s_{ij}}{\sigma_m'} \tag{5.176}$$

The subscript (0) indicates the state at the end of consolidation. In the normal consolidated (NC) region, we have

$$\tilde{M}^{*}=\text{constant} \tag{5.177}$$

and in the overconsolidated (OC) region, we have

$$\begin{aligned}\tilde{M}^{*}&=M_m^{*} && f_b\geq 0\\ &=-\frac{(\eta_{ij}^{*}\eta_{ij}^{*})^{1/2}}{\ln(\sigma_m'/\sigma_{mc}')} && f_b<0\end{aligned} \tag{5.178}$$

Thus, $f = 0$ implies no viscoplastic deformation. The viscoplastic potential is defined similarly to (5.175) as

$$g = \bar{\eta}^*_{(0)} + \tilde{M}^* \ln(\frac{\sigma'_m}{\sigma'_{mp}}) \tag{5.179}$$

In this model, it was assumed that there exists a boundary between the NC region and the OC region. This boundary is depicted by $f_b = 0$ with

$$f_b = \bar{\eta}^*_{(0)} + \tilde{M}^* \ln(\frac{\sigma'_m}{\sigma'_{mb}}) \tag{5.180}$$

As shown in (5.178), the NC and OC regions are defined, respectively, by

$$f_b \geq 0, \quad \text{and} \quad f_b < 0 \tag{5.181}$$

The physical meaning of the yield function, viscoplastic potential, and OC boundary used in the model is illustrated in Figs. 5.23 and 5.24 (Kimoto et al., 2004).

These functions eventually lead to the following viscoplastic strain (Kimoto et al., 2004):

$$\dot{\varepsilon}^{vp}_{ij} = c_0 \exp\left\{ m'[\bar{\eta}^*_{(0)} + \tilde{M}^* \ln(\frac{\sigma'_m}{\sigma'_{mb}})] \right\} \frac{(\eta^*_{ij} - \eta^*_{ij(0)})}{\bar{\eta}^*_{(0)}} \tag{5.182}$$

where c_0 and m' are constants to be determined by experiments. Note that in this model the stress state always exists outside the static yield function, and this is the feature of the so-called overstress model discussed earlier.

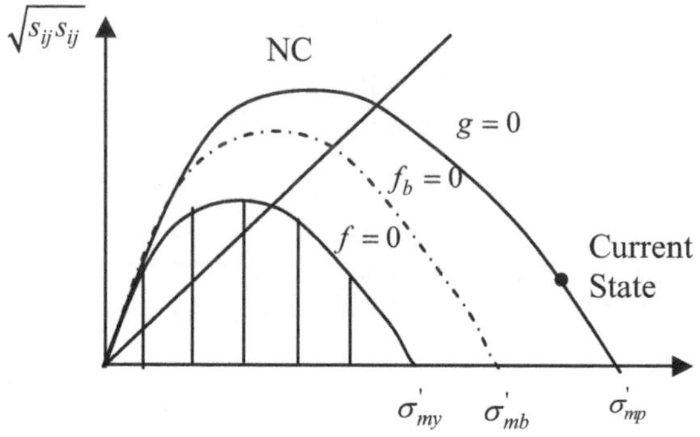

Figure 5.23 OC boundary, static yield function, and potential function in the NC region (after Kimoto et al. (2004) with permission from Elsevier)

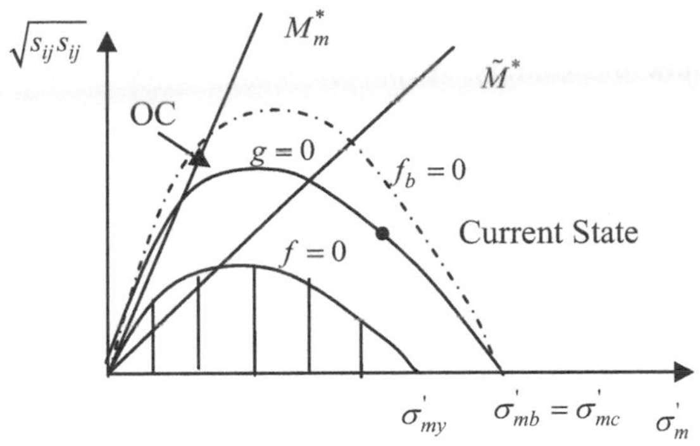

Figure 5.24 OC boundary, static yield function, and potential function in the OC region (after Kimoto et al. (2004) with permission from Elsevier)

5.18 SUMMARY AND FURTHER READING

One of the classic books on the mathematical theory of plasticity is by Hill (1950). The most up-to-date book on the mathematical theory of plasticity theory is by Lubliner (1990). These books do not pay special attention to geomaterials. There are a number of excellent and comprehensive books on soil plasticity. The most notable ones are by Chen (1975), Salencon (1974), Chen and Baladi (1985), Chen and Mizuno (1990), Yu (2006), Chen and Saleeb (1988), Davis and Selvadurai (2002), Pietruszczak (2010), Nakai (2012). The application of viscoplasticity to geomaterials is discussed by Cristescu (1994).

Some important topics in soil plasticity have not been included in this chapter. They are limit analysis for perfectly plastic materials, slip line theory, and numerical implementation. These topics have been covered in many of these books and outside of the scope of the present chapter. For example, Chen and Liu (1990) write extensively on limit analysis in soil mechanics.

Another important topic related to soil and rock plasticity is material instability and strain localization (see Section 9.8.2 for a brief discussion on the equivalence of strain localization and stationary acceleration wave). In soil, this strain localization is typically observed in the form of shear banding. A more recently recognized form of localization is called compaction band in porous rocks and has prompted considerable interest in this problem again (Issen and Rudnicki, 2000; Rudnicki, 2002; Wong and Baud, 1999; Wong et al., 2001). A tremendous amount of work has been conducted since the classic work of Rudnicki and Rice (1975). Instability of geomaterials due to geometry and boundary conditions can also appear in the form of diffuse mode bifurcations, including barreling and surface instability (Chau and Rudnicki, 1990; Chau, 1992, 1993, 1994a, 1995a; Vardoulakis, 1979, 1983). For bifurcation analysis in geomechanics, the readers are referred to the comprehensive book by Vardoulakis and Sulem (1996) and the review paper by Besuelle and Rudnicki (2004).

5.19 PROBLEMS

Problem 5.1 Show the validity of the Sherman–Morrison equation given in (5.16).

Problem 5.2 Show the validity of (5.21) for the Rudnicki–Rice (1975) model.

Problem 5.3 Show that (5.47)–(5.49) are the solution of (5.41).

Problem 5.4 Show the validity of (5.117) for P_{ij} in Yatomi's (1989) model by using the chain rule of differentiation.

Problem 5.5 Show the validity of (5.120). (*Hint*: it is a scalar equation and the effective Cauchy stress is symmetric.)

Problem 5.6 Show the validity of (5.121) for Λ in Yatomi's (1989) model (*Hint*: see the technique used in Section 5.4.)

Problem 5.7 The principal stress states at three different points (*A*, *B*, and *C*) in a solid are given as $(\sigma_1, \sigma_2, \sigma_3) = (12, 2, 4), (2, 4, 12), (2, 12, 4)$. They are given in the stress space in Fig. 5.25. Use (5.42) to find $I_1 (= \sigma_1 + \sigma_2 + \sigma_3)$, J_2, and J_3 for all stress points *A*, *B*, and *C*.

Problem 5.8 Show the validity of the following vectors in the π-plane shown in Fig. 5.26:

$$\overrightarrow{OD} = (6,6,6) \tag{5.183}$$

$$\overrightarrow{DA} = \overrightarrow{OA} - \overrightarrow{OD} = (6,-4,-2) \tag{5.184}$$

$$\overrightarrow{DB} = \overrightarrow{OB} - \overrightarrow{OD} = (-4,-2,6) \tag{5.185}$$

$$\overrightarrow{DC} = \overrightarrow{OC} - \overrightarrow{OD} = (-4,6,-2) \tag{5.186}$$

Problem 5.9 This problem continues from Problem 5.7. Prove that the distance of stress points A', B', and C' from the origin in the π-plane are the same. Use (5.42) and (5.70) to show ρ and θ in Fig. 5.26 as

$$\rho = 2\sqrt{14}\,, \quad \theta = 10.89^\circ \tag{5.187}$$

Problem 5.10 The principal stress states at three different points (*E*, *F*, and *G*) in a solid are given as $(\sigma_1, \sigma_2, \sigma_3) = (12, 9, 3), (9, 3, 12), (9, 12, 3)$. Use (5.42) to find $I_1 (= \sigma_1 + \sigma_2 + \sigma_3)$, J_2 and J_3 for all stress points *E*, *F*, and *G*.

Problem 5.11 Show the validity of the following vectors in the π-plane illustrated in Fig. 5.27:

$$\overrightarrow{HE} = (-4,1,-5) \tag{5.188}$$

$$\overrightarrow{HF} = (1,-5,4) \tag{5.189}$$

$$\overrightarrow{HG} = (1,4,-5) \tag{5.190}$$

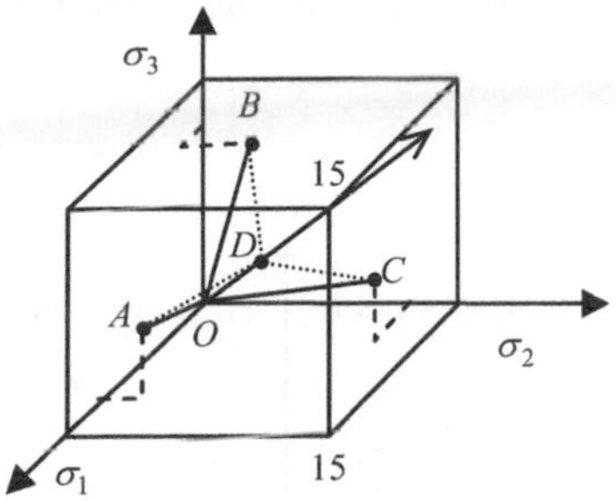

Figure 5.25 Locations of three points in stress space

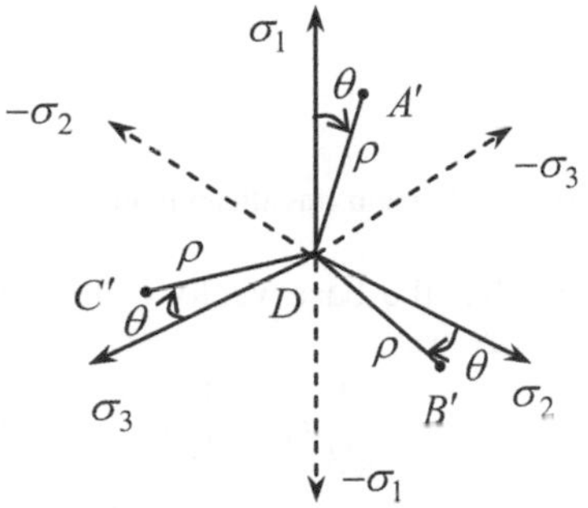

Figure 5.26 Stress points *A, B,* and *C* on the π-plane

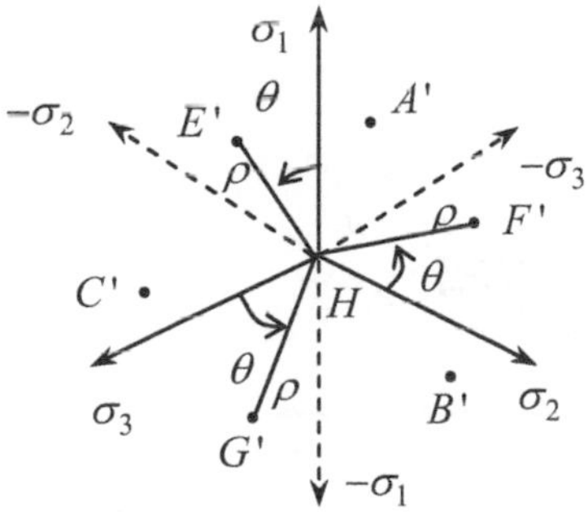

Figure 5.27 Stress points *A, B,* and *C* on the π-plane

Problem 5.12 Prove that the distance of stress points E', F', and G' from the origin in the π-plane are the same. Use (5.42) and (5.70) to show ρ and θ in Fig. 5.27 as

$$\rho = \sqrt{42}, \quad \theta = 40.89° \tag{5.191}$$

Problem 5.13 From the results of Problems 5.7–5.12 show that the stress points A', B', C', E', F', and G' satisfy the domain classification shown in Fig. 5.28.

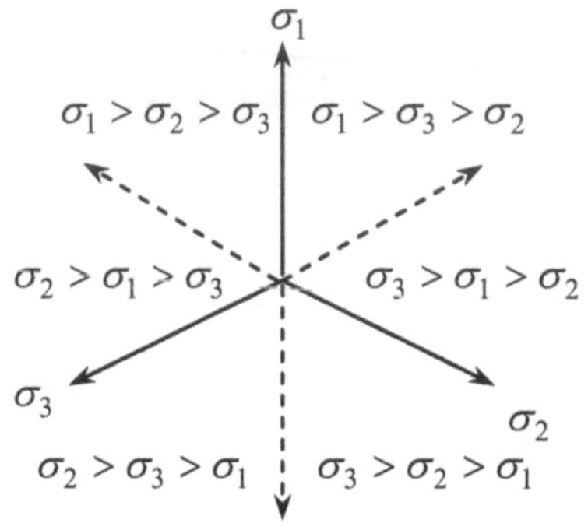

Figure 5.28 Domain classification of stress on the π-plane

Problem 5.14 Show that the base vectors along the X- and Y-axes of the π-plane as shown in Fig. 5.29 are

$$\boldsymbol{e}_y = \frac{2}{\sqrt{3}}\boldsymbol{e}_1 - \frac{1}{\sqrt{6}}\boldsymbol{e}_2 - \frac{1}{\sqrt{6}}\boldsymbol{e}_3 \tag{5.192}$$

$$\boldsymbol{e}_x = -\frac{1}{\sqrt{2}}\boldsymbol{e}_2 + \frac{1}{\sqrt{2}}\boldsymbol{e}_3 \tag{5.193}$$

(*Hint*: the base vector for $\boldsymbol{n}$ is given in (5.61).)

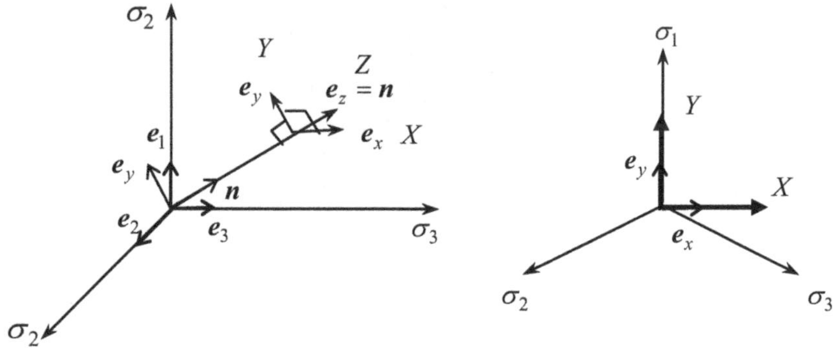

Figure 5.29 Domain classification of stress on the π-plane

Problem 5.15 With the base vectors found in Problem 5.14 show that the following transformation between stresses X, Y, and Z (note that $Z = 0$ on the π-plane) and σ_1, σ_2, and σ_3 are

$$\begin{Bmatrix} X \\ Y \\ Z \end{Bmatrix} = \begin{bmatrix} 0 & -\frac{1}{\sqrt{2}} & \frac{1}{\sqrt{2}} \\ \frac{2}{\sqrt{6}} & -\frac{1}{\sqrt{6}} & -\frac{1}{\sqrt{6}} \\ \frac{1}{\sqrt{3}} & \frac{1}{\sqrt{3}} & \frac{1}{\sqrt{3}} \end{bmatrix} \begin{Bmatrix} \sigma_1 \\ \sigma_2 \\ \sigma_3 \end{Bmatrix} \tag{5.194}$$

Problem 5.16 Show that the inverse of (5.194) is

$$\begin{Bmatrix} \sigma_1 \\ \sigma_2 \\ \sigma_3 \end{Bmatrix} = \begin{bmatrix} 0 & \frac{2}{\sqrt{6}} & \frac{1}{\sqrt{3}} \\ -\frac{1}{\sqrt{2}} & -\frac{1}{\sqrt{6}} & \frac{1}{\sqrt{3}} \\ \frac{1}{\sqrt{2}} & -\frac{1}{\sqrt{6}} & \frac{1}{\sqrt{3}} \end{bmatrix} \begin{Bmatrix} X \\ Y \\ Z \end{Bmatrix} \tag{5.195}$$

Problem 5.17 Assuming $\sigma_1 > \sigma_2 > \sigma_3$, use the result in Problem 5.16 to show that the following equation of the Mohr–Coulomb failure criterion:

$$\sin\phi = \frac{\frac{1}{2}(\sigma_1 - \sigma_3)}{c\cot\phi + \frac{1}{2}(\sigma_1 + \sigma_3)} \tag{5.196}$$

can be expressed in the π-plane as

$$Y = \frac{\sqrt{3}(1+\sin\phi)}{(3-\sin\phi)} X + \frac{2\sqrt{6}c\cos\phi}{(3-\sin\phi)} \tag{5.197}$$

Problem 5.18 Assuming $\sigma_1 > \sigma_3 > \sigma_2$, use the result in Problem 5.16 to show that Mohr-Coulomb failure criterion can be expressed in the π-plane as:

$$Y = -\frac{\sqrt{3}(1+\sin\phi)}{(3-\sin\phi)} X + \frac{2\sqrt{6}c\cos\phi}{(3-\sin\phi)} \tag{5.198}$$

Problem 5.19 Find the coordinates at the vertex points 1, 2, and 3 on the Mohr–Coulomb yield surface shown in Fig. 5.30.

Answer:

$$X_1 = 0, \quad Y_1 = \frac{2\sqrt{6}c\cos\phi}{(3-\sin\phi)} \tag{5.199}$$

$$X_2 = -\frac{3\sqrt{2}c\cos\phi}{(3-\sin\phi)}, \quad Y_2 = \frac{\sqrt{6}c\cos\phi}{(3-\sin\phi)} \tag{5.200}$$

$$X_3 = \frac{3\sqrt{2}c\cos\phi}{(3-\sin\phi)}, \quad Y_3 = \frac{\sqrt{6}c\cos\phi}{(3-\sin\phi)} \tag{5.201}$$

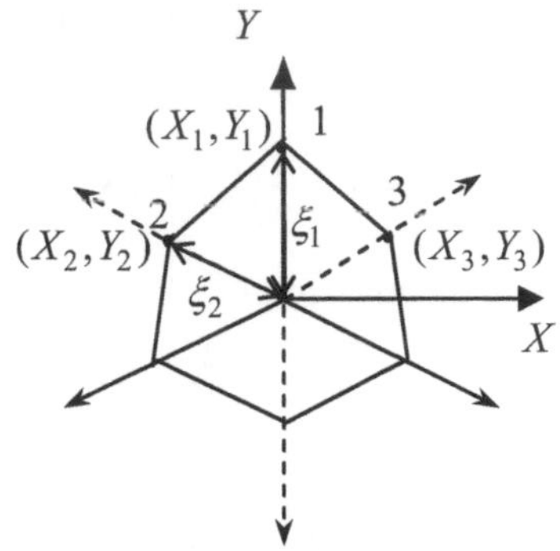

Figure 5.30 Domain classification of stress on the π-plane

Problem 5.20 Find the ratio between ξ_1 and ξ_2 shown in Fig. 5.30.

Answer: $\sqrt{2}$

Problem 5.21 Use symmetric properties to show the validity of the yield surface for the Mohr–Coulomb failure criterion shown in Fig. 5.9.

CHAPTER SIX

Fracture Mechanics and Its Applications

6.1 INTRODUCTION

The first analytical result for fracture mechanics is the stress concentration at crack tips and sharp corners obtained by Inglis (1913). The main result of Inglis (1913) is that the stress singularity near the crack tip is $1/r^{1/2}$, where r is the distance from the crack tip. Based upon this result, Griffith (1920) presented his celebrated fracture criteria for ideally brittle solids using the concept of "minimum potential energy", in which surface energy was incorporated. In this classic paper, Griffith used his results to explain why glass and quartz crystals have tensile strength that is much smaller than that for the perfect solids containing no cracks. Although Griffith's classic paper was originally motivated by its applications to brittle materials, most of the developments in fracture mechanics were, however, fostered by its application to ductile metallic solids. For example, the concept of *energy release rate*, G, was originally proposed by Irwin (1956) for steel and aluminum alloys. The application of fracture mechanics to rock-like solids remains a relatively new area of research; and many fundamental issues remain to be resolved. The main difficulty in applying classical fracture mechanics to rocks is due to the fact that rock masses are normally subjected to all-round compressions when tensile fractures start to grow. It is usually believed that the deviatoric stress and stress difference will cause a local tensile stress field near the crack tip. However, the relation between the local tensile stress and the far field compressions remains uncertain. Although various wing crack models have been commonly adopted to account for the micromechanics of tensile cracks under compression (e.g., Nemat-Nasser and Obata, 1988; Ashby and Hallam, 1986), actual wing cracks are seldom observed in real rocks (e.g., Tapponnier and Brace, 1976; Kranz, 1979).

We will start the chapter by following the tradition approach. We will first consider the stress concentration at an elliptical hole subject to tensile stress, then specialize the solution to crack geometry (i.e., the size of the minor axis is very small compared to the major axis). This consideration then extends to shear crack and tearing crack. The universality of the order of stress singularity at the crack tip by Williams (1957, 1959) will be discussed for all mode I, mode II, and mode III. The idea of energy release rate is then introduced, followed by the discussion on the J-integral. The method of superposition using the Westergaard stress function is summarized. The concept of cohesive crack is introduced through the growth of slip surface in slopes and in fault zones. The wing crack model is demonstrated by considering the local tensile stress induced by compression. Bažant's model of size effect on compressive strength is formulated via the application of the J-integral.

The last topic in this chapter is the application of fracture mechanics to damages and microcrack models.

6.2 STRESS CONCENTRATION AT AN ELLIPTICAL HOLE

We apply here the Muskhelishvili method discussed in Chapter 2 to obtain the stress concentration at the boundary of an elliptical hole. This solution will then be specialized to model the crack problem in the next section. The following discussion follows closely the presentation by Xu (1982) and Savin (1962).

In particular, we consider the following conformal mapping:

$$z=\omega(\zeta)=R(\frac{1}{\zeta}+m\zeta) \tag{6.1}$$

where $z = x_1+ix_2$ and $\zeta = \rho e^{i\theta}$. Substitution of these definitions into (6.1) and comparing the real and imaginary parts give

$$x_1=R(\frac{1}{\rho}+m\rho)\cos\theta,\quad x_2=-R(\frac{1}{\rho}-m\rho)\sin\theta. \tag{6.2}$$

Elimination of θ from (6.2) gives

$$\frac{x_1^2}{R^2(\frac{1}{\rho}+m\rho)^2}+\frac{x_2^2}{R^2(\frac{1}{\rho}-m\rho)^2}=1 \tag{6.3}$$

which depicts the locus of an ellipse on the physical plane. When $\theta = 0$, we have $x_1 = a$ and when $\theta = \pi/2$, we have $x_2 = b$; therefore, the major and minor axes of the ellipse are

$$a=R(1+m),\quad b=R(1-m) \tag{6.4}$$

Rewriting R and m in terms of a and b, we have

$$R=\frac{a+b}{2},\quad m=\frac{a-b}{a+b} \tag{6.5}$$

Therefore, as expected, the boundary of an elliptical hole is mapped onto a circle on $\zeta = \sigma = e^{i\theta}$ (with $\rho = 1$). Conversely, if we eliminate ρ from (6.2), we have

$$\frac{x_1^2}{4R^2m\cos^2\theta}-\frac{x_2^2}{4R^2m\sin^2\theta}=1 \tag{6.6}$$

which is a hyperbola for the θ = const. on the z-plane. By setting $x_2 = 0$, the focal length of the ellipse is $2R\sqrt{m}\cos\theta$. As shown in Fig. 6.1, the points A, B, C and D in the z-plane are mapped to A', B', C', and D' in the ζ-plane, respectively. These points are tabulated in Table 6.1 below.

The following derivatives and their conjugates for the mapping function in (6.1) are given here for later reference:

$$\omega(\sigma)=R(\frac{1}{\sigma}+m\sigma),\quad \bar{\omega}(\bar{\sigma})=R(\sigma+\frac{m}{\sigma}) \tag{6.7}$$

$$\omega'(\sigma)=R(m-\frac{1}{\sigma^2}),\quad \bar{\omega}'(\bar{\sigma})=R(m-\sigma^2) \tag{6.8}$$

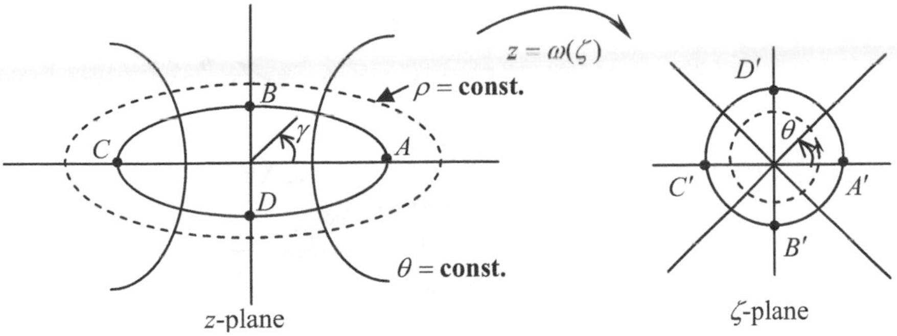

Figure 6.1 The mapping of an ellipse to a unit circle

Table 6.1

θ	x_1	x_2	γ
0	$R(1+m)$	0	0
$\pi/2$	0	$-R(1-m)$	$-\pi/2$
π	$-R(1+m)$	0	π
$3\pi/2$	0	$R(1-m)$	$\pi/2$

$$\frac{\omega(\sigma)}{\overline{\omega}'(\overline{\sigma})}=-\frac{1}{\sigma}\left(\frac{m\sigma^2+1}{\sigma^2-m}\right),\qquad \frac{\overline{\omega}(\overline{\sigma})}{\omega'(\sigma)}=\sigma\left(\frac{\sigma^2+m}{m\sigma^2-1}\right) \tag{6.9}$$

Recalling (3.158) and (3.141) of Chapter 3, we have

$$\phi_0(\zeta)+\frac{1}{2\pi i}\int_\sigma\left[-\frac{1}{\sigma}\left(\frac{m\sigma^2+1}{\sigma^2-m}\right)\right]\left(\overline{\alpha}_1+2\frac{\overline{\alpha}_2}{\sigma}+3\frac{\overline{\alpha}_3}{\sigma^2}+\ldots\right)\frac{d\sigma}{\sigma-\zeta}=\frac{1}{2\pi i}\int_\sigma\frac{f_0 d\sigma}{\sigma-\zeta} \tag{6.10}$$

It is straightforward to see that the kernel for the Cauchy integral on the left-hand side is analytic outside the unit circle $\zeta=\sigma$. In view of (3.152) of Chapter 3, we obtain

$$\phi_0(\zeta)=\frac{1}{2\pi i}\int_\sigma\frac{f_0\, d\sigma}{\sigma-\zeta} \tag{6.11}$$

Recalling (3.159) in Chapter 3, we have

$$\psi_0(\zeta)+\frac{1}{2\pi i}\int_\sigma\sigma\left(\frac{\sigma^2+m}{m\sigma^2-1}\right)(\alpha_1+2\alpha_2\sigma+3\alpha_3\sigma^2+\ldots)\frac{d\sigma}{\sigma-\zeta}=\frac{1}{2\pi i}\int_\sigma\frac{\overline{f}_0 d\sigma}{\sigma-\zeta} \tag{6.12}$$

The kernel inside the Cauchy integral on the left-hand side is analytic inside the unit circle $\zeta=\sigma$. Therefore, we have

$$\psi_0(\zeta)=\frac{1}{2\pi i}\int_\sigma\frac{\overline{f}_0 d\sigma}{\sigma-\zeta}-\zeta\left(\frac{\zeta^{-2}+m}{m\zeta^2-1}\right)\phi_0'(\zeta) \tag{6.13}$$

Now, we consider an elliptical hole subject to far field tension q with an inclination α measured from the x_1-axis, as shown in Fig. 6.2. Consequently, we have $T_1=T_2=X=Y=0$ and $\sigma_1=q$ and $\sigma_2=0$. Hence, similar to the discussion in Section 3.3, we obtain

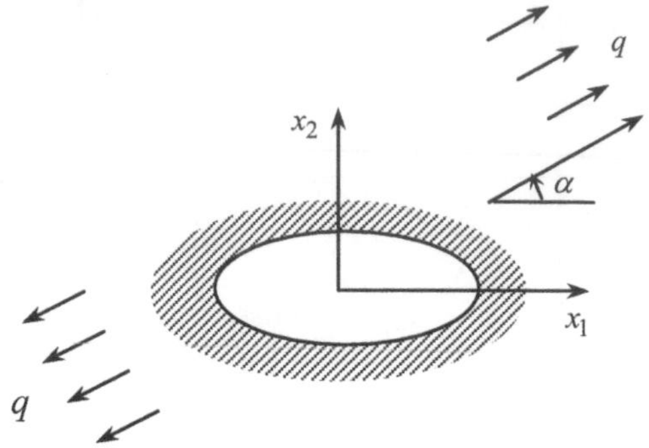

Figure 6.2 An elliptical hole subject to far field tension inclined at α from x_1

$$B=\frac{1}{4}(\sigma_1+\sigma_2)=\frac{q}{4},\quad B'+iC'=-\frac{1}{2}(\sigma_1-\sigma_2)e^{-2i\alpha}=-\frac{q}{2}e^{-2i\alpha} \tag{6.14}$$

From the definition of f_0 given in (3.148), we get

$$\begin{aligned} f_0 &= -2B\omega(\sigma)-(B'-iC')\bar{\omega}(\bar{\sigma}) \\ &= \frac{qR}{2}[-(\frac{1}{\sigma}+m\sigma)+(\sigma+\frac{m}{\sigma})e^{2i\alpha}] \end{aligned} \tag{6.15}$$

Substitution of (6.15) into (6.11) yields

$$\phi_0(\zeta)=\frac{qR}{2}(e^{2i\alpha}-m)\zeta \tag{6.16}$$

Similarly, substitution of (6.15) into (6.13) yields

$$\begin{aligned} \psi_0(\zeta) &= \frac{qR}{2}(me^{-2i\alpha}-1)\zeta-\frac{qR}{2}(e^{2i\alpha}-m)\zeta(\frac{\zeta^{-2}+m}{m\zeta^{-2}-1}) \\ &= -\frac{qR}{2}[(1-me^{-2i\alpha})\zeta+(m-e^{2i\alpha})(\frac{m+\zeta^{-2}}{1-m\zeta^{-2}})\zeta] \end{aligned} \tag{6.17}$$

Back-substitution of (6.16) and (6.17) into (3.139) and (3.140) of Chapter 3, respectively; we find

$$\phi(\zeta)=\frac{qR}{4}[\frac{1}{\zeta}+(2e^{2i\alpha}-m)\zeta] \tag{6.18}$$

and

$$\psi(\zeta)=-\frac{qR}{2}[\frac{1}{\zeta}e^{-2i\alpha}+\frac{\zeta^{3}e^{2i\alpha}+(me^{2i\alpha}-m^2-1)\zeta}{m\zeta^2-1}] \tag{6.19}$$

Note that $\zeta\bar{\zeta}=\rho^2$, thus we get

$$\omega'(\zeta)=R(m-\frac{1}{\zeta^2}) \tag{6.20}$$

$$\bar{\omega}'(\bar{\zeta}) = R(m - \frac{\zeta^2}{\rho^4}), \tag{6.21}$$

$$\bar{\omega}(\bar{\zeta}) = R(\frac{\zeta}{\rho^2} + \frac{m\rho^2}{\zeta}) \tag{6.22}$$

Differentiating (6.18) and (6.19) with respect to ζ gives

$$\phi'(\zeta) = \frac{qR}{4}[(2e^{2i\alpha} - m) - \frac{1}{\zeta^2}] \tag{6.23}$$

and

$$\psi'(\zeta) = -\frac{qR}{2}[-\frac{1}{\zeta}e^{-2i\alpha} + \frac{3\zeta^2 e^{2i\alpha} + (me^{2i\alpha} - m^2 - 1)}{m\zeta^2 - 1} - \frac{\zeta^2 e^{2i\alpha} + (me^{2i\alpha} - m^2 - 1)}{(m\zeta^2 - 1)^2}2m\zeta^2] \tag{6.24}$$

Substituting (6.20) and (6.23) into (3.122) of Chapter 3, we finally have

$$\Phi(\zeta) = \frac{\varphi'(\zeta)}{\omega'(\zeta)} = \frac{q}{4}[\frac{(2e^{2i\alpha} - m)\zeta^2 - 1}{m\zeta^2 - 1}] \tag{6.25}$$

Substituting (6.21) and (6.24) into (3.123) of Chapter 3, we get

$$\Psi(\zeta) = -\frac{q}{2}[-\frac{e^{-2i\alpha}}{m\zeta^2 - 1} + \frac{3\zeta^2 e^{2i\alpha} + (me^{2i\alpha} - m^2 - 1)}{(m\zeta^2 - 1)^2}\zeta^2 - \frac{\zeta^2 e^{2i\alpha} + (me^{2i\alpha} - m^2 - 1)}{(m\zeta^2 - 1)^3}2m\zeta^4] \tag{6.26}$$

Differentiating (6.25) with respect to the argument yields

$$\Phi'(\zeta) = \frac{q}{4}[\frac{(2e^{2i\alpha} - m)2\zeta}{m\zeta^2 - 1} - \frac{(2e^{2i\alpha} - m)\zeta^2 - 1}{(m\zeta^2 - 1)^2}2m\zeta] \tag{6.27}$$

Substituting (6.25) (6.27) into (3.128) and (3.129) of Chapter 3, we get

$$\sigma_{\theta\theta} + \sigma_{\rho\rho} = q\ \mathrm{Re}[\frac{(2e^{2i\alpha} - m)\zeta^2 - 1}{m\zeta^2 - 1}] \tag{6.28}$$

$$\sigma_{\theta\theta} - \sigma_{\rho\rho} + 2i\sigma_{\rho\theta} = \frac{q\zeta^2(\zeta^2 + m\rho^4)}{(\rho^4 m - \zeta^2)(m\zeta^2 - 1)}[2e^{2i\alpha} - m + m\frac{1 - (2e^{2i\alpha} - m)\zeta^2}{m\zeta^2 - 1}]$$
$$+ \frac{q\rho^2}{(\rho^4 m - \zeta^2)}[e^{-2i\alpha} - \frac{3e^{2i\alpha}\zeta^2 + (me^{2i\alpha} - m^2 - 1)}{m\zeta^2 - 1}\zeta^2 + \frac{e^{2i\alpha}\zeta^2 + me^{2i\alpha} - m^2 - 1}{(m\zeta^2 - 1)^2}2m\zeta^4] \tag{6.29}$$

The expression for $\sigma_{\rho\rho}$, $\sigma_{\theta\theta}$, and $\sigma_{\rho\theta}$ can be obtained explicitly from (6.28) and (6.29); however, these expressions are very lengthy. Of particular interest is the tangential stress on the hole's boundary; for this case, we set $\sigma_{\rho\rho}$ and $\sigma_{\rho\theta} = 0$ on $\zeta = \sigma$. Thus, we have the following form of stress concentration:

$$\begin{aligned}\sigma_{\theta\theta}(\zeta = \sigma) &= q\ \mathrm{Re}[\frac{(2e^{2i\alpha} - m)\sigma^2 - 1}{m\sigma^2 - 1}] \\ &= q[\frac{1 - m^2 + 2m\cos 2\alpha - 2\cos 2(\alpha + \theta)}{1 + m^2 - 2m\cos 2\theta}]\end{aligned} \tag{6.30}$$

The largest compression and tension occur either at $\theta = 0, \pi$ or at $\theta = \pm\pi/2$, as summarised in Table 6.2.

6.3 STRESS CONCENTRATION AT A TENSILE CRACK

The result given in the last section can easily be specialized to the crack problem. Consider the limit that $b \to 0$, we have $m = 1$ and $R = a/2$. That is, the mapping function becomes

$$z = \omega(\zeta) = \frac{a}{2}\left(\frac{1}{\zeta} + \zeta\right) \tag{6.31}$$

Muskhelishvili's analytic functions become, in view of (6.18) and (6.19),

$$\phi(\zeta) = \frac{qa}{8}\left[\frac{1}{\zeta} + (2e^{2i\alpha} - 1)\zeta\right] \tag{6.32}$$

$$\psi(\zeta) = -\frac{qa}{4}\left[\frac{1}{\zeta}e^{-2i\alpha} + \frac{\zeta^3 e^{2i\alpha} + (e^{2i\alpha} - 2)\zeta}{\zeta^2 - 1}\right] \tag{6.33}$$

For the inclined tension shown in Fig. 6.3, both tensile and shear modes of deformations exist (as will be discussed in the later section these are normally referred as mode I and II cracks in fracture mechanics). To focus on the stress field near the crack tip, these modes will be considered separately here.

For the opening mode of deformation at the crack tip under tensile far field stress, we set $\alpha = \pi/2$ into (6.32) and (6.33); thus, they become

$$\phi(\zeta) = \frac{qa}{8}\left[\frac{1}{\zeta} - 3\zeta\right] \tag{6.34}$$

Table 6.2 Stress concentration around the elliptical hole

θ	$\pm\pi/2$	$0, \pi$
$\alpha = 0$	$q(1 + 2b/a)$	$-q$
$\alpha = \pi/2$	$-q$	$q(1 + 2a/b)$

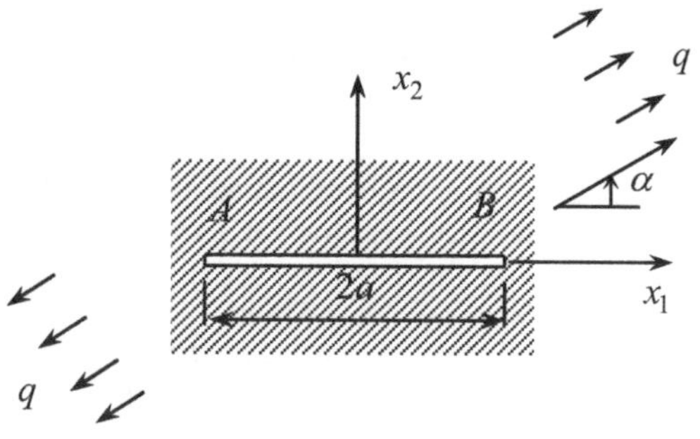

Figure 6.3 A 2-D crack subject to far field tension *q* inclined at *α* from horizontal

$$\psi(\zeta) = \frac{qa}{4}[\frac{1}{\zeta} - \frac{\zeta^3 + 3\zeta}{1-\zeta^2}] \tag{6.35}$$

Since we are interested in the stress field in the physical plane, it is advantageous to rewrite (6.31) as

$$\zeta = \frac{z}{a} - \sqrt{\frac{z^2}{a^2} - 1} \tag{6.36}$$

The negative sign in front of the square root is chosen to reflect the fact that $z \to \infty$ is being mapped to $\zeta \to 0$. By noting that

$$\sqrt{z^2 - a^2} = \frac{a(1-\zeta^2)}{2\zeta} \tag{6.37}$$

it is straightforward to show, in view of (3.120) and (3.121) of Chapter 3, that

$$\phi_1(z) = \phi(\zeta) = \frac{q}{4}(2\sqrt{z^2 - a^2} - z) \tag{6.38}$$

$$\psi_1(z) = \psi(\zeta) = \frac{q}{2}(z - \frac{a^2}{\sqrt{z^2 - a^2}}) \tag{6.39}$$

With these expressions, (3.34) and (3.35) of Chapter 3 can be used to evaluate the stress field as

$$\sigma_{11} + \sigma_{22} = 4\,\mathrm{Re}[\phi_1'(z)] = q\left[2\,\mathrm{Re}(\frac{z}{\sqrt{z^2 - a^2}}) - 1\right] \tag{6.40}$$

$$\sigma_{22} - \sigma_{11} + 2i\sigma_{12} = 2[\bar{z}\phi_1''(z) + \psi_1'(z)] = q\left[\frac{2ia^2 x_2}{(z^2 - a^2)^{3/2}} + 1\right] \tag{6.41}$$

We can now examine the stress field near the crack tip. As shown in Fig. 6.4, first we introduce a polar coordinate (r, θ) at the crack tip B such that

$$z = a + re^{i\theta} = a + r\cos\theta + ir\sin\theta, \qquad x_2 = r\sin\theta \tag{6.42}$$

Substitution of (6.42) into (6.40) and (6.41) with the approximation that $r/a \ll 1$, we obtain the following dominant terms near the crack tip:

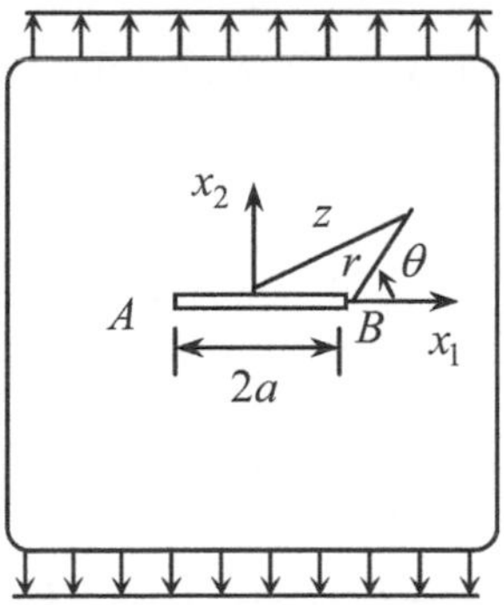

Figure 6.4 The polar coordinates at a crack subject to far field tension

$$\sigma_{11}+\sigma_{22}=q(\frac{2a}{r})^{1/2}\cos\frac{\theta}{2}+O(1) \tag{6.43}$$

$$\sigma_{22}-\sigma_{11}+2i\sigma_{12}=q(\frac{2a}{r})^{1/2}\sin\frac{\theta}{2}\cos\frac{\theta}{2}(\sin\frac{3\theta}{2}+i\cos\frac{3\theta}{2})+O(1) \tag{6.44}$$

Rearranging (6.43) and (6.44), we finally have the crack-tip field

$$\sigma_{11}=\frac{K_I}{(2\pi r)^{1/2}}\cos\frac{\theta}{2}(1-\sin\frac{\theta}{2}\sin\frac{3\theta}{2}) \tag{6.45}$$

$$\sigma_{22}=\frac{K_I}{(2\pi r)^{1/2}}\cos\frac{\theta}{2}(1+\sin\frac{\theta}{2}\sin\frac{3\theta}{2}) \tag{6.46}$$

$$\sigma_{12}=\frac{K_I}{(2\pi r)^{1/2}}\sin\frac{\theta}{2}\cos\frac{\theta}{2}\cos\frac{3\theta}{2} \tag{6.47}$$

where K_I is the so-called *stress intensity factor* at the crack tip. For the crack geometry and loading given in Fig. 6.4, we have

$$K_I=q(\pi a)^{1/2} \tag{6.48}$$

More generally, K_I depends on the crack geometry and loading type for different crack problems. Although the θ-variation for $\boldsymbol{\sigma}$ given in (6.45)–(6.47) is derived for the particular problem shown in Fig. 6.4, it is important to note that they are the universal forms of crack-tip field and apply to all cracks regardless of the crack geometry and loading (as long as the crack tip is in opening mode of deformation). To honor the original contribution by Griffith (1920) on fracture mechanics, the problem given in Fig. 6.4 is usually referred as the *Griffith crack problem.*

The stress field predicted by (6.45)–(6.47) are singular at the crack tip (i.e., becoming unbounded at $r\to 0$). Physically, all materials will either yield or be damaged at the crack tip; therefore, an inelastic zone around the crack tip is expected. Normally, we assume that a *small-scale yielding zone* exists, that is, r remains small outside the inelastic zone such that the dominant terms remain those predicted by (6.45)–(6.47).

Substituting (6.38) and (6.39) into (3.36) of Chapter 3, and dropping the rigid body motion but retaining the most dominant terms, we have

$$2\mu(u_1+iu_2)=q(\frac{ar}{2})^{1/2}[\cos\frac{\theta}{2}(\kappa-\cos\theta)+i\sin\frac{\theta}{2}(\kappa-\cos\theta)] \tag{6.49}$$

Noting the definition given in (6.48) for K_I, we finally get

$$u_1=\frac{K_I}{2\mu}(\frac{r}{2\pi})^{1/2}\cos\frac{\theta}{2}[\kappa-1+2\sin^2\frac{\theta}{2}] \tag{6.50}$$

$$u_2=\frac{K_I}{2\mu}(\frac{r}{2\pi})^{1/2}\sin\frac{\theta}{2}[\kappa+1-2\cos^2\frac{\theta}{2}] \tag{6.51}$$

In terms of polar coordinates, (3.52) of Chapter 3 can be applied to show that

$$u_r=\frac{K_I}{4\mu}(\frac{r}{2\pi})^{1/2}[(2\kappa-1)\cos\frac{\theta}{2}-\cos\frac{3\theta}{2}] \tag{6.52}$$

$$u_\theta=-\frac{K_I}{4\mu}(\frac{r}{2\pi})^{1/2}[(2\kappa+1)\sin\frac{\theta}{2}-\sin\frac{3\theta}{2}] \tag{6.53}$$

Similar to (6.45)–(6.47), the displacement field near the crack tips are also universal.

6.4 STRESS FIELD NEAR A SHEAR CRACK

We now consider a 2-D crack subject to a far field shear stress q, as shown in Fig. 6.5. For this shear crack, we can superimpose a far field tension q at $\alpha = \pi/4$ and a far field compression $-q$ at $\alpha = -\pi/4$. The following Muskhelishvili's analytic functions were obtained:

$$\begin{aligned}\phi(\zeta) &= \frac{qa}{8}[\frac{1}{\zeta} + (2e^{2i\alpha} - 1)\zeta]_{\alpha=\pi/4} - \frac{qa}{8}[\frac{1}{\zeta} + (2e^{2i\alpha} - 1)\zeta]_{\alpha=-\pi/4} \\ &= \frac{iqa}{2}\zeta\end{aligned} \tag{6.54}$$

$$\psi(\zeta) = \frac{iqa}{2}[\frac{1}{\zeta} + \frac{\zeta^3 + \zeta}{1-\zeta^2}] \tag{6.55}$$

In terms of z, these analytic functions become

$$\phi_1(z) = \phi(\zeta) = \frac{iq}{4}(z - \sqrt{z^2 - a^2}) \tag{6.56}$$

$$\psi_1(z) = \psi(\zeta) = \frac{iq}{2}(2\sqrt{z^2 - a^2} + \frac{a^2}{\sqrt{z^2 - a^2}}) \tag{6.57}$$

Substituting (6.56) and (6.57) into (3.34) and (3.35) of Chapter 3, we get

$$\sigma_{11} + \sigma_{22} = 4\,\mathrm{Re}[\phi_1'(z)] = -2q\,\mathrm{Re}(\frac{iz}{\sqrt{z^2 - a^2}}) \tag{6.58}$$

$$\begin{aligned}\sigma_{22} - \sigma_{11} + 2i\sigma_{12} &= 2[\bar{z}\phi_1''(z) + \psi_1'(z)] \\ &= q\left[\frac{ia^2\bar{z} + iz(2z^2 - 3a^2)}{(z^2 - a^2)^{3/2}}\right]\end{aligned} \tag{6.59}$$

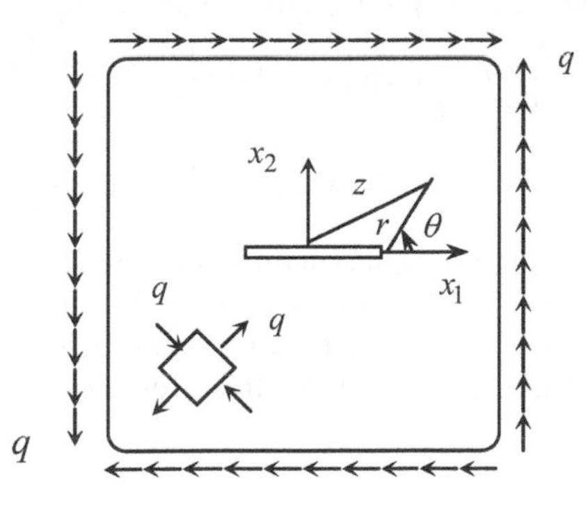

Figure 6.5 A 2-D crack subject to far field shear *q*

Again, we consider the crack-tip field using $z = a + re^{i\theta}$ as $r \to 0$ and $r/a \ll 1$. Retaining only the dominant terms in (6.58) and (6.59), we find

$$\sigma_{11} + \sigma_{22} = (\frac{2a}{r})^{1/2} q \sin\frac{\theta}{2} \tag{6.60}$$

$$\begin{aligned}\sigma_{22} - \sigma_{11} + 2i\sigma_{12} = q(\frac{2a}{r})^{1/2}\{&\sin\frac{\theta}{2}[1+\cos\frac{\theta}{2}\cos\frac{3\theta}{2}] \\ &+ i\cos\frac{\theta}{2}[1-\sin\frac{\theta}{2}\sin\frac{3\theta}{2}]\}\end{aligned} \tag{6.61}$$

Consequently, the stress field near the crack tip is

$$\sigma_{11} = -\frac{K_{II}}{(2\pi r)^{1/2}} \sin\frac{\theta}{2}(2+\cos\frac{\theta}{2}\cos\frac{3\theta}{2}) \tag{6.62}$$

$$\sigma_{22} = \frac{K_{II}}{(2\pi r)^{1/2}} \sin\frac{\theta}{2}\cos\frac{\theta}{2}\cos\frac{3\theta}{2} \tag{6.63}$$

$$\sigma_{12} = \frac{K_{II}}{(2\pi r)^{1/2}} \cos\frac{\theta}{2}[1-\sin\frac{\theta}{2}\sin\frac{3\theta}{2}] \tag{6.64}$$

where the shear (or mode II) stress intensity factor is

$$K_{II} = q(\pi a)^{1/2} \tag{6.65}$$

Substituting (6.56) and (6.57) into (3.36) of Chapter 3, and retaining the most dominant terms, we have

$$2\mu(u_1 + iu_2) = q(\frac{ar}{2})^{1/2}[\sin\frac{\theta}{2}(2+\kappa+\cos\theta) + i\cos\frac{\theta}{2}(2-\kappa-\cos\theta)) \tag{6.66}$$

Noting the definition (6.65) for K_{II}, we get

$$u_1 = \frac{K_{II}}{2\mu}(\frac{r}{2\pi})^{1/2} \sin\frac{\theta}{2}[\kappa+1+2\cos^2\frac{\theta}{2}] \tag{6.67}$$

$$u_2 = -\frac{K_{II}}{2\mu}(\frac{r}{2\pi})^{1/2} \cos\frac{\theta}{2}[\kappa-1-2\sin^2\frac{\theta}{2}] \tag{6.68}$$

In terms of polar coordinates, (3.52) of Chapter 3 can be applied to show that

$$u_r = -\frac{K_{II}}{4\mu}(\frac{r}{2\pi})^{1/2}[(2\kappa-1)\sin\frac{\theta}{2} - 3\sin\frac{3\theta}{2}] \tag{6.69}$$

$$u_\theta = -\frac{K_{II}}{4\mu}(\frac{r}{2\pi})^{1/2}[(2\kappa+1)\cos\frac{\theta}{2} - 3\cos\frac{3\theta}{2}] \tag{6.70}$$

In most fracture mechanics problems, the main issue is to find the appropriate form of the stress intensity factor for different cracks under different loading types.

In the next section, we will consider the stress field around crack tips by using the eigenfunction expansions on Airy stress function proposed by Williams (1957).

6.5 THE GENERAL STRESS AND DISPLACEMENT FIELD FOR TENSILE CRACKS

Before we consider the general stress field at crack tips, we first classify here three modes of crack deformations, namely: mode I (tensile mode), mode II (sliding or shear mode), and mode III (tearing mode) cracks. Figure 6.6 illustrates the types of loading and the corresponding deformation of each mode.

We now reconsider the crack-tip field following an approach similar to those adopted by Williams (1957) and subsequently modified by Hellan (1984). First, we recall here the Airy stress function introduced in Section 2.17. In particular, the stress components in polar coordinates can be expressed in terms of a 2-D function φ:

$$\boldsymbol{\sigma}=[(\frac{1}{r}\frac{\partial\varphi}{\partial r}+\frac{1}{r^2}\frac{\partial^2\varphi}{\partial\theta^2})\boldsymbol{e}_r\boldsymbol{e}_r+(\frac{\partial^2\varphi}{\partial r^2})\boldsymbol{e}_\theta\boldsymbol{e}_\theta-\frac{\partial}{\partial r}(\frac{1}{r}\frac{\partial\varphi}{\partial\theta})(\boldsymbol{e}_r\boldsymbol{e}_\theta+\boldsymbol{e}_\theta\boldsymbol{e}_r)] \tag{6.71}$$

which satisfies the following biharmonic equation:

$$(\frac{\partial^2}{\partial r^2}+\frac{1}{r}\frac{\partial}{\partial r}+\frac{1}{r^2}\frac{\partial^2}{\partial\theta^2})(\frac{\partial^2}{\partial r^2}+\frac{1}{r}\frac{\partial}{\partial r}+\frac{1}{r^2}\frac{\partial^2}{\partial\theta^2})\varphi=0 \tag{6.72}$$

Recalling the Almansi theorem given in (4.342) of Chapter 4, the general solution of the biharmonic equation can be expressed in terms of two harmonic functions f and g as

$$\varphi=r^2 f(r,\theta)+g(r,\theta) \tag{6.73}$$

That is, both f and g satisfy the Laplace equation:

$$\nabla^2 f=0, \qquad \nabla^2 g=0 \tag{6.74}$$

It is known from the Cauchy–Riemann relation that harmonic function can form both real and imaginary parts of an analytic function (e.g., Spiegel). Note that if

$$f=\mathrm{Re}(z^\alpha) \quad or \quad f=\mathrm{Im}(z^\alpha) \tag{6.75}$$

where Re( ) and Im( ) stand for the real and imaginary parts of (…), respectively, and $z = x_1 + ix_2$, then f is harmonic (since z^α is analytic). In polar form, z^α becomes

$$z^\alpha=(re^{i\theta})^\alpha=r^\alpha e^{i\alpha\theta} \tag{6.76}$$

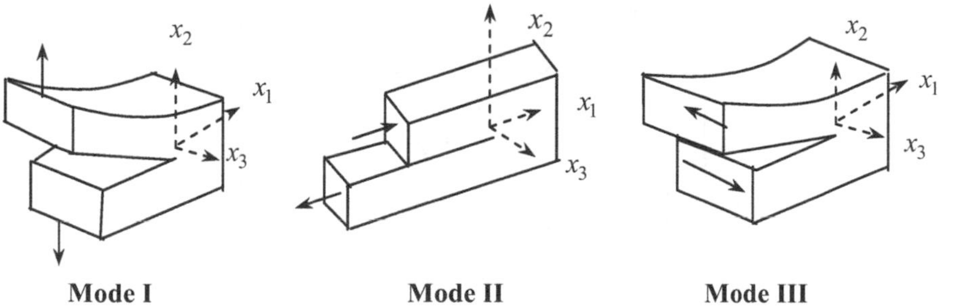

Figure 6.6 Three modes of crack deformation

This suggests that a possible solution form is in terms of series:

$$f(r,\theta)=\sum_{\lambda=1}^{\infty}[A_\lambda r^\lambda \cos\lambda\theta + C_\lambda r^\lambda \sin\lambda\theta] \tag{6.77}$$

$$g(r,\theta)=\sum_{\lambda=1}^{\infty}[B_\lambda r^{\lambda+2} \cos(\lambda+2)\theta + D_\lambda r^{\lambda+2} \sin(\lambda+2)\theta] \tag{6.78}$$

We now apply this series solution to our crack problems. First, we consider the mode I crack deformation shown in Fig. 6.7. For such an opening mode, it is obvious that the solution for $\sigma_{\theta\theta}$ must be symmetric about the x_1-axis, i.e., $\sigma_{\theta\theta}(\theta) = \sigma_{\theta\theta}(-\theta)$; consequently, the terms involving $\sin\theta$ must vanish in the solution. Therefore, for mode I crack we set

$$C_\lambda = D_\lambda = 0 \tag{6.79}$$

Substitution of (6.77) and (6.78) into (6.73) and (6.71) yields the following expressions for φ and the stresses

$$\varphi(r,\theta)=\sum_{\lambda=1}^{\infty} r^{\lambda+2}[A_\lambda \cos\lambda\theta + B_\lambda \cos(\lambda+2)\theta] \tag{6.80}$$

$$\sigma_{\theta\theta}=\sum_{\lambda=1}^{\infty}(\lambda+2)(\lambda+1)r^\lambda[A_\lambda \cos\lambda\theta + B_\lambda \cos(\lambda+2)\theta] \tag{6.81}$$

$$\sigma_{r\theta}=\sum_{\lambda=1}^{\infty}(\lambda+1)r^\lambda[A_\lambda \lambda \sin\lambda\theta + B_\lambda(\lambda+2)\sin(\lambda+2)\theta] \tag{6.82}$$

$$\sigma_{rr}=\sum_{\lambda=1}^{\infty} r^\lambda[A_\lambda[(\lambda+2)-\lambda^2]\cos\lambda\theta - B_\lambda(\lambda+2)(1+\lambda)\cos(\lambda+2)\theta] \tag{6.83}$$

Applying the following boundary conditions on the crack face ($\theta = \pi$)

$$\sigma_{\theta\theta}(r,\theta=\pi)=0, \qquad \sigma_{r\theta}(r,\theta=\pi)=0, \tag{6.84}$$

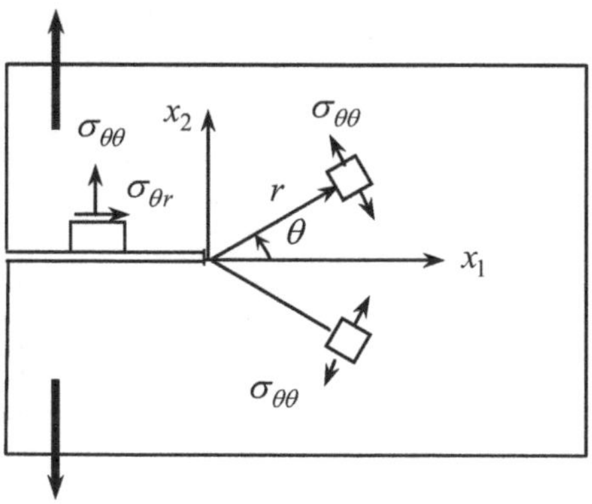

Figure 6.7 The symmetry stress field in mode I crack

we obtain two equations for A_λ and B_λ

$$(A_\lambda + B_\lambda)\cos\lambda\pi = 0 \tag{6.85}$$

$$[A_\lambda \lambda + (\lambda+2)B_\lambda]\sin\lambda\pi = 0 \tag{6.86}$$

There are two sets of solutions for (6.85) and (6.86): the first set is

$$\sin\lambda\pi = 0 \quad and \quad A_\lambda = -B_\lambda \tag{6.87}$$

and the second one is

$$\cos\lambda\pi = 0 \quad and \quad B_\lambda = -A_\lambda \frac{\lambda}{\lambda+2} \tag{6.88}$$

The solutions for λ satisfying (6.87) are

$$\lambda = 0, \pm 1, \pm 2, \ldots \tag{6.89}$$

and the solutions for λ satisfying (6.88) are

$$\lambda = \pm\frac{1}{2}, \pm\frac{3}{2}, \pm\frac{5}{2}, \ldots \tag{6.90}$$

Since we are interested in the most dominant stress field in the vicinity of the crack tip, only the negative of λ needed to be considered (note that the stresses are proportional to r^λ). There are, however, infinite possibilities of negative powers in (6.89) and (6.90); thus, we further impose the boundedness of the strain energy in any finite volume surrounding the crack tip. In particular, we require

$$\int_0^R \int_{-\pi}^{\pi} \mathcal{E}(r,\theta) r dr d\theta = \frac{1}{2\mu} \int_0^R \int_{-\pi}^{\pi} (\sigma_{rz}^2 + \sigma_{z\theta}^2) r dr d\theta < \infty \tag{6.91}$$

where R is an arbitrary finite distance measured from the crack tip. For linear elastic solids, the strain energy function $\mathcal{E}$ is the sum of products of an elastic compliance multiplied by the stress squared; therefore $\mathcal{E}$ is proportional to $r^{2\lambda}$. Thus, the boundedness condition (6.91) becomes

$$\int_0^R r^{2\lambda+1} dr < \infty \tag{6.92}$$

Therefore, we must have $2\lambda+1 > -1$ or

$$\lambda > -1 \tag{6.93}$$

From (6.90), the only singular term with $\lambda > -1$ is $\lambda = -1/2$. Therefore, we have

$$B_{-1/2} = A_{-1/2} \frac{(-1/2)}{(-1/2)+2} = \frac{1}{3} A_{-1/2} \tag{6.94}$$

In addition, it is customary to write

$$A_{-1/2} = \frac{K_I}{(2\pi)^{1/2}} \tag{6.95}$$

where K_I is the *mode I stress intensity factor*. Substitution of this singular term into (6.81)–(6.83), we obtain

$$\sigma_{\theta\theta} = \frac{K_I}{(2\pi r)^{1/2}} [\frac{3}{4}\cos\frac{\theta}{2} + \frac{1}{4}\cos\frac{3\theta}{2}] = \frac{K_I}{(2\pi r)^{1/2}} \cos^3\frac{\theta}{2} \tag{6.96}$$

$$\sigma_{r\theta} = \frac{K_I}{(2\pi r)^{1/2}} [\frac{1}{4}\sin\frac{\theta}{2} + \frac{1}{4}\sin\frac{3\theta}{2}] \tag{6.97}$$

$$\sigma_{rr} = \frac{K_I}{(2\pi r)^{1/2}}[\frac{5}{4}\cos\frac{\theta}{2} - \frac{1}{4}\cos\frac{3\theta}{2}] \tag{6.98}$$

Note that only the singular terms are retained in (6.96)–(6.98), and all other terms that vanish at the crack tip are not displayed.

To transform these expressions to Cartesian coordinates, we recall (3.37) of Chapter 3:

$$\begin{aligned}\sigma_{22} - \sigma_{11} + 2i\sigma_{12} &= (\sigma_{\theta\theta} - \sigma_{rr} + 2i\sigma_{r\theta})e^{-2i\theta} \\ &= (\sigma_{\theta\theta} - \sigma_{rr})\cos 2\theta + 2\sigma_{r\theta}\sin 2\theta + i[2\sigma_{r\theta}\cos 2\theta - (\sigma_{\theta\theta} - \sigma_{rr})\sin 2\theta]\end{aligned} \tag{6.99}$$

Summing (6.96) and (6.98) gives:

$$\sigma_{11} + \sigma_{22} = \sigma_{\theta\theta} + \sigma_{rr} = \frac{K_I}{(2\pi r)^{1/2}} 2\cos\frac{\theta}{2} \tag{6.100}$$

Now substitution of (6.96) to (6.98) into (6.99) gives

$$\sigma_{22} - \sigma_{11} = \frac{K_I}{(2\pi r)^{1/2}} 2\sin\frac{3\theta}{2}\sin\frac{\theta}{2}\cos\frac{\theta}{2} \tag{6.101}$$

$$\sigma_{12} = \frac{K_I}{(2\pi r)^{1/2}}\sin\frac{\theta}{2}\cos\frac{\theta}{2}\cos\frac{3\theta}{2} \tag{6.102}$$

Solving (6.100) and (6.101) gives

$$\sigma_{11} = \frac{K_I}{(2\pi r)^{1/2}}\cos\frac{\theta}{2}(1 - \sin\frac{\theta}{2}\sin\frac{3\theta}{2}) \tag{6.103}$$

$$\sigma_{22} = \frac{K_I}{(2\pi r)^{1/2}}\cos\frac{\theta}{2}(1 + \sin\frac{\theta}{2}\sin\frac{3\theta}{2}) \tag{6.104}$$

It is important to note that (6.102)–(6.104) are exactly the same as (6.45)–(6.47) for the Griffith crack. But in this section we do not make any assumption regarding the loading and the crack geometry of the mode I crack. Therefore, (6.102)–(6.104) give the universal spatial dependence at the crack tip for mode I cracks.

To examine the crack-tip deformation field, we first recall the two-dimensional Hooke's law given in (2.112) of Chapter 2.

$$\varepsilon_{rr} = \frac{\partial u_r}{\partial r} = \frac{1}{2\mu}[\sigma_{rr} - \frac{3-\kappa}{4}(\sigma_{rr} + \sigma_{\theta\theta})] \tag{6.105}$$

$$\varepsilon_{\theta\theta} = \frac{1}{r}(\frac{\partial u_\theta}{\partial \theta} + u_r) = \frac{1}{2\mu}[\sigma_{\theta\theta} - \frac{3-\kappa}{4}(\sigma_{rr} + \sigma_{\theta\theta})] \tag{6.106}$$

Substituting (6.96)–(6.98) into (6.105) and (6.106) and integrating the resulting expressions, we obtain

$$u_r = \frac{K_I}{4\mu}(\frac{r}{2\pi})^{1/2}[(2\kappa - 1)\cos\frac{\theta}{2} - \cos\frac{3\theta}{2}] \tag{6.107}$$

$$u_\theta = -\frac{K_I}{4\mu}(\frac{r}{2\pi})^{1/2}[(2\kappa + 1)\sin\frac{\theta}{2} - \sin\frac{3\theta}{2}] \tag{6.108}$$

As expected, they are consistent with (6.52) and (6.53) for the near field displacement for the Griffith crack. Thus, (6.107) and (6.108) also provide the universal spatial dependency of the crack-tip displacement field.

The stress singularity around a crack-tip can also be examined using complex variable technique (e.g., Wang and Chau, 1998, 2001; Chau and Wang, 1998a,b).

6.6 THE GENERAL STRESS AND DISPLACEMENT FIELD FOR MODE II CRACKS

For mode II cracks, the shear stress $\sigma_{r\theta}$ is symmetric about the x_1-axis but $\sigma_{\theta\theta}$ becomes antisymmetric about the crackling. We therefore retain only the sine terms in (6.77) and (6.78). Consequently, we have

$$\varphi(r,\theta) = \sum_{\lambda=1}^{\infty} r^{\lambda+2}[C_\lambda \sin\lambda\theta + D_\lambda \sin(\lambda+2)\theta] \tag{6.109}$$

$$\sigma_{\theta\theta} = \sum_{\lambda=1}^{\infty} (\lambda+2)(\lambda+1)r^{\lambda}[C_\lambda \sin\lambda\theta + D\sin(\lambda+2)\theta] \tag{6.110}$$

$$\sigma_{r\theta} = -\sum_{\lambda=1}^{\infty} (\lambda+1)r^{\lambda}[C_\lambda \lambda\cos\lambda\theta + D_\lambda(\lambda+2)\cos(\lambda+2)\theta] \tag{6.111}$$

$$\sigma_{rr} = \sum_{\lambda=1}^{\infty} r^{\lambda}\{C_\lambda[(\lambda+2)-\lambda^2]\sin\lambda\theta - D_\lambda(\lambda+2)(1+\lambda)\sin(\lambda+2)\theta\} \tag{6.112}$$

With these stresses, the boundary condition (6.84) yields

$$(C_\lambda + D_\lambda)\sin\lambda\pi = 0 \tag{6.113}$$

$$[\lambda C_\lambda + (\lambda+2)D_\lambda]\cos\lambda\pi = 0 \tag{6.114}$$

Following the same argument used in the previous section, we have

$$\lambda = -\frac{1}{2}, \quad and \quad C_{-1/2} = -D_{-1/2} \tag{6.115}$$

The corresponding stress field near the crack tip can be shown to be

$$\sigma_{\theta\theta} = -\frac{K_{II}}{(2\pi r)^{1/2}}[\frac{3}{4}\sin\frac{\theta}{2} + \frac{3}{4}\sin\frac{3\theta}{2}] \tag{6.116}$$

$$\sigma_{r\theta} = \frac{K_{II}}{(2\pi r)^{1/2}}[\frac{1}{4}\cos\frac{\theta}{2} + \frac{3}{4}\cos\frac{3\theta}{2}] \tag{6.117}$$

$$\sigma_{rr} = -\frac{K_{II}}{(2\pi r)^{1/2}}[\frac{3}{4}\sin\frac{\theta}{2} + \frac{3}{4}\sin\frac{3\theta}{2}] \tag{6.118}$$

Substituting these stresses into (6.99) we obtain

$$\sigma_{11} = -\frac{K_{II}}{(2\pi r)^{1/2}}\sin\frac{\theta}{2}[2 + \cos\frac{\theta}{2}\cos\frac{3\theta}{2}] \tag{6.119}$$

$$\sigma_{22} = \frac{K_{II}}{(2\pi r)^{1/2}} \sin\frac{\theta}{2}\cos\frac{\theta}{2}\cos\frac{3\theta}{2} \tag{6.120}$$

$$\sigma_{12} = \frac{K_{II}}{(2\pi r)^{1/2}} \cos\frac{\theta}{2}[1-\sin\frac{\theta}{2}\sin\frac{3\theta}{2}] \tag{6.121}$$

where K_{II} is the mode II stress intensity factor. Again (6.119)–(6.121) are the same as (6.62) and (6.64) of the Griffith crack. Thus, (6.119)–(6.121) are the universal spatial dependency of stresses of mode II crack. Similar to the mode I case, all loading and crack geometry enters the problem only through the calculation of K_{II}.

For the deformation field, substitution of (6.116) and (6.118) into (6.105) and (6.106) yields the following expressions for u_r and u_θ:

$$u_r = -\frac{K_{II}}{4\mu}(\frac{r}{2\pi})^{1/2}[(2\kappa-1)\sin\frac{\theta}{2} - 3\sin\frac{3\theta}{2}] \tag{6.122}$$

$$u_\theta = -\frac{K_{II}}{4\mu}(\frac{r}{2\pi})^{1/2}[(2\kappa+1)\cos\frac{\theta}{2} - 3\cos\frac{3\theta}{2}] \tag{6.123}$$

As expected, these displacements are the same as those given in (6.69) and (6.70). Therefore, they are the universal spatial dependency for the crack-tip displacement field.

6.7 THE GENERAL STRESS AND DISPLACEMENT FIELD FOR MODE III CRACKS

For tearing modes, we have in polar coordinates $u_r = u_\theta = 0$. The only nonzero displacement component is u_z, and the only nonzero stresses are σ_{rz} and $\sigma_{\theta z}$ which can be expressed in terms of u_z as

$$\sigma_{rz} = \mu\frac{\partial u_z}{\partial r}, \qquad \sigma_{\theta z} = \frac{\mu}{r}\frac{\partial u_z}{\partial \theta} \tag{6.124}$$

The only equilibrium equation needed to be satisfied is (compare (1.74) of Chapter 1)

$$\frac{\partial(r\sigma_{rz})}{\partial r} + \frac{\partial \sigma_{\theta z}}{\partial \theta} = 0 \tag{6.125}$$

Substituting (6.124) into (6.125), we obtain

$$\frac{\partial}{\partial r}(r\frac{\partial u_z}{\partial r}) + \frac{1}{r}\frac{\partial^2 u_z}{\partial \theta^2} = 0 \tag{6.126}$$

which is the Laplacian equation in the polar plane [compare (1.72) of Chapter 1]. This special form of elasticity is normally referred to as the *anti-plane problem*, which was coined by L.N.G. Filon (Milne-Thomson, 1962). This feature of anti-plane problems renders a different but simpler method of analysis. The stress field for mode III crack problems is demonstrated in Fig. 6.8. The displacement boundary conditions along the crack line are

$$u_z(r,\theta=0) = 0, \qquad \frac{\partial u_z(r,\theta=\pi)}{\partial \theta} = 0 \tag{6.127}$$

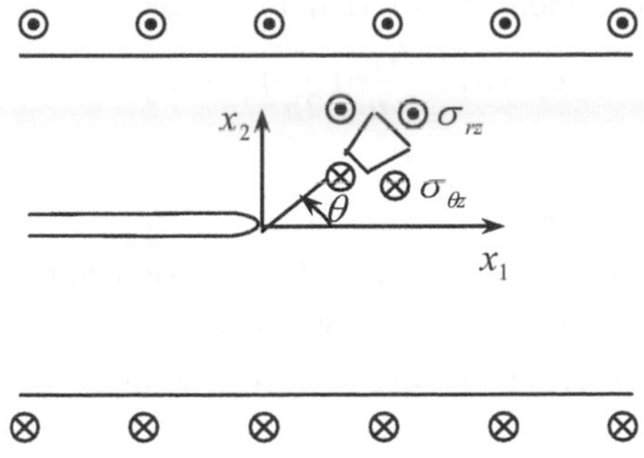

Figure 6.8 The stress field in mode III crack

Since we are interested in the asymptotic solution near $r = 0$, we look for the following general solution for u_z:

$$u_z = r^{\lambda} f(\theta) \tag{6.128}$$

Substitution of (6.128) into (6.126) yields the following governing equation for f:

$$f''(\theta) + \lambda^2 f(\theta) = 0 \tag{6.129}$$

Therefore, the general solution for u_z is

$$u_z = r^{\lambda}[A \sin \lambda\theta + B \cos \lambda\theta] \tag{6.130}$$

where A and B are real constants to be determined from boundary conditions. Substitution of (6.130) into the boundary condition (6.127) gives

$$B = 0, \qquad \cos \lambda\pi = 0 \tag{6.131}$$

Therefore, the possible values for λ are

$$\lambda = \pm\frac{1}{2}, \pm\frac{3}{2}, \pm\frac{5}{2}, \ldots \tag{6.132}$$

Since the shear stresses are proportional to $r^{\lambda-1}$ all $\lambda < 1$ will give stress singularity at the crack tip. Again, the boundedness of the strain energy for a finite body around the crack tip must be imposed to render a unique solution. In particular, we require

$$\int_0^R \int_{-\pi}^{\pi} \mathcal{E}(r,\theta) r dr d\theta = \frac{1}{2\mu} \int_0^R \int_{-\pi}^{\pi} (\sigma_{rz}^2 + \sigma_{z\theta}^2) r dr d\theta < \infty \tag{6.133}$$

where R is an arbitrary finite distance measured from the crack tip. Therefore, since $\mathcal{E}$ is proportional to $r^{2\lambda-2}$, we have

$$\int_0^R r^{2\lambda-1} dr < \infty \tag{6.134}$$

This implies that $2\lambda - 1 > -1$ or simply

$$\lambda > 0 \tag{6.135}$$

Therefore, the only λ gives both finite strain energy and stress singularity at the crack tip is $\lambda = 1/2$. Before we evaluate the stresses and displacement, it is customary to redefine

$$A = \frac{K_{III}}{\mu} \left(\frac{2}{\pi}\right)^{1/2} \tag{6.136}$$

Consequently, the displacement and stress fields become

$$u_z = u_3 = \frac{K_{III}}{\mu}(\frac{r}{2\pi})^{1/2} \sin\frac{\theta}{2} \tag{6.137}$$

$$\sigma_{zr} = \frac{K_{III}}{(2\pi r)^{1/2}} \sin\frac{\theta}{2}, \qquad \sigma_{\theta z} = \frac{K_{III}}{(2\pi r)^{1/2}} \cos\frac{\theta}{2} \tag{6.138}$$

To express the stress field in Cartesian coordinates, we note that $\boldsymbol{e}_3 = \boldsymbol{e}_z$ and

$$\begin{aligned} \boldsymbol{\sigma} &= \sigma_{32}\mathbf{e}_3\mathbf{e}_2 + \sigma_{31}\mathbf{e}_3\mathbf{e}_1 = \sigma_{zr}\mathbf{e}_z\mathbf{e}_r + \sigma_{z\theta}\mathbf{e}_z\mathbf{e}_\theta \\ &= \sigma_{zr}\mathbf{e}_3(\mathbf{e}_1\cos\theta + \mathbf{e}_2\sin\theta) + \sigma_{z\theta}\mathbf{e}_3(-\mathbf{e}_1\sin\theta + \mathbf{e}_2\cos\theta) \end{aligned} \tag{6.139}$$

Therefore, we find

$$\sigma_{31} = -\frac{K_{III}}{(2\pi r)^{1/2}} \sin\frac{\theta}{2}, \qquad \sigma_{32} = \frac{K_{III}}{(2\pi r)^{1/2}} \cos\frac{\theta}{2} \tag{6.140}$$

Physically, mode III cracking resembles the strike-slip faulting in the earth whereas mode II cracking is analogous to thrust faulting (e.g., Scholz, 1990).

To find the stress intensity factor for modes I, II, and III, one can first consult stress intensity factor handbooks (e.g., Tada et al., 1973; Sih, 1973; Rooke and Cartwright, 1976; Murakami et al., 1987). Before we discuss other methods in estimating the stress intensity factors, we will present the energy release rate.

6.8 ENERGY RELEASE RATE AT CRACK TIPS

When a crack propagates either in mode I, II, or III in a solid, the energy inside the body will be released. For example, consider the idealized shear experiment of displacement control shown in Fig. 6.9. The presentation here closely follows that of Rudnicki (1988).

Let $Q = \sigma_{12}A$ be the net force applied to the upper surface and q be the displacement of the upper surface. The work done due to an increment in the applied load is then Qdq (per unit thickness). If Φ is the strain energy, then for elastic materials (not necessarily linear)

$$d\Phi = Qdq \tag{6.141}$$

at fixed crack length l. We now define the crack-tip energy release rate as

$$G = -(\frac{\partial\Phi}{\partial l})_q \tag{6.142}$$

Some authors believed that the symbol G is used in honor of Griffith (Kannien and Popelar, 1985). But, incidentally, G is also the first letter of George, the first name of Irwin, who was the first one to use this symbol. A brief biography of G.R. Irwin is given at the end of this book. As shown in Fig. 6.9, the strain energy is the area under the Q–q curve, thus, $G\Delta l$ is the area between the curves of Q versus q at crack length of l to $l + \Delta l$. Note that the energy release rate defined in (6.142) is for a fixed load point displacement. However, intuitively the energy release rate should also be independent of the loading type (i.e., either load or displacement control).

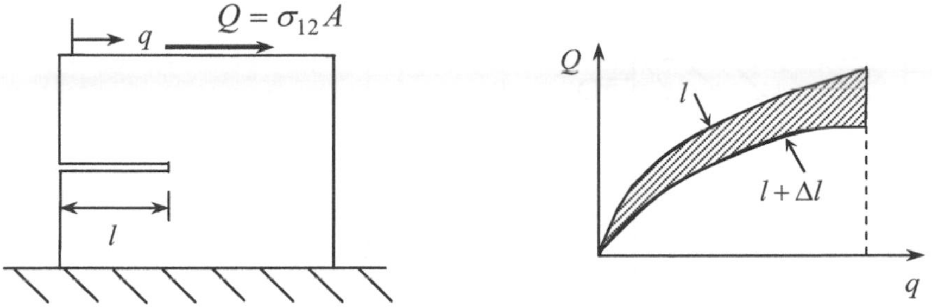

Figure 6.9 An idealized shear experiment and its load-displacement curve

In particular, to derive an expression for G for the case of the dead load case, as shown in Fig. 6.10, we consider the following total differential of the strain energy:

$$d\Phi(q,l) = (\frac{\partial \Phi}{\partial q})_l \, dq + (\frac{\partial \Phi}{\partial l})_q \, dl = Qdq - Gdl \tag{6.143}$$

We now change the variable from q to Q using the Legendre transformation (see Appendix C):

$$-d\,U = d[\Phi(q,l) - Qq] = d\Phi - d(Q\,q) = -q\,dQ - Gdl \tag{6.144}$$

where $U = U(Q,l)$ is the complementary strain energy and $\Phi - Qq$ is the total energy, which is the sum of the strain energy and the energy of the applied load. Therefore, for a fixed load ($dQ = 0$), we have the energy release rate defined as

$$G = -[\frac{\partial}{\partial l}(\Phi - Qq)]_Q \tag{6.145}$$

Similarly, the change in total energy $\Delta[\Phi - Qq]$ is the area between the curves for l and $l + \Delta l$ of a fixed Q, as shown in Fig. 6.10.

Within our definition of energy release rate, it can be shown that (6.142) and (6.145) are the same. In particular, the total energy released by the solid due to a crack increment of Δl is

$$\Delta\Phi = (\frac{\partial \Phi}{\partial l})_q \Delta l + \frac{1}{2!}(\frac{\partial^2 \Phi}{\partial l^2})_q (\Delta l)^2 + ... = -G\,\Delta l + O[(\Delta l)^2] \tag{6.146}$$

Therefore, the energy release rate can be interpreted as the linear order term of the Taylor's series expansion of $\Delta\Phi$. Similarly, $\Delta(\Phi - Qq)$ can be interpreted as the first-order term of the increment of the total energy. The difference between $\Delta\Phi$ and $\Delta(\Phi - Qq)$ equals the shaded area in Fig. 6.11. However, each dimension of the shaded area is proportional to Δl; consequently, the shaded area is of the order of $(\Delta l)^2$ and hence can be neglected. Thus, the energy release rate defined by (6.142) and (6.145) must be the same.

In general, it is customary to express G in terms of the stress intensity factor discussed in the earlier sections. In particular, consider a small advance of a shear crack from length l to $l + \Delta l$, as shown in Fig. 6.12. The energy release by the shear crack growth can be computed alternatively as the work necessary to restore the state "after" propagation to the original state "before" the propagation. This is the negative of the potential energy change, that is

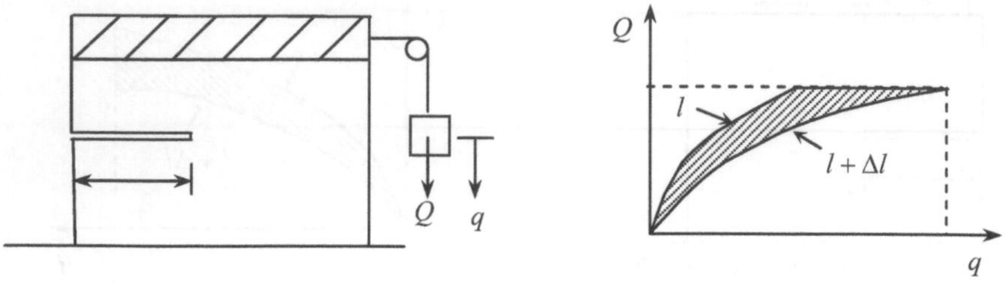

Figure 6.10 A idealized shear experiment subject to dead load

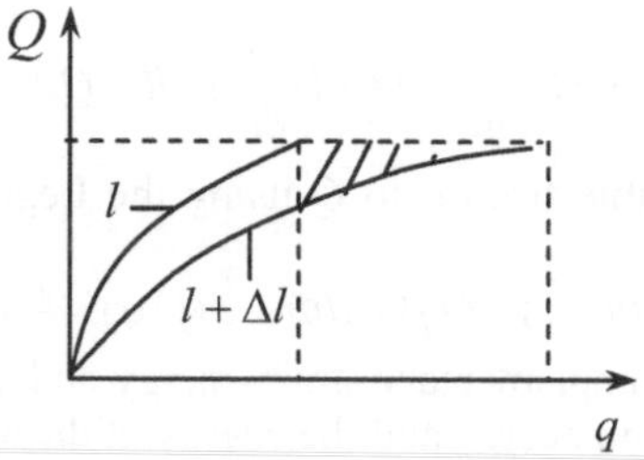

Figure 6.11 Difference between the changes in the strain energy and the total energy

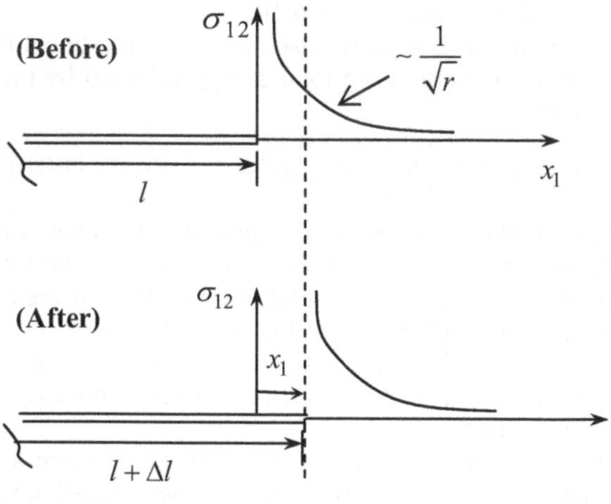

Figure 6.12 The crack geometry before and after crack propagation

$$\Delta\Phi = -\frac{1}{2}\int_0^{\Delta l} \sigma_{12}^{before}(u_1^{top} - u_1^{bot})^{after}\, dx_1 \qquad (6.147)$$

For small Δl, we can adopt the universal spatial dependence of both stress and displacement obtained earlier. More specifically, we set $\theta = 0$ and $r = x_1$ into (6.64) and $\theta = \pi$ and $r = (\Delta l - x_1)$ into (6.67), we get

$$\sigma_{12}^{before} = \frac{K_{II}(l)}{(2\pi x_1)^{1/2}} + ... \qquad (6.148)$$

$$u_1^{after} = \frac{[K_{II}(l) + \Delta K_{II}]}{2\mu}(\kappa+1)\left(\frac{\Delta l - x_1}{2\pi}\right)^{1/2} + ... \qquad (6.149)$$

Substitution of (6.148) and (6.149) into (6.147) yields

$$\Delta\Phi = -\int_0^{\Delta l} \frac{(\kappa+1)}{4\pi\mu} K_{II}(l)[K_{II}(l) + \Delta K_{II}]\left(\frac{\Delta l - x_1}{x_1}\right)^{1/2} dx_1 \qquad (6.150)$$

The energy release rate is

$$G = \lim_{\Delta l \to 0}\left[-\frac{\Delta\Phi}{\Delta l}\right] = \left(\frac{\kappa+1}{4\pi\mu}\right)K_{II}^2 \lim_{\Delta l \to 0}\left[\frac{1}{\Delta l}\int_0^{\Delta l}\left(\frac{\Delta l - x_1}{x_1}\right)^{1/2} dx_1\right] = \frac{\kappa+1}{8\mu}K_{II}^2 \qquad (6.151)$$

This expression can easily be generalized to include mode II and mode III simultaneously. The final result for the energy release rate is

$$G = \frac{\kappa+1}{8\mu}(K_I^2 + K_{II}^2) + \frac{1}{2\mu}K_{III}^2 \qquad (6.152)$$

Alternatively, it is customary to write the energy release rate as

$$G = \frac{(1-\nu^2)}{E}(K_I^2 + K_{II}^2) + \frac{1}{2\mu}K_{III}^2 \qquad (6.153)$$

for plane strain and

$$G = \frac{1}{E}(K_I^2 + K_{II}^2) + \frac{1}{2\mu}K_{III}^2 \qquad (6.154)$$

for plane stress. This result was first obtained by Irwin (1956). Therefore, a commonly used fracture criterion is

$$G = G_{crit} \qquad (6.155)$$

where G_{crit} is commonly regarded as a material constant, and it reflects the resistance of a material to fracture growth. For pure mode cracks, this is equivalent to assuming the existence of a fracture toughness K_{crit}, which indicates crack growth when

$$K = K_{crit} \qquad (6.156)$$

For mode I, K_{crit} is normally referred as K_{IC} or the mode I critical stress intensity factor. It is also known as mode I fracture toughness. In terms of Griffith's (1920) original fracture criterion, the crack growth occurs when

$$G = 2\gamma \qquad (6.157)$$

where γ is the effective surface energy and all energy dissipations at the crack tip should be incorporated. It is the energy required to create the new surfaces of the crack.

6.9 FRACTURE TOUGHNESS FOR ROCKS

For rock materials, two types of standard testing technique were proposed by the International Society for Rock Mechanics (ISRM, 1988): the *Chevron bend test* and *short rod test*. The details are referred to in ISRM (1988). Another fracture toughness test getting popularity is the semi-circular bend test which was proposed in 1984 (Kuruppu and Chong, 2012). Fracture toughness or the critical energy release rates for various rocks have been compiled by various authors (e.g., Rudnicki, 1980; Atkinson, 1984; Li, 1987; ISRM, 1988).

For the shear rupture zone or slip-weakening shear model, Li (1987) compiled the typical value for G_{crit}. The main results are tabulated in Table 6.3.

Table 6.3 Critical energy release rate G_{crit} for slipping zones or shear cracks

Rock types	G_{crit} (J/m^2)
Intact rocks	$0.3\sim0.5\times10^4$
Sawcut rocks	0.1~2.4
Clay	50~270
Natural rock joints	9~1000
Natural crustal faults	$10^6 \sim 10^8$

Table 6.4 Fracture toughness K_{IC} for various rock types

Rock types	K_{IC} (No. of samples) (MNm$^{-1.5}$)	
	ISRM (1988)	Atkinson (1984)
Andesite	1.26~1.68 (24)	
Basalt		2.5~2.58
Dolerite	2.86~3.26 (12)	3.28
Dunit		1.39~3.74
Gabbro	2.22~3.23 (54)	0.84~2.88
Granite	0.65~2.8 (249)	0.59~2.5
Granodiorite	2.95~3.35 (6)	
Limestone	1.31~1.87 (28)	0.37~2.01
Marble	1.26~2.62 (74)	0.64~1.39
Micrite		1.01
Norite	2.23~2.69 (17)	
Quartzite		2.1~2.65
Sandstone	0.73~2.56 (56)	0.28~2.53
Shale	0.25~1.02 (14)	0.61~1.37
Siltstone		1.04~1.37
Tholeiite		0.87
Tuff	1.02~1.08 (64)	
	Lim *et al.* (1994)	
Coal	0.02~0.15	
Johnstone	0.07	
Schist	0.9	
Syenite	1.36~1.86	

The fracture toughness for various rocks was compiled by ISRM (1988), Atkinson (1984), and Lim et al. (1994); Table 6.4 presents only the ranges of K_{IC} for different rock groups. For example, for the value of K_{IC} for granite from a specific location refer to the original paper. In general, for most rocks the order of K_{IC} is about 1 MPa$\sqrt{m}$; and for artificial rock, such as Johnstone, the K_{IC} is of the order of 0.1 MPa$\sqrt{m}$.

For different minerals, Atkinson (1984) also compiled a table. The results are summarized briefly in Table 6.5. In general, the fracture toughness is smaller in minerals than in rocks, except for diamonds.

Whether G_{crit} or K_{IC} is given, we can transform one to the other according to (6.154). For example, if μ = 20 GPa and ν = 0.2, K_{IC} values of 1, 2.5 and 4 $MNm^{-3/2}$ yield G_{crit} = 20, 125 and 320 Jm^{-2}, respectively.

For modes II and III fracture toughness, the values for K_{IIC} and K_{IIIC} is not readily available for most rocks. Some data for K_{IIC} are compiled by Lim et al. (1994). In general, as remarked by Li (1987), the order of this fracture toughness is again about 1 $MNm^{-3/2}$.

Table 6.5 Fracture toughness K_{IC} for various minerals (after Atkinson, 1984)

Minerals	K_{IC} ($MNm^{-1.5}$)
Calcite	0.16~0.2
Diamond	3.4~3.9
Feldspar	0.39~1.3
Fluorite	0.3
Galena	0.18
Halite	0.18
Muscovite	0.21
Olivine	0.59~0.73
Periclase	0.79
Pyrite	0.96
Quartz	0.28~2.4
Spinel	1.27~1.9

6.10 *J*-INTEGRAL AND THE ENERGY RELEASE RATE

The energy release rate G discussed in Section 6.8 can be expressed in terms of a path-independent integral called the *J-integral*. The J-integral was derived independently by Rice (1968a) and Cherepanov (1969) for crack problems, although the same integral had been derived much earlier by Eshelby (1957) as a way of calculating forces on heterogeneities and dislocations. However, Rice's approach is surely the most popular one, thus his work has also received most of the credit. Except for the classic paper by Griffith (1920), Rice's paper is probably the most cited paper in the fracture mechanics literature. The derivation of the J-integral is indeed a giant step and milestone in the development of fracture mechanics because it opens up a whole new field of *elastic-plastic fracture mechanics*. The most attractive feature of the J-integral is the path-independency and the energy release rate interpretation of it. Both of these will be discussed later in the section in terms of nonlinear elastic fracture mechanics.

We will discuss here a slightly modified version of the *J*-integral that incorporates the effect of body forces. In particular, the total potential energy of a linear elastic body containing a crack, as shown in Fig. 6.13, is

$$\Pi(a) = \int_A W dA - \int_\Gamma T_i u_i ds - \int_A f_i u_i dA \tag{6.158}$$

where W is the strain energy density, u_i is the displacement field, Γ is the contour of the body, A is the area of the two-dimensional body, T_i is the traction field applied on the boundary of the body, and f_i is the body force component. The strain energy density can be defined as

$$W(\varepsilon_{pq}) = \int_0^{\varepsilon_{pq}} \sigma_{ij} d\varepsilon_{ij} \tag{6.159}$$

Conversely, the stress can be expressed as

$$\sigma_{ij} = \frac{\partial W}{\partial \varepsilon_{ij}} \tag{6.160}$$

We now consider the change of the potential energy with respect to an increment of the crack length as

$$\frac{d\Pi(a)}{da} = \int_A \frac{dW}{da} dA - \int_\Gamma T_i \frac{du_i}{da} ds - \int_A f_i \frac{du_i}{da} dA \tag{6.161}$$

In obtaining (6.161), we have assumed that both the body force and surface traction remain constant when the crack propagation takes place. We now introduce a moving coordinate, which is located at the crack tip, such that

$$X_1 = x_1 - a, \quad X_2 = x_2 \tag{6.162}$$

Thus, it is straightforward to see that

$$\frac{d}{da} = \frac{\partial}{\partial a} + \frac{\partial}{\partial X_1}\frac{\partial X_1}{\partial a} = \frac{\partial}{\partial a} - \frac{\partial}{\partial X_1} = \frac{\partial}{\partial a} - \frac{\partial}{\partial x_1} \tag{6.163}$$

since $\partial X_1/\partial a = -1$ and $\partial/\partial X_1 = \partial/\partial x_1$. Substitution of (6.163) into (6.161) yields

$$\frac{d\Pi(a)}{da} = \int_A \left(\frac{\partial W}{\partial a} - \frac{\partial W}{\partial x_1}\right) dA - \int_\Gamma T_i \left(\frac{\partial u_i}{\partial a} - \frac{\partial u_i}{\partial x_1}\right) ds - \int_A f_i \left(\frac{\partial u_i}{\partial a} - \frac{\partial u_i}{\partial x_1}\right) dA \tag{6.164}$$

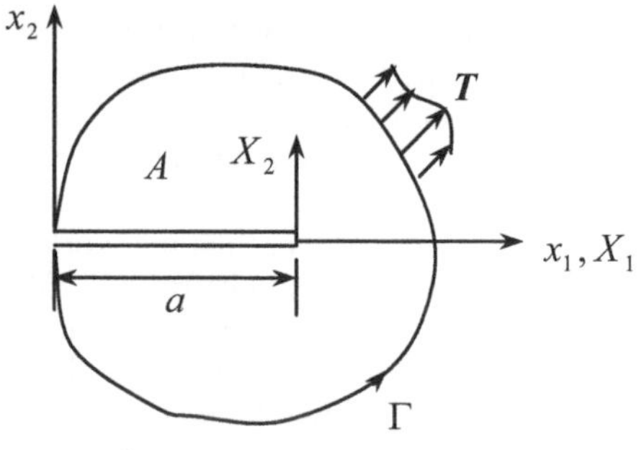

Figure 6.13 A cracked nonlinear elastic solid with contour Γ

Note that

$$\frac{\partial W}{\partial a}=\frac{\partial W}{\partial \varepsilon_{ij}}\frac{\partial \varepsilon_{ij}}{\partial a}=\sigma_{ij}\frac{1}{2}\frac{\partial}{\partial a}\left(\frac{\partial u_i}{\partial x_j}+\frac{\partial u_j}{\partial x_i}\right)=\sigma_{ij}\frac{\partial}{\partial x_j}\frac{\partial u_i}{\partial a} \tag{6.165}$$

Therefore, the following area integral becomes

$$\begin{aligned}\int_A \frac{\partial W}{\partial a}dA &= \int_A \sigma_{ij}\frac{\partial}{\partial x_j}\left(\frac{\partial u_i}{\partial a}\right)dA=\int_A \frac{\partial}{\partial x_j}\left(\sigma_{ij}\frac{\partial u_i}{\partial a}\right)dA-\int_A \frac{\partial \sigma_{ij}}{\partial x_j}\frac{\partial u_i}{\partial a}dA \\ &= \int_A \frac{\partial}{\partial x_j}\left(\sigma_{ij}\frac{\partial u_i}{\partial a}\right)dA+\int_A f_i\frac{\partial u_i}{\partial a}dA\end{aligned} \tag{6.166}$$

The last of (6.166) is due to the substitution of the equilibrium equation. We further apply the Gauss theorem (or divergence theorem) given in Section 1.6.1 to the first term of the last of (6.166), and the result is

$$\int_A \frac{\partial W}{\partial a}dA=\int_\Gamma \sigma_{ij}n_j\left(\frac{\partial u_i}{\partial a}\right)ds+\int_A f_i\frac{\partial u_i}{\partial a}dA=\int_\Gamma T_i\frac{\partial u_i}{\partial a}ds+\int_A f_i\frac{\partial u_i}{\partial a}dA \tag{6.167}$$

Finally, substitution of (6.167) into (6.164) yields

$$\frac{d\Pi}{da}=-\int_A\left(\frac{\partial W}{\partial x_1}-f_i\frac{\partial u_i}{\partial x_1}\right)dA+\int_\Gamma T_i\left(\frac{\partial u_i}{\partial x_1}\right)ds \tag{6.168}$$

The first integral on the right-hand side can obviously be expressed in terms of the contour integral. The J-integral is defined as

$$J=-\frac{d\Pi}{da}=\int_\Gamma\left[(W-f_iu_i)dx_2-T_i\frac{\partial u_i}{\partial x_1}ds\right] \tag{6.169}$$

By noting that $n_1ds=dx_2$, the J-integral can alternatively be written as

$$J=\int_\Gamma\left[(W-\boldsymbol{f}\bullet\boldsymbol{u})n_1-\boldsymbol{T}\bullet\frac{\partial \boldsymbol{u}}{\partial x_1}\right]ds \tag{6.170}$$

When $f_i=0$, the original J-integral obtained by Rice (1968a) is recovered as a special case.

To examine the path-independency of the J-integral, we consider the J-integral around a contour Γ_t, which equals $\Gamma_1+\Gamma_2+\Gamma+\Gamma_3$, as shown in Fig. 6.14:

$$J_t=\int_{\Gamma_t}\left[(W-\boldsymbol{f}\bullet\boldsymbol{u})n_1-\boldsymbol{T}\bullet\frac{\partial \boldsymbol{u}}{\partial x_1}\right]ds \tag{6.171}$$

Note that we have $n_1=0$ and $T_i=0$ on both the sub-contours Γ_2 and Γ_3. Therefore, (6.171) can be simplified to

$$J_t=J_1+J=\int_{\Gamma_1+\Gamma}\left[(W-\boldsymbol{f}\bullet\boldsymbol{u})n_1-\boldsymbol{T}\bullet\frac{\partial \boldsymbol{u}}{\partial x_1}\right]ds \tag{6.172}$$

Applying the divergence or Gauss theorem back to (6.172), we get

$$J_1+J=\int_A[(\frac{\partial W}{\partial x_1}-f_i\frac{\partial u_i}{\partial x_1}-\frac{\partial}{\partial x_j}(\sigma_{ij}\frac{\partial u_i}{\partial x_1})]dA$$
$$=\int_A[\frac{\partial W}{\partial \varepsilon_{ij}}\frac{\partial \varepsilon_{ij}}{\partial x_1}-f_i\frac{\partial u_i}{\partial x_1}-\frac{\partial \sigma_{ij}}{\partial x_j}\frac{\partial u_i}{\partial x_1}-\sigma_{ij}\frac{\partial}{\partial x_j}(\frac{\partial u_i}{\partial x_1})]dA \tag{6.173}$$
$$=\int_A[\sigma_{ij}\frac{\partial \varepsilon_{ij}}{\partial x_1}-\sigma_{ij}\frac{\partial \varepsilon_{ij}}{\partial x_1}]dA=0$$

Therefore, finally we get

$$J_1=-J \tag{6.174}$$

Thus, if both contours Γ and Γ_1 are taken as counter-clockwise, we must have $J_1 = J$. Therefore, the J-integral must be path-independent.

The relation between the J-integral and the energy release rate G can be considered by taking the limit of the contour approaching the crack tip, as shown in Fig. 6.15. For mathematical simplicity, we will consider the case of mode III crack. In particular, we can apply the results in (6.137) and (6.138) here, that is,

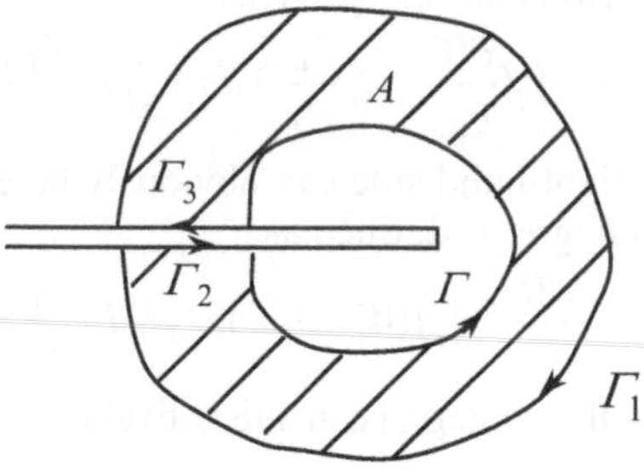

$$\Gamma_t=\Gamma_1+\Gamma_2+\Gamma+\Gamma_3$$

Figure 6.14 The path $\Gamma_t = \Gamma_1+\Gamma_2+\Gamma+\Gamma_3$ for the J-integral

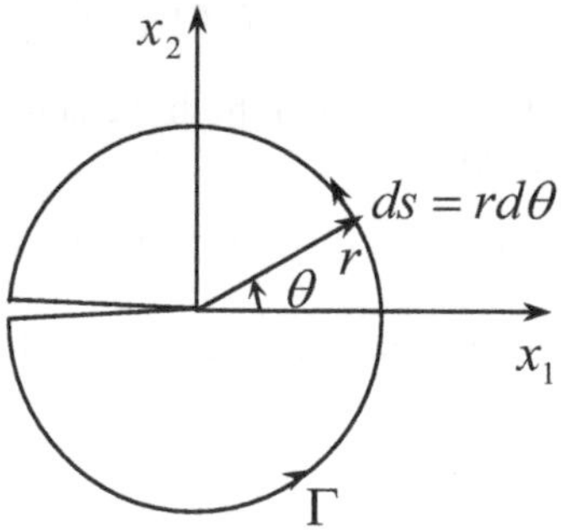

Figure 6.15 The J-integral around the crack-tip field

$$W = \frac{1}{2\mu}(\sigma_{rz}^2 + \sigma_{\theta z}^2) = \frac{K_{III}^2}{2\mu}\frac{1}{(2\pi r)}[\sin^2(\frac{\theta}{2}) + \cos^2(\frac{\theta}{2})] = \frac{K_{III}^2}{4\pi\mu r} \tag{6.175}$$

$$T_i \frac{\partial u_i}{\partial x_1} = \sigma_{rz}\frac{\partial u_z}{\partial x_1} = \frac{K_{III}^2}{\pi\mu}\frac{1}{r^{1/2}}\sin(\frac{\theta}{2})\frac{\partial}{\partial x_1}[r^{1/2}\sin(\frac{\theta}{2})] \tag{6.176}$$

Noting that

$$\frac{\partial r}{\partial x_1} = \cos\theta, \quad \frac{\partial\theta}{\partial x_1} = -\frac{\sin\theta}{r} \tag{6.177}$$

(6.176) reduces to

$$T_i \frac{\partial u_i}{\partial x_1} = \frac{K_{III}^2}{2\pi\mu r}\sin^2(\frac{\theta}{2}) \tag{6.178}$$

Substitution of (6.175) and (6.178) into (6.169), noting that $ds = rd\theta$ and $dx_2 = r\cos\theta \, d\theta$, yields

$$J = \frac{K_{III}^2}{4\pi\mu}\int_{-\pi}^{\pi}\cos\theta d\theta + \frac{K_{III}^2}{2\pi\mu}\int_{-\pi}^{\pi}\sin^2(\frac{\theta}{2})d\theta = \frac{K_{III}^2}{2\mu} \tag{6.179}$$

More generally, if all three modes of crack displacement are taken into account, we have

$$J = \frac{\kappa+1}{8\mu}(K_I^2 + K_{II}^2) + \frac{1}{2\mu}K_{III}^2 = G \tag{6.180}$$

That is, the J-integral is physically the energy release rate at the crack tip, at least for nonlinear elastic solids. For elastic-plastic solids, although the J-integral does not correspond exactly to G, it provides a useful way to estimate the crack-tip field because of its path-independent property.

Although we only demonstrate the validity of (6.180) using mode III crack, the reader can follow the procedure given here for both mode I and II crack tip fields.

6.11 WESTERGAARD STRESS FUNCTION AND SUPERPOSITION

The modes I and II stress intensity factors for some two-dimensional crack problems can be found analytically in terms of a stress function called the *Westergaard stress function* (Westergaard, 1939). The Westergaard stress function is actually a special form of Muskhelishvili's analytic functions, and this function has been found very useful in calculating the stress intensity factors for symmetric and antisymmetric crack problems subject to near field loadings.

In particular, for mode I cracks there is a symmetric property that $\sigma_{12} = 0$ on $x_2 = 0$. By virtue of this fact we can set the following special form of Muskhelishvili's analytic functions:

$$\psi(z) = -z\phi'(z) + \phi(z) \tag{6.181}$$

We first rewrite (3.34) and (3.35) of Chapter 3 as

$$\sigma_{22} = 2\,\mathrm{Re}[\phi' + \tfrac{1}{2}\bar{z}\phi'' + \tfrac{1}{2}\psi'] \tag{6.182}$$

$$\sigma_{11} = 2\,\mathrm{Re}[\phi' - \frac{1}{2}\bar{z}\phi'' - \frac{1}{2}\psi'] \tag{6.183}$$

$$\sigma_{12} = \mathrm{Im}\left[\bar{z}\phi'' + \psi'\right] \tag{6.184}$$

Substitution of (6.181) into (6.182)–(6.184) yields

$$\sigma_{11} = 2\,\mathrm{Re}\left[\phi'(z)\right] - 2x_2\,\mathrm{Im}\left[\phi''(z)\right] \tag{6.185}$$

$$\sigma_{22} = 2\,\mathrm{Re}\left[\phi'(z)\right] + 2x_2\,\mathrm{Im}\left[\phi''(z)\right] \tag{6.186}$$

$$\sigma_{12} = -2x_2\,\mathrm{Re}\left[\phi''(z)\right] \tag{6.187}$$

In obtaining these expressions, we have used the following identities

$$\mathrm{Re}\{if(z)\} = -\mathrm{Im}\left[f(z)\right], \quad \mathrm{Im}\{if(z)\} = \mathrm{Re}\left[f(z)\right] \tag{6.188}$$

for any analytic function $f(z)$. Similarly, substitution of (6.181) into (3.36) of Chapter 3 gives

$$2\mu u_1 = (\kappa - 1)\,\mathrm{Re}\left[\phi(z)\right] - 2x_2\,\mathrm{Im}\left[\phi'(z)\right] \tag{6.189}$$

$$2\mu u_2 = (\kappa + 1)\,\mathrm{Im}\left[\phi(z)\right] - 2x_2\,\mathrm{Re}\left[\phi'(z)\right] \tag{6.190}$$

It is customary to use the following notation for the Westergaard stress function Z:

$$Z(z) = 2\phi'(z), \quad \frac{d\bar{Z}}{dz} = Z \tag{6.191}$$

Consequently, the stresses and displacements become

$$\sigma_{11} = \mathrm{Re}\,Z - x_2\,\mathrm{Im}\,Z' \tag{6.192}$$

$$\sigma_{22} = \mathrm{Re}\,Z + x_2\,\mathrm{Im}\,Z' \tag{6.193}$$

$$\sigma_{12} = -x_2\,\mathrm{Re}\,Z' \tag{6.194}$$

$$2\mu u_1 = \left(\frac{\kappa - 1}{2}\right)\mathrm{Re}\,\bar{Z} - x_2\,\mathrm{Im}\,Z \tag{6.195}$$

$$2\mu u_2 = \left(\frac{\kappa + 1}{2}\right)\mathrm{Im}\,\bar{Z} - x_2\,\mathrm{Re}\,Z \tag{6.196}$$

We can solve Griffith crack problem by using superposition. In particular, the original problem can be considered the sum of the uniform stress solution in an infinite domain without crack subject to far field tension σ and the solution for an isolated crack subject to internal pressure σ (see Fig. 6.16). The solution for the first auxiliary problem is trivial and the solution for the second auxiliary problem can be solved using the following Westergaard stress function (Westergaard, 1939):

$$Z(z) = \frac{\sigma}{\sqrt{1 - (a/z)^2}} \tag{6.197}$$

In terms of the polar coordinates given in Fig. 6.16, we have

$$z = re^{i\theta}, \quad z - a = r_1 e^{i\theta_1}, \quad z + a = r_2 e^{i\theta_2} \tag{6.198}$$

where all θ, θ_1, and θ_2 are between π and −π. Substitution of (6.198) into (6.197) gives

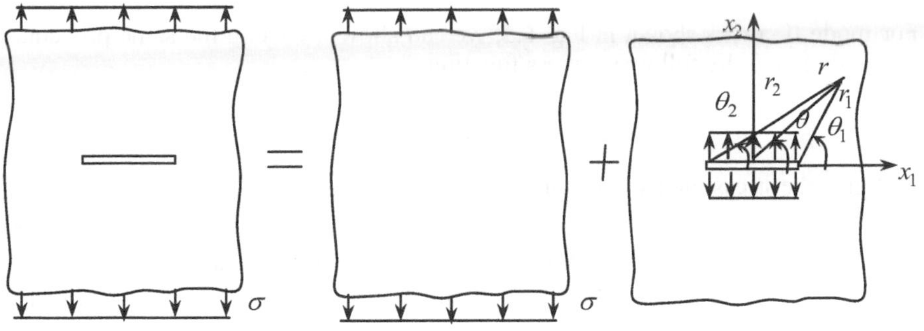

Figure 6.16 Method of superposition for crack

$$Z = \frac{\sigma r}{\sqrt{r_1 r_2}} e^{i\left[\theta-(\theta_1+\theta_2)/2\right]} \tag{6.199}$$

The derivative of (6.197) is

$$Z' = -\frac{\sigma a^2}{\left(z^2-a^2\right)^{3/2}} = -\frac{\sigma a^2}{(r_1 r_2)^{3/2}} e^{-i\frac{3}{2}(\theta_1+\theta_2)} \tag{6.200}$$

On the other hand, integration of (6.197) gives

$$\bar{Z} = \sigma\sqrt{z^2-a^2} = \sigma\sqrt{r_1 r_2}\, e^{i(\theta_1+\theta_2)/2} \tag{6.201}$$

Along the crack surfaces (i.e., $x_2 = 0$ and $x_1 < a$), it is not difficult to show that ReZ = x_2ReZ'= x_2ImZ'= 0. We now consider the stress field near the crack tip; that is, we set

$$\frac{r_1}{a} \to 0, \quad r \to a, \quad r_2 \to 2a, \quad \theta \to 0, \quad \theta_2 \to 0 \tag{6.202}$$

into (6.199). The resulting stress function is approximately

$$Z = \frac{\sigma\sqrt{a}}{\sqrt{2r_1}} e^{-i\theta_1/2} \tag{6.203}$$

In terms of polar coordinates with the origin at the right crack tip, i.e.,

$$\zeta = r_1 e^{i\theta_1} = z - a \tag{6.204}$$

(6.203) becomes

$$Z(\zeta) = \frac{\sigma\sqrt{a}}{\sqrt{2\zeta}} = \frac{K_I}{\sqrt{2\pi\zeta}} \tag{6.205}$$

where $K_I = \sigma(\pi a)^{1/2}$ is the mode I stress intensity factor defined in (6.48) of Section 6.3.

For mode II cracks, it is not difficult to show that the following form of the Westergaard stress function applies:

$$\sigma_{11} = 2\,\mathrm{Im}\,Z + x_2\,\mathrm{Re}\,Z', \quad \sigma_{22} = -x_2\,\mathrm{Re}\,Z', \quad \sigma_{12} = \mathrm{Re}\,Z - x_2\,\mathrm{Im}\,Z' \tag{6.206}$$

$$2\mu u_1 = (\frac{\kappa-1}{2}) \operatorname{Im} \bar{Z} + x_2 \operatorname{Re} Z', \quad 2\mu u_2 = -(\frac{\kappa+1}{2}) \operatorname{Re} \bar{Z} - x_2 \operatorname{Im} Z \quad (6.207)$$

For mode II cracks shown in Fig. 6.5, we can show, by using the same procedure for mode I, that the following stress function

$$Z(z) = \frac{\tau}{\sqrt{1-(a/z)^2}} \quad (6.208)$$

will give the stress intensity factor as

$$K_{II} = \tau\sqrt{\pi a} \quad (6.209)$$

For mode III cracks, the Westergaard stress function becomes

$$\mu u_3 = \operatorname{Im} \bar{Z}, \quad \sigma_{23} = \operatorname{Re} Z, \quad \sigma_{13} = \operatorname{Im} Z \quad (6.210)$$

For mode III crack subject to far field uniform shear, a stress function same as those given in (6.208) can be used; the K_{III} is found equal to $\tau(\pi a)^{1/2}$. The detailed steps will, however, be omitted here. Therefore, the stress intensity factor can be defined as

$$\begin{Bmatrix} K_I \\ K_{II} \\ K_{II} \end{Bmatrix} = \lim_{r\to 0} \sqrt{2\pi r} \begin{Bmatrix} \sigma_{22} \\ \sigma_{12} \\ \sigma_{13} \end{Bmatrix} \quad (6.211)$$

Since superposition applies to the linear-elastic stress field, it is evident from (6.211) that the K values are additive. For example, the Westergaard stress function for a 2-D cracks of length $2a$ subject to a pair point forces P applied on $x_1 = b$ (as shown in Fig. 6.17) is (Tada et al., 1973, p. 5.9):

$$Z(z) = \frac{P}{\pi(z-b)} \frac{\sqrt{a^2-b^2}}{\sqrt{z^2-a^2}} \quad (6.212)$$

Again, in terms of the polar coordinates,

$$z - a = \zeta = r_1 e^{i\theta_1}, \quad z + a = r_2 e^{i\theta_2} \quad (6.213)$$

we have

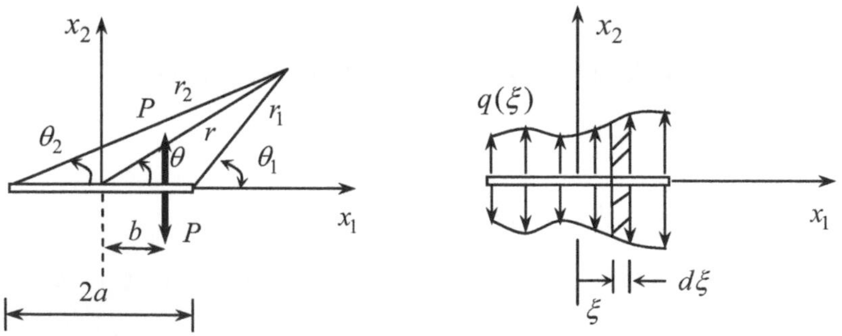

Figure 6.17 Cracks subject to pair point forces P and general internal pressure

$$Z(z) = \frac{P}{\pi(z-b)} \frac{\sqrt{a^2-b^2}}{\sqrt{r_1 r_2}} e^{-i(\theta_1+\theta_2)/2} \tag{6.214}$$

Consider the crack tip by letting $r_1/a \to 0$, $\theta \to 0$, $r_2 \to 2a$ and $z \to a$, we find

$$Z(z) = \frac{P}{\pi(a-b)} \frac{\sqrt{a^2-b^2}}{\sqrt{2a\zeta}} \tag{6.215}$$

Recalling the second of (6.205), we have

$$K_I = Z\sqrt{2\pi\zeta} = \frac{P}{a-b} \frac{\sqrt{a^2-b^2}}{\sqrt{\pi a}} = \frac{P}{\sqrt{\pi a}} \sqrt{\frac{a+b}{a-b}} \tag{6.216}$$

This result can readily be used to find the Westergaard stress function and stress intensity factor for a crack subject to a varying internal pressure $g(\xi)$, shown also in Fig. 6.17. In particular, we put the following identifications

$$b \leftarrow \xi, \qquad P \leftarrow g(\xi)d\xi \tag{6.217}$$

into (6.212) and carry out the integration from $-a$ to a. This gives

$$Z(z) = \frac{1}{\pi\sqrt{z^2-a^2}} \int_{-a}^{a} \frac{g(\xi)\sqrt{a^2-\xi^2}}{z-\xi} d\xi \tag{6.218}$$

If we consider the crack-tip field and multiply the stress function by $(2\pi\zeta)^{1/2}$, we get

$$K_I = \frac{1}{\sqrt{\pi a}} \int_{-a}^{a} \frac{g(\xi)\sqrt{a^2-\xi^2}}{a-\xi} d\xi \tag{6.219}$$

Note that a similar superposition procedure applies equally to mode II and III cases.

Chau et al. (2000, 2002) and Chau and Shao (2006) applied such a superposition technique to derive 2-D and 3-D crack models subject to loading of a center of dilatation.

6.12 GROWTH OF SLIP SURFACE IN SLOPES

In this section, we will present an application of fracture mechanics to the progressive failure of slopes. The presentation follows closely those of Palmer and Rice (1973). In particular, referring to Fig. 6.18, we consider a long flat slope of inclination angle α into which a step of height h has been cut. Thus, a man-made surface is formed as the toe of the natural slope. A slip surface or shear band of length l emanates from the base of the cut in a direction paralleling the slope surface. Based upon the J-integral, we will derive the driving force on the band and the propagation criteria for the slip surface. To make the problem mathematically tractable, we assume that the length of the slip surface l from the toe is large compared to the layer thickness h and to the size of the end region ω, which is defined in Fig. 6.19. The end region is defined as the region near the tip of the shear band where the shear strength is larger than the *residual shear strength,* as shown in Fig. 6.19. Under this assumption, the energy transfer during the shear band extension will be due to the gravitational work on the down slope movements of the layer and to deformations of the layer from changes in the normal stress acting parallel to the slope surface.

The stress state σ^0_{ij} existing *before* the cut is made is supposed to depend only on the depth from the slope surface; thus the initial stresses are

$$\sigma^0_{22} = -\gamma x_2 \cos\alpha, \quad \sigma^0_{21} = \gamma x_2 \sin\alpha, \quad \sigma^0_{11} = f(x_2) \tag{6.220}$$

where γ is the average unit weight over the depth h of the layer. Note that σ^0_{11} is undetermined by equilibrium consideration alone. Since h is small, we will consider only the average value of σ_{11} over the depth as

$$\bar{\sigma}_{11} = \frac{1}{h}\int_0^h \sigma_{11} dx_2 \tag{6.221}$$

The average lateral earth pressure existing before the introduction of the cut is denoted by p_0, i.e.,

$$p_0 = -\bar{\sigma}^0_{11} \tag{6.222}$$

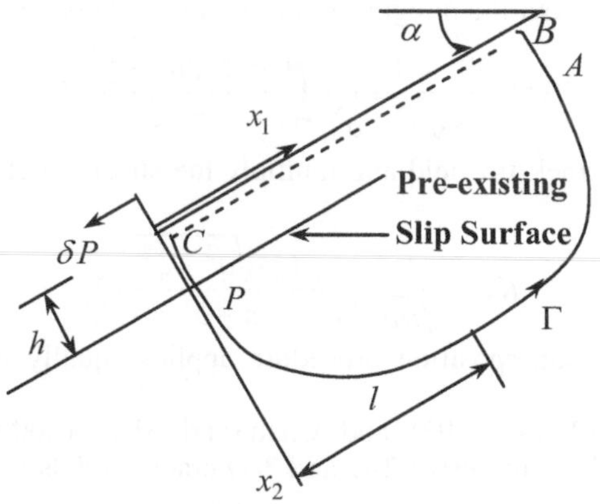

Figure 6.18 Propagation of a slip from the step of a slope (after Palmer and Rice (1973) with permission from the Royal Society)

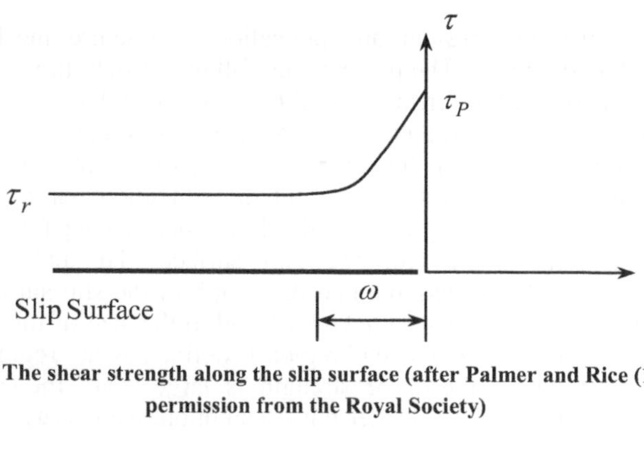

Figure 6.19 The shear strength along the slip surface (after Palmer and Rice (1973) with permission from the Royal Society)

The gravitationally induced shear stress on the prospective failure plane is

$$\tau_g = (\sigma_{21}^0)_{x_2=h} = \gamma h \sin\alpha \tag{6.223}$$

All displacements and strains will be measured from zero in the prestressed state.

We now apply the J-integral around the tip of the shear band along a contour Γ shown in Fig. 6.18. The point P is chosen as the starting and ending points of the contour Γ. All the displacements and straining in the base material below the slip surface (i.e., $x_2 > h$) can be neglected since the dominant deformation is mainly due to the sliding layer (i.e., $0 < x_2 < h$). Thus, the J-integral vanishes for the portion of Γ which is below the slip surface (along PA is zero). The displacements and straining far up the slope can also be neglected since it is simply the prestressed state. Therefore, the J-integral for path AB also vanishes. Since on the slope surface along the path BC, we have $dx_2 = 0$ and traction is also zero; and along the cut surface we also have the traction being zero. Consequently, we are left with the following integral along the path CP:

$$J_p \approx \int_{\Gamma_{CP}} (W - f_i u_i) dx_2 = -\int_0^h (W + \gamma \sin\alpha u_1 - \gamma \cos\alpha u_2)_{x_1=0} dx_2 \tag{6.224}$$

since $f_1 = -\gamma \sin\alpha$ and $f_2 = \gamma\cos\alpha$. When the layer is long compared to the depth h, we must have one-dimensional displacement (i.e., $u_2 = 0$ on $x_1 = 0$). The displacement along the negative x_1-direction must also equal the relative sliding $\delta(x_1)$ along the slip plane. That is,

$$u_1(x_1) = -\delta(x_1) \tag{6.225}$$

Therefore, (6.224) reduces to

$$J_p = -\overline{W}h + (\gamma h \sin\alpha)\delta_P = -\overline{W}h + \tau_g \delta_P \tag{6.226}$$

where $\overline{W}$ is the thickness average of the energy density at the end of the slope. The last part of (6.226) is obtained in view of (6.223). Interpreted by the stress-strain curve relating the average stress $\overline{\sigma}_{11}$ and the average strain $\overline{\varepsilon}$ in the layer, $\overline{W}$ is

$$\overline{W} = \int_{\overline{\sigma}_{11}=-p_0}^{\overline{\sigma}_{11}=0} \overline{\sigma}_{11}(\overline{\varepsilon}_{11}) d\overline{\varepsilon}_{11} \tag{6.227}$$

or is the negative of the hatched area in Fig. 6.20 below $\overline{\sigma}_{11} = 0$.

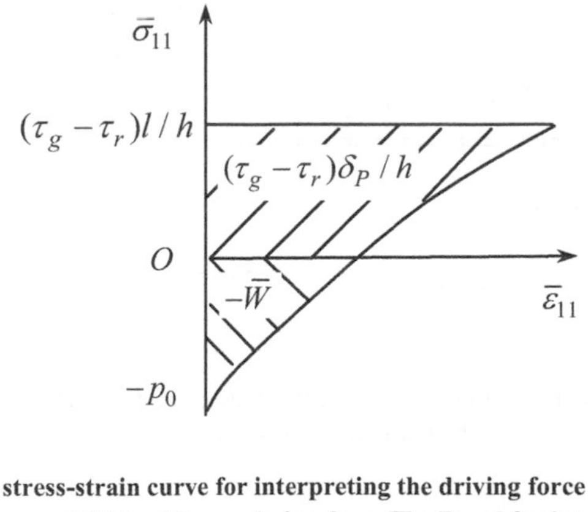

Figure 6.20 The stress-strain curve for interpreting the driving force (after Palmer and Rice (1973) with permission from The Royal Society)

Before we continue to determine the driving force, we first want to establish the following path-independent quantity:

$$J_P - \tau_r \delta_P = \int (\tau - \tau_r) d\delta = (\tau_P - \tau_r)\overline{\delta} \tag{6.228}$$

where the second term is clearly the hatched area shown in Fig. 6.21, and it is the path-independent quantity of P as long as P is outside the end region near the tip of the shear band.

To show (6.228), we consider a special path P^-TP^+ around the tip of the shear band, as shown in Fig. 6.22. We now consider the J-integral along P^-TP^+ which follows the lower surface of the slip band from P^- to the tip T of the band, and returns to P^+ along the upper surface. Since $dx_2 = 0$ along the whole Γ, the first term of the J-integral disappears. Along the band, we have u_2 to be continuous across the band and therefore $\partial u_2/\partial x_1$ is also continuous, but T_2 on the upper surface is equal and opposite to the T_2 on the lower surface, and thus $T_2\partial u_2/\partial x_1$ will make no contribution to the J-integral. Therefore, we have

$$J_P = \int_\Gamma \sigma_{21} \frac{\partial u_1}{\partial x_1} dx_1 = \int_T^P \sigma_{21} \frac{\partial}{\partial x_1}(u_1^+ - u_1^-) dx_1 = \int_T^P \tau \frac{\partial \delta}{\partial x_1} dx_1 = \int_0^{\delta_P} \tau(\delta) d\delta \tag{6.229}$$

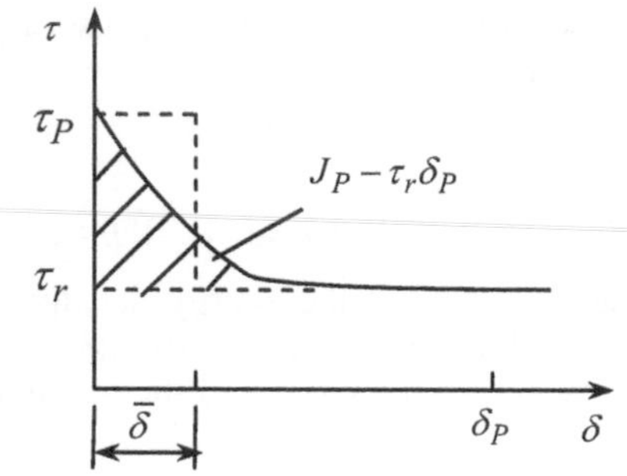

Figure 6.21 The relation between the shear stress τ and relative displacement δ

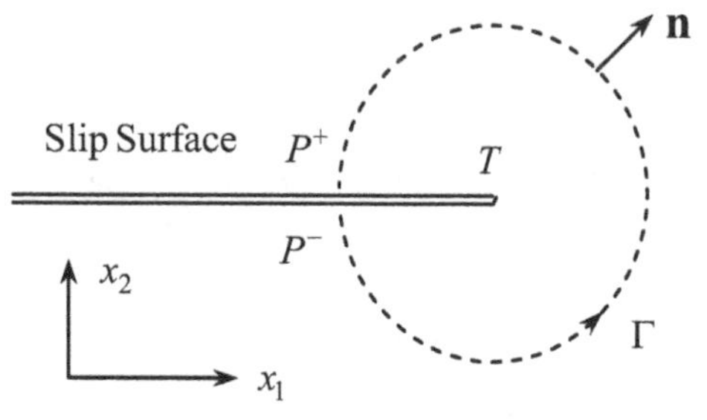

Figure 6.22 Integration path for the J-integral (after Palmer and Rice (1973) with permission from the Royal Society)

where τ is the shear stress along the band, and u_1^+ and u_1^- are the displacements along the upper and lower slip surfaces (their difference equals δ the relative sliding along the band). Then, J_P becomes the area under the curve in Fig. 6.21. However, at the same time it is obvious from Fig. 6.21 that J_P is not a path-independent quantity and it increases with δ_P. But if we can subtract the contribution from the residual part of strength from J_P we have

$$J_P - \tau_r \delta_P = \int (\tau - \tau_r) d\delta = (\tau_P - \tau_r)\bar{\delta} \tag{6.230}$$

Therefore, (6.228) is established. As discussed by Palmer and Rice (1973), $J_p - \tau_r \delta_p$ can be interpreted as the energy surplus made available per unit advance of the band. Therefore, (6.228) implies that for propagation to occur this net energy surplus must just balance the additional dissipation in the end region against shear strengths in excess of the residual one.

We now return to (6.226) and subtract from it by $\tau_r \delta_p$; therefore, we have

$$J_P - \tau_r \delta_P = -\bar{W}h + (\tau_g - \tau_r)\delta_p \tag{6.231}$$

We want to show further that the second term on the right of (6.231) can be interpreted as the hatched area above $\bar{\sigma}_{11} = 0$ in Fig. 6.20. To do so, we consider the force equilibrium along the x_1-direction for the free body of the upper layer, shown in Fig. 6.23. In particular, we find

$$\bar{\sigma}_{11}(x_1)h = (\tau_g - \tau_r)x_1 \tag{6.232}$$

Thus

$$(\tau_g - \tau_r)\delta_P = (\tau_g - \tau_r)\int_0^l \bar{\varepsilon}_{11} dx_1 = h\int_0^{(\tau_g - \tau_r)l/h} \bar{\varepsilon}_{11} d\bar{\sigma}_{11} \tag{6.233}$$

which corresponds to the hatched area above $\bar{\sigma}_{11} = 0$ in Fig. 6.20, as we expect. The last of (6.233) is the result of using the change of variable given in (6.232) which is from force equilibrium. Substitution of (6.233) and (6.227) into (6.231) yields

$$J_P - \tau_r \delta_P \approx h\int_{-p_0}^{(\tau_g - \tau_r)l/h} \bar{\varepsilon}_{11}(\bar{\sigma}_{11}) d\sigma_{11} \tag{6.234}$$

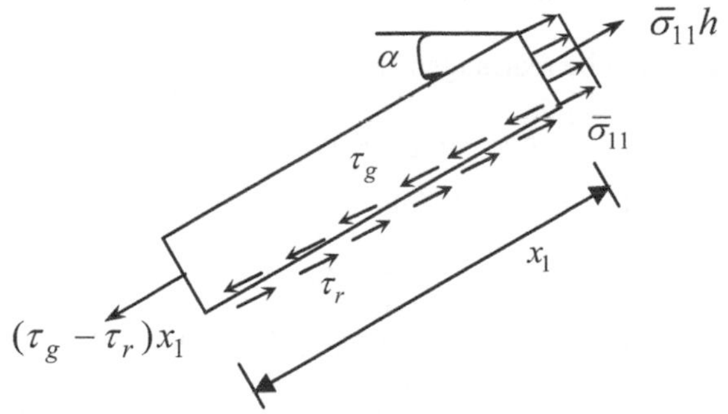

Figure 6.23 The force equilibrium along x_1 for a free body of the upper layer

From the energetic point of view, the lower hatched area is the energy recovered due to unit advance of the shear band resulting from the relief of the lateral pressure p_0, whereas the upper hatched area is the excess of work done input by the gravity pull over the dissipation against the residual shear strength.

To simplify the problem further, a linear stress-strain relation is assumed:

$$\bar{\sigma}_{11} = -p_0 + E'\bar{\varepsilon}_{11} \tag{6.235}$$

where E' is the overall elastic modulus for the layer under the assumed plane strain condition [or $E' = E/(1-\nu^2)$ as given in (1.2-33) of Kannien and Popelar, 1985], then the integral in (6.234) becomes

$$\begin{aligned}\int_{-p_0}^{(\tau_g-\tau_r)l/h} \bar{\varepsilon}_{11}(\bar{\sigma}_{11})d\bar{\sigma}_{11} &= \frac{1}{E'}\left[\frac{1}{2}\bar{\sigma}_{11}^2 + p_0\bar{\sigma}_{11}\right]_{-p_0}^{(\tau_g-\tau_r)l/h} \\ &= \frac{1}{2E'}\left[(\tau_g-\tau_r)\frac{l}{h} + p_0\right]^2\end{aligned} \tag{6.236}$$

Finally, (6.234) becomes

$$J_P - \tau_r\delta_P = \frac{h}{2E'}\left[(\tau_g-\tau_r)\frac{l}{h} + p_0\right]^2 = G = \frac{1-\nu^2}{E}K_{II}^2 \tag{6.237}$$

where G is the energy release at the tip of the shear band for unit advance of the slip surface. The last of (6.237) is a consequence of (6.153). We have assumed that the end region (ω in Fig. 6.19) is small such that the stress singularity predicted by the linear elastic theory applies equally here. This is equivalent to the small-scale yielding concept discussed earlier. The crack-tip field will, however, become

$$\begin{Bmatrix}\sigma_{12}\\ \sigma_{22}\\ \sigma_{11}\end{Bmatrix} = \frac{K_{II}}{(2\pi r)^{1/2}}\begin{Bmatrix}\cos\frac{\theta}{2}[1-\sin\frac{\theta}{2}\sin\frac{3\theta}{2}]\\ \sin\frac{\theta}{2}\cos\frac{\theta}{2}\cos\frac{3\theta}{2}\\ -\sin\frac{\theta}{2}[2+\cos\frac{\theta}{2}\cos\frac{3\theta}{2}]\end{Bmatrix} + \begin{Bmatrix}\tau_r\\ \sigma_n\\ \sigma_t\end{Bmatrix} + O(r^{1/2}) \tag{6.238}$$

where τ_r is the residual stress strength of the band, σ_n is the normal stress transmitted across the band, and σ_t is the transverse or lateral stress acting along the line directly ahead of the band. Further discussion can be referred to Palmer and Rice (1973).

Thus, (6.237) can be rearranged as:

$$K_{II} = \left[(\tau_g-\tau_r)\frac{l}{h} + p_0\right]\sqrt{\frac{h}{2}} \tag{6.239}$$

However, (6.228) and (6.237) will yield

$$K_{IIC} = \left[\frac{E}{1-\nu^2}(\tau_g-\tau_r)\bar{\delta}\right]^{1/2} \tag{6.240}$$

Therefore, the slip surface will start to propagate as long as (6.239) equals (6.240). However, much remains to be done to verify the applicability of this approach to real problems. In addition, the effect of pore water pressure should also be examined (see Rice, 1973). Nevertheless, this problem does illustrate the prospect of using fracture mechanics in geotechnical problems.

6.13 ENERGY RELEASE RATE FOR EARTHQUAKES

This section illustrates how to use fracture mechanics in estimating the amount energy being released during earthquakes. The discussion follows those employed by Palmer and Rice (1973) and Rudnicki (1980). We idealize the 1857 California earthquake at Cholame as a long strike-fault being loaded by displacement not too far from the strike line, as shown in Fig. 6.24. In particular, the fault is assumed to be locked at point T but is undergoing steady relative displacement to the left of T. At point P, the relative displacement is δ_p. The fault is loaded by displacements u_b at a distance $h/2$ from the fault. These displacements can be viewed as those displacements being imposed by large-scale tectonic plate movement.

To calculate the energy release rate, we again use the J-integral. To the right far ahead of T, the Earth's crust is in a state of homogeneous shear strain and stress:

$$2\varepsilon_{12} = \gamma_0 = \frac{u_b}{h}, \qquad \sigma_{12} = \tau_0 < \tau_P \tag{6.241}$$

Again, as defined in the last section τ_p is the peak shear strength of the fault zone. Likewise, far to the left of the tip T, there is also a homogeneous state in the Earth's crust on both sides of the fault zone:

$$2\varepsilon_{12} = \gamma_r, \qquad \sigma_{12} = \tau_r \tag{6.242}$$

At point P shown in Fig. 6.24, the relative displacement is

$$\delta_P = \gamma_0 h - \frac{1}{2}\gamma_r h - \frac{1}{2}\gamma_r h = h(\gamma_0 - \gamma_r) \tag{6.243}$$

where the first term is the imposed boundary displacement and the two subtracted terms $1/2\gamma_r h$ represent that portion of the imposed boundary displacement taken up by crustal rock deformation in the regions above and below the fault shown in Fig. 6.24. Along the rigid boundaries AB and DC all $dx_2 = 0$ and $\partial u_i/\partial x_1 = 0$, thus both will not contribute to the J-integral. Likewise, $\partial u_i/\partial x_1$ vanishes in the homogeneously strained regions far to the right and left of the tip T, so that for path Γ

$$J_P = \int_\Gamma W dx_2 = hW(\gamma_0) - hW(\gamma_r) \tag{6.244}$$

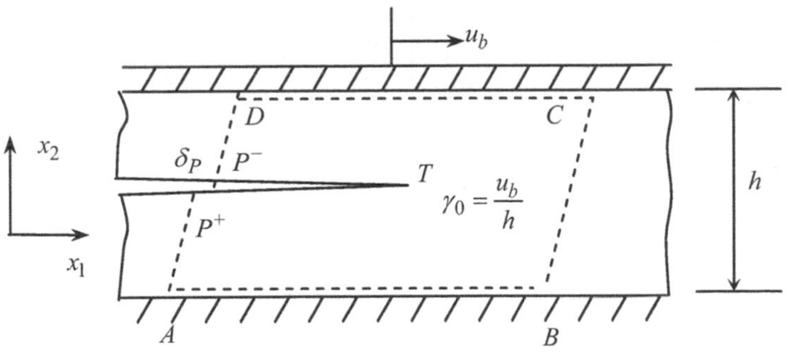

Figure 6.24 Idealization of a long strike-slip fault loaded by displacement

where $W(\gamma)$ is the energy density in a region under homogeneous shear strain γ. As discussed in the previous section the driving force term is

$$J_P - \tau_r\delta_P = h[W(\gamma_0) - W(\gamma_r) - \tau_r(\gamma_0 - \gamma_r)] \tag{6.245}$$

To interpret the driving force in terms of the shear stress-strain curve $\tau = \tau(\gamma)$, we note from Fig. 6.25 that

$$W(\gamma_0) - W(\gamma_r) = \int_{\gamma_r}^{\gamma_0} \tau(\gamma)d\gamma \tag{6.246}$$

Substitution of (6.246) into (6.245) yields

$$J_P - \tau_r\delta_P = h\int_{\gamma_r}^{\gamma_0} [\tau(\gamma) - \tau_r]d\gamma \tag{6.247}$$

To simplify the problem, a linear stress-strain curve is assumed:

$$\tau(\gamma) - \tau_r = \mu(\gamma - \gamma_r) \tag{6.248}$$

Substitution of (6.248) into (6.247) and integration give

$$J_P - \tau_r\delta_P = \frac{h}{2\mu}(\tau_0 - \tau_r)^2 = \frac{1}{2}h\mu(\gamma_0 - \gamma_r)^2 = \frac{1-\nu^2}{E}K_{II}^2 \tag{6.249}$$

From (6.249), the corresponding stress intensity factor is obtained as (Palmer and Rice, 1973)

$$K_{II} = (\tau_0 - \tau_r)\sqrt{\frac{h}{1-\nu}} \tag{6.250}$$

Earthquake occurs when the fault starts to propagate; and, as usual, the criterion for propagation can be related to the critical energy release rate:

$$J_P - \tau_r\delta_P = G_{crit} = \frac{\mu}{2h}\delta_P^2 \tag{6.251}$$

Along the creeping portion of the San Andreas fault the observed displacement is about 3 cm/year. For return period of 160 years and for a $M_s = 8.25$ earthquake, $\delta_P \sim 160$ (year) × 3(cm/year) = 480 cm. Taking $\mu = 20$ GPa and $h = 60$ km, we have $G_{crit} = 3.84\times 10^6$ Jm^{-2} from (6.251). It should be emphasized that this value is comparable to the estimation by other researchers using different methods. Field data show that rupture length along the fault zone during the earthquake is about 275 km and taking the depth to 10 km, we find that the total energy released during the 1857 California earthquake was about 10^{16} J. Using the empirical Gutenberg –Richter relationship (Rudnicki, 1980)

$$\log E = 11.8 + 1.5M_S \tag{6.252}$$

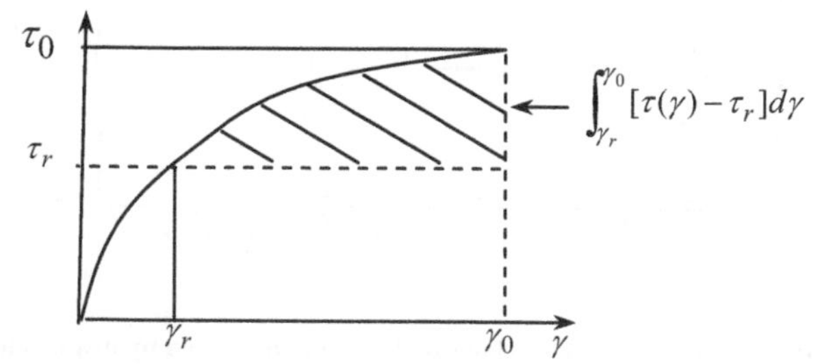

Figure 6.25 The interpretation of the driving force term in the propagation criterion

where M_s is the surface wave magnitude (Scholz, 1990; Aki and Richards, 1980), alternatively we find the seismic energy E is about 10^{24} J. However, we do not expect the Gutenberg-Richter relationship formula to be very accurate, especially given the fact that none of the local crystal rock parameters is incorporated into the relation. Nevertheless, the fracture mechanics model considered here does provide a simple means of calculating the seismic energy being released. We, of course, do not expect our calculation to be very accurate either, but at least our approach gives more insight into the qualitative effect of each parameter compared to the black-box formula (6.252).

6.14 WING CRACK MODEL UNDER COMPRESSIONS

One of the distinct features of brittle solids, including rocks, ceramics, and concrete, under uniaxial compression is that crack growth is normally observed in a direction parallel to the compression field. According to classical elastic fracture mechanics, a vertical pre-existing crack will not be activated under uniaxial compression; thus, conventional fracture mechanics is unable to explain this observation. The extent of crack growth is also a function of confining stress. In order to activate cracking, it is observed experimentally that vertical stress has to be larger than compressive confining stress. In view of this observation, it has been proposed that vertical cracking is controlled by the deviatoric stress component normal to the crack front (Costin, 1983, 1985). See Section 6.18 for further application of Costin's (1985) model. One of the most popular models to explain vertical cracking of brittle solids under compression is the wing crack model. As shown in Fig. 6.26, it is assumed that vertical cracking (dotted lines in the figure) is induced by frictional sliding on pre-existing inclined cracks. In particular, the wing cracks are wedged open by frictional sliding at both the upper and lower tips. The existence of wing cracks in PMMA is well documented by Horii and Nemat-Nasser (1985). In terms of experimental observations in real rocks, the existence of sliding wing cracks is inconclusive (e.g., Brace, 1960; Tapponnier and Brace, 1976; Kranz, 1979). Nevertheless, the wing crack model is among the most successful models for cracking in rocks (Brace, 1960; Hoek and Bieniawski, 1965; Nemat-Nasser and Obata, 1988; Jeyakumaran and Rudnicki 1995; Wong et al. 1996; Wong and Chau, 1998). It has also been used in modeling hysteresis of dilatancy (Scholz and Kranz, 1974).

In this section, we will not go into the details of various wing crack models, such as Nemat-Nasser and Obata (1988) and Ashby and Hallam (1986). Instead, we will simply illustrate the idea of wing crack by considering a simple model by Muhlhaus et al. (1996). This approach was later extended by Chau and Wang (2001) to consider bifurcation in the growth of arrays of en echelon cracks. In particular, Muhlhaus et al. (1996) estimated the pop-up force F for an equivalent vertical crack by projecting the resolved shear and normal stresses on the crack face, as shown in Fig. 6.27. The local resolved stresses on the crack face are:

$$\tau_s = \boldsymbol{n} \bullet \boldsymbol{\sigma} \bullet \boldsymbol{s}, \quad \sigma_n = \boldsymbol{n} \bullet \boldsymbol{\sigma} \bullet \boldsymbol{n} \tag{6.253}$$

In component forms, we have

$$\tau_s = -\frac{\sin 2\theta}{2}(\sigma_{11}^{\infty} - \sigma_{22}^{\infty}), \quad \sigma_n = -(\sigma_{11}^{\infty}\cos^2\theta + \sigma_{22}^{\infty}\sin^2\theta) \tag{6.254}$$

where $-\pi/2 < \theta < \pi/2$. The shear stress driving the sliding crack is

$$\tau = \tau_s \pm \mu\sigma_n \tag{6.255}$$

where μ is the frictional coefficient. The plus or minus sign must be chosen such that the frictional stress is always against sliding. The equivalent point force (per unit length) is

$$F = 2c_0\tau\cos\theta = c_0(\psi_1\sigma_{11}^{\infty} - \psi_2\sigma_{22}^{\infty}) \tag{6.256}$$

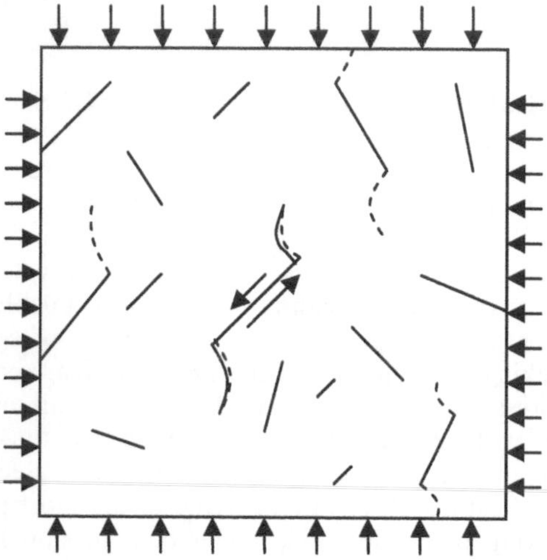

Figure 6.26 Crack growth under compression in brittle rocks

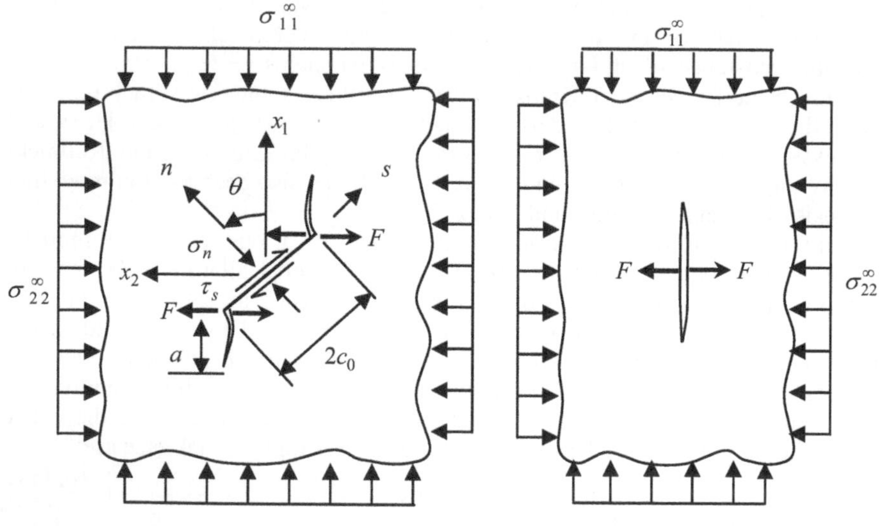

Figure 6.27 Wing crack model in estimating the pop-up force F

where

$$\psi_1 = \cos\theta(|\sin 2\theta| - 2\mu\cos^2\theta), \quad \psi_2 = \cos\theta(|\sin 2\theta| + 2\mu\sin^2\theta) \quad (6.257)$$

for $-\pi/2 < \theta < \pi/2$. The angle θ should be chosen as θ_{max} which can be evaluated as:

$$\frac{dF}{d\theta}(\theta_{\max}) = c_0\left(\frac{d\psi_1}{d\theta}\sigma_{11}^{\infty} - \frac{d\psi_2}{d\theta}\sigma_{22}^{\infty}\right) = 0 \quad (6.258)$$

The prediction of this θ_{max} has been compared to the modified Griffith crack model (e.g., McClintock and Walsh, 1962; Jaeger and Cook, 1979):

$$\theta_{\max} = \frac{\pi}{2} - \frac{1}{2}\tan^{-1}\left(\frac{1}{\mu}\right) \quad (6.259)$$

This comparison is also reported in Fig. 6.28. The prediction of θ_{max} is a function of stress ratio whereas the prediction by the modified Griffith crack model is independent of the stress ratio. Therefore, we expect the prediction by (6.258) to be more reliable than (6.259).

6. 15 BAŽANT'S SIZE EFFECT LAW VIA *J*-INTEGRAL

According to Bažant and Rajapakse (1991), scaling and size effect in fracture strength date back to the time of Leonardo da Vinci. The qualitative foundation of statistical size effect of strength was laid down by Marriotte in the 17th century. Bažant and Planas (1998) showed that the strength dependence from plastic yielding and from linear elastic fracture mechanics behavior can be linked together by a simple "one over square root law" of size effect. For large-size quasi-brittle solids, the strength approaches the plastic yielding, whereas for small-size solids, the strength is controlled by linear elastic fracture mechanics.

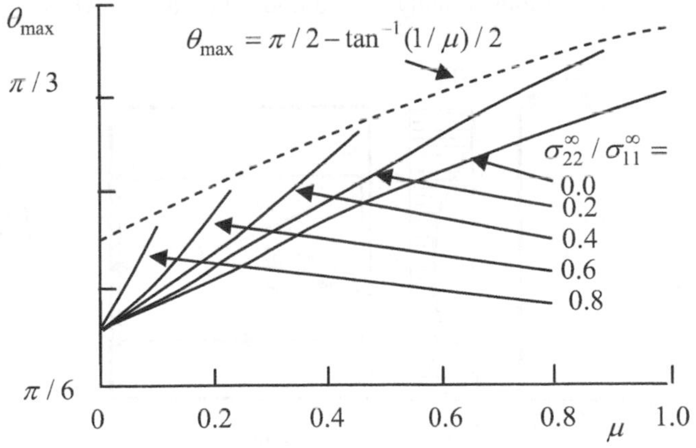

Figure 6.28 Prediction of most optimal angle for activating sliding cracks (after Muhlhaus et al. (1996) with permission from Springer)

In this section, we adopt a particular cohesive crack model proposed by Bažant et al. (1999) for a finite solid under compression. For more comprehensive discussion of cohesive zone models, we refer to Sun and Yin (2012). As shown in Fig. 6.29, the *J*-integral can be employed to find such size or scale effect. In particular, a finite specimen of width D contains an edge crack of length a and a fracture process zone (FPZ) c is subject to compression σ_N. Two triangular energy release zones OFGO and OCBO are assumed behind the crack front as shown. The stress level in these zones is assumed to drop to a residual level σ_r. The fracture process zone (FPZ) in front of the crack tip is modeled by a cohesive crack model.

In particular, for the case of zero body force we would evaluate the energy release rate for the contour ABCDEFGH as (see (6.170))

$$J = \int_{\Gamma}[Wn_1 - \boldsymbol{T}\bullet\frac{\partial \boldsymbol{u}}{\partial x_1}]ds = \int_{\Gamma}[Wdx_2 - \boldsymbol{T}\bullet\frac{\partial \boldsymbol{u}}{\partial x_1}ds] \tag{6.260}$$

The length of FC is

$$\overline{FC} = 2k(a_0 + c) \tag{6.261}$$

The first part of the *J*-integral on the right side of (6.260) can be evaluated by following the closed contour ABCDEFGH. On the horizontal contours EF and CD, the contribution to the first term are zeros as $dx_2 = 0$. The contour integrals of AB and GH balance with each other. The contour integral on DE can be assumed subject to uniform stress σ_N without disturbance from the crack, whereas contours BC and FG are both subject to uniform residual stress σ_r. These contributions can be summed as:

$$\int_{\Gamma} Wdx_2 = 2k(a_0 + c)[\frac{\sigma_N^2}{2E} - \frac{\sigma_r^2}{2E}] \tag{6.262}$$

The contribution to the second term on the right side of (6.258) can be broken down as the sum of contours AB, BC, CD, DE, EF, FG, and GH. The integrals for EF, FG, and BC are zeros as dx = 0 on them. The displacement field on both EF and CD can be assumed to be uniform (i.e., undisturbed by the crack and FPZ), leading to zero contributions. Finally, the contour integrals on AB and GH are

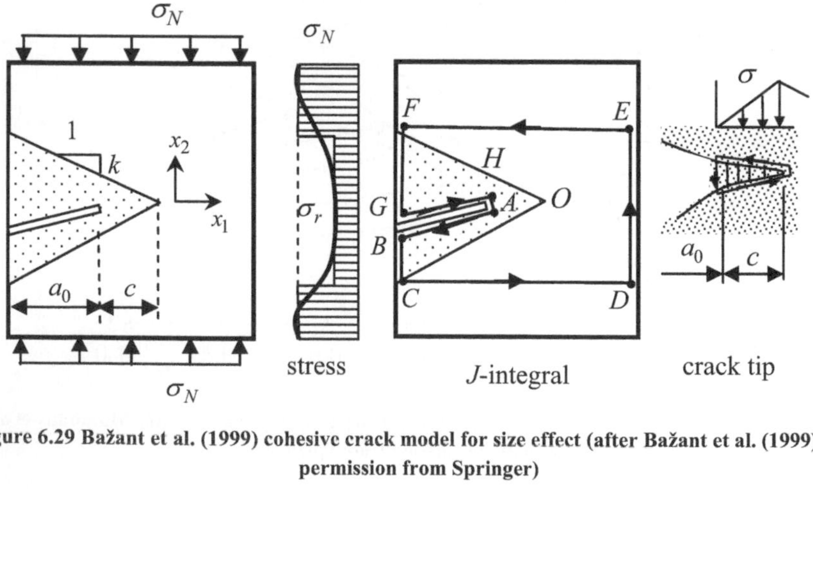

Figure 6.29 Bažant et al. (1999) cohesive crack model for size effect (after Bažant et al. (1999) with permission from Springer)

$$\int_{\Gamma} \boldsymbol{T} \bullet \frac{\partial \boldsymbol{u}}{\partial x_1} ds = \int_{AB} \sigma_r \frac{d}{dx_1}[\frac{1}{2}\delta(x_1)]dx_1 - \int_{GH} \sigma_r \frac{d}{dx_1}[\frac{1}{2}\delta(x_1)]dx_1$$
$$= \int_0^{a_0} \sigma_r d\delta(x_1) = \sigma_r \int_0^{a_0} d\delta(x_1) = \sigma_r \delta_{BG} \tag{6.263}$$

where δ_{BG} is the relative displacement between points B and G. This displacement can be estimated from the difference between the lengths of ED and FC as

$$\delta_{BG} = 2k(a_0 + c)\frac{\sigma_N}{E} - 2k(a_0 + c)\frac{\sigma_r}{E} = 2k(a_0 + c)\frac{\sigma_N - \sigma_r}{E} \tag{6.264}$$

Finally, combining (6.262)–(6.264) gives

$$J = \int_\Gamma [W dx_2 - \boldsymbol{T} \bullet \frac{\partial \boldsymbol{u}}{\partial x_1} ds] = \frac{k}{E}(a_0 + c)(\sigma_N - \sigma_r)^2 \tag{6.265}$$

On the other hand, the critical value of the *J*-integral at the crack tip can be evaluated by the contour around the tip shown in Fig. 6.29 together with the cohesive model shown in Fig. 6.30.

$$J = \int_{\Gamma} \boldsymbol{T} \bullet \frac{\partial \boldsymbol{u}}{\partial x_1} dx_1$$
$$= -\int_{a_0}^{a_0+c} f[\delta(x_1)]\frac{d}{dx}[\frac{1}{2}\delta(x_1)]dx_1 + \int_{a_0+c}^{a_0} f[\delta(x_1)]\frac{d}{dx_1}[\frac{1}{2}\delta(x_1)]dx_1$$
$$= -\int_{a_0}^{a_0+c} f[\delta(x_1)]\frac{d\delta(x_1)}{dx_1}dx_1 \tag{6.266}$$
$$= \int_0^{\delta_r} f[\delta(x_1)]d\delta(x_1) = G_b + \sigma_r \delta_r$$

The last term is obvious from the hatched areas in Fig. 6.30. Equating (6.265) to (6.266), we have

$$\sigma_N = \sigma_r + \sqrt{\frac{E(G_b + \sigma_r \delta_r)/(kc)}{1 + D/D_0}} = \sigma_r + \frac{\sigma_0}{\sqrt{1 + D/D_0}} \tag{6.267}$$

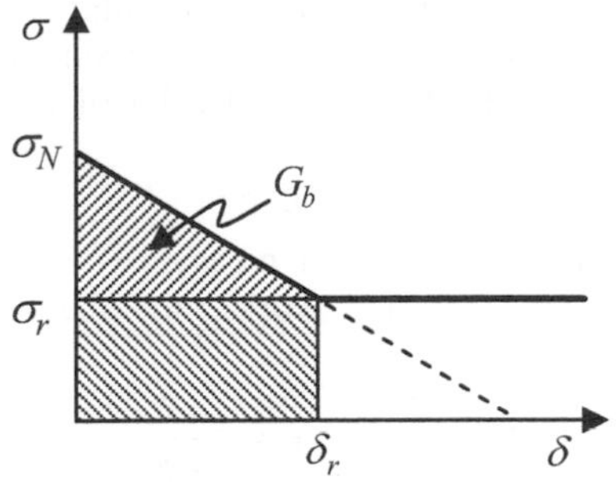

Figure 6.30 shear resistance versus slipping displacement in cohesive model (after Bažant et al. (1999) with permission from Springer)

where

$$D_0 = \frac{c}{\alpha_0}, \quad D = \frac{a}{\alpha_0}, \quad \sigma_N = \sqrt{\frac{E(G_b + \sigma_r \delta_r)}{kc}} \tag{6.268}$$

The size of the FPZ is characterized by the value of D_0 whereas the size of the specimen is characterized by D. The size effect law is given in (6.267) and Bažant and Planas (1998) found that this scale or size effect law is universal and results from many crack problems. Equation (6.267) is illustrated in Fig. 6.31. The asymptote for small D is controlled by plastic yielding and the asymptote for large D is controlled by linear elastic fracture mechanics. Therefore, this size effect links the plasticity for small specimens to linear elastic fracture mechanics for large specimens.

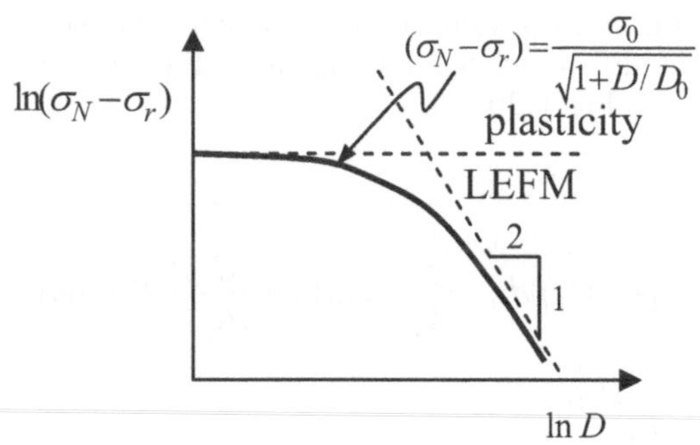

Figure 6.31 Bažant size effect law (after Bažant et al. (1999) with permission from Springer)

6. 16 CONTINUUM DAMAGE MECHANICS

In this section, we will extend the analysis given in Section 5.16 for internal variables to model continuum damage mechanics. In this approach, the fracture process of individual microcracks will not be considered explicitly, but instead the growth of microcracks will be considered damages, and mathematically they were modeled by the evolution of internal variables. Similar to Section 5.16, we start with the following free energy potential, which is assumed to be a function of a strain measure $\boldsymbol{E}$ (a second-order tensor), and a set of internal variables or as a damage vector $\boldsymbol{\xi}$ (a first-order tensor or vector):

$$\Psi = \Psi(\boldsymbol{E}, \boldsymbol{\xi}) \tag{6.269}$$

Then, the stress $\boldsymbol{S}$ conjugate to the strain measure (i.e., the double dot product between $d\boldsymbol{E}$ and $\boldsymbol{S}$ equals the external work), and the thermodynamic force $\boldsymbol{\Lambda}$ associated with the internal variable $\boldsymbol{\xi}$ are

$$\boldsymbol{S} = \frac{\partial \Psi}{\partial \boldsymbol{E}}, \quad \boldsymbol{\Lambda} = -\frac{\partial \Psi}{\partial \boldsymbol{\xi}} \tag{6.270}$$

Then, we introduce an initial damage surface $f(\boldsymbol{\Lambda}, \boldsymbol{\xi})$ such that $f < 0$ implies a reversible process (i.e., no damage), $f = 0$ implies initiation of damage, and $f > 0$ implies irreversible process of damages. The evolution of the internal variable can be determined from another potential $F(\boldsymbol{\Lambda}, \boldsymbol{\xi})$ as

$$d\boldsymbol{\xi} = \lambda \frac{\partial F}{\partial \boldsymbol{\Lambda}} \tag{6.271}$$

We have continuous damage if $\lambda > 0$ and λ can be determined from the consistency condition (similar to the plasticity formulation discussed in Chapter 5) as

$$df = \frac{\partial f}{\partial \boldsymbol{\Lambda}} \bullet d\boldsymbol{\Lambda} + \frac{\partial f}{\partial \boldsymbol{\xi}} \bullet d\boldsymbol{\xi} = 0 \tag{6.272}$$

The change of the thermodynamic force given in (6.272) can be determined from the second part of (6.270) as:

$$\begin{aligned} d\boldsymbol{\Lambda} &= -\frac{\partial}{\partial \boldsymbol{\xi}}(d\Psi) = -\frac{\partial}{\partial \boldsymbol{\xi}}\left(\frac{\partial \Psi}{\partial \boldsymbol{E}} : d\boldsymbol{E} + \frac{\partial \Psi}{\partial \boldsymbol{\xi}} \bullet d\boldsymbol{\xi}\right) \\ &= -\frac{\partial^2 \Psi}{\partial \boldsymbol{\xi} \partial \boldsymbol{E}} : d\boldsymbol{E} - \frac{\partial^2 \Psi}{\partial \boldsymbol{\xi} \partial \boldsymbol{\xi}} \bullet d\boldsymbol{\xi} \end{aligned} \tag{6.273}$$

Substitution of (6.273) into (6.272) leads to

$$\left(\frac{\partial f}{\partial \boldsymbol{\xi}} - \frac{\partial f}{\partial \boldsymbol{\Lambda}} \bullet \frac{\partial^2 \Psi}{\partial \boldsymbol{\xi} \partial \boldsymbol{\xi}}\right) \bullet d\boldsymbol{\xi} = \frac{\partial f}{\partial \boldsymbol{\Lambda}} \bullet \frac{\partial^2 \Psi}{\partial \boldsymbol{\xi} \partial \boldsymbol{E}} : d\boldsymbol{E} \tag{6.274}$$

In view of (6.271), (6.274) becomes

$$\left(\frac{\partial f}{\partial \boldsymbol{\xi}} - \frac{\partial f}{\partial \boldsymbol{\Lambda}} \bullet \frac{\partial^2 \Psi}{\partial \boldsymbol{\xi} \partial \boldsymbol{\xi}}\right) \bullet \lambda \frac{\partial F}{\partial \boldsymbol{\Lambda}} = \frac{\partial f}{\partial \boldsymbol{\Lambda}} \bullet \frac{\partial^2 \Psi}{\partial \boldsymbol{\xi} \partial \boldsymbol{E}} : d\boldsymbol{E} \tag{6.275}$$

Solving for λ from the scalar equation (6.275) gives

$$\lambda = \frac{\dfrac{\partial f}{\partial \boldsymbol{\Lambda}} \bullet \dfrac{\partial^2 \Psi}{\partial \boldsymbol{\xi} \partial \boldsymbol{E}} : d\boldsymbol{E}}{\dfrac{\partial f}{\partial \boldsymbol{\xi}} \bullet \dfrac{\partial F}{\partial \boldsymbol{\Lambda}} - \dfrac{\partial f}{\partial \boldsymbol{\Lambda}} \bullet \dfrac{\partial^2 \Psi}{\partial \boldsymbol{\xi} \partial \boldsymbol{\xi}} \bullet \dfrac{\partial F}{\partial \boldsymbol{\Lambda}}} \tag{6.276}$$

In view of the first part of (6.270), the change of stress tensor $d\boldsymbol{E}$ can now be written as:

$$d\boldsymbol{S} = \frac{\partial}{\partial \boldsymbol{E}}\left(\frac{\partial \Psi}{\partial \boldsymbol{E}} : d\boldsymbol{E} + \frac{\partial \Psi}{\partial \boldsymbol{\xi}} \bullet d\boldsymbol{\xi}\right) \tag{6.277}$$

Substitution of (6.271) and (6.276) into (6.277) gives

$$d\boldsymbol{S} = \left\{ \frac{\partial^2 \Psi}{\partial \boldsymbol{E} \partial \boldsymbol{E}} + \frac{\dfrac{\partial \Psi}{\partial \boldsymbol{\xi}} \bullet \dfrac{\partial F}{\partial \boldsymbol{\Lambda}} \dfrac{\partial f}{\partial \boldsymbol{\Lambda}} \bullet \dfrac{\partial^2 \Psi}{\partial \boldsymbol{\xi} \partial \boldsymbol{E}}}{\dfrac{\partial f}{\partial \boldsymbol{\xi}} \bullet \dfrac{\partial F}{\partial \boldsymbol{\Lambda}} - \dfrac{\partial f}{\partial \boldsymbol{\Lambda}} \bullet \dfrac{\partial^2 \Psi}{\partial \boldsymbol{\xi} \partial \boldsymbol{\xi}} \bullet \dfrac{\partial F}{\partial \boldsymbol{\Lambda}}} \right\} : d\boldsymbol{E} \tag{6.278}$$

Alternatively, (6.278) can be written in rate form as:

$$\dot{\boldsymbol{S}} = \left\{ \boldsymbol{L} - \frac{\boldsymbol{\Omega}_{\xi} \bullet \boldsymbol{N}_{\alpha} \boldsymbol{N}_{\beta} \bullet \boldsymbol{\Omega}_{\xi E}}{\boldsymbol{N}_{\beta} \bullet \boldsymbol{\Omega}_{\xi\xi} \bullet \boldsymbol{N}_{\alpha} - \boldsymbol{R}_{\beta} \bullet \boldsymbol{N}_{\alpha}} \right\} : \dot{\boldsymbol{E}} \tag{6.279}$$

where

$$\boldsymbol{\Omega}_\xi = \frac{\partial \Psi}{\partial \boldsymbol{\xi}}, \quad \boldsymbol{\Omega}_{\xi E} = \frac{\partial^2 \Psi}{\partial \boldsymbol{\xi} \partial \boldsymbol{E}} \quad \boldsymbol{\Omega}_{\xi\xi} = \frac{\partial^2 \Psi}{\partial \boldsymbol{\xi} \partial \boldsymbol{\xi}}$$

$$\boldsymbol{L} = \frac{\partial^2 \Psi}{\partial \boldsymbol{E} \partial \boldsymbol{E}}, \quad \boldsymbol{N}_\alpha = \frac{\partial F}{\partial \boldsymbol{\Lambda}}, \quad \boldsymbol{N}_\beta = \frac{\partial f}{\partial \boldsymbol{\Lambda}}, \quad \boldsymbol{R}_\beta = \frac{\partial f}{\partial \boldsymbol{\xi}}, \tag{6.280}$$

Note that $\boldsymbol{\Omega}_\xi$, $\boldsymbol{\Omega}_{\xi\xi}$, $\boldsymbol{\Omega}_{\xi E}$, and $\boldsymbol{L}$ are vector, second-order tensor, third-order tensor, and fourth-order tensors, respectively. Comparison of (6.280) with (5.21) of Chapter 5 shows that the framework for plasticity differs fundamentally from the continuum damage mechanics formulation. For more comprehensive coverage on continuum damage mechanics, we refer to Krajcinovic (1986).

6. 17 SOLIDS CONTAINING MICROCRACKS

In this section, we will summarize the essence of the effective mechanical properties of solids containing microcracks. There are a few books on damage mechanics (e.g., Kachanov, 1986; Lemaître, 1996; Krajcinovic, 1996; Nemat-Nasser and Hori, 1993), but the following presentation follows from Chau (1991) and Chau and Wong (1997).

6.17.1 Compliance Change due to a Single Crack

We consider an isotropic, homogeneous, elastic solid containing a penny-shaped microcrack of radius a and with unit normal $\boldsymbol{n}$. The resolved normal and shear stresses on the surface of the penny-shaped crack are denoted by σ and τ, respectively. According to Budiansky and O'Connell (1976), the energy change due to the introduction of the crack is

$$\Delta U = -\int_0^{2\pi} \int_0^a \frac{1}{2} \left[\tau(u_1^+ - u_1^-) + \sigma(u_2^+ - u_2^-) \right] r dr d\theta \tag{6.281}$$

where the superscripts + and − indicate the shear displacement u_1 and normal displacement u_2 on the top and bottom of the crack face, respectively. These displacements are given by Rice (1968b) as

$$u_1^\pm = \pm \frac{4(1-\nu)}{(2-\nu)\pi} \frac{\tau}{\mu} (a^2 - r^2)^{1/2} \tag{6.282}$$

$$u_2^\pm = \pm \frac{4(1-\nu)}{(2-\nu)\pi} \frac{\sigma}{\mu} (a^2 - r^2)^{1/2} \tag{6.283}$$

where μ and ν are the shear modulus and Poisson's ratio, and r is the distance measured from the center of the crack. Substitution of (6.282) and (6.283) into (6.281) and integration over the crack face give

$$\Delta U = -\frac{4(1-\nu)}{3\mu} \left[\frac{2}{(2-\nu)} \tau^2 + \sigma^2 \right] a^3 \tag{6.284}$$

The resolved shear and normal stresses can be expressed in terms of the far-field Cauchy stress tensor as

$$\sigma = n_p n_q \sigma_{pq} \tag{6.285}$$

$$\tau^2 = n_p \sigma_{pq} n_r \sigma_{rq} - n_p \sigma_{pq} n_q n_r \sigma_{rs} n_s \tag{6.286}$$

The change of compliance due to the introduction of the penny-shaped crack is given by Rice (1975):

$$\sigma_{ij} \Delta C_{ijkl} = -\frac{1}{2V}\left[\frac{\partial(\Delta U)}{\partial \sigma_{kl}} + \frac{\partial(\Delta U)}{\partial \sigma_{lk}}\right] \tag{6.287}$$

where V is the volume of the solid. Substitution of (6.285), (6.286), and (6.284) into (6.287) gives

$$V\sigma_{ij} \Delta C_{ijkl} = \frac{8(1-\nu)}{3\mu(2-\nu)} a^3 (n_p \sigma_{pl} n_k + n_p \sigma_{pk} n_l - \nu n_k n_l n_r n_s \sigma_{rs}) \tag{6.288}$$

Differentiating (6.288) with respect to σ_{ij} and σ_{ji} and taking the average of their results, we have

$$\Delta C_{ijkl} = \frac{1}{V} \frac{4(1-\nu)}{3\mu(2-\nu)} a^3 \Phi_{ijkl} \tag{6.289}$$

where

$$\Phi_{ijkl} = n_i n_k \delta_{lj} + n_j n_k \delta_{li} + n_i n_l \delta_{kj} + n_j n_l \delta_{ki} - 2\nu n_i n_j n_k n_l \tag{6.290}$$

This fourth-order tensor was first derived by Salganik (1973).

6.17.2 Effective Compliance for Cracked Bodies

Consider now a solid containing N penny-shaped cracks of varying size and varying orientation $\boldsymbol{n}$. However, spatial variations of these cracks are assumed random. Then, (6.289) and (6.290) can be used to estimate the change of compliance as

$$\Delta C_{ijkl} = \frac{4(1-\nu)}{3\mu(2-\nu)} \frac{N}{2\pi V} \int_0^{2\pi} \int_0^{\pi/2} a^3(\phi,\theta) \Phi_{ijkl} \sin\phi \, d\theta \, d\phi \tag{6.291}$$

where unit normal $\boldsymbol{n}$ = ($\sin\phi\cos\theta$, $\sin\phi\sin\theta$, $\cos\phi$) and μ and ν are the uncracked moduli. The overall or effective compliance of any cracked body can then be evaluated as

$$C_{ijkl} = C^0_{ijkl} + \Delta C_{ijkl} \tag{6.292}$$

where the superscript 0 denotes the compliance tensor of the uncracked body.

6.17.3 Noninteracting result for Planar Transverse Isotropy

For the planar transverse isotropy (PTI) case, all cracks have a normal perpendicular to the x_1-x_2 plane or all normals are parallel to the x_3-axis (Lekhnitskii, 1963). The spatial distribution is again random. In this case, no integration is needed, and we have

$$\Delta C_{ijkl} = \frac{4(1-\nu)}{3\mu(2-\nu)} \Phi_{ijkl} \varepsilon \tag{6.293}$$

By setting $n_1 = n_2 = 0$ and $n_3 = 1$ and recognizing the following formulas for the effective moduli

$$C_{3333} = 1/\tilde{E}, \quad 4C_{2323} = 4C_{1313} = 1/\tilde{\mu} \tag{6.294}$$

and

$$C^0_{ijkl} = \frac{1}{E}[\frac{1}{2}(1+\nu)(\delta_{ik}\delta_{jl} + \delta_{jk}\delta_{il}) - \nu\delta_{ij}\delta_{kl}], \tag{6.295}$$

we find that

$$\frac{\tilde{E}}{E} = \frac{1}{1+\dfrac{16}{3}(1-\nu^2)\varepsilon}, \tag{6.296}$$

$$\frac{\tilde{\mu}}{\mu} = \frac{1}{1+\dfrac{16}{3}\dfrac{(1-\nu)}{(2-\nu)}\varepsilon} \tag{6.297}$$

We can also consider a hydrostatic compression ($\sigma_{11} = \sigma_{22} = \sigma_{33} = -p$) and define the effective bulk modulus as

$$\tilde{K} = -\frac{p}{\varepsilon_{kk}} \tag{6.298}$$

where ε_{ij} is the strain tensor. We obtain

$$\frac{1}{\tilde{K}} = \frac{1}{E}(2 - 6\nu + \frac{E}{\tilde{E}}) \tag{6.299}$$

Substitution of (6.296) into (6.299) leads to

$$\frac{\tilde{K}}{K} = \frac{1}{1+\dfrac{16(1-\nu^2)}{9(1-2\nu)}\varepsilon} \tag{6.300}$$

In these formulas, we have adopted the definition of crack density as defined by Budiansky and O'Connell (1976)

$$\varepsilon = \frac{N}{V}\langle a^3 \rangle \tag{6.301}$$

where the bracket indicates average. These results can be found in Nemat-Nasser and Hori (1993). Since interactions between cracks have been ignored, as discussed by Willis (1980) these results correspond to the upper Hashin–Shtrikman bounds. Taya (1981) obtained these results by Mori-Tanaka's method.

6.17.4 Planar Transverse Isotropy by Self-Consistent Method

The self-consistent method is a simple way to estimate the crack interaction by assuming the moduli used in (6.291) and (6.293) as the final unknown moduli of the crack body. In particular, we have

$$\Delta C_{ijkl} = \frac{4(1-\nu)}{3\tilde{\mu}(2-\nu)}\Phi_{ijkl}\varepsilon = \frac{8(1-\nu)(1+\nu)}{3\tilde{E}(2-\nu)}\Phi_{ijkl}\varepsilon \tag{6.302}$$

Putting (6.294), (6.295), and (6.293) into (6.292), we obtain

$$\frac{\tilde{E}}{E} = 1 - \frac{16}{3}(1-\nu^2)\varepsilon , \tag{6.303}$$

$$\frac{\tilde{\mu}}{\mu} = 1 - \frac{16}{3}\frac{(1-\nu)}{(2-\nu)}\varepsilon \tag{6.304}$$

These self-consistent results clearly overestimate the crack interaction because both moduli become zero for finite values of crack density. Therefore, the self-consistent method has been criticized by many researchers (e.g., Henyey and Pomphery, 1982; Chatterjee et al., 1978; Hudson, 1980; Horii and Sahasakmontri, 1989).

6.17.5 Planar Transverse Isotropy by Differential Scheme

Another method of estimating crack interactions is called the differential scheme (Salganik, 1973). In this method, a system of differential equations for the moduli is obtained by successively introducing crack groups with increasing sizes into a material with the effective moduli at the current cracked stage. Therefore, we expect that this approach gives a more accurate result than that obtained by the self-consistent method.

For the case of planar transverse isotropy, we can make the following substitution in (6.303) and (6.304):

$$\tilde{E} \leftarrow E + dE , \; \tilde{\mu} \leftarrow \mu + d\mu \quad \text{and } \varepsilon \leftarrow d\varepsilon \tag{6.305}$$

The resulting differential equations are

$$\frac{dE}{d\varepsilon} = -\frac{16}{3}(1-\nu^2)E , \tag{6.306}$$

$$\frac{d\mu}{d\varepsilon} = -\frac{16}{3}\frac{(1-\nu)}{(2-\nu)}\mu \tag{6.307}$$

In this approach, we end up integrating two differential equations. This approach has been used by Henyey and Pomphery (1982), Salganik (1973), and Hashin (1988). Approximate solutions for (6.306) and (6.307) are given as Problem 6.12.

6.17.6 Noninteracting Result for Cylindrical Transverse Isotropy

Another commonly encountered isotropy is called cylindrical transverse isotropy (Lekhnitskii, 1963). In this case, normals of all cracks will lie randomly on planes parallel to the x_1-x_2 plane and thus $\phi = \pi/2$. The unit normals become $\boldsymbol{n} = (\cos\theta, \sin\theta, 0)$ and (6.291) becomes

$$\Delta C_{ijkl} = \frac{4(1-\nu)}{3\mu(2-\nu)}\frac{\varepsilon}{2\pi}\int_0^{2\pi} \Phi_{ijkl}(\theta)d\theta \tag{6.308}$$

It is straightforward to show that only the following three moduli are affected by the cracks:

$$C_{1111} = 1/\tilde{E} , \; 4C_{2323} = 4C_{1313} = 1/\tilde{\mu} , \; 4C_{1212} = 1/\mu^* \tag{6.309}$$

After integration, we find that

$$\frac{\tilde{E}}{E} = \frac{1}{1 + \dfrac{2(1-\nu^2)(8-3\nu)}{3(2-\nu)}\varepsilon}, \tag{6.310}$$

$$\frac{\tilde{\mu}}{\mu} = \frac{1}{1 + \dfrac{8}{3}\dfrac{(1-\nu)}{(2-\nu)}\varepsilon} \tag{6.311}$$

$$\frac{\mu^*}{\mu} = \frac{1}{1 + \dfrac{4(1-\nu)(4-\nu)}{3(2-\nu)}\varepsilon} \tag{6.312}$$

In addition, if we define

$$C_{1122} = C_{2211} = -\tilde{\nu} / E \tag{6.313}$$

we have

$$\tilde{\nu} = \nu[1 + \frac{2(1-\nu^2)}{3(2-\nu)}\varepsilon] \tag{6.314}$$

Using the same definition defined in (6.298) we have

$$\frac{1}{\tilde{K}} = \frac{1}{E}([1 + 2\frac{E}{\tilde{E}} - 2(\tilde{\nu} + 2\nu)] \tag{6.315}$$

Substituting (6.314) and (6.310) into (6.315), we again obtain (6.300). Thus, unlike the Young's and shear moduli the effective bulk modulus is independent of the crack distribution. Both self-consistent results and differential scheme results can be obtained similar to the case of PTI considered in Sections 6.17.4 and 6.17.5 (see Problems 6.13 and 6.14).

6.17.7 Noninteracting Result for Isotropically Cracked Solids

For the case of isotropy, no simplification can be made to (6.291). After integration, we can obtain

$$\frac{\tilde{E}}{E} = \frac{1}{1 + \dfrac{16(1-\nu^2)(10-3\nu)}{45(2-\nu)}\varepsilon}, \tag{6.316}$$

$$\frac{\tilde{\mu}}{\mu} = \frac{1}{1 + \dfrac{32(1-\nu)(5-\nu)}{45(2-\nu)}\varepsilon} \tag{6.317}$$

$$\varepsilon = \frac{45}{16}\frac{(\nu - \tilde{\nu})(2-\nu)}{(1-\nu^2)(10\tilde{\nu} - 3\tilde{\nu}\nu - \nu)} \tag{6.318}$$

For isotropically cracked solids, we have

$$3\tilde{K}(1 - 2\tilde{\nu}) = \tilde{E} \tag{6.319}$$

Substitution of (6.316)–(6.318) into (6.319) gives again (6.300).

Therefore, the change of bulk moduli is independent of the nature of crack distribution, but only a function of the number of cracks. However, this result is restricted to noninteracting cases. The corresponding results for self-consistent results and differential scheme results can be obtained similarly and the results will not be given here (see Problems 6.15 and 6.16).

6.18 RUDNICKI–CHAU (1996) MULTIAXIAL MICROCRACK MODEL

The local resolved normal stress σ given in (6.285) however does not predict the normally observed phenomenon that horizontal tensile stress is developed on a vertical crack under uniaxial compression. This local tensile stress can be viewed as a consequence of microscale inhomogeneities (Rudnicki and Chau, 1996). In view of this, Costin (1985) proposed the following form of local tensile stress:

$$\sigma_t = n_p n_q \sigma_{pq} + f' g(a) n_p S_{pq} n_q \tag{6.320}$$

where $\boldsymbol{S}$ is the deviatoric stress tensor defined as:

$$\boldsymbol{\sigma} = \boldsymbol{S} + \frac{1}{3} tr(\boldsymbol{\sigma}) \boldsymbol{I} \tag{6.321}$$

and

$$f' = f \frac{d / a_0}{\sqrt{1-(a_0/d_1)^2}} - 1 \tag{6.322}$$

$$g(a) = \frac{1}{a/a_0} \sqrt{\frac{1-(a_0/d_1)^2}{1-(a/d_1)^2}} \tag{6.323}$$

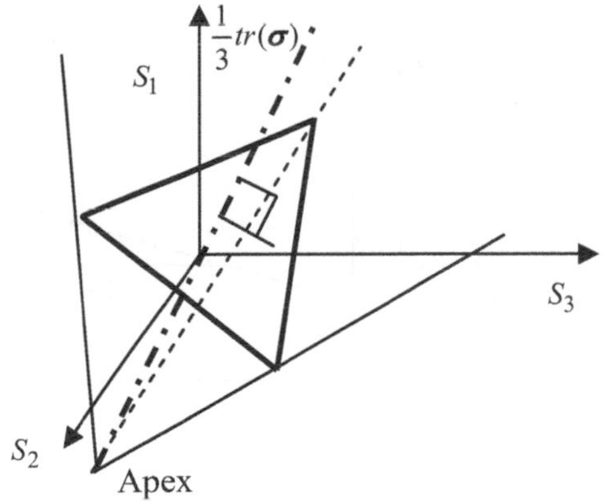

Figure 6.32 The initial damage surface in deviatoric space

where a is the current crack length, a_0 is the initial crack length, d_1 is a measure of the distance between cracks, d is a measure of the size of local tensile zone, and f is a controlling parameter on the magnitude of tensile stress. Then, the stress intensity factor at the penny-shaped crack is

$$K = \frac{2}{\pi}\sqrt{\pi a}\sigma_t \tag{6.324}$$

where σ_t is given in (6.320). Crack growth initiation will start when K is equal to K_{IC} and $a = a_0$, and we have

$$\boldsymbol{n}\bullet\boldsymbol{S}\bullet\boldsymbol{n} = \frac{\pi}{2}\frac{K_{IC}}{(\pi a_0)^{1/2}}\frac{1}{1+f'} - \frac{1}{1+f'}\frac{1}{3}tr(\boldsymbol{\sigma}) \tag{6.325}$$

As shown in Fig. 6.32, the initial damage surface is a cone with a base of an equilateral triangle. At the apex, we have

$$\frac{1}{3}tr(\boldsymbol{\sigma}) = \frac{\pi}{2}\frac{K_{IC}}{(\pi a_0)^{1/2}} \tag{6.326}$$

At zero hydrostatic stress, the intercept on Fig. 6.33 is

$$\boldsymbol{n}\bullet\boldsymbol{S}\bullet\boldsymbol{n} = \frac{\pi}{2}\frac{K_{IC}}{(\pi a_0)^{1/2}}\frac{1}{1+f'} \tag{6.327}$$

For the growth of micrcocrack from a_0 to a, we can rewrite (6.284) as

$$dU = -\frac{4(1-\nu)}{\mu}\left[\frac{2}{(2-\nu)}\tau^2 + \sigma^2\right]a^2 da \tag{6.328}$$

$$dC_{ijkl} = \frac{8(1-\nu)}{\mu(2-\nu)}\int_{a_0}^{a}\int_{n} a^2\tilde{\Phi}_{ijkl}(\boldsymbol{n})dnda \tag{6.329}$$

where

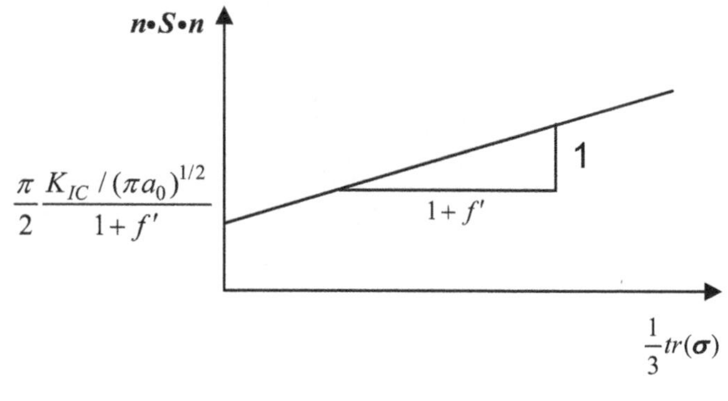

Figure 6.33 The initial damage surface in the deviatoric stress-hydrostatic stress space

$$\tilde{\Phi}_{ijkl} = (2-\nu)\{n_l n_k[1+f'g(a)] - \frac{1}{3}\delta_{kl} f'g(a)\}\{n_i n_j[1+f'g(a)] - \frac{1}{3}\delta_{ij} f'g(a)\}$$
$$+\frac{1}{2}(n_i n_k \delta_{lj} + n_j n_k \delta_{li} + n_i n_l \delta_{kj} + n_j n_l \delta_{ki}) - 2n_i n_j n_k n_l \tag{6.330}$$

When we consider $f' = 0$, we recover the classical result without local tensile stress induced by compressive stress (see (6.289) and (6.290)).

For the case of axisymmetric compression, Rudnicki and Chau (1996) found that the change of compliance is caused by vertical cracks (with horizontal unit normal of $\phi = 90°$) and cracks close to vertical with $\phi = \phi_{\min}$ which is given by

$$\cos\phi_{\min} = \sqrt{\frac{f'(\Sigma_a - 1)}{3\Sigma_a(1+f')}} \tag{6.331}$$

where

$$\Sigma_a = \frac{\sigma_a - \sigma_c}{(\sigma_a - \sigma_c)_0} \tag{6.332}$$

This is the stress at current stress level comparing to the stress at microcrack initiation. For numerical results, we refer to Rudnicki and Chau (1996).

6. 19 SUMMARY AND FURTHER READING

In this chapter, we start with the stress concentration at an elliptic hole and take the minor axis to major axis ratio approaching zero to consider the crack limit. The stress singularity at the crack tip is examined. More generally, the universal stress singularities at the crack tip under all mode I, mode II, and mode III are considered by using William's (1957) eigenfunction expansion. Energy release and the *J*-integral concept are then introduced. Superposition using the Westergaard stress function is discussed. The application of the *J*-integral to the slope problem and earthquake mechanics are summarized, following by discussions on the wing crack model and size effect. Continuum damage mechanics and compliance of microcracked solids are reviewed, including the noninteracting method, the self-consistent method, and the differential scheme. Finally, we also introduce the concept of local tensile stress under deviatoic compression proposed by Costin (1983, 1985).

There are many good textbooks and reference books on fracture mechanics (e.g., Hellan, 1984). Kannien and Popelar (1985), Broberg (1999), and Slepyan (2002) compiled many advanced topics and problems in a single volume. The book by Shah (1995) is devoted exclusively to fracture mechanics applied to concrete and rocks. Atkinson (1987) is an edited book on various topics in rock fracture mechanics. Li (1987), Rice (1980), Dmowska and Rice (1986), and Rudnicki (1980) applied fracture mechanics to earthquakes and faulting in the Earth's crust, and they are excellent review articles on earthquake mechanics. To solve practical fracture mechanics problems, the readers are recommended to first look up the problems in a number of stress intensity factor handbooks, including Sih (1973), Tada et al. (1973), Murakami (1987), and Rooke and Cartwright (1976). Only when such solutions are not available, should one continue to solve the problem analytically or numerically. The numerical simulations of failure process of actual solids have been done assuming the Weibull distribution of material property variations and the elastic-damage finite element model, and the simulations compare favorably with experiments (Tang et al., 2001; Wong et al., 2001, 2002). Wong et al. (2006) theoretically established the Weibull parameters in terms of the microstructures of rocks.

Fracture mechanics has also been applied to crack problems in poroelastic solids. More notable references include Rudnicki (1987, 1991, 1996), Rudnicki et al. (1993), Simons (1977), and Rice and Cleary (1976). The effect of friction has been incorporated into crack analysis by Qian and Sun (1998), Sun and Qian (1998), Chau and Wang (1998b), Chau and Wong (2009), and Chau et al. (2000, 2002).

6. 20 PROBLEMS

Problem 6.1 Starting from (6.34) and (6.35), show the validity of (6.38) and (6.39).

Problem 6.2 Starting from (6.38) and (6.39), show the validity of (6.40) and (6.41).

Problem 6.3 Use the Westergaard stress function given in (6.212) for a crack of length $2a$ subject to a pair of point forces P shown in Fig. 6.17 to show that the stress intensity factor of a crack problem with two pairs of point forces P applied at $x_1 = b$ and $-b$, as shown in Fig. 6.34 below, is

$$K_I = \frac{2P}{\sqrt{\pi a}} \frac{a}{\sqrt{a^2 - b^2}} \tag{6.333}$$

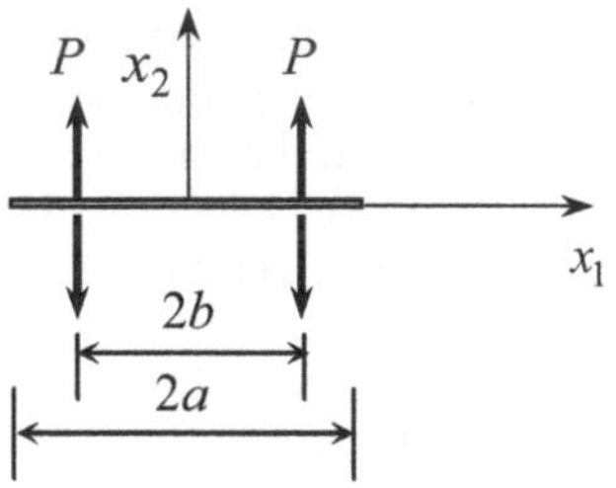

Figure 6.34 A central of length $2a$ subject to two pairs of point forces at $x_1 = b$ and $-b$

Problem 6.4 Referring to Fig. 6.35, this problem is similar to the analysis in Section 6.13, except that the fault is now driven by mode III motion. The upper and lower faces are bonded to two rigid bodies and are displaced by w_0 along the positive x_3-direction on the upper surface and by w_0 along the negative x_3-direction on the lower surface. Find the mode III stress intensity factor.

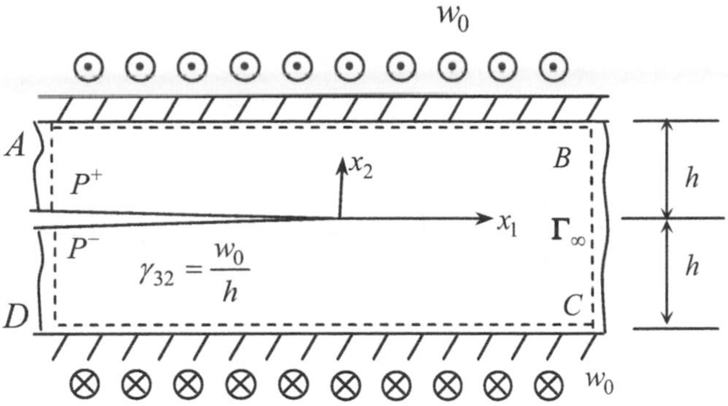

Figure 6.35 An elastic slab of thickness $2h$ and infinite dimensions in the x_1- and x_3-directions and containing a semi-infinite crack

Answer:

$$K_{III} = \mu w_0 \sqrt{\frac{2}{h}} \tag{6.334}$$

Problem 6.5 Referring to Fig. 6.36, this problem is similar to the analysis in Section 6.13, except that the fault is now driven by mode I motion. The elastic layer is under plane strain condition. The upper and lower faces are bonded to two rigid bodies and are displaced by u_2 along the positive x_2-direction on the upper surface and by u_2 along the negative x_2-direction on the lower surface. Find the mode I stress intensity factor.

Answer:

$$K_I = \frac{Eu_2}{h(1+\nu)} \sqrt{\frac{h}{(1-2\nu)}} \tag{6.335}$$

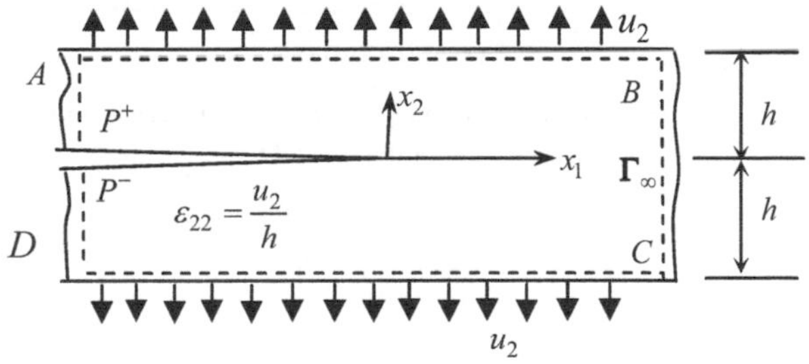

Figure 6.36 An elastic slab of thickness $2h$ and containing a semi-infinite crack; uniform displacement is applied on the upper and lower faces

Problem 6.6 This is a modification to Problem 6.5. Instead of being driven by a uniform displacement over the top and bottom surfaces, a uniform pressure σ is applied on the face of the semi-infinite crack, as shown in Fig. 6.37. The elastic layer is under plane strain condition. Find the mode I stress intensity factor.

Answer:
$$K_I = \frac{\sqrt{1-2\nu}}{1-\nu}\sigma\sqrt{h} \tag{6.336}$$

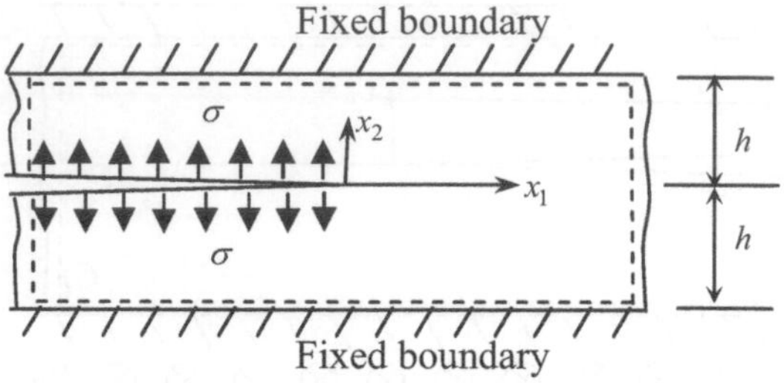

Figure 6.37 An elastic slab of thickness 2h containing a semi-infinite crack subject to pressure

Problem 6.7 Now consider that an elastic layer of thickness h contains two semi-infinite cracks and both of them are subject to internal pressure σ (see Fig. 6.38). The elastic layer is under plane strain condition. Find the mode I stress intensity factor at the crack tip for both the upper and lower cracks.

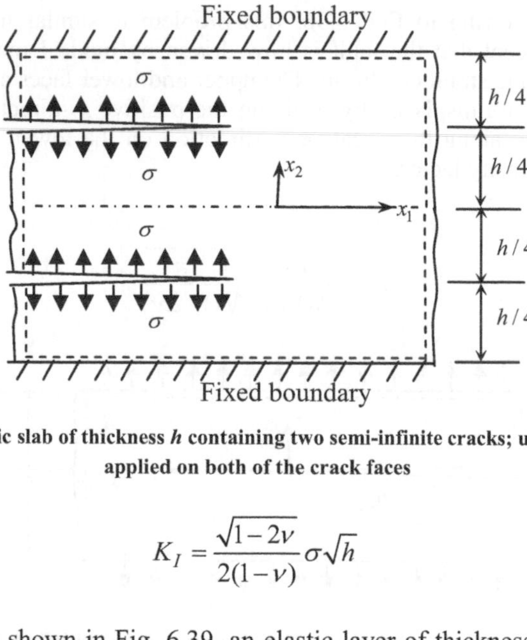

Figure 6.38 An elastic slab of thickness h containing two semi-infinite cracks; uniform tension σ is applied on both of the crack faces

Answer:
$$K_I = \frac{\sqrt{1-2\nu}}{2(1-\nu)}\sigma\sqrt{h} \tag{6.337}$$

Problem 6.8 As shown in Fig. 6.39, an elastic layer of thickness h contains two semi-infinite cracks. The elastic layer is under plane strain condition. The upper and lower faces are bonded to two rigid bodies and are displaced by w_0 along the positive x_3-direction on the upper surface and by w_0 along the negative x_3-direction on the lower surface. Find the mode III stress intensity factor.

Answer:
$$K_{III} = \mu w_0\sqrt{\frac{2}{h}} \tag{6.338}$$

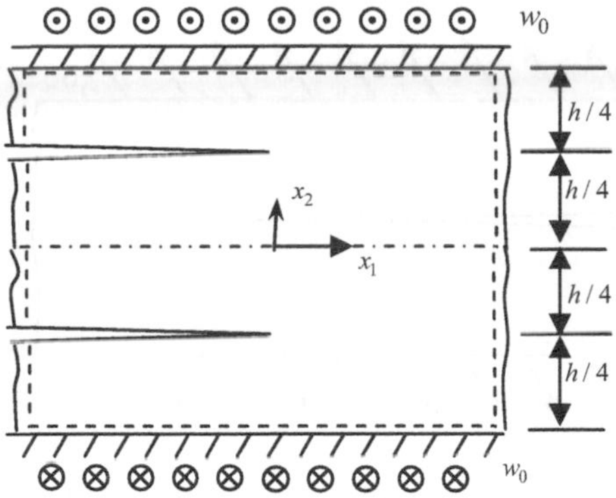

Figure 6.39 An elastic slab of thickness *h* containing two semi-infinite cracks

Problem 6.9 Consider an elastic layer of thickness $2h$ containing a semi-infinite crack, as shown in Fig. 6.40. The elastic layer is under plane strain condition. The upper and lower faces are bonded to two rigid bodies and are displaced by u_0 along the positive x_1-axis on the top and along the negative x_1-axis at the bottom as shown in Fig. 6.40. Find the mode II intensity factor K_{II}.

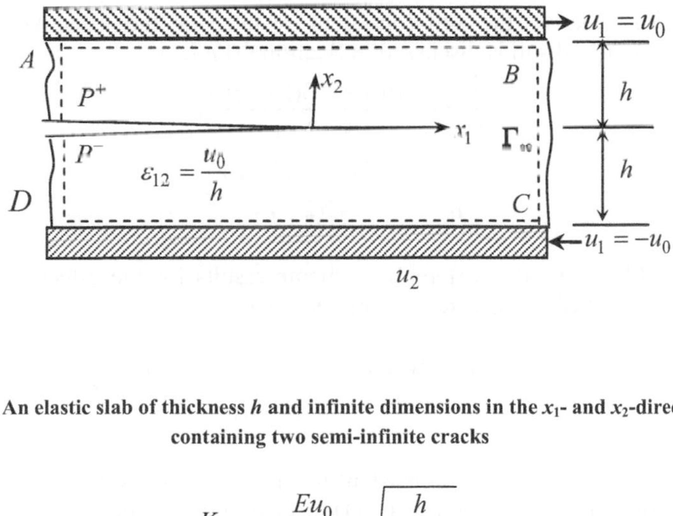

Figure 6.40 An elastic slab of thickness *h* and infinite dimensions in the x_1- and x_2-directions and containing two semi-infinite cracks

Answer:

$$K_{II} = \frac{Eu_0}{2h(1+\nu)}\sqrt{\frac{h}{(1-\nu)}} \tag{6.339}$$

Problem 6.10 Consider an elastic layer of thickness $2h$ containing a semi-infinite crack, as shown in Fig. 6.41. The elastic layer is under plane strain condition. The upper and lower faces are bonded to two rigid bodies and are displaced by u_0 at 60° with the x_1-axis as shown. Find a relation between K_I and K_{II}.

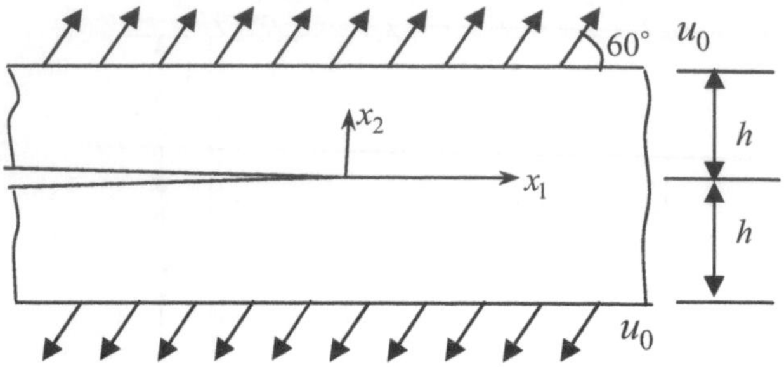

Figure 6.41 The displacement u_0 applied at 60° with the x_1-axis

Answer:
$$K_I^2 + K_{II}^2 = \frac{(5-7\nu)}{(1-\nu)(1-2\nu)}[\frac{Eu_0}{2h(1+\nu)}]^2 \tag{6.340}$$

Problem 6.11 Show that the solutions for the differential scheme given in (6.306) and (6.307) can be approximated by

$$\mu = \mu_0 e^{-\frac{16(1-\nu)}{3(2-\nu)}\varepsilon}, \quad E = E_0 e^{-\frac{16}{3}(1-\nu^2)\varepsilon} \tag{6.341}$$

Problem 6.12 Derive the self-consistent results for the effective moduli of isotropically cracked solids discussed in Section 6.17.7.

Answer:
$$\frac{\tilde{E}}{E} = 1 - \frac{16(1-\nu^2)(10-3\nu)}{45(2-\nu)}\varepsilon, \tag{6.342}$$

$$\frac{\tilde{\mu}}{\mu} = 1 - \frac{32(1-\nu)(5-\nu)}{45(2-\nu)}\varepsilon \tag{6.343}$$

Problem 6.13 Derive the differential scheme results for the effective moduli of isotropically cracked solids discussed in Section 6.17.7.

Answer:
$$E = E_0 e^{-\frac{16(1-\nu^2)(10-3\nu)}{45(2-\nu)}\varepsilon}, \quad \mu = \mu_0 e^{-\frac{32(1-\nu)(5-\nu)}{45(2-\nu)}\varepsilon} \tag{6.344}$$

Problem 6.14 Derive the self-consistent results for the effective moduli of solids with cylindrical transverse isotropy (CTI) discussed in Section 6.17.6.

Answer:
$$\frac{\tilde{E}}{E} = 1 - \frac{2(1-\nu^2)(8-3\nu)}{3(2-\nu)}\varepsilon, \tag{6.345}$$

$$\frac{\tilde{\mu}}{\mu} = 1 - \frac{8}{3}\frac{(1-\nu)}{(2-\nu)}\varepsilon \tag{6.346}$$

$$\frac{\mu^*}{\mu} = 1 - \frac{4(1-\nu)(4-\nu)}{3(2-\nu)}\varepsilon \tag{6.347}$$

Problem 6.15 Derive the differential scheme results for the effective moduli of solids with cylindrical transverse isotropy (CTI) discussed in Section 6.17.6.

Answer:

$$E = E_0 e^{-\frac{2(1-\nu^2)(8-3\nu)}{3(2-\nu)}\varepsilon} \tag{6.348}$$

$$\mu = \mu_0 e^{-\frac{8(1-\nu)}{3(2-\nu)}\varepsilon}, \quad \mu^* = \mu_0^* e^{-\frac{4(1-\nu)(4-\nu)}{3(2-\nu)}\varepsilon} \tag{6.349}$$

Problem 6.16 Show the validity of the following equation for isotropically cracked solids discussed in Section 6.17.7.

$$\Delta C_{ijkl} = \frac{16(1-\nu^2)}{45E(2-\nu)}[(5-\nu)(\delta_{ik}\delta_{jl} + \delta_{il}\delta_{jk}) - \nu\delta_{ij}\delta_{kl}]\varepsilon \tag{6.350}$$

CHAPTER SEVEN

Viscoelasticity and Its Applications

7.1 INTRODUCTION

Deformation and stress in certain materials are known to vary with time even though the external excitations, regardless of displacement or loads, are constant with time. In terms of dynamic problems, vibrations in solids are known to damp with time. In a sense, there appears a viscous effect in the solid. A solid can exhibit such a time effect but remains elastic. In reality, there may be some permanent deformation remains in the body, depending on the magnitude of the excitations. The theory of viscoplasticity was introduced in Section 5.17. However, if such permanent deformation is relatively small, we can model it as viscoelastic. That is, when the excitation is removed, the body returns to its original shape and size. Theoretical formulation that deals with such a viscoelastic body is called *viscoelasticity*. There are two extremes of viscoelasticity: if viscous response is negligible, the solid is purely elastic; if the elastic response is negligible, the material is a viscous fluid.

Time-dependent creeping has been reported in both rock and soil slopes, and deformation in excavated tunnels is often found to increase with time. There have been many examples of delayed geomechanical failure after loadings have been applied. Therefore, viscoelasticity finds its application in many applications in geotechnical problems.

A special feature of viscoelastic solids is that their present state of deformation cannot be determined if their entire loading history is not known. In other words, a viscoelastic body appears to have memory of its entire past. Because of this the deformation of viscoelastic solid at time t must be summed from its total loading history. In the case of stress relaxation (i.e., imposing strain as a controlling parameter), the current stress is a function of the current strain as well as its entire strain history. In the case of creeping (i.e., imposing stress as a controlling parameter), the current deformation is a function of the current stress as well as its entire stress history. If the loading is applied at a different rate, clearly because of this memory effect, the response of a viscoelastic solid will also change. Therefore, viscoelastic solids should also be considered rate sensitive. The actual micromechanism for such a time-dependent effect is still a mystery in most materials. If a solid is purely elastic, its response should not depend on how its current state is attained through its loading history. In a sense, there must be some irreversible processes involved in the deformation process. Energy must have been dissipated because as viscous effect is involved. Therefore, viscoelasticity has been linked to entropy evolution through irreversible thermodynamics (Fung, 1965). These irreversible processes have been modeled by using hidden state variables and their associated generalized forces. However, such models will not be considered here.

7.2 BOLTZMANN'S INTEGRAL FORM OF STRESS AND STRAIN

As mentioned above, the present state of deformation of a viscoelectic solid cannot be determined if its entire loading history is not known. Because of this the deformation of a body at time t must be summed from the total loading history. In particular, stress at time t can be evaluated as:

$$\sigma_{ij}(x,t) = \int_{-\infty}^{t} G_{ijkl}(x,t-\tau)\frac{\partial \varepsilon_{kl}}{\partial \tau}d\tau \tag{7.1}$$

where G_{ijkl} is the relaxation tensor. Since stress is a function of time, (7.1) can be considered as a relaxation type of stress-strain law. Alternatively, the inverse of (7.1) can expressed as

$$\varepsilon_{ij}(x,t) = \int_{-\infty}^{t} J_{ijkl}(x,t-\tau)\frac{\partial \sigma_{kl}}{\partial \tau}d\tau \tag{7.2}$$

where J_{ijkl} is the creeping tensor. Now the strain is a function of time, and (7.2) can be considered as a creeping type of stress-strain law. Clearly, G_{ijkl} and J_{ijkl} are not independent. Gurtin and Sternberg (1962) showed that if G_{ijkl} is twice differentiable and the initial value of it is not zero, the inverse of (7.1) given in (7.2) exists and is unique.

Solids satisfying the constitutive law in the integral form of (7.1) and (7.2) are referred to as Boltzmann solids or, following Volterra's terminology, the viscoelastic solid is the linear heredity solid (Fung, 1965). The integrals in (7.1) and (7.2) can also be interpreted as Duhamel integrals. Boltzmann was the founder of statistical mechanics in physics. However, he committed suicide because his theory on atom structure was opposed by others. This theory became widely accepted after his tragic death (see biography section).

Because of the symmetric properties of stress and strain, we must have

$$J_{ijkl} = J_{jikl} = J_{ijlk} \tag{7.3}$$

$$G_{ijkl} = G_{jikl} = G_{ijlk} \tag{7.4}$$

In addition, if the deformation starts at $t = 0$, we also have

$$G_{ijkl} = 0\,, \quad J_{ijkl} = 0 \tag{7.5}$$

for $-\infty < t < 0$. For isothermal viscoelasticity, a counterpart to Drucker's postulate for plasticity has been proposed by Gurtin and Herrera (1965):

$$\int_{0}^{t} \sigma_{ij}(\tau)\frac{\partial \varepsilon_{ij}(\tau)}{\partial \tau}d\tau \geq 0 \tag{7.6}$$

This is called the non-negative work requirement. Gurtin and Herrera (1965) deduced from this requirement that

$$G_{ijkl}(0)\gamma_{ij}\gamma_{kl} \geq 0\,, \quad G_{ijkl}(0) = G_{klij}(0) \tag{7.7}$$

and

$$G_{ijkl}(\infty)\gamma_{ij}\gamma_{kl} \geq 0\,, \quad G_{ijkl}(\infty) = G_{klij}(\infty) \tag{7.8}$$

for any symmetric tensor γ_{ij}. As shown by Christensen (1971), if the Helmholtz free energy per unit mass is non-negative, we can have a more restrictive requirement:

$$G_{ijkl}(t) \geq 0 \tag{7.9}$$

The requirements given by (7.7) have been found consistent with experiments.

Regarding the memory effect, we expect that the effect of recent history is more profound than more distant history. This common sense is called fading memory hypothesis (Christensen, 1971). Mathematically, we can write

$$\left|\frac{dG_{ijkl}(t)}{dt}\right|_{t=t_1} \leq \left|\frac{dG_{ijkl}(t)}{dt}\right|_{t=t_2} \qquad t_1 > t_2 > 0 \tag{7.10}$$

Similarly, we also have

$$\left|\frac{dJ_{ijkl}(t)}{dt}\right|_{t=t_1} \leq \left|\frac{dJ_{ijkl}(t)}{dt}\right|_{t=t_2} \qquad t_1 > t_2 > 0 \tag{7.11}$$

Mathematically, we can also write (7.10) and (7.11) as

$$\frac{d^2 G_{ijkl}(t)}{dt^2} \geq 0, \quad \frac{d^2 J_{ijkl}(t)}{dt^2} \geq 0 \tag{7.12}$$

In fact, all experimentally established G_{ijkl} and J_{ijkl} agree with this fading memory hypothesis, as summarized by Christensen (1971).

As illustrated in Fig. 7.1, when there is a jump in the strain at $t = 0$, and with all zero strain and stress for $t < 0$, (7.1) can be written as

$$\sigma_{ij}(x,t) = \varepsilon_{kl}(x,0^+)G_{ijkl}(x,t) + \int_0^t G_{ijkl}(x,t-\tau)\frac{\partial \varepsilon_{kl}(x,\tau)}{\partial \tau} d\tau \tag{7.13}$$

This integral can be written into another form. By considering a change of variable $s = t - \tau$, we have (Fung, 1965)

$$\sigma_{ij}(x,t) = \varepsilon_{kl}(x,0^+)G_{ijkl}(x,t) + \int_t^0 G_{ijkl}(x,s)\frac{\partial \varepsilon_{kl}(x,t-s)}{\partial s} ds \tag{7.14}$$

We now apply integration by parts to the last integral in (7.14) as

$$\begin{aligned} \int_t^0 G_{ijkl}(x,s)\frac{\partial \varepsilon_{kl}(x,t-s)}{\partial s} ds &= \varepsilon_{kl}(x,t)G_{ijkl}(x,0) \\ &- \varepsilon_{kl}(x,0)G_{ijkl}(x,t) + \int_0^t \varepsilon_{kl}(x,t-\tau)\frac{\partial G_{ijkl}(x,\tau)}{\partial \tau} ds \end{aligned} \tag{7.15}$$

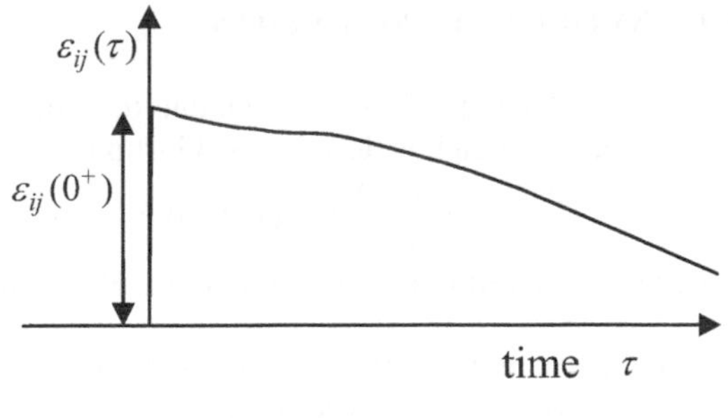

Figure 7.1. Illustration of strain history with jump at initial time

Note that we have reversed the limits of integration of the last integral and changed back the dummy variable from s to τ. Substitution of (7.15) into (7.14) gives

$$\sigma_{ij} = \varepsilon_{kl}(x,t)G_{ijkl}(x,0) + \int_0^t \varepsilon_{kl}(x,t-\tau)\frac{\partial G_{ijkl}(x,\tau)}{\partial \tau} ds \tag{7.16}$$

Note that (7.16) is equivalent to (7.13). This is the so-called commutative property of Stieltjes convolution (see below).

For isotropic solids, it can be shown that the following tensor forms apply (Lubliner, 1990):

$$G_{ijkl}(t) = \frac{G_2 - G_1}{3}\delta_{ij}\delta_{kl} + \frac{G_1}{3}(\delta_{ik}\delta_{jl} + \delta_{il}\delta_{jk}) \tag{7.17}$$

$$J_{ijkl}(t) = \frac{J_2 - J_1}{3}\delta_{ij}\delta_{kl} + \frac{J_1}{3}(\delta_{ik}\delta_{jl} + \delta_{il}\delta_{jk}) \tag{7.18}$$

Substituting (7.17) into (7.13), we have the following explicit forms:

$$s_{ij} = \int_0^t G_1(t-\tau)\frac{\partial e_{ij}(\tau)}{\partial \tau} d\tau \tag{7.19}$$

$$\sigma_{kk} = \int_0^t G_2(t-\tau)\frac{\partial \varepsilon_{kk}(\tau)}{\partial \tau} d\tau \tag{7.20}$$

where the deviatoric stress and strain are defined as

$$s_{ij} = \sigma_{ij} - (1/3)\delta_{ij}\sigma_{kk}, \quad e_{ij} = \varepsilon_{ij} - (1/3)\delta_{ij}\varepsilon_{kk} \tag{7.21}$$

Therefore, physically G_1 is the relaxation function under shear while G_2 is the relaxation function under dilatation or contraction. Similarly, the strain integral can also be expressed as

$$e_{ij} = \int_0^t J_1(t-\tau)\frac{\partial s_{ij}(\tau)}{\partial \tau} d\tau \tag{7.22}$$

$$\varepsilon_{kk} = \int_0^t J_2(t-\tau)\frac{\partial \sigma_{kk}(\tau)}{\partial \tau} d\tau \tag{7.23}$$

Therefore, physically J_1 is the creep function under shear while J_2 is the creep function under dilatation or contraction.

7.3 STIELTJES CONVOLUTION NOTATION

In the treatment of viscoelasticity, it is convenient to write these integrals in Stieltjes convolution notation, which is defined as (Gurtin and Sternberg, 1962)

$$S(t) = \int_0^t \phi(t-\tau)\frac{\partial \psi(\tau)}{\partial \tau} d\tau + \phi(t)\psi(0) = \phi * d\psi \tag{7.24}$$

The integral is called the convolution of ϕ and ψ and is denoted by a composition product form.

We first note the following properties of convolution:

$$\phi * d\psi = \psi * d\phi \quad \text{(cummutativity)} \tag{7.25}$$

$$\phi * (d\psi * d\theta) = (\phi * d\psi) * d\theta = \phi * d\psi * d\theta \quad \text{(associativity)} \tag{7.26}$$

$$\phi * d(\psi + \theta) = \phi * d\psi + \phi * d\theta \quad \text{(distributivity)} \tag{7.27}$$

$$\phi * d\psi = 0 \text{ implies } \phi \equiv 0 \text{ or } \psi \equiv 0 \tag{7.28}$$

Note that the scalar functions ϕ and ψ can be extended to tensors. Clearly, the commutative property given in (7.25) was proved in the last section by the change of variable and integration by parts (compare (7.13) and (7.16)). With the Stieltjes convolution notation, the stress and strain integrals now become

$$\sigma_{ij} = G_{ijkl} * d\varepsilon_{kl} = \varepsilon_{kl} * dG_{ijkl} \tag{7.29}$$

$$\varepsilon_{ij} = J_{ijkl} * d\sigma_{kl} = \sigma_{kl} * dJ_{ijkl} \tag{7.30}$$

For isotropic solids,

$$s_{ij} = G_1 * de_{ij} = e_{ij} * dG_1 \tag{7.31}$$

$$\sigma_{kk} = G_2 * d\varepsilon_{kk} = \varepsilon_{kk} * dG_2 \tag{7.32}$$

$$e_{ij} = J_1 * ds_{ij} = s_{ij} * dJ_1 \tag{7.33}$$

$$\varepsilon_{kk} = J_2 * d\sigma_{kk} = \sigma_{kk} * dJ_2 \tag{7.34}$$

If G_{ijkl}, J_{ijkl}, G_1 and G_2 are step functions, these convolution integrals can be reduced to the elastic Hooke's law.

In the next section, we present constitutive models for viscoelastic solids in terms of the differential equation form.

7.4 STRESS-STRAIN RELATION IN DIFFERENTIAL EQUATION FORM

For the convenience of comparison with viscoelastic material, the constitutive relations of isotropic linear elastic materials are first given in terms of deviatoric and hydrostatic stresses as

$$s_{ij} = 2\mu e_{ij}, \quad \sigma_{kk} = 3K\varepsilon_{kk} \tag{7.35}$$

Note that for fluids under small pressure and undergoing small deformation gradient, we have approximately

$$s_{ij} = 2\mu^* \frac{\partial e_{ij}}{\partial t}, \quad \sigma_{kk} = 3\kappa^* \frac{\partial \varepsilon_{kk}}{\partial t} \tag{7.36}$$

where μ^* and κ^* are the coefficients of viscosity of the fluid. Comparison of (7.35) and (7.36) shows that the constitutive laws for solids and fluids differ only by the rates of deviatoric strain.

For linear isotropic viscoelastic materials, the time-dependent constitutive relation can, in general, be expressed in a differential form as (Flugge, 1967)

$$P_1(t)s_{ij}(t) = Q_1(t)e_{ij}(t), \tag{7.37}$$

$$P_2(t)\sigma_{kk}(t) = Q_2(t)\varepsilon_{kk}(t) \tag{7.38}$$

where P_1, P_2, Q_1, and Q_2 are polynomials of the time differential operator $\partial/\partial t$. In explicit form, (7.37) can be expressed as

$$p_0 s_{ij}(t) + p_1 \frac{\partial s_{ij}(t)}{\partial t} + p_2 \frac{\partial^2 s_{ij}(t)}{\partial t^2} + \dots = q_0 e_{ij}(t) + q_1 \frac{\partial e_{ij}(t)}{\partial t} + q_2 \frac{\partial^2 e_{ij}(t)}{\partial t^2} + \dots \tag{7.39}$$

Similarly, (7.38) can also be expanded in differential form like (7.39). Note also that we have dropped the explicit spatial dependence of stress and strain, but they are implicitly understood as functions of position $\boldsymbol{x}$.

We can see that (7.37) and (7.38) are natural generalizations of (7.35) and (7.36) to include both elastic and viscous behavior. In the process, we have also included higher derivatives as well as adding differential operators on the deviatoric stress. Therefore, both elastic deformation and viscous flow can be obtained as special cases of viscoelastic models. Another motivation for having the differential form given in (7.37) and (7.38) probably originated from equivalent mechanical models for viscoelastic solids proposed by Maxwell, Voigt, and Kelvin. Figure 7.2 shows some of the commonly adopted models in viscoelasticity. The applied stress σ is shear stress and the deformation is engineering shear strain γ. They were made by various combinations of elastic spring and viscous dashpot. Clearly, the instantaneous elastic deformation is modeled by the spring whereas the time-dependent viscous flow is modeled by the dashpot. For example, the Maxwell model put a spring and a dashpot in series, whereas the Voigt or Kelvin model put a spring and a dashpot in parallel, as shown in Fig. 7.2.

7.4.1 Maxwell Model

Referring to the Maxwell model shown in Fig. 7.2(a), the shear stresses for both dashpot and spring are the same and equal:

$$\sigma = \mu\gamma_1 = \eta\frac{\partial\gamma_2}{\partial t} \tag{7.40}$$

The total shear strain is the sum of the strain from dashpot and spring and leads to

$$\eta\frac{\partial\gamma}{\partial t} = \eta\frac{\partial\gamma_1}{\partial t} + \eta\frac{\partial\gamma_2}{\partial t} = (1+\frac{\eta}{\mu}\frac{\partial}{\partial t})\sigma = (1+\tau\frac{\partial}{\partial t})\sigma \tag{7.41}$$

where $\tau = \eta/\mu$ is the relaxation time. The third part of (7.41) is a consequence of applying (7.40). Therefore, for Maxwell solids the polynomial differential operators are, by comparing (7.41) and (7.39),

$$P_1(t) = 1+\tau\frac{\partial}{\partial t}, \quad Q_1(t) = 2\eta\frac{\partial}{\partial t} \tag{7.42}$$

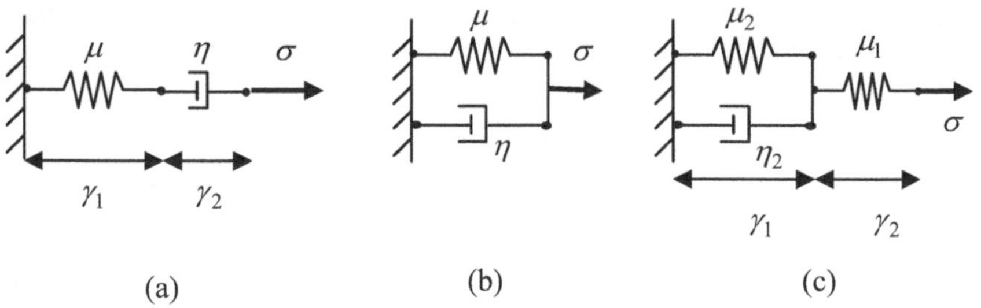

Figure 7.2. Mechanical models: (a) Maxwell, (b) Voigt–Kelvin, (c) standard linear solid

Note that the Maxwell solid will sustain a permanent deformation even if an applied load is removed permanently. This is because the dashpot is in series with the spring. Because of this, the Maxwell solid is sometimes referred to as the Maxwell fluid.

7.4.2 Kelvin–Voigt Model

Similarly, by considering the force equilibrium and strain compatibility the corresponding differential form of the Voigt or Kelvin model shown in Fig. 7.2(b) is:

$$\sigma = (\mu + \eta \frac{\partial}{\partial t})\gamma \tag{7.43}$$

Thus, we have the differential operator for the Voigt or Kelvin model as:

$$P_1(t) = 1, \quad Q_1(t) = 2(\mu + \eta \frac{\partial}{\partial t}) \tag{7.44}$$

Note that Kelvin–Voigt model does not allow instantaneous deformation because the dashpot is in parallel with the spring.

7.4.3 Three-Parameter Model

In the standard linear model, the viscoelastic behavior of materials can be represented by an elastic spring (with shear stiffness of μ_1) connected in series with a Kelvin or Voigt solid (an elastic spring of shear stiffness μ_2 in parallel with a dashpot of viscoelastic constant η_2). The standard linear solid was also called the "three-parameter body" (Flugge, 1967) or "generalized Kelvin body" (Lama and Vutukuri, 1978; Goodman, 1989). For the standard linear solid or three-paramater model A shown in Fig. 7.2(c), the differential form is

$$(1 + \frac{\eta_2}{\mu_1 + \mu_2}\frac{\partial}{\partial t})\sigma = \frac{\mu_1}{\mu_1 + \mu_2}(\mu_2 + \eta_2 \frac{\partial}{\partial t})\gamma \tag{7.45}$$

$$P_1(t) = (1 + \frac{\eta_2}{\mu_1 + \mu_2}\frac{\partial}{\partial t}), \quad Q_1(t) = \frac{2\mu_1}{\mu_1 + \mu_2}(\mu_2 + \eta_2 \frac{\partial}{\partial t}) \tag{7.46}$$

For the three-parameter models, we have four possible combinations. The one shown in Fig. 7.2(c) is called the standard linear solid, and the three other combinations are shown in Fig. 7.3(b–d). Here we called them three-parameter model A, model B, model C and model D. Note that Fung (1965) called the model B shown in Figure 7.3(b) standard linear solid instead of model A. However, model A is more often called the standard linear solid (Haddad, 1995).

The corresponding constitutive law for the three-parameter model B is:

$$(1 + \tau_2 \frac{\partial}{\partial t})\sigma = [(\tau_2 \mu_1 + 1)\frac{\partial}{\partial t} + \mu_1]\gamma \tag{7.47}$$

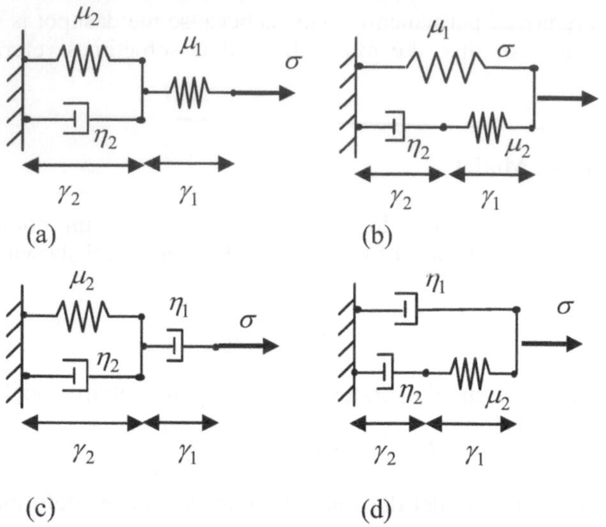

Figure 7.3. Three-parameter models: (a) standard linear solid A (b) model B, (c) model C, (d) model D

$$P_1(t) = (1 + \tau_2 \frac{\partial}{\partial t}) \ , \ Q_1(t) = 2[(\tau_2 \mu_1 + 1)\frac{\partial}{\partial t} + \mu_1] \tag{7.48}$$

where $\tau_2 = \eta_2/\mu_2$ is the relaxation time. The differential form of the three-parameter model C is

$$(1 + \frac{\eta_1 + \eta_2}{\mu_2}\frac{\partial}{\partial t})\sigma = \eta_1(\frac{\partial}{\partial t} + \tau_2 \frac{\partial^2}{\partial t^2})\gamma \tag{7.49}$$

$$P_1(t) = (1 + \frac{\eta_1 + \eta_2}{\mu_2}\frac{\partial}{\partial t}) \ , \ Q_1(t) = 2\eta_1(\frac{\partial}{\partial t} + \tau_2 \frac{\partial^2}{\partial t^2}) \tag{7.50}$$

The differential form of the three-parameter model D is

$$(1 + \tau_2 \frac{\partial}{\partial t})\sigma = [(\eta_2 + \eta_1)\frac{\partial}{\partial t} + \tau_2 \eta_1 \frac{\partial^2}{\partial t^2}]\gamma \tag{7.51}$$

$$P_1(t) = (1 + \tau_2 \frac{\partial}{\partial t}) \ , \ Q_1(t) = 2[(\eta_2 + \eta_1)\frac{\partial}{\partial t} + \tau_2 \eta_1 \frac{\partial^2}{\partial t^2}] \tag{7.52}$$

The proof of these equations is considered in Problem 7.12.

7.4.4 Generalized Maxwell and Kelvin Models

The more general models are given in Fig. 7.4 as the generalized Maxwell model and generalized Kelvin model. By noting that the strain is the same for each of the Maxwell elements shown in Fig. 7.4(a), we find the following differential form of the stress-strain law for the generalized Maxwell model:

$$\left(\mu_0 + \frac{\mu_1\eta_1(\partial/\partial t)}{\mu_1 + \eta_1(\partial/\partial t)} + \ldots + \frac{\mu_n\eta_n(\partial/\partial t)}{\mu_n + \eta_n(\partial/\partial t)} + \eta_{n+1}(\partial/\partial t)\right)\gamma = \sigma \tag{7.53}$$

By noting that the stress through each Kelvin–Voigt element shown in Fig. 7.4(b) is the same, the differential form of the stress-strain law for the generalized Kelvin model becomes

$$\left(\frac{1}{\mu_0} + \frac{1}{\mu_1 + \eta_1(\partial/\partial t)} + \ldots + \frac{1}{\mu_n + \eta_n(\partial/\partial t)} + \frac{1}{\eta_{n+1}(\partial/\partial t)}\right)\sigma = \gamma \tag{7.54}$$

Although there are some concerns about the mechanical base models presented in this section (e.g., Christensen, 1971), Section 13.8 of Fung (1965) illustrated that the irreversible thermodynamics-based constitutive model between generalized force and generalized displacement has the same mathematical form as the generalized Maxwell model. Thus, spring-dashpot type models did provide a simple way to interpret viscoelastic solids.

7.5 STRESS-STRAIN RELATION IN LAPLACE TRANSFORM SPACE

Viscoelastic problems modeled by constitutive law (7.37) and (7.38) can be solved by using integral transforms. As mentioned by Christensen (1971), either the Laplace transform or the Fourier transform can be applied to consider the time dependence of the viscoelastic problems. We only consider the Laplace transform in this section.

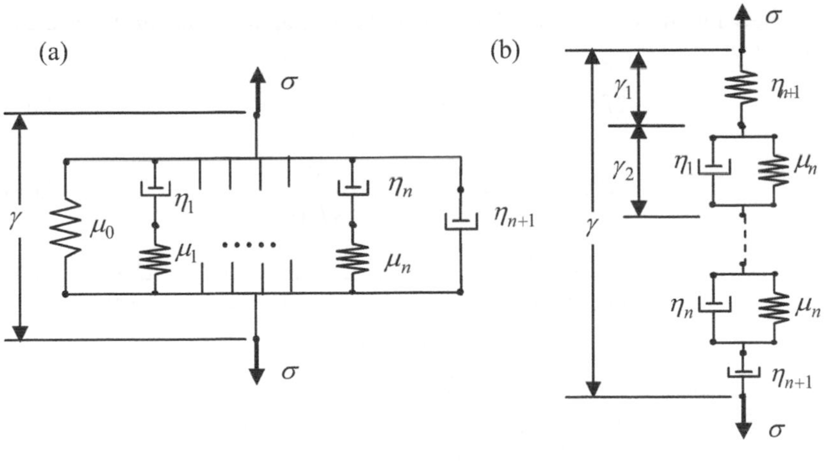

Figure 7.4. Generalized Maxwell (a) and Kelvin (b) models

Taking the Laplace transform of (7.37) and (7.38) and assuming a zero stress and strain state prior to time $t = 0$, we have

$$\hat{P}_1(s)\hat{s}_{ij}(s) = \hat{Q}_1(s)\hat{e}_{ij}(s) \tag{7.55}$$

$$\hat{P}_2(s)\hat{\sigma}_{kk}(s) = \hat{Q}_2(s)\hat{\varepsilon}_{kk}(s) \tag{7.56}$$

where $\hat{s}_{ij}(s)$ represents the Laplace transform of $s_{ij}(t)$ as

$$\hat{s}_{ij}(s) = \mathfrak{L}\left[s_{ij}(t)\right] = \int_0^\infty s_{ij}(t)e^{-st}dt \tag{7.57}$$

$$s_{ij}(t) = \mathfrak{L}^{-1}[\hat{s}_{ij}(s)] = \frac{1}{2\pi i}\int_{\gamma-i\infty}^{\gamma+i\infty} e^{st}\hat{s}_{ij}(s)ds \tag{7.58}$$

More details on the Laplace transform can be found in Appendix B. We can also apply the Laplace transform to (7.31)–(7.34), and obtain the following constitutive law in terms of creep function and relaxation function:

$$\hat{s}_{ij}(s) = s\hat{G}_1(s)\hat{e}_{ij}(s) \tag{7.59}$$

$$\hat{\sigma}_{kk}(s) = s\hat{G}_2(s)\hat{\varepsilon}_{kk}(s) \tag{7.60}$$

$$\hat{e}_{ij}(s) = s\hat{J}_1(s)\hat{s}_{ij}(s) \tag{7.61}$$

$$\hat{\varepsilon}_{kk}(s) = s\hat{J}_2(s)\hat{\sigma}_{kk}(s) \tag{7.62}$$

To obtain the above results, we apply the following Faltung or convolution theorem of the Laplace transform (Sneddon, 1951):

$$\mathfrak{L}\left[\int_0^t g(\tau)f(t-\tau)d\tau\right] = \hat{g}(s)\hat{f}(s) \tag{7.63}$$

The proof of this is given in Appendix B. We can also rewrite this constitutive law in the Laplace transform space by defining equivalent shear modulus, bulk modulus, Poisson's ratio, and Lamé's constants as

$$\hat{s}_{ij}(s) = 2\tilde{\mu}(s)\hat{e}_{ij}(s), \quad \hat{\sigma}_{kk}(s) = 3\tilde{K}(s)\hat{\varepsilon}_{kk}(s) \tag{7.64}$$

where the new moduli of the viscoelastic material in Laplace transform space can be expressed as

$$2\tilde{\mu}(s) = \frac{\hat{Q}_1(s)}{\hat{P}_1(s)} = s\hat{G}_1(s) = \frac{1}{s\hat{J}_1(s)} \tag{7.65}$$

$$3\tilde{K}(s) = \frac{\hat{Q}_2(s)}{\hat{P}_2(s)} = s\hat{G}_2(s) = \frac{1}{s\hat{J}_2(s)} \tag{7.66}$$

$$\tilde{\nu}(s) = \frac{\hat{P}_1(s)\hat{Q}_2(s) - \hat{Q}_1(s)\hat{P}_2(s)}{2\hat{P}_1(s)\hat{Q}_2(s) + \hat{P}_2(s)\hat{Q}_1(s)} = \frac{\hat{G}_2(s) - \hat{G}_1(s)}{2\hat{G}_2(s) + \hat{G}_1(s)} = \frac{\hat{J}_1(s) - \hat{J}_2(s)}{2\hat{J}_1(s) + \hat{J}_2(s)} \tag{7.67}$$

$$\tilde{E}(s) = \frac{3\hat{Q}_2(s)\hat{Q}_1(s)}{2\hat{P}_1(s)\hat{Q}_2(s) + \hat{P}_2(s)\hat{Q}_1(s)} = \frac{3s\hat{G}_2(s)\hat{G}_1(s)}{2\hat{G}_2(s) + \hat{G}_1(s)} = \frac{3}{s[2\hat{J}_1(s) + \hat{J}_2(s)]} \tag{7.68}$$

$$\tilde{\lambda}(s) = \frac{\hat{P}_1(s)\hat{Q}_2(s) - \hat{P}_2(s)\hat{Q}_1(s)}{3\hat{P}_2(s)\hat{P}_1(s)} = \frac{1}{3}s[\hat{G}_2(s) - \hat{G}_1(s)] = \frac{1}{3s}\left[\frac{1}{J_2(s)} - \frac{1}{J_1(s)}\right] \tag{7.69}$$

7.5.1 Viscoelastic Solids with Elastic Bulk Modulus

For engineering applications, it is normally assumed that there is no viscoelastic effect under isotropic compression and the viscous effect only appears under shear stress as is assumed in the spring-dashpot-based models shown in the previous section. More specifically, we can set

$$\hat{Q}_2(s)/\hat{P}_2(s) = 3K \tag{7.70}$$

where K is not a function of s. For such a case, equations (7.65)–(7.69) can be simplified to

$$2\tilde{\mu}(s) = \frac{\hat{Q}_1(s)}{\hat{P}_1(s)} = s\hat{G}_1(s) = \frac{1}{s\hat{J}_1(s)} \tag{7.71}$$

$$\tilde{\nu}(s) = \frac{3K\hat{P}_1(s) - \hat{Q}_1(s)}{6K\hat{P}_1(s) + \hat{Q}_1(s)} = \frac{3K - sG_1(s)}{6K + sG_1(s)} = \frac{3KsJ_1(s) - 1}{6KsJ_1(s) + 1} \tag{7.72}$$

$$\tilde{E}(s) = \frac{9K\hat{Q}_1(s)}{6K\hat{P}_1(s) + \hat{Q}_1(s)} = \frac{9Ks\hat{G}_1(s)}{6K + s\hat{G}_1(s)} = \frac{9K}{6Ks\hat{J}_1(s) + 1} \tag{7.73}$$

$$\tilde{\lambda}(s) = \frac{3K\hat{P}_1(s) - \hat{Q}_1(s)}{3\hat{P}_1(s)} = K - \frac{s\hat{G}_1(s)}{3} = K - \frac{1}{3s\hat{J}_1(s)} \tag{7.74}$$

Note of course that G_2 and J_2 do not exist for this case as the solid does not exhibit viscoelastic responses.

7.5.2 Maxwell Solids

For Maxwell solids, the Laplace transform of (7.42) gives

$$P_1(s) = 1 + \tau s \tag{7.75}$$

$$Q_1(s) = 2\eta s \tag{7.76}$$

Substitution of (7.75) and (7.76) into (7.71)–(7.74) gives the following equivalent moduli of the Maxwell solids in the transformed space:

$$2\tilde{\mu}(s) = \frac{2\eta s}{1 + \tau s} \tag{7.77}$$

$$3\tilde{K} = 3K \tag{7.78}$$

$$\tilde{\nu}(s) = \frac{3K + (3K\tau - 2\eta)s}{6K + (6K\tau + 2\eta)s} \tag{7.79}$$

$$\tilde{E}(s) = \frac{18K\eta s}{6K + (6K\tau + 2\eta)s} \tag{7.80}$$

$$\tilde{\lambda}(s) = \frac{3K + (3K\tau - 2\eta)s}{3(1 + \tau s)} \tag{7.81}$$

As remarked earlier, Maxwell solids are also referred to as Maxwell fluids since there are continuous creeping displacements.

7.5.3 Kelvin-Voigt Solids

For Kelvin–Voigt solids, the Laplace transform of (7.44) gives

$$P_1(s) = 1 \tag{7.82}$$

$$Q_1(s) = 2(\mu + \eta s) \tag{7.83}$$

The equivalent moduli of the Kelvin–Voigt solids in the transformed space become

$$2\tilde{\mu}(s) = 2(\mu + \eta s) \tag{7.84}$$

$$3\tilde{K} = 3K \tag{7.85}$$

$$\tilde{\nu}(s) = \frac{3K - 2\mu - 2\eta s}{6K + 2\mu + 2\eta s} \tag{7.86}$$

$$\tilde{E}(s) = \frac{18K(\mu + \eta s)}{6K + 2\mu + 2\eta s} \tag{7.87}$$

$$\tilde{\lambda}(s) = \frac{3K - 2\mu - 2\eta s}{3} \tag{7.88}$$

7.5.4 Standard Linear Solid and Three-Parameter Models

For a standard linear solid or the three-parameter model A, the Laplace transform of (7.46) gives (Bland, 1960)

$$P_1(s) = \frac{\eta_2 s}{\mu_1 + \mu_2} + 1 \tag{7.89}$$

$$Q_1(s) = \frac{2\mu_1\mu_2}{\mu_1 + \mu_2}\left(\frac{\eta_2}{\mu_2} s + 1\right), \tag{7.90}$$

These results are obvious in view of (B.11) of Appendix B.3. For the three-parameter model A, the equivalent moduli in the transformed space can be obtained by substituting (7.89) and (7.90) into (7.71)–(7.74):

$$2\tilde{\mu}(s) = \frac{2\mu_1(\eta_2 s + \mu_2)}{(\eta_2 s + \mu_1 + \mu_2)} \tag{7.91}$$

$$3\tilde{K} = 3K \tag{7.92}$$

$$\tilde{\nu}(s) = \frac{3K(\eta_2 s + \mu_1 + \mu_2) - 2\mu_1(\eta_2 s + \mu_2)}{6K(\eta_2 s + \mu_1 + \mu_2) + 2\mu_1(\eta_2 s + \mu_2)} \tag{7.93}$$

$$\tilde{E}(s) = \frac{18K\mu_1(\eta_2 s + \mu_2)}{6K(\eta_2 s + \mu_1 + \mu_2) + 2\mu_1(\eta_2 s + \mu_2)} \tag{7.94}$$

$$\tilde{\lambda}(s) = \frac{3K(\eta_2 s + \mu_1 + \mu_2) - 2\mu_1(\eta_2 s + \mu_2)}{3(\eta_2 s + \mu_1 + \mu_2)} \tag{7.95}$$

For the three-parameter model B, the Laplace transform of (7.48) gives

$$P_1(s) = 1 + \tau_2 s \tag{7.96}$$

$$Q_1(s) = 2[(\mu_1\tau_2 + 1)s + \mu_1] \tag{7.97}$$

Similarly, it is straightforward to show that the equivalent moduli for the three-parameter model B in the transformed space become

$$2\tilde{\mu}(s)=\frac{2[(\mu_1\tau_2+1)s+\mu_1]}{(1+\tau_2 s)} \tag{7.98}$$

$$3\tilde{K}=3K \tag{7.99}$$

$$\tilde{\nu}(s)=\frac{3K(1+\tau_2 s)-2[(\mu_1\tau_2+1)s+\mu_1]}{6K(1+\tau_2 s)+2[(\mu_1\tau_2+1)s+\mu_1]} \tag{7.100}$$

$$\tilde{E}(s)=\frac{18K[(\mu_1\tau_2+1)s+\mu_1]}{6K(1+\tau_2 s)+2[(\mu_1\tau_2+1)s+\mu_1]} \tag{7.101}$$

$$\tilde{\lambda}(s)=\frac{3K(1+\tau_2 s)-2[(\mu_1\tau_2+1)s+\mu_1]}{3(1+\tau_2 s)} \tag{7.102}$$

For the three-parameter model C, by virtue of (B.11) and (B.12) of Appendix B.3 the Laplace transform of (7.50) gives

$$P_1(s)=1+(\frac{\eta_1+\eta_2}{\mu_2})s \tag{7.103}$$

$$Q_1(s)=2(\frac{\eta_1\eta_2}{\mu_2}s^2+\eta_1 s) \tag{7.104}$$

In view of (7.103) and (7.104), the equivalent moduli for the three-parameter model C in the transformed space become

$$2\tilde{\mu}(s)=\frac{2(\eta_1\eta_2 s^2+\mu_2\eta_1 s)}{\mu_2+(\eta_1+\eta_2)s} \tag{7.105}$$

$$3\tilde{K}=3K \tag{7.106}$$

$$\tilde{\nu}(s)=\frac{3K[\mu_2+(\eta_1+\eta_2)s]-2(\eta_1\eta_2 s^2+\mu_2\eta_1 s)}{6K[\mu_2+(\eta_1+\eta_2)s]+2(\eta_1\eta_2 s^2+\mu_2\eta_1 s)} \tag{7.107}$$

$$\tilde{E}(s)=\frac{18K(\eta_1\eta_2 s^2+\mu_2\eta_1 s)}{6K[\mu_2+(\eta_1+\eta_2)s]+2(\eta_1\eta_2 s^2+\mu_2\eta_1 s)} \tag{7.108}$$

$$\tilde{\lambda}(s)=\frac{3K[\mu_2+(\eta_1+\eta_2)s]-2(\eta_1\eta_2 s^2+\mu_2\eta_1 s)}{3[\mu_2+(\eta_1+\eta_2)s]} \tag{7.109}$$

Following the procedure used for model C above, the Laplace transform of (7.52) for the three-parameter model D gives

$$P_1(s)=1+\frac{\eta_2}{\mu_2}s \tag{7.110}$$

$$Q_1(s)=2[\frac{\eta_1\eta_2}{\mu_2}s^2+(\eta_1+\eta_2)s] \tag{7.111}$$

Subsequently, the following equivalent moduli for the three-parameter model D in the transformed space are obtained:

$$2\tilde{\mu}(s)=\frac{2[\eta_1\eta_2 s^2+\mu_2(\eta_1+\eta_2)s]}{\mu_2+\eta_2 s} \tag{7.112}$$

$$3\tilde{K}=3K \tag{7.113}$$

$$\tilde{\nu}(s)=\frac{3K(\mu_2+\eta_2 s)-2[\eta_1\eta_2 s^2+\mu_2(\eta_1+\eta_2)s]}{6K(\mu_2+\eta_2 s)+2[\eta_1\eta_2 s^2+\mu_2(\eta_1+\eta_2)s]} \tag{7.114}$$

$$\tilde{E}(s) = \frac{18K[\eta_1\eta_2 s^2 + \mu_2(\eta_1+\eta_2)s]}{6K(\mu_2+\eta_2 s)+2[\eta_1\eta_2 s^2 + \mu_2(\eta_1+\eta_2)s]} \tag{7.115}$$

$$\tilde{\lambda}(s) = \frac{3K(\mu_2+\eta_2 s)-2[\eta_1\eta_2 s^2 + \mu_2(\eta_1+\eta_2)s]}{3(\mu_2+\eta_2 s)} \tag{7.116}$$

7.6 CORRESPONDENCE PRINCIPLE

When inertia effects are not negligible in solving elasticity problems, dynamic terms need to be included. The Laplace transform is a useful mathematical tool to solve dynamic problems involving time derivatives. The transformation translates the time derivative to an algebraic problem with respect to parameter s. After the algebraic equation in s is solved, the solution is translated back to the time domain. But, of course, the inverse transform may not be easily obtained analytically and a numerical or approximate method may be needed. In a sense, viscoelastic problems are somehow similar.

The corresponding principle was first deduced by Lee (1955) for viscoelastic solids subject to proportional loadings, and Lee was awarded the Timoshenko medal in 1976 (see biography section). For proportional loadings, he observed that spatial dependence of a transformed viscoelastic solution is the same as that of a geometrically similar elastic solid if the spatial dependence of the prescribed boundary is the same for both problems. Laplace transform can be applied to viscoelastic problems and the transformed problem involving parameter s can normally be put into the same mathematical form as linear elasticity problems. Note from the last section that constitutive law in viscoelasticity in the Laplace transform space is also a function s. The Laplace parameter s can simply be considered a constant in solving the associated elastic problems. Therefore, if the associated elastic problems can be solved, the viscoelasticity solution can be obtained by taking the inverse transform back to the time domain. This identification of a problem in linear elasticity with one in viscoelasticity in the transform space s is called *correspondence principle*.

Note that this method applies as long as the space and time dependence of the prescribed loading and displacements appear as separate factors with a common time factor (i.e., separation of variables). In principle, the Fourier transform can also be used to solve viscoelastic problems. It is particularly useful in tackling problems with loading applied in a harmonic manner (e.g., see Section 9.9 of Chapter 9). However, for suddenly applied load (or load applied as a unit-step function) the Fourier transform is not applicable, and instead the Laplace transform should be used.

In the following discussion, we will restrict ourselves to quasi-static viscoelastic problems (i.e., problems in which the inertia effect can be neglected). Table 7.1 summarized the similarity and correspondence between the static elastic problem and the quasi-static viscoelastic problems. In Table 7.1, S_σ and S_u are the traction and displacement boundaries of the solid, respectively. The last row indicates that the elastic solution can be used to obtain the viscoelastic solution in the transform space by replacing the elastic moduli by the equivalent moduli in terms of s and the loading by transformed loading.

Table 7.1 Correspondence of elastic and viscoelastic problems

Parameter	Elastic		Viscoelastic in Laplace Space
Displacement	$\boldsymbol{u}(\boldsymbol{x})$		$\hat{\boldsymbol{u}}(\boldsymbol{x},s)$
Stress	$\boldsymbol{\sigma}(\boldsymbol{x})$		$\hat{\boldsymbol{\sigma}}(\boldsymbol{x},s)$
Equilibrium	$\nabla\bullet\boldsymbol{\sigma}(\boldsymbol{x})+\boldsymbol{F}=0$		$\nabla\bullet\hat{\boldsymbol{\sigma}}(\boldsymbol{x},s)+\boldsymbol{F}(s)=0$
Strain	$\boldsymbol{\varepsilon}(\boldsymbol{x})$		$\hat{\boldsymbol{\varepsilon}}(\boldsymbol{x},s)$
Constitutive Law	$\boldsymbol{s}(\boldsymbol{x})=2\mu\boldsymbol{e}(\boldsymbol{x})$ $\sigma_{kk}(\boldsymbol{x})=3K\varepsilon_{kk}(\boldsymbol{x})$		$\hat{\boldsymbol{s}}(\boldsymbol{x},s)=2\hat{\mu}(s)\hat{\boldsymbol{e}}(\boldsymbol{x},s)$ $\hat{\sigma}_{kk}(\boldsymbol{x},s)=3\hat{K}(s)\hat{\varepsilon}_{kk}(\boldsymbol{x},s)$
Boundary Condition	$\boldsymbol{\sigma}\bullet\boldsymbol{n}=\boldsymbol{f}$ on S_σ $\boldsymbol{u}=\boldsymbol{g}$ on S_u		$\boldsymbol{\sigma}(s)\bullet\boldsymbol{n}=\boldsymbol{f}(s)$ on S_σ $\boldsymbol{u}(s)=\boldsymbol{g}(s)$ on S_u
Solution	Available elastic solution: $\boldsymbol{u}=\boldsymbol{u}(\boldsymbol{x},\boldsymbol{g},\boldsymbol{f},\mu,\nu)$	substitutions $\hat{\nu}(s)\to\nu$ $\hat{\mu}(s)\to\mu$ $\hat{\boldsymbol{f}}(s)\to\boldsymbol{f}$ $\hat{\boldsymbol{g}}(s)\to\boldsymbol{g}$	Solution in transform space: $\hat{\boldsymbol{u}}=\hat{\boldsymbol{u}}[\boldsymbol{x},\boldsymbol{g}(s),\boldsymbol{f}(s),\hat{\mu}(s),\hat{\nu}(s)]$ Solution in time domain: $\boldsymbol{u}=\mathfrak{L}^{-1}\{\hat{\boldsymbol{u}}(\boldsymbol{x},s)\}$

Regarding the boundary conditions, if a suddenly applied load is imposed on the traction boundary, we have

$$\mathfrak{L}[f(t)]=\mathfrak{L}[f_0H(t)]=\frac{1}{s}f_0 \tag{7.117}$$

$$\mathfrak{L}[f(t)]=\mathfrak{L}[f_0\delta(t)]=f_0 \tag{7.118}$$

The proof of the last equation in (7.117) and (7.118) on the Laplace transform of Heaviside step function and the Dirac delta function is given in Appendix B.4.

Finally, the viscoelastic solution in transform space can be converted back to time domain by applying the inverse Laplace transform. Some useful formulas for the inverse Laplace transform are given in Table B.1 in Appendix B.5.

7.6.1 Boussinesq Problem for Maxwell Half-Space

Consider the problem of a concentrated point force applied on the surface of a half-space or the so-called Boussinesq problem considered in Chapter 4. The elastic solution is given in (4.152) and (4.153) of Chapter 4. Applying the correspondence principle, the solutions in transform space become

$$\hat{u}_r(s)=\frac{\hat{P}(s)}{4\pi\hat{\mu}(s)R}\{\frac{rz}{R^2}-\frac{[1-2\hat{\nu}(s)]r}{R+z}\},\quad \hat{u}_z(s)=\frac{\hat{P}(s)}{4\pi\hat{\mu}(s)R}\{2[1-\hat{\nu}(s)]+\frac{z^2}{R^2}\} \tag{7.119}$$

$$\hat{\sigma}_{rr}(s)=\frac{\hat{P}(s)}{2\pi R^2}\{\frac{[1-2\hat{v}(s)]R}{R+z}-\frac{3r^2z}{R^3}\},\quad \hat{\sigma}_{\theta\theta}(s)=\frac{[1-2\hat{v}(s)]\hat{P}(s)}{2\pi R^2}(\frac{z}{R}-\frac{R}{R+z})$$

$$\hat{\sigma}_{zz}(s)=-\frac{3\hat{P}(s)z^3}{2\pi R^5},\quad \hat{\sigma}_{rz}(s)=-\frac{3\hat{P}(s)z^2r}{2\pi R^5} \tag{7.120}$$

The inversion of the Laplace transform for stresses only involves:

$$\mathfrak{L}^{-1}[\hat{P}(s)]=P(t),\quad \mathfrak{L}^{-1}\{\hat{P}(s)[1-2\hat{v}(s)]\}=\mathbf{\Lambda}_1(t),\quad \mathfrak{L}^{-1}\{\hat{P}(s)[1-\hat{v}(s)]\}=\mathbf{\Lambda}_2(t) \tag{7.121}$$

for the case that the point load is applied suddenly as $P_0H(t)$, such that

$$\hat{P}(s)=\mathfrak{L}[P_0H(t)]=\frac{1}{s}P_0 \tag{7.122}$$

where $H(t)$ is a Heaviside step function. By further assuming that the viscoelastic half-space is a Maxwell solid, we have

$$1-2\bar{v}(s)=1-2[\frac{3K+(3K\tau-2\eta)s}{6K+(6K\tau+2\eta)s}] \tag{7.123}$$

Combining (7.122) and (7.123), and simplifying the result by partial fraction yields

$$\hat{P}(s)[1-2\bar{v}(s)]=P_0\alpha_0[\frac{1}{\alpha_1+s}] \tag{7.124}$$

where

$$\alpha_0=\frac{6\eta}{6K\tau+2\eta},\quad \alpha_1=\frac{6K}{6K\tau+2\eta} \tag{7.125}$$

By referring to Appendix B, (7.124) can be inverted to give

$$\mathfrak{L}^{-1}\{\hat{P}(s)[1-2\bar{v}(s)]\}=P_0\alpha_0e^{-\alpha_1t}H(t) \tag{7.126}$$

Similarly, the inversion for u_z involves the following inversion:

$$\mathfrak{L}^{-1}\{\hat{P}(s)[1-\bar{v}(s)]\}=\frac{P_0}{2}\{1+\alpha_0e^{-\alpha_1t}\}H(t) \tag{7.127}$$

The inversion is now complete and the final results are

$$\sigma_{rr}(t)=\frac{P_0}{2\pi R^2}\{\frac{R}{R+z}\alpha_0e^{-\alpha_1t}-\frac{3r^2z}{R^3}\}H(t),\quad \sigma_{\theta\theta}(t)=\frac{P_0}{2\pi R^2}(\frac{z}{R}-\frac{R}{R+z})e^{-\alpha_1t}H(t)$$

$$\sigma_{zz}(t)=-\frac{3z^3P_0}{2\pi R^5}H(t),\quad \sigma_{rz}(t)=-\frac{3P_0z^2r}{2\pi R^5}H(t) \tag{7.128}$$

Note that the elastic solution cannot be recovered as a special case either for $t\to\infty$ or $t\to 0$ because the dashpot movement is not restricted by the spring. Similarly, the displacement can also be obtained as

$$u_r(t)=\frac{P_0}{4\pi\mu R}\{\frac{rz}{R^2}f_1(t)-\frac{r}{R+z}f_2(t)\}H(t),\quad u_z(t)=\frac{P_0}{4\pi\mu R}\{2f_3(t)+\frac{z^2}{R^2}f_1(t)\}H(t) \tag{7.129}$$

where time functions f_1, f_2, and f_3 are given in Problem 7.13. The details of the analysis are left as a problem for the reader.

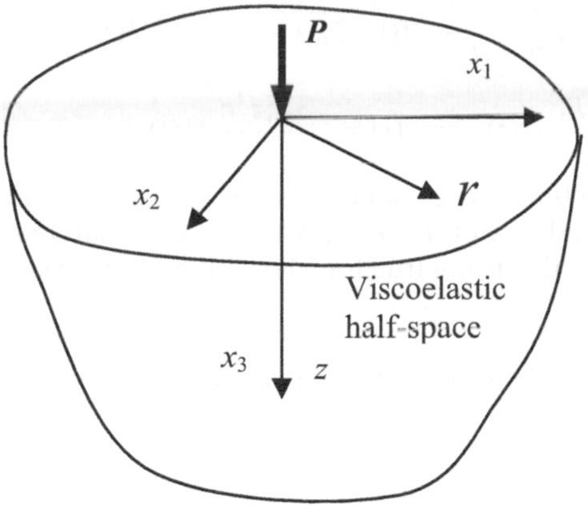

Figure 7.5. Boussinesq problem on viscoelastic half-space

7.6.2 Boussinesq Problem for Kelvin–Voigt Half-Space

For the case that the point load is applied suddenly as $P_0H(t)$ on the surface of a Kelvin-Voigt half-space, we have

$$1-2\tilde{\nu}(s)=1-2[\frac{3K-2\mu-2\eta s}{6K+2\mu+2\eta s}] \tag{7.130}$$

Combining (7.122) and (7.130), and simplifying it by partial fraction yields

$$\hat{P}(s)[1-2\tilde{\nu}(s)]=P_0\left\{\frac{1-2\bar{\beta}}{s}+\beta_1(\frac{1}{\beta_2+s})\right\} \tag{7.131}$$

where

$$\bar{\beta}=\frac{3K-2\mu}{6K+2\mu},\quad \beta_1=\frac{9K}{3K+\mu},\quad \beta_2=\frac{6K+2\mu}{2\eta} \tag{7.132}$$

By referring to Appendix B, (7.131) can be inverted to give

$$\mathfrak{L}^{-1}\{\hat{P}(s)[1-2\tilde{\nu}(s)]\}=P_0\{(1-2\bar{\beta})+\beta_1e^{-\beta_2t}\}H(t) \tag{7.133}$$

Similarly, the inversion for u_z involves the following inversion:

$$\mathfrak{L}^{-1}\{\hat{P}(s)[1-\tilde{\nu}(s)]\}=P_0\{(1-\bar{\beta})+\frac{1}{2}\beta_1e^{-\beta_2t}\}H(t) \tag{7.134}$$

The inversion gives the displacements and stresses as

$$u_r(t)=\frac{P_0}{4\pi\mu R}\{\frac{rz}{R^2}g_1(t)-\frac{r}{R+z}g_2(t)\}H(t) \tag{7.135}$$

$$u_z(t)=\frac{P_0}{4\pi\mu R}\{2g_3(t)+\frac{z^2}{R^2}g_1(t)\}H(t) \tag{7.136}$$

$$\sigma_{rr}(t)=\frac{P_0}{2\pi R^2}\{\frac{R}{R+z}[(1-2\bar{\beta})+\beta_1e^{-\beta_2t}]-\frac{3r^2z}{R^3}\}H(t) \tag{7.137}$$

$$\sigma_{\theta\theta}(t) = \frac{P_0}{2\pi R^2}\left(\frac{z}{R} - \frac{R}{R+z}\right)[(1-2\bar{\beta}) + \beta_1 e^{-\beta_2 t}]H(t) \tag{7.138}$$

$$\sigma_{zz}(t) = -\frac{3z^3 P_0}{2\pi R^5} H(t), \quad \sigma_{rz}(t) = -\frac{3P_0 z^2 r}{2\pi R^5} H(t) \tag{7.139}$$

where the time functions g_1, g_2, and g_3 are given in Problem 7.14 and left as an exercise for the reader. For $t \to \infty$, we can see that the elastic stresses are recovered. However, this is not true for Maxwell half-spaces.

7.6.3 Boussinesq Problem for Three-Parameter Model A

For the case that the point load is applied suddenly as $P_0 H(t)$ on the surface of a three-parameter model A or standard linear solid half-space, we have

$$1 - 2\tilde{\nu}(s) = 1 - 2\left[\frac{3K(\mu_1+\mu_2) - 2\mu_1\mu_2 + \eta_2(3K - 2\mu_1)s}{6K(\mu_1+\mu_2) + 2\mu_1\mu_2 + \eta_2(6K + 2\mu_1)s}\right] \tag{7.140}$$

$$\hat{P}(s)[1 - 2\tilde{\nu}(s)] = P_0 \left\{ \frac{1-2\bar{\beta}}{s} + \beta_1\left(\frac{1}{\beta_2 + s}\right) \right\} \tag{7.141}$$

where

$$\bar{\beta} = \frac{3K(\mu_1+\mu_2) - 2\mu_1\mu_2}{6K(\mu_1+\mu_2) + 2\mu_1\mu_2}, \quad \beta_1 = \frac{9K\mu_1^2}{(3K+\mu_1)[3K(\mu_1+\mu_2) + \mu_1\mu_2]},$$

$$\beta_2 = \frac{3K(\mu_1+\mu_2) + \mu_1\mu_2}{\eta_2(3K+\mu_1)} \tag{7.142}$$

Note that mathematical forms for the Laplace inverses for stresses are exactly the same as those for the Kelvin–Voigt half-space. Therefore, the final solutions will not be repeated here. For $t \to \infty$, similar to the Kelvin–Voigt model, the elastic solution is recovered as a special case, provided that the following identification is made:

$$\mu_\infty = \frac{\mu_1\mu_2}{(\mu_1+\mu_2)} \tag{7.143}$$

where μ_∞ is naturally called the long-term shear modulus. From Fig. 7.2(c), we can see that initial shear deformation is only proportional to μ_1 since the spring within the Kelvin–Voigt unit cannot respond instantaneously because of the viscous dashpot. Therefore, naturally we have

$$\mu_0 = \mu_1 \tag{7.144}$$

where μ_0 is naturally called the initial shear modulus. Another parameter commonly used is the relaxation time, defined as:

$$\tau_G = \frac{\eta_2}{\mu_1+\mu_2} \tag{7.145}$$

Since their physically meanings are well defined, it is sometimes more convenient to write the parameters of the standard linear solid in terms of these newly defined parameters as

$$\mu_1 = \mu_0, \quad \mu_2 = \frac{\mu_\infty\mu_0}{\mu_0 - \mu_\infty}, \quad \eta_2 = \left(\frac{\mu_0^2}{\mu_0 - \mu_\infty}\right)\tau_G \tag{7.146}$$

The displacements are given in Problem 7.15 and left as an exercise for the reader.

7.7 CREEPING AND RELAXATION TESTS

To calibrate the material constants in the viscoelastic models, either a creeping test or a relaxation test has to be conducted. Creeping compliance and the relaxation modulus for various models are considered in this section.

7.7.1 Maxwell Material

7.7.1.1 Creeping Test

In creeping tests, a constant stress is suddenly applied at $t = 0$:

$$\sigma = \sigma_0 H(t) \tag{7.147}$$

Then the evolution of the strain is measured as a function of time. This measured function can then be used to calibrate the material constants for any viscoelastic models. For the case of the Maxwell model, we have (see (7.41))

$$\sigma + \tau \frac{d\sigma}{dt} = \eta \frac{d\gamma}{dt} \tag{7.148}$$

where $\tau = \eta/\mu$ is the relaxation time. Because of the jump at $t = 0$, there is a singularity of the time derivative of stress at $t = 0$. Integration of (7.148) around the vicinity of $t = 0$ gives

$$\int_{-\tau}^{+\tau} \sigma dt + \tau[\sigma(\tau) - \sigma(-\tau)] = \eta[\gamma(\tau) - \gamma(-\tau)] \tag{7.149}$$

Considering the limit that $\tau \to 0$, and noting that the first integral shrinks to zero, we have

$$\tau\sigma(0^+) = \eta\gamma(0^+)\,, \text{ or } \tau\sigma_0 = \eta\gamma_0 \tag{7.150}$$

This equation gives an initial condition for the stress and strain. We now return to the creeping test situation by setting $\sigma = \sigma_0$ and $d\sigma/(dt) = 0$ in (7.148) and integrate the resulting equation once with respect to time to obtain

$$\eta\gamma = \sigma_0 t + C \tag{7.151}$$

Setting $t = 0$ in (7.151) and using the initial condition (7.150), the integration constant can be determined, and the final representation for the strain under creeping test is

$$\gamma(t) = \frac{\sigma_0}{\eta}(t + \tau) \tag{7.152}$$

The creeping compliance can be obtained as

$$J_1(t) = \frac{\gamma(t)}{2\sigma_0} = \frac{1}{2\eta}(t + \tau) \tag{7.153}$$

This creeping prediction is shown in Fig. 7.6(a). This result shows that there is an instantaneous strain response at $t = 0$, followed by a linear increasing strain. The material is more like fluid than solid. This appears not to be agreeable with creeping experiments on most geomaterials, unless under high temperature and pressure.

7.7.1.2 Relaxation Test

For the stress relaxation test, we impose

$$\gamma = \gamma_0 H(t) \qquad \text{at } t = 0 \tag{7.154}$$

By substituting (1.754) into (7.148), we get

$$\sigma + \tau \frac{d\sigma}{dt} = 0 \tag{7.155}$$

Integrating (7.155) once and setting $\sigma = \sigma_0$ at $t = 0$ results in

$$\sigma(t) = \sigma_0 e^{-t/\tau} \tag{7.156}$$

Substitution of the initial condition (7.150) into (7.156) gives the relaxation modulus:

$$G_1(t) = \frac{\sigma(t)}{2\gamma_0} = \mu e^{-t/\tau} \tag{7.157}$$

This relaxation behavior is illustrated in Fig. 7.7(a).

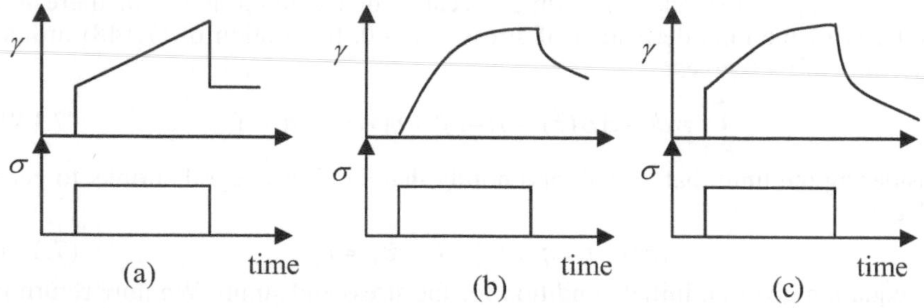

Figure 7.6. Creep function of (a) Maxwell, (b) Kelvin–Voigt, (c) standard linear solid (after Fung, 1965)

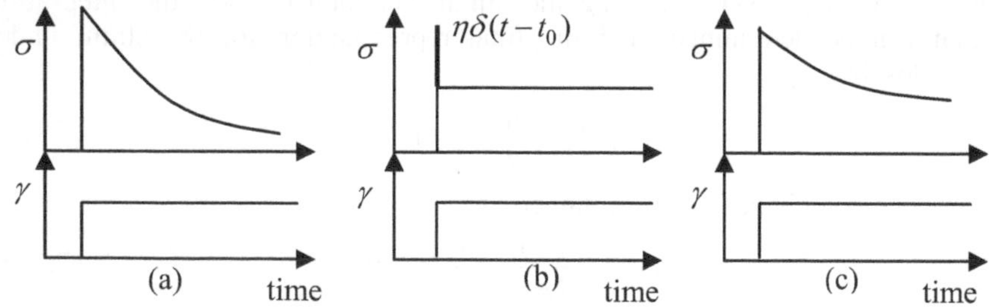

Figure 7.7. Relaxation function of (a) Maxwell, (b) Kelvin–Voigt, (c) standard linear solid (after Fung, 1965)

7.7.2 Kelvin-Voigt Material

7.7.2.1 Creeping Test

From (7.43), the stress-strain law for the Kelvin–Voigt model is

$$\sigma = (\mu\gamma + \eta\frac{d\gamma}{dt}) \tag{7.158}$$

Using the creeping test condition in (7.158), we have

$$\frac{\sigma_0}{\mu} = \gamma + \tau\frac{d\gamma}{dt} \tag{7.159}$$

where τ is again the relaxation time. It is straightforward to show, by back-substitution, that γ has the following solution form:

$$\gamma = \frac{\sigma_0}{\mu} + Ce^{-t/\tau} \tag{7.160}$$

Substitution of the initial condition $\gamma(0^+) = 0$ gives the final solution

$$J_1(t) = \frac{\gamma(t)}{2\sigma_0} = \frac{1}{2\mu}(1 - e^{-t/\tau}) \tag{7.161}$$

This function is plotted in Fig. 7.6(b). According to the creeping test, a Kelvin–Voigt solid does not undergo an instantaneous strain response but instead strain grows exponentially with time.

7.7.2.2 Relaxation Test

Imagine that after we started the creeping test considered in the previous section, we suddenly stop the strain at $t = t_1$ and keep it at a constant value for $t > t_1$:

$$\gamma = \gamma_0 \qquad \text{for} \quad t > t_1 \tag{7.162}$$

By substituting (7.162) into (7.161) and (7.160), we get

$$\sigma(t_1) = \sigma_0(1 - e^{-t_1/\tau}) \tag{7.163}$$

and from (7.161) with a corresponding strain of

$$\gamma(t_1) = \frac{\sigma_0}{\mu}(1 - e^{-t_1/\tau}) \tag{7.164}$$

We now define the Dirac delta function and observe its relation with the Heaviside step function before we proceed. As shown in Fig. 7.8, the derivative of the Heaviside step function is everywhere zero except at the origin where its differentiation becomes infinite. This is precisely the definition of the Dirac delta function. Mathematically, the Dirac delta function is defined as

$$\delta(t) = 0 \quad \text{for } t \neq 0 \tag{7.165}$$

$$\delta(t) = +\infty \quad \text{for } t = 0 \tag{7.166}$$

$$\int_{-\infty}^{+\infty} \delta(t)dt = \int_{0^-}^{0^+} \delta(t)dt = 1 \tag{7.167}$$

$$\frac{dH(t)}{dt} = \delta(t) \tag{7.168}$$

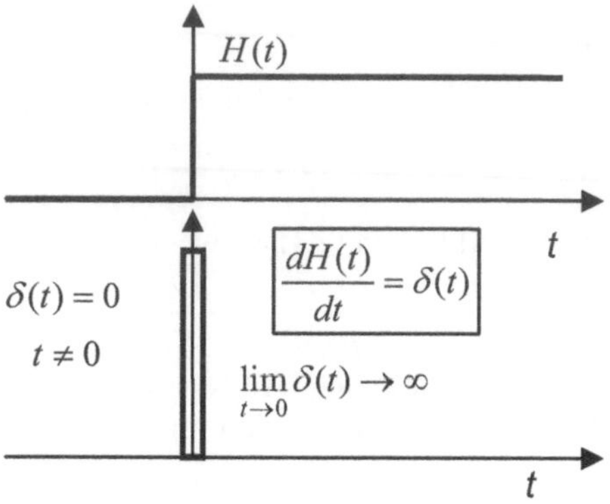

Figure 7.8. Dirac delta function and Heaviside step function

For any arbitrary continuous function *f(t)* at $t = 0$, we have

$$\int_{-\varepsilon}^{+\varepsilon} f(t)\delta(t)dt = f(0) \tag{7.169}$$

where $\varepsilon > 0$. The distribution sense of the Dirac delta function is defined in (7.167) while its relation with the Heaviside step function is given in (7.168).

We now return to (7.164) and consider the following limit as $t_1 \to 0$:

$$\gamma_1 = \gamma(t_1) = \frac{\sigma_0}{\mu}(1 - e^{-t_1/\tau}) = \frac{\sigma_0}{\mu}[1 - 1 + \frac{t_1}{\tau} - \frac{1}{2}(\frac{t_1}{\tau})^2 + ...] \approx \frac{\sigma_0 t_1}{\eta} \tag{7.170}$$

This gives

$$\gamma_1 = \frac{\sigma_0 t_1}{\eta} \tag{7.171}$$

Therefore, when $t_1 \to 0$, we must have $\sigma_0 \to \infty$. However, right after t_1 we have imposed a constant strain of γ_1. Subsequently, we can set the strain

$$\sigma(t) = \eta\gamma_1\delta(t) + \mu\gamma_1 H(t) \tag{7.172}$$

Finally, we have

$$G_1(t) = \frac{\sigma(t)}{2\gamma_1} = \eta\delta(t) + \mu H(t) \tag{7.173}$$

This relaxation function is plotted in Fig. 7.7(b). This again may not be realistic since the load cell used in the laboratory has finite capacity, and thus (7.172) cannot be verified easily by experiments.

7.7.3 Three-Parameter Model A or Standard Linear Solid

7.7.3.1 Creeping Test

From (7.45), the stress-strain law for the three-parameter model A or standard linear solid is

$$(1+\tau_G \frac{\partial}{\partial t})\sigma = \frac{\mu_1 \mu_2}{\mu_1 + \mu_2}\gamma + \mu_1 \tau_G \frac{\partial \gamma}{\partial t} \tag{7.174}$$

where τ_G is defined in (7.145). Using (7.143)–(7.145), we first rewrite (7.174) in the following form:

$$(1+\tau_G \frac{\partial}{\partial t})\sigma = \mu_\infty \gamma + \mu_0 \tau_G \frac{\partial \gamma}{\partial t} \tag{7.175}$$

Taking the Laplace transform of the boundary condition (7.147) for the creeping test, we have

$$\hat{\sigma}(s) = \frac{\sigma_0}{s} \tag{7.176}$$

Taking the Laplace transform of (7.175), we have

$$(1+\tau_G s)\hat{\sigma} = (\mu_\infty + \mu_0 \tau_G s)\hat{\gamma} \tag{7.177}$$

Then, substitution of (7.176) into (7.177) leads to

$$(\frac{1}{s} + \tau_G)\sigma_0 = (\mu_\infty + \mu_0 \tau_G s)\hat{\gamma} \tag{7.178}$$

Rearrangement of (7.178) results in

$$\hat{\gamma} = \sigma_0 \frac{(1+s\tau_G)}{s(\mu_\infty + \mu_0 \tau_G s)} = (\frac{\sigma_0}{\mu_0 \tau_G}) \frac{(1+s\tau_G)}{s[\mu_\infty / (\mu_0 \tau_G) + s]} \tag{7.179}$$

Now, we define another relaxation time as:

$$\tau_2 = \frac{\eta_2}{\mu_2} = \frac{\mu_0 \tau_G}{\mu_\infty} \tag{7.180}$$

With this second relaxation time, (7.179) becomes

$$\hat{\gamma} = (\frac{\sigma_0}{\mu_0 \tau_G}) \frac{(1+s\tau_G)}{s(1/\tau_2 + s)} = (\frac{\sigma_0}{\mu_0 \tau_G}) \left[\frac{1}{s(1/\tau_2 + s)} + \frac{\tau_G}{1/\tau_2 + s} \right] \tag{7.181}$$

Using the formulas given in Appendix B, we obtain the inversion of the Laplace transform of (7.181) as

$$\begin{aligned} \gamma(t) &= (\frac{\sigma_0}{\mu_0 \tau_G}) \left[\tau_2 (1 - e^{-t/\tau_2}) + \tau_G e^{-t/\tau_2} \right] \\ &= (\frac{\sigma_0 \tau_2}{\mu_0 \tau_G}) \left[1 - (1 - \frac{\tau_G}{\tau_2}) e^{-t/\tau_2} \right] \end{aligned} \tag{7.182}$$

Considering the limit case of $t \to 0$ and $t \to \infty$, we have

$$\gamma(0^+) = \frac{\sigma_0}{\mu_0} \tag{7.183}$$

$$\gamma(\infty) = \frac{\sigma_0}{\mu_\infty} \tag{7.184}$$

These physical meanings of μ_0 and μ_∞ again agree with our observations from the result of the Boussinesq problem for viscoelastic half-space considered in Section 7.6.3. The creeping compliance becomes

$$J_1(t)=\frac{\gamma(t)}{2\sigma_0}=\frac{1}{2\mu_\infty}\left[1-(1-\frac{\tau_G}{\tau_2})e^{-t/\tau_2}\right]$$
$$=\frac{1}{2\mu_\infty}+\frac{1}{2\mu_0}(1-\frac{\mu_0}{\mu_\infty})e^{-t/\tau_2} \tag{7.185}$$
$$=\frac{1}{2\mu_1}+\frac{1}{2\mu_2}(1-e^{-t/\tau_2}).$$

The first equation in (7.185) is similar to the form given by Fung (1965), whereas the last one in (7.185) was derived by Chau and Wong (2009) for the standard linear solid. All the equations given in (7.185) are equivalent, depending on what parameters we use to define the constitutive model. The evolution of (7.185) is plotted in Fig. 7.6(c).

7.7.3.2 Relaxation Test

Taking the Laplace transform of the boundary condition (7.154) for the relaxation test, we have

$$\hat{\gamma}(s)=\frac{\gamma_0}{s} \tag{7.186}$$

Substitution of this condition in (7.177) gives the following differential equation:

$$(1+\tau_G s)\hat{\sigma}=(\mu_\infty+\mu_0\tau_G s)\frac{\gamma_0}{s} \tag{7.187}$$

Rearrangement of (7.187) results in

$$\hat{\sigma}(s)=\frac{\gamma_0}{s}\frac{(\mu_\infty+\mu_0\tau_G s)}{(1+s\tau_G)}=\frac{\gamma_0\mu_\infty}{\tau_G}[\frac{(1+\tau_2 s)}{s(1/\tau_G+s)}]$$
$$=\frac{\gamma_0\mu_\infty}{\tau_G}\left[\frac{1}{s(1/\tau_G+s)}+\frac{\tau_2}{(1/\tau_G+s)}\right] \tag{7.188}$$

Using the formulas given in Appendix A, we obtain the inversion of the Laplace transform of (7.188) as

$$\sigma(t)=(\frac{\gamma_0\mu_\infty}{\tau_G})\left[\tau_G(1-e^{-t/\tau_G})+\tau_2 e^{-t/\tau_G}\right]$$
$$=\gamma_0\mu_\infty\left[1-(1-\frac{\tau_2}{\tau_G})e^{-t/\tau_G}\right] \tag{7.189}$$

Again, considering the limit case of $t\to 0$ and $t\to\infty$, we have

$$\sigma(0^+)=\gamma_0\frac{\mu_\infty\tau_2}{\tau_G}=\gamma_0\mu_0 \tag{7.190}$$

$$\sigma(\infty)=\mu_\infty\gamma_0 \tag{7.191}$$

The physical meanings of μ_0 and μ_∞ in (7.190) and (7.191) are obvious and are similar to those by (7.183) and (7.184). From (7.189), the relaxation modulus can be obtained as

$$G_1(t) = \frac{2\sigma(t)}{\gamma_0} = 2\mu_\infty\left[1-(1-\frac{\tau_2}{\tau_G})e^{-t/\tau_G}\right] = 2\mu_\infty\left[1-(1-\frac{\mu_0}{\mu_\infty})e^{-t/\tau_G}\right]$$
$$= 2\mu_1\left[1-\frac{\mu_1}{(\mu_1+\mu_2)}(1-e^{-t/\tau_G})\right] \tag{7.192}$$

The first equation in (7.192) is the same as that given by Fung (1965), whereas the last one in (7.192) is the same as that given by Flugge (1967). This function is plotted in Fig. 7.7(c).

7.7.3.3 Relaxation Test in Compression

Although most of the formulas available in textbooks are for shear relaxation or creeping test, in reality shear test is seldom conducted because of its complexity. More often, the uniaxial compression test is conducted, for creeping or for relaxation. Therefore, it is important to relate the uniaxial compression results with our shear model. For the standard linear solid or generalized Kelvin solid, the uniaxial relaxation modulus is given as (Lama and Vutukuri, 1978)

$$E(t) = E_0 - (E_0 - E_\infty)[1-\exp(-t/\tau_E)] \tag{7.193}$$

where

$$\tau_E = (E_\infty / E_0)\tau_2 \tag{7.194}$$

Taking the Laplace transform of (7.193) yields

$$\hat{E}(s) = \frac{(1+s\tau_E)E_0 - E_0 + E_\infty}{s(1+s\tau_E)} \tag{7.195}$$

Using the standard relation between the relaxation modulus and creep compliance in the transform space (Flugge, 1967 or see (7.71)),

$$\hat{C}(s)\hat{E}(s) = \frac{1}{s^2}, \tag{7.196}$$

we can find that

$$\hat{C}(s) = \frac{1}{E_0 s}\left\{1+\frac{1-E_\infty/E_0}{E_\infty/E_0 + s\tau_E}\right\} \tag{7.197}$$

Taking the inverse Laplace transform of (7.197), we obtain the following results:

$$C(t) = \frac{1}{E_0} + \left(\frac{1}{E_\infty} - \frac{1}{E_0}\right)[1-\exp(-t/\tau_2)] \tag{7.198}$$

where τ_2 is defined in (7.180). This result agrees with that of Chau and Wong (2009).

7.8. CALIBRATION OF THE VISCOELASTIC MODEL

There is little published data on the creeping and relaxation tests on rocks. Renner et al. (2000) conducted creeping tests on the foliated shales from the Swiss Central Alps. These data were used to calibrate the standard linear viscoelastic solid by

Chau and Wong (2009). In this section, we basically follow the approach proposed by Chau and Wong (2009). The complete set of mechanical properties of the so-called S samples drilled perpendicular to the foliation is obtained by uniaxial and triaxial compressions, and by creeping and relaxation tests. This data set includes Poisson's ratio $\nu = 0.18 \pm 0.07$ (13); Young's modulus $E_0 = 16 \pm 7$ GPa (18); and uniaxial compressive strength $\sigma_{max} = 73 \pm 33$ MPa (18) (the value following the "±" is the standard deviation, and the number in the bracket is the number of samples tested by Renner et al., 2000). Therefore, it is evident that elastic properties in different layers of shale differ drastically. To calibrate the standard linear solid, Figure 9(b) of Renner et al. (2000) is used, which is reproduced in Fig. 7.9.

In particular, we first recall from (7.198) that the creep compliance for the standard linear solid under compression is

$$C(t) = \frac{1}{E_0} + \left(\frac{1}{E_\infty} - \frac{1}{E_0} \right) [1 - \exp(-t / \tau_2)] \tag{7.199}$$

where

$$\tau_2 = (\mu_0 / \mu_\infty) \tau_G \tag{7.200}$$

By virtue of (7.199), it is straightforward to show that the difference between any two creeping strain data ($\Delta\varepsilon_1$ and $\Delta\varepsilon_2$) recorded at two different times (t_1 and t_2) can be written as

$$\Delta\varepsilon_2 - \Delta\varepsilon_1 = \left(\frac{1}{E_\infty} - \frac{1}{E_0} \right) \Delta\sigma \left[\exp(-t_1 / \tau_2) - \exp(-t_2 / \tau_2) \right] \tag{7.201}$$

where $\Delta\sigma$ is the stress increment applied in the creeping test. Therefore, if three creeping data are extracted from the curve of Fig. 7.9, we can establish the following equation for the unknown relaxation time τ_2:

$$\gamma_{21}[\exp(-t_1 / \tau_2) - \exp(-t_3 / \tau_2)] - \gamma_{31}[\exp(-t_1 / \tau_2) - \exp(-t_2 / \tau_2)] = 0 \tag{7.202}$$

where

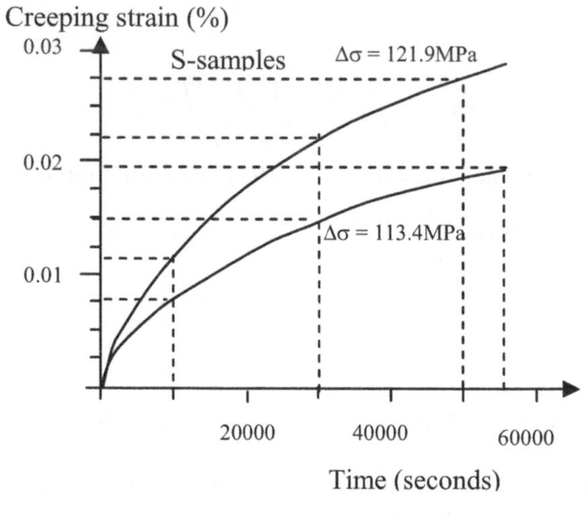

Figure 7.9. Creeping test results for foliated shales from Swiss Central Alps (after Renner et al. (2000) with permission from Springer)

$$\gamma_{21} = (\Delta\varepsilon_2 - \Delta\varepsilon_1) / \Delta\sigma\,, \quad \gamma_{31} = (\Delta\varepsilon_3 - \Delta\varepsilon_1) / \Delta\sigma \tag{7.203}$$

The data for creeping tests with Δσ =113.4 MPa and 121.9 MPa have been extracted for data fitting, and these values are given in Table 7.2. The value of τ_2 can be solved from (7.202) using the standard numerical approach (Press et al., 1992), and then E_∞ can be found from (7.201). Subsequently,

$$\mu_\infty = 3K_0 E_\infty / (9K_0 - E_\infty)\,, \quad \tau_G = \tau_2(\mu_\infty / \mu_0) \tag{7.204}$$

can be found. Since E_0 has been obtained from the non-creeping test, all three parameters of the three-parameter model have been found. Table 7.3 summarizes the calibrated viscoelastic parameters for shales from the Swiss Central Alps for various initial material parameters (E_0 and ν_0).

These combinations of mechanical parameters are motivated by the variations of the basic properties reported in Table 1 of Renner et al. (2000). As illustrated in Table 7.3 for the shales with horizontal foliations from the Swiss Central Alps, the instantaneous shear modulus may vary as much as 2.9 times, the ratio of instantaneous/long-term shear modulus 2.8 times, Poisson's ratio 2.3 times, and the shear relaxation time 2.3 times.

Table 7.2. The actual strain data extracted from Fig. 7.9 for calibration

$\Delta\sigma$ (MPa)	ε	Time (seconds)
113.4	0.0075	10,000
	0.015	30,000
	0.019	55,000
121.9	0.012	10,000
	0.0218	30,000
	0.027	50,000

Table 7.3 Viscoelastic parameters of the standard linear viscoelastic solid calibrated for shales of the Swiss Central Alps from Fig.9(b) of Renner et al. (2000)

$\Delta\sigma$ (MPa)	E_0 (GPa)	ν_0	μ_0 (GPa)	τ_G (s)	μ_∞ / μ_0
113.4	16	0.18	6.78	5552	0.21357
	9	0.11	4.05	8119	0.3123
	9	0.25	3.60	8797	0.3384
	23	0.25	9.20	4335	0.1668
	23	0.11	10.36	3923	0.1509
121.9	16	0.18	6.78	5458	0.1729
	9	0.11	4.05	8176	0.2591
	9	0.25	3.60	8917	0.2825
	23	0.25	9.20	4214	0.1335
	23	0.11	10.36	3798	0.1204

Chau and Wong (2009) adopted these viscoelastic data for their analysis of the time-dependent cracking in Colorado shales. They derived analytically the stress intensity factors (SIFs) of a frictional interfacial crack in a viscoelastic bimaterial subject to the action of a center of dilatation, which modeled the effect of cyclic steam injection during oil sand extraction. This problem will be summarized briefly in the next section.

7.9 VISCOELASTIC CRACK MODELS FOR STEAM INJECTION

Cyclic steam stimulation (CSS), which was invented by accident in Venezuela in 1959, is one of the viable thermal recovery methods for extracting bitumen from the oil sand ores buried in deep overburden of up to 500 m. In Canada, this CSS technique had been widely adopted in Alberta in oil sand extraction. However, at Cold Lake oil sands area in eastern Alberta, cyclic steam injection is known to have caused over 250 well failures (Dusseault et al., 2001). Some of these failures occurred at the base of the Colorado shale, at where natural horizontal fissures or fractures are commonly found (Williams and Burk, 1970; Wong, 1998). These casing impairments have been largely found to be associated with horizontal shear along pre-existing planar features, such as interfaces between two materials of different stiffness, discontinuities, natural fissures, weakness planes, bedding or foliation planes which have previously slipped, the shale–sandstone interfaces, the thin bed of clay shale lying between stiffer beds, clayey zones, joints, or faults (Talebi et al., 1998; Dusseault et al., 2001). The thickness of these planar structures measure from a few millimeters to a few meters.

Field monitoring data show that there may have continuous and substantial creeping movements in the horizontal shale layers (Talebi et al., 1998). It was also reported that some of the failures in casing occurred 2 days after the end of the steaming operation; therefore, there is clearly a time delay process. The elastic theoretical crack model by Chau et al. (2000, 2002) would not model such a time-dependent effect. In view of this, Chau and Wong (2009) derived a simple analytical model to incorporate the effect of viscoelastic behavior of shales and the presence of a bimaterial crack, as shown in Fig. 7.10. This section will briefly summarize their work.

7.9.1 Superposition of Auxiliary Problems I and II

The associated elastic problem is solved by superimposing the solutions of two auxiliary problems as shown in Fig. 7.11: (I) a center of dilatation in a two-dimensional bimaterial; and (II) a bimaterial crack subject to tractions on the crack face that cancel out the stresses induced on the position of the crack by Auxiliary Problem I.

7.9.2 Center of Dilatation in Two-Dimensional Bimaterial

The elastic solution of a center of dilatation in a bimaterial has been given by Carvalho and Curran (1992) in terms of the matrix-vector formulation proposed by

Vijayakumar and Cormack (1987). Chau and Wong (2009) had written their solutions in the Laplace transform by virtue of the correspondence principle of Lee discussed in Section 7.6. More specifically, if the center of dilatation is of the magnitude

$$c_0 = \zeta_0 H(t), \tag{7.205}$$

where $H(t)$ is the Heaviside step function, that is suddenly applied at (h_1, h_2) in medium 1, the nonzero stress and displacement fields can be given in the Laplace transform space:

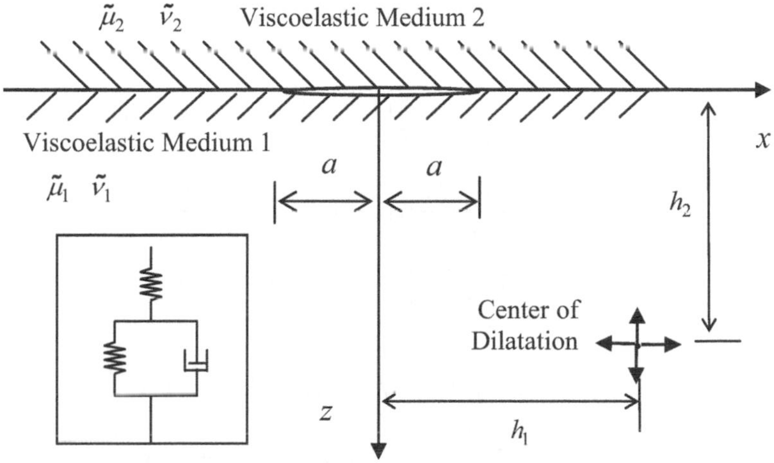

Figure 7.10. Bimaterial crack subject to center of dilatation (after Chau and Wong (2009) with permission from Elsevier)

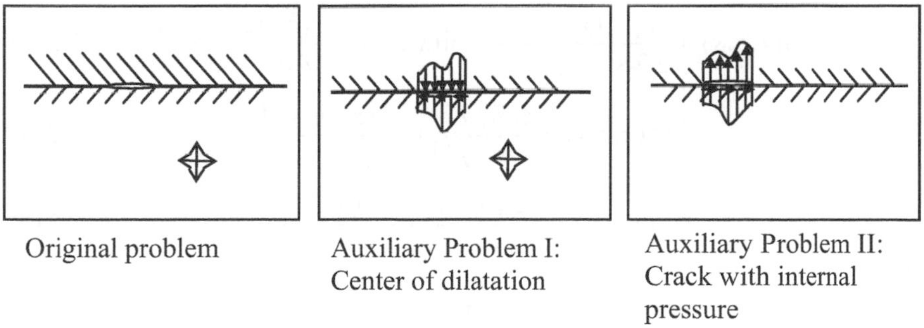

Figure 7.11. Superposition of Auxiliary Problems I and II (after Chau and Wong (2009) with permission from Elsevier)

Medium 1

$$\hat{u}(x,z,s) = \frac{\zeta_0 (x-h_1)h_2}{2\tilde{\mu}_1 s}\left\{\frac{1}{r^2} + \frac{\tilde{p}_1 (x-h_1)h_2}{\bar{r}^2}\left[\tilde{\kappa}_1 - \frac{4(z+h_2)z}{\bar{r}^2}\right]\right\} \tag{7.206}$$

$$\hat{w}(x,z,s) = \frac{\zeta_0 h_2}{2\tilde{\mu}_1 s}\left\{\frac{z-h_2}{r^2} - \frac{\tilde{p}_1}{\bar{r}^2}\left[(z+h_2)(1-4\tilde{\nu}_1) + 2h_2 + \frac{4(z+h_2)^2 z}{\bar{r}^2}\right]\right\} \tag{7.207}$$

$$\begin{aligned}\hat{\sigma}_{zz}(x,z,s) = \frac{\zeta_0 h_2}{s}\Bigg\{&\frac{1}{r^2}\left[1 - \frac{2(z-h_2)^2}{r^2}\right] \\ &+ \frac{\tilde{p}_1}{\bar{r}^2}\left[\frac{2(z+h_2)}{\bar{r}^2}\left(h_2 - 5z + \frac{8z(z+h_2)^2}{\bar{r}^2}\right) - 1\right]\Bigg\}\end{aligned} \tag{7.208}$$

$$\hat{\sigma}_{xz}(x,z,s) = \frac{2\zeta_0 h_2 (x-h_1)}{s}\left\{-\frac{(z-h_2)}{r^2} + \frac{\tilde{p}_1}{\bar{r}^4}\left[\frac{8z(z+h_2)^2}{\bar{r}^2} - (3z+h_2)\right]\right\} \tag{7.209}$$

$$\begin{aligned}\hat{\sigma}_{yy}(x,z,s) = \frac{\zeta_0 \tilde{\nu}_1 \tilde{p}_1 h_2}{s\bar{r}^2}\Bigg\{&2(1-4\tilde{\nu}_1) - \frac{2\tilde{\kappa}_1 (x-h_1)^2}{\bar{r}^2} \\ &+ \frac{2(z+h_2)}{\bar{r}^2}\left[3z + h_2 + 4\tilde{\nu}_1 (z+h_2)\right]\Bigg\}\end{aligned} \tag{7.210}$$

where

$$r = \left[(x-h_1)^2 + (z-h_2)^2\right]^{1/2}, \quad \bar{r} = \left[(x-h_1)^2 + (z+h_2)^2\right]^{1/2} \tag{7.211}$$

$$\tilde{p}_1 = \frac{\tilde{m}-1}{\tilde{m}+\tilde{\kappa}_1}, \quad \tilde{m} = \frac{\tilde{\mu}_1}{\tilde{\mu}_2}, \quad \tilde{\kappa}_i = 3 - 4\tilde{\nu}_i \quad (i=1,2). \tag{7.212}$$

Note that the unit of ζ_0 is N/m, and μ_i and ν_i denote the shear modulus and Poisson ratio of the i-th medium, respectively.

Medium 2

$$\hat{u}(x,z,s) = \frac{\zeta_0 \tilde{p}_2}{2\tilde{\mu}_2 s}\frac{(x-h_1)h_2}{r^2}, \quad \hat{w}(x,z,s) = \frac{\zeta_0 \tilde{p}_2}{2\tilde{\mu}_2 s}\frac{(z-h_2)h_2}{r^2} \tag{7.213}$$

$$\hat{\sigma}_{zz}(x,z,s) = \frac{\zeta_0 \tilde{p}_2}{s}\frac{h_2}{r^2}\left[1 - \frac{2(z-h_2)^2}{r^2}\right] \tag{7.214}$$

$$\hat{\sigma}_{xz}(x,z,s) = -\frac{\zeta_0 \tilde{p}_2}{s}\frac{2h_2 (x-h_1)(z-h_2)}{r^4} \tag{7.215}$$

where

$$\tilde{p}_2 = \frac{1+\tilde{\kappa}_1}{\tilde{m}+\tilde{\kappa}_1} \tag{7.216}$$

It is straightforward to show that these solutions of both media 1 and 2 are continuous on the interface $z = 0$. The solutions for a center of dilatation in a half-plane with a rigidly constrained boundary and with a free surface can be obtained by setting $m = 0$ and $m \to \infty$, respectively. The solution of a center of dilatation in an infinite plane can be obtained by setting m=1. For example, the stresses given in (6) and (7) of Chau et al. (2002) can be recovered as a special case by setting $m \to \infty$.

7.9.3 Stress Intensity Factor of Auxiliary Problem II

The fundamental or point force solution on a finite crack of length $2a$ on the interface of a bimaterial (or so-called dissimilar media) was given by Rice and Sih (1965) and was also compiled in Section 8.3 of Murakami (1987). Applying this fundamental solution, we obtain the solution of Auxiliary Problem II as:

$$\begin{Bmatrix} \hat{K}_I(s) \\ \hat{K}_{II}(s) \end{Bmatrix} = \frac{1}{\sqrt{\pi a}} \int_{-a}^{a} \begin{Bmatrix} \hat{\sigma} H(\hat{\sigma}) \cos \tilde{B} + \hat{\tau} \sin \tilde{B} \\ \hat{\sigma} H(\hat{\sigma}) \sin \tilde{B} - \hat{\tau} \cos \tilde{B} \end{Bmatrix} \sqrt{\frac{a+\xi}{a-\xi}} d\xi \tag{7.217}$$

$$\tilde{B} = \tilde{\varepsilon} \ln \left[\frac{2a(a-\xi)}{a+\xi} \right], \quad \tilde{\varepsilon} = \frac{1}{2\pi} \ln \left[\frac{\tilde{\mu}_2 \tilde{\kappa}_1 + \tilde{\mu}_1}{\tilde{\mu}_1 \tilde{\kappa}_2 + \tilde{\mu}_2} \right] \tag{7.218}$$

The normal and shear stresses are defined as

$$\hat{\sigma}(\xi) = \hat{\sigma}_{zz}(\xi, 0, s) - \gamma d \tag{7.219}$$

$$\begin{aligned} \hat{\tau}(\xi) &= \hat{\sigma}_{xz}(\xi,0,s) - H(\hat{\sigma}_{xz}) |\hat{\sigma}| \tan\phi && \text{if } \hat{\sigma} < 0 \text{ and } |\hat{\sigma}_{xz}(\xi,0,s)| > |\hat{\sigma}| \tan\phi \\ &= \hat{\sigma}_{xz}(\xi,0,s) && \text{if } \hat{\sigma} > 0 \\ &= 0 && \text{if } \hat{\sigma} < 0 \text{ and } |\hat{\sigma}_{xz}(\xi,0,s)| < |\hat{\sigma}| \tan\phi \end{aligned} \tag{7.220}$$

where γ is the unit weight of medium 2, d is the depth of the bimaterial crack from the ground surface, and ϕ is the friction angle on the crack (note that d is typically a few hundred meters, thus the free ground surface has been neglected in the formulation).

7.9.4 Inverse Laplace Transform

To obtain the stress intensity factor as a function of time, we have to perform the inverse Laplace transform. Books that contain tables of inverse Laplace transform include McCollum and Brown (1965), Nixon (1965), Abramowitz and Stegun (1964), Erdelyi (1954), Spiegel (1965), and Gradshteyn and Ryzhik (1980); and the numerical inversion of the Laplace transform is discussed in length by Bellman et al. (1966). For the solution given in (7.217) in the Laplace transform space, Chau and Wong (2009) used Schapery's (1961, 1967) approximate inverse method to estimate the stress intensity factor:

$$\begin{Bmatrix} K_I(t) \\ K_{II}(t) \end{Bmatrix} = \mathfrak{L}^{-1} \begin{Bmatrix} \hat{K}_I(s) \\ \hat{K}_{II}(s) \end{Bmatrix} \approx \begin{Bmatrix} s\hat{K}_I(s) \\ s\hat{K}_{II}(s) \end{Bmatrix}_{s=\gamma_E/t} \tag{7.221}$$

where $\gamma_E = \exp(-C) \approx 0.56158$ (C is the Euler's constant). The derivation of this formula is given in Appendix B.6.

7.9.5 Numerical Results

Here we only report the effect of overburden, and for other numerical results we refer to Chau and Wong (2009). Figure 7.12 plots the normalized mode I and II stress intensity factors versus the angle θ for various overburden pressure ratios β. The dimensionless parameter β is a ratio of the overburden stress normalized with respect to the strength of the center of dilatation defined as:

$$\beta = \frac{\gamma d}{\zeta_0 / a} \tag{7.222}$$

The material parameters used in the plot are summarized in Table 7.4 below. In addition, the friction angle is $\phi = 15°$ and $r/a = 1$.

Figure 7.12 is for a stiff medium 1 (which contains the center of dilatation) and a soft medium 2. The dash lines are for instantaneous stress intensity factors whereas the solid lines are for the long-term stress intensity factors. If the medium with the center of dilatation is stiffer than the one without the center of dilatation, the long-term stress intensity factors are always larger than those of instantaneous ones. This conclusion is of profound importance because the long-term behavior of the solid will determine whether there is crack propagation (assuming the fracture toughness is independent of time). Therefore, this solution potentially can explain why there is a delay in the failure after the steam injection is stopped, as remarked earlier.

Inclinometer surveys indicate that the localized shear displacements on weak bedding planes are on the order of 10 cm and in some cases larger than 20cm (Dusseault et al., 2001). Such monitoring can also provide a check on the potential problem with future fracture propagation. By virtue of the present bimaterial crack model, the relative crack surface sliding displacement can be estimated as (e.g., Sun and Qian, 1998)

$$\Delta u(r) = \left[\frac{K_{II} \gamma_0 \sin(\lambda \pi)}{2(1-\lambda)(2\pi)^{\lambda}} \right] r^{1-\lambda} \tag{7.223}$$

where λ is roughly 1/2 (as we have neglected the frictional effect) and γ_0 is

$$\gamma_0 = \left[\frac{\kappa_1(1-\beta_0)+(1+\beta_0)}{2\mu_1} \right] + \left[\frac{\kappa_2(1+\beta_0)+(1-\beta_0)}{2\mu_2} \right] \tag{7.224}$$

where β_0 is the Dundurs parameter defined as (Dundurs, 1969)

$$\beta_0 = \frac{\mu_1(\kappa_2 - 1) - \mu_2(\kappa_1 - 1)}{\mu_1(\kappa_2 + 1) + \mu_2(\kappa_1 + 1)} \tag{7.225}$$

$$\kappa_i = 3 - 4\nu_i \tag{7.226}$$

Table 7.4 Material parameters for Fig. 7.12

Material	ν_0	μ_0 (GPa)	τ_G (s)	μ_∞ / μ_0
Medium 1	0.11	10.36	3798	0.1204
Medium 2	0.25	3.6	8925.3	0.3384

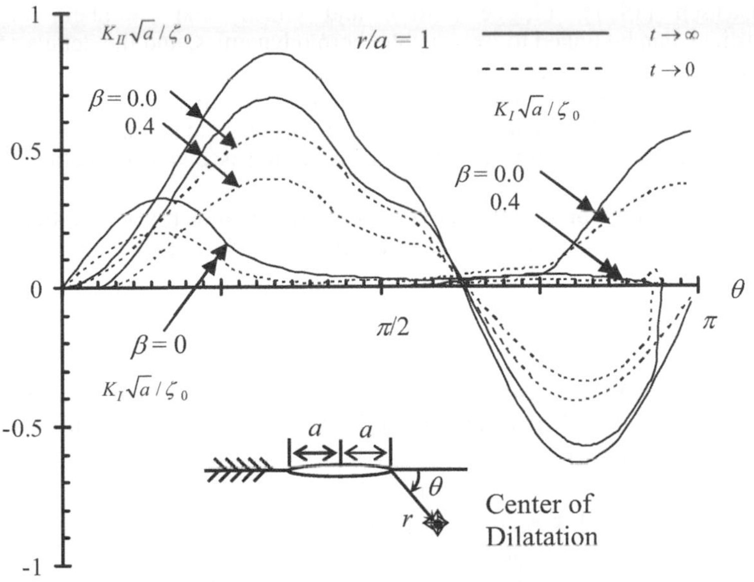

Figure 7.12. The normalized mode I and II stress intensity factors vs. the angle θ for various overburden pressure ratios β (after Chau and Wong (2009) with permission from Elsevier)

The maximum sliding displacement that can be sustained in the bimaterial crack before crack propagation can be estimated by setting K_{II} equal to K_{IIC} (the mode II fracture toughness) and r equal to a in (7.223). This is of course an approximate solution, since, strictly speaking, (7.223) is valid only near the crack tip. Nevertheless, for the case of nearly incompressible soft rocks with $\mu_1 = \mu_2 = 0.8$ GPa, $a = 100$ m, $\nu_1 = \nu_2 = 0.49$, shear fracture toughness being 1 MPa$\sqrt{\text{m}}$, Δu is in the order of 10 mm, which may be small but should not be too difficult for a sensitive inclinometer to pick up.

7.10 SUMMARY AND FURTHER READING

In this chapter, we present a brief summary of the theory of viscoelasticity. The application of viscoelasticity to time-dependent cracking induced by steam stimulations is considered. In particular, we discuss viscoelastic formulation in terms of Boltzmann's integral, differential form, and mechanical string-dashpot models. Spring-dashpot models include the Maxwell model, the Voigt–Kelvin model, and four three-parameter models. The use of Laplace transform and correspondence principle is discussed by considering the Boussinesq problem for various viscous elastic half-spaces. Experimental calibration is discussed in terms of creeping tests and relaxation tests. A real example of parameter calibration is given for shales of the Swiss Central Alps by using the data of Renner et al. (2000).

We refer beginning readers to the introductory discussion on viscoelasticity by Mase (1964) and Mase and Mase (1999). More in-depth coverage can be found in Christensen (1971), Flugge (1967), and Gurtin and Sternberg (1962). Viscoelasticity can be linked to irreversible thermodynamics, and the details of this theory can be found in Fung (1965). Regarding the approximate technique of taking the inverse Laplace transform, we refer to Schapery (1961, 1967) and Schapery and Park (1999). Wave propagation in viscoelastic solids is discussed in full detail in Borcherdt (2009), and a brief discussion is given in Section 9.9 of Chapter 9.

More advanced topics on thermoviscoelasticity, nonlinear viscoelasticity, and other general theorems can be found in Christensen (1971) and Haddad (1995).

7.11 PROBLEMS

Problem 7.1 Derive the constitutive model in terms of stress σ and strain γ and find the following equivalent moduli of the Kelvin–Voigt and Maxwell models in series (shown in Fig. 7.13):

$$2\tilde{\mu},\quad \tilde{\nu},\quad \tilde{E},\quad \tilde{\lambda} \tag{7.227}$$

This model is also called Burgers viscoelastic material.

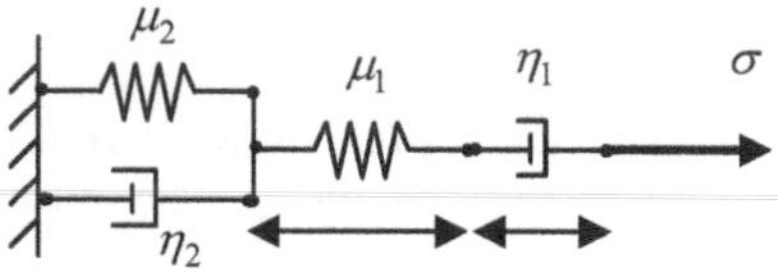

Figure 7.13 Kelvin and Maxwell models in series (Burgers material)

Problem 7.2 Find the creeping compliance $J_1(t)$ of the model shown in Fig. 7.13.

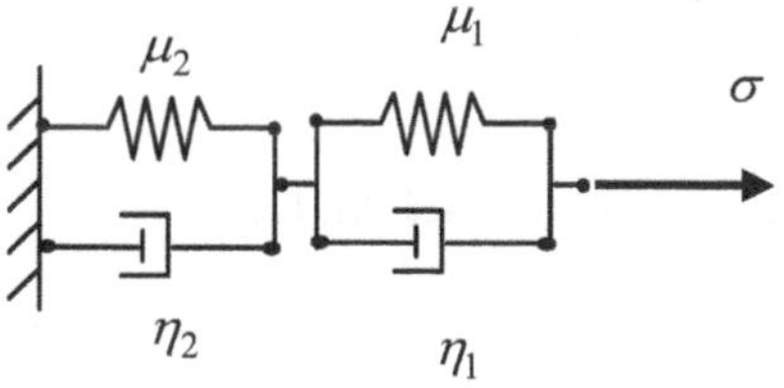

Figure 7.14 Double Kelvin model

Problem 7.3 Derive the constitutive model in terms of stress σ and strain γ and find the following equivalent moduli of the double Kelvin model shown in Fig. 7.14:

$$2\tilde{\mu},\quad \tilde{\nu},\quad \tilde{E},\quad \tilde{\lambda} \tag{7.228}$$

Problem 7.4 Find the relaxation function $G_1(t)$ of the model shown in Fig. 7.14.

Problem 7.5 Use the correspondence principle to find the solutions of the decaying point force applied on the surface of the viscoelastic half-space shown in Fig. 7.15. The elastic solution is given in (4.152) and (4.153) of Chapter 4. Assume the viscoelastic solid is a Kelvin–Voigt solid.

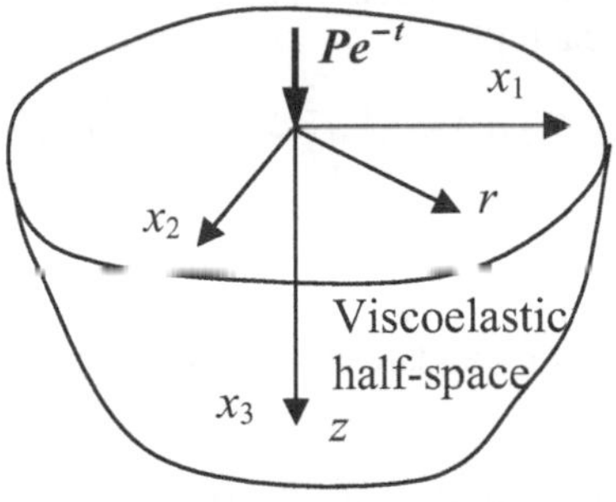

Figure 7.15 A viscoelastic half-space subject to time decaying force

Problem 7.6 Solve the problem shown in Fig. 7.15 again for a three-parameter model A (or standard linear solid).

Problem 7.7 Solve the problem shown in Fig. 7.15 again for a three-parameter model B.

Problem 7.8 Solve the problem shown in Fig. 7.15 again for a three-parameter model C.

Problem 7.9 Solve the problem shown in Fig. 7.15 again for a three-parameter model D.

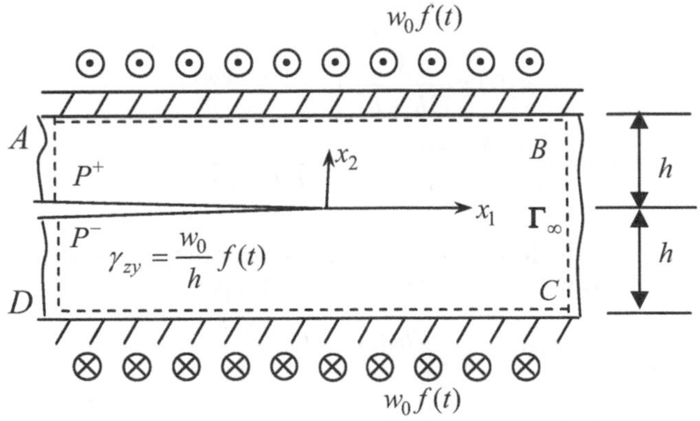

Figure 7.16 A viscoelastic slab of thickness $2h$ and infinite dimensions in the x_1- and x_3-directions and containing a semi-infinite crack

Problem 7.10 This is an extension of Problem 6.4. The layer is now viscoelastic with a relaxation function of $G_1(t)$ under shear deformation. The upper and lower faces are bonded to two rigid bodies and are displaced by $w_0 f(t)$ along the positive x_3-direction on the upper surface and by $w_0 f(t)$ along negative x_3 direction on the lower surface. The time function starts at $t = 0$. What is the expression for the mode III stress intensity factor?

Answer:
$$K_{III}(t) = \frac{w_0}{2}\sqrt{\frac{2}{h}}\mathfrak{L}^{-1}[s\hat{G}_1(s)\hat{f}(s)] \tag{7.229}$$

Problem 7.11 For Problem 7.10, the displacement is applied as a sudden applied function as $w_0 H(t)$ (where $H(t)$ is the Heaviside step function). The rock is modeled by the model A (or standard linear solid). Find the time-dependent stress intensity factor.

Problem 7.12 Show the validity of (7.47), (7.49), and (7.51) for the three parameter models B, C, and D shown in Fig. 7.3(b–d).

Problem 7.13 Derive the following functions involved in the displacements given in (7.129) for the Boussinesq problem of a Maxwell half-space:

$$f_1(t) = \frac{1}{\eta}(t+\tau) \tag{7.230}$$

$$f_2(t) = \frac{1}{(3K\tau+\eta)K}\{(3K\tau+\eta) - \eta e^{-3Kt/(3K\tau+\eta)}\} \tag{7.231}$$

$$f_3(t) = \frac{1}{2\eta(3K\tau+\eta)K}\{K(3K\tau+\eta)t + (3K\tau+\eta)(K\tau+\eta) - \eta^2 e^{-3Kt/(3K\tau+\eta)}\} \tag{7.232}$$

Problem 7.14 Derive the following functions involved in the displacements given in (7.135) and (7.136) for the Boussinesq problem of a Kelvin–Voigt half-space.

$$g_1(t) = \frac{1}{\mu}(1 - e^{-\mu t/\eta}) \tag{7.233}$$

$$g_2(t) = \frac{3}{(3K+\mu)}\{1 - e^{-(3K+\mu)t/\eta}\} \tag{7.234}$$

$$g_3(t) = \frac{1}{2\mu}\{\frac{3K+4\mu}{(3K+\mu)} - e^{-\mu t/\eta} - \frac{3\mu}{(3K+\mu)}e^{-(3K+\mu)t/\eta}\} \tag{7.235}$$

Problem 7.15 Derive the following displacements for the Boussinesq problem of a standard linear solid (three-paramter model A) half-space:

$$u_r(t) = \frac{P_0}{4\pi\mu R}\{\frac{rz}{R^2}h_1(t) - \frac{r}{R+z}h_2(t)\}H(t) \tag{7.236}$$

$$u_z(t) = \frac{P_0}{4\pi\mu R}\{2h_3(t) + \frac{z^2}{R^2}h_1(t)\}H(t) \tag{7.237}$$

$$h_1(t) = \frac{\mu_1 + \mu_2}{\mu_1 \mu_2} - \frac{1}{\mu_2} e^{-\mu_2 t/\eta_2} \tag{7.238}$$

$$h_2(t) = \frac{3}{3K(\mu_1 + \mu_2) + \mu_1 \mu_2} \{(\mu_1 + \mu_2) - \frac{\mu_1^2}{(3K + \mu_1)} e^{-[3K(\mu_1 + \mu_2) + \mu_1 \mu_2]t/[\eta_2 (3K + \mu_1)]}\} \tag{7.239}$$

$$h_3(t) = \frac{3K + 4\mu_1}{2\mu_1 (3K + \mu_1)} \{X_1 - X_2 e^{-\mu_2 t/\eta_2} + X_3 e^{-[3K(\mu_1 + \mu_2) + \mu_1 \mu_2]t/[\eta_2 (3K + \mu_1)]}\} \tag{7.240}$$

$$X_1 = (\frac{\mu_1 + \mu_2}{\mu_2})(\frac{3K + \mu_1}{3K + 4\mu_1})[\frac{3K(\mu_1 + \mu_2) + 4\mu_2 \mu_1}{3K(\mu_1 + \mu_2) + \mu_2 \mu_1}] \tag{7.241}$$

$$X_2 = \frac{\mu_2}{\mu_1}(\frac{3K + \mu_1}{3K + 4\mu_1}) \tag{7.242}$$

$$X_3 = 1 - (\frac{\mu_1 + \mu_2}{\mu_2})(\frac{3K + \mu_1}{3K + 4\mu_1})[\frac{3K(\mu_1 + \mu_2) + 4\mu_2 \mu_1}{3K(\mu_1 + \mu_2) + \mu_2 \mu_1}] + \frac{\mu_2}{\mu_1}(\frac{3K + \mu_1}{3K + 4\mu_1}) \tag{7.243}$$

CHAPTER EIGHT

Linear Elastic Fluid-Infiltrated Solids and Poroelasticity

8.1 INTRODUCTION

Stress-induced flow of interstitial fluid in porous solids and the subsequent solid deformations have been used to explain a variety of phenomena observed in geophysics and in engineering practice. The one-dimensional consolidation theory of Terzaghi has been used successfully for predicting subsidence in clay under surface loading (Terzaghi, 1943). However, when the surface loading is of finite size, the clay layer is not thin and not confined, or the clay is of nonuniform thickness, three-dimensional deformation becomes inevitable. In these situations, Terzaghi's 1-D theory is inapplicable. Biot (1941) extended the formulation for poroelastic solids to three-dimensional and retained the coupling between solid deformation and pore-space pressure. The solutions of Biot theory have been studied by Derski (1964, 1965), McNamee and Gibson (1960a,b), and Schiffman and Fungaroli (1965). Although there have been formulations of mixture theory to model porous geomaterials and they were formulated following more rigorous procedure from thermodynamics (e.g., de Boer, 2000), in practice they do not offer any advantage over Biot's theory (Detourney and Cheng, 1993).

Mathematically, Biot's theory of poroelasticity is precisely analogous to linear coupled thermoelasticty (e.g., Carslaw and Jaeger, 1959), which is a very well-developed area with an abundance of available solutions in the literature. Unfortunately, in thermoelasticity the adiabatic (no heat loss in solids) and isothermal (no temperature change in solids) Poisson's ratios of solids under thermal effects are typically indistinguishable, so that the coupling terms between deformation fields and heat conduction are commonly and justifiably dropped. Most of the available solutions are obtained for uncoupled thermelasticity; that is, heat diffusion first can be solved independent of the deformation field of the solids. However, in poroelasticity the undrained and drained Poisson's ratios differ considerably. Unless the pore fluid is highly compressible, the diffusion process of fluid cannot be uncoupled from the deformation of geomaterials. Therefore, most of the solutions from thermoelasticity cannot be converted to poroelasticity. On the contrary, all solutions for poroelasticity can be employed to consider thermal effect in solids. To solve the coupled equations of Biot's poroelasticity, displacement functions have been proposed by McNamee and Gibson (1960a,b) for axisymmetric problems, and by Schiffman and Fungaroli (1965) for anti-symmetric problems. Such a displacement function approach will be discussed in this chapter, in conjunction with the Laplace–Hankel transform method given by Chau (1996) and others.

In Biot's original formulation, there are two moduli, one relating the fluid mass variation by pore-pressure change and the other relating the fluid mass variation with the mean stress of the solid phase. These parameters are not readily observed in standard experiments. Rice and Cleary (1976) reinterpreted these moduli in terms of the well-understood parameters of the undrained Poisson's ratio and the Skempton pore pressure coefficient B. Rudnicki (1985, 1986) presented a similar constitutive form but in favor of the undrained Lamé's constant and a pore water pressure coefficient. These various forms of Biot's theory will be summarized. The fundamental point forces and fluid point source solution by Cleary (1977) will be summarized (with corrections by Rudnicki, 1981) and another more compact form derived by Rudnicki (1986) will also be discussed in detail. This solution forms a useful basis for generating the center of dilatation, the center of shear, concentrated couples, and double couples, with the latter related to slip on the fault plane in fluid-infiltrated porous rocks during earthquakes.

Figure 8.1 illustrates schematically the pore distributions of a porous rock with specified volume V and S. The porous solid is assumed fully saturated. The total volume of the porous solid consists of volume of solid and volume of pore (usually referred to as volume of void in soil mechanics). The pore volume can further be decomposed into interstitial pores and isolated pores. The solid behavior should also include the effect of fully saturated isolated pores depicted as V^I in Fig. 8.1. The other interstitial pore space V^F is assumed completely connected so that fluid diffusion can occur. The porosity intersects the surface S at points shown as hatched lines in Fig. 8.1, the union of which is S^F. Rice and Cleary (1976) also interpreted undrained deformation to mean the loading is applied over a time scale which is short enough not to allow loss of fluid due to diffusion in a global sense, but not too short to allow local pore pressure equilibrium within a typical "point element" within the continuum. In doing so, there is no need to consider another theory for the nonequilibrating situation of pore pressures within pore networks locally.

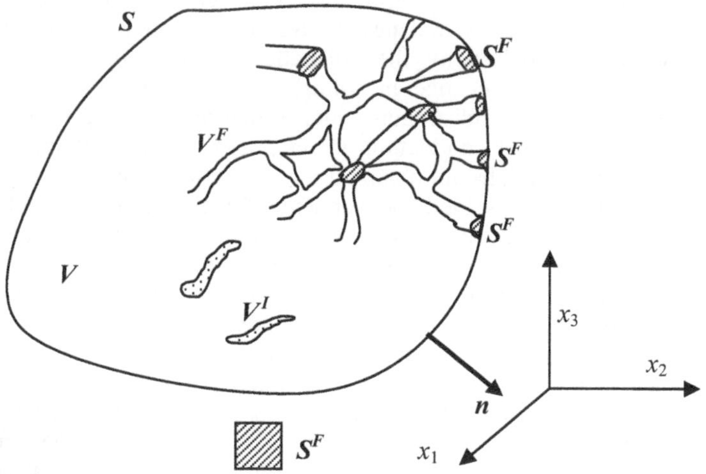

Figure 8.1 Schematic diagram of porous rock of volume V and surface S. The porosity of interconnected pore space is V^F and isolated pore-space is V^I

Figure 8.2 illustrates schematically the pore distributions of a typical granular soil, composed of particles of various sizes. The pore is again fully interconnected and the solid is fully saturated. The formulation to be discussed applies equally to porous rocks shown in Fig. 8.1 and to saturated soils shown in Fig. 8.2.

The schematic diagram of a fully saturated clay is illustrated in Fig. 8.3, showing both dispersed and flocculated clay plate-like particles, and clay assemblages in bookhouse forms popping up the pore-spaces between plate-like particles and assemblages. This kind of soft clay is highly deformable and small strain linear poroelasticity may not be applicable, although Terzaghi's classical formulation was actually motivated by such applications.

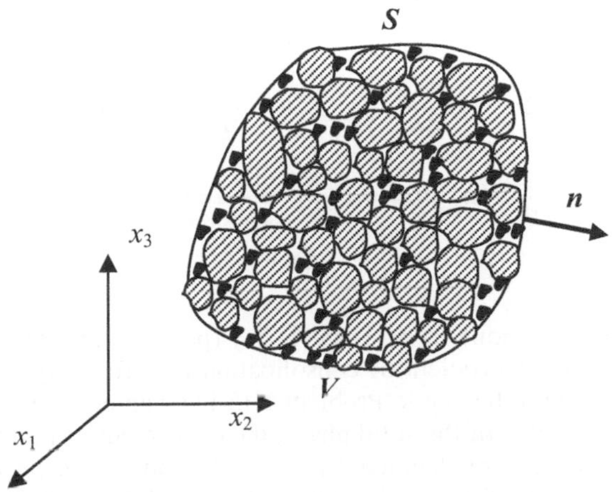

Figure 8.2 Schematic diagram of porous soil of volume *V* with surface *S* and normal *n*; The porosity between contacting particles is interconnected

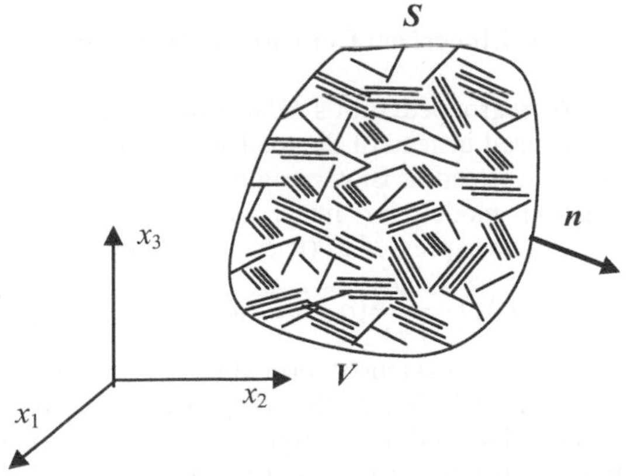

Figure 8.3 Schematic diagram of saturated clay of volume *V* with surface *S*

8.2 BIOT'S THEORY OF POROELASTICITY

8.2.1 McNamee and Gibson's Cylindrical Coordinates Form

Biot's (1941) three-dimensional consolidation theory has been written in various different forms by others, including McNamee and Gibson (1960a,b), Rice and Cleary (1976), Detournay and Cheng (1993), and Rudnicki (1985, 1986). In cylindrical coordinates, the following form is given by McNamee and Gibson (1960a,b).

In particular, the equilibrium equations in terms of displacements ($\boldsymbol{u}$) and excess pore water pressure (p) are

$$\nabla^2 u_r + (2\eta-1)\frac{\partial e}{\partial r} - \frac{1}{r}\left(\frac{2}{r}\frac{\partial u_\theta}{\partial \theta} + \frac{u_r}{r}\right) + \frac{\partial p}{G\partial r} = 0 \tag{8.1}$$

$$\nabla^2 u_\theta + (2\eta-1)\frac{\partial e}{r\partial \theta} - \frac{1}{r^2}\left(u_\theta - 2\frac{\partial u_r}{\partial \theta}\right) + \frac{\partial p}{Gr\partial \theta} = 0 \tag{8.2}$$

$$\nabla^2 u_z + (2\eta-1)\frac{\partial e}{\partial z} + \frac{\partial p}{G\partial z} = 0 \tag{8.3}$$

$$c\nabla^2 e = \frac{\partial e}{\partial t} \tag{8.4}$$

where $e = \boldsymbol{\nabla}\cdot\boldsymbol{u}$ (the dilatation), $\eta = (1-\nu)/(1-2\nu)$ (poroelastic stress coefficient), and $c = (2G\eta k)/\gamma_w$ (the coefficient of consolidation or diffusivity). The proof of (8.1)–(8.4) is obvious by referring to Problem 1.13 in Chapter 1. Poisson's ratio of the skeleton, shear modulus of the solid phase, the coefficient of permeability, and the unit weight of water are denoted by ν, G, k, and γ_w, respectively. Since Poisson's ratio ranges from 0 to 1/2 for linear elastic solids, we have $1 \le \eta < \infty$. Note that for axisymmetric deformation, we have $u_\theta = 0$, $u_r = u_r(r,z)$, and $u_z = u_z(r,z)$.

Various forms of the extended Biot's theory, including the compressibilities of fluid and solid constituents, are presented by Rice and Cleary (1976), Rudnicki (1985, 1986), and Detournay and Cheng (1993). They will be introduced next.

8.2.2. Rice-Cleary (1976) Linearized Constitutive Relation

Rice and Cleary (1976) regrouped Biot's (1941) formulation such that effective stress (i.e., $\sigma_{kk} + p\delta_{ij}$) is used instead of the total stress. They further proposed two new material parameters, instead of Biot's original moduli. In particular, the strain tensor of Biot's (1941) model for linear isotropic poroelastic solid under isothermal conditions can be expressed as (Rice and Cleary, 1976)

$$2\mu\varepsilon_{ij} = (\sigma_{ij} + p\delta_{ij}) - \frac{\nu}{1+\nu}(\sigma_{kk} + 3p)\delta_{ij} + \frac{2G}{3}\left(\frac{1}{H} - \frac{1}{K}\right)p\delta_{ij} \tag{8.5}$$

where K is the bulk modulus. This tensor form of (8.5) is given by Rice and Cleary (1976), which is of course equivalent to the component form given in Eq. (2.4) of Biot (1941). In addition, Biot (1941) introduced another modulus, H_1, relating water content change with mean stress and pore water pressure. In the terminology of Rice and Cleary (1976), the water content change can be expressed as

$$v - v_0 = \frac{\sigma_{kk}}{3H_1} + \frac{p}{R} = \frac{1}{3H_1}(\sigma_{kk} + 3p) - \frac{v_0}{K_s''} p \tag{8.6}$$

The first part of (8.6) is from Biot (1941) whereas the second part of (8.6) is the form proposed by Rice and Cleary (1976).

Actually, it can be shown that H in (8.5) is equal to H_1 in (8.6). By following Biot's (1941) argument of the existence of potential energy of the soil, Rice and Cleary (1976) assumed that the following are exact differentials:

$$dW = \sigma_{ij} d\varepsilon_{ij} + p dv = \varepsilon_{ij} d\sigma_{ij} + v dp \tag{8.7}$$

This equation implies the reversibility of the deformation process (recall that we are using poroelasticity instead of poroplasticity). Because they are exact differentials, (8.7) can be written as

$$dW = \frac{\partial W}{\partial \varepsilon_{ij}} d\varepsilon_{ij} + \frac{\partial W}{\partial v} dv = \frac{\partial W}{\partial \sigma_{ij}} d\sigma_{ij} + \frac{\partial W}{\partial p} dp \tag{8.8}$$

Comparing the first part of (8.7) and the first part of (8.8), we have

$$\sigma_{ij} = \frac{\partial W}{\partial \varepsilon_{ij}}, \quad p = \frac{\partial W}{\partial v} \tag{8.9}$$

Taking the partial derivative of the first part of (8.9) with respect to v and the second part of (8.9) with respect to ε_{ij} we have the following equality:

$$\frac{\partial^2 W}{\partial \varepsilon_{ij} \partial v} = \frac{\partial \sigma_{ij}}{\partial v} = \frac{\partial p}{\partial \varepsilon_{ij}} \tag{8.10}$$

Following the same procedure for the second equations of (8.8) and (8.9), alternatively we can have

$$\frac{\partial^2 W}{\partial \sigma_{ij} \partial p} = \frac{\partial v}{\partial \sigma_{ij}} = \frac{\partial \varepsilon_{ij}}{\partial p} \tag{8.11}$$

For linear elastic solids, (8.10) and (8.11) are identical. For the case of isotropic compressions, we can take the trace of (8.5) and then apply either (8.10) or (8.11) to show that

$$H = H_1 \tag{8.12}$$

In view of the formulation in effective stress, Rice and Cleary (1976) introduced the following moduli instead of H and R given in (8.5) and (8.6):

$$\frac{1}{K_s'} = \frac{1}{K} - \frac{1}{H}, \quad \frac{v_0}{K_s''} = \frac{1}{H} - \frac{1}{R} \tag{8.13}$$

For special circumstances stated by Rice and Cleary (1976) both K_s'' and K_s' can be interpreted as bulk moduli of solid constituents.

In addition, Rice and Cleary (1976) regrouped (8.6) in a linearized fashion to give the fluid mass change per volume as

$$\begin{aligned} m - m_0 &= (\rho - \rho_0) v_0 + \rho_0 (v - v_0) \\ &= \rho_0 \frac{v_0}{K_f} p + \frac{\rho_0}{3}\left(\frac{1}{K} - \frac{1}{K_s'}\right)(\sigma_{kk} + 3p) - \rho_0 \frac{v_0}{K_s''} p \end{aligned} \tag{8.14}$$

where the bulk modulus K_f is defined as

$$K_f = \frac{\rho_0 p}{(\rho - \rho_0)} \tag{8.15}$$

Using Skempton's definition of pore pressure coefficient B (Skempton, 1954)

$$\Delta p = -B\frac{\Delta\sigma_{kk}}{3} \tag{8.16}$$

for the "undrained situation" (i.e., with no loss of fluid by diffusion but with local pressure equilibrium), we can set Δm to zero in (8.14) to obtain

$$B = \frac{1/K - 1/K_s'}{v_0 / K_f + 1/K - 1/K_s' - v_0 / K_s''} \tag{8.17}$$

For the undrained Poisson's ratio, we can substitute (8.17) back into (8.5) and observe the following definition of ν_u

$$2\mu\Delta\varepsilon_{ij} \equiv \Delta\sigma_{ij} - \frac{\nu_u}{1+\nu_u}\Delta\sigma_{kk}\delta_{ij} \tag{8.18}$$

to obtain the final result as

$$\nu_u = \frac{3\nu + B(1-2\nu)(1-K/K_s')}{3 - B(1-2\nu)(1-K/K_s')} \tag{8.19}$$

It is obvious that $1/2 \geq \nu_u \geq \nu$. The upper limit corresponds to $B = 1$ and $K/K_s' = 0$, where the lower limit corresponds to highly compressible fluid with $K_f \ll v_0 K$ or $B \approx 0$. With these new terms of B and ν_u, it is more advantageous to rewrite the constitution law (8.5) and mass dependence on stress (8.14) as (Rice and Cleary, 1976)

$$2\mu\varepsilon_{ij} = \sigma_{ij} - \frac{\nu}{1+\nu}\sigma_{kk}\delta_{ij} + \frac{3(\nu_u - \nu)}{B(1+\nu)(1+\nu_u)}p\delta_{ij} \tag{8.20}$$

$$m - m_0 = \frac{3\rho_0(\nu_u - \nu)}{2\mu B(1+\nu)(1+\nu_u)}(\sigma_{kk} + \frac{3}{B}p) \tag{8.21}$$

This particular form can now be easily identified with thermoelasticity. The pore pressure corresponds to a multiple of temperature fluctuation and fluid mass with some multiple of specific entropy per volume. The analogue of undrained response is the adiabatic (or isentropic) deformation and drained response is the isothermal condition. The linear Fourier law of heat conduction corresponds to the following Darcy's law:

$$q_i = -\rho_0\kappa\frac{\partial p}{\partial x_i} \tag{8.22}$$

Substituting (8.20) in the equilibrium equation without body force, we can obtain the compatibility condition:

$$\nabla^2[(1+\nu)\sigma_{ij} - \nu\sigma_{kk}\delta_{ij}] + \sigma_{kk},_{ij} + \frac{3(\nu_u - \nu)}{B(1+\nu_u)}[\nabla^2 p\delta_{ij} + p_{,ij}] = 0 \tag{8.23}$$

Taking the trace of (8.23) provides a useful formula between isotropic compression and pore pressure:

$$\nabla^2[\sigma_{kk} + \frac{6(\nu_u - \nu)}{B(1-\nu)(1+\nu_u)}p] = 0 \tag{8.24}$$

The effect of pressure looks like body force term, but actually it couples with the stress of the solid through mass conservation as:

$$\frac{\partial q_i}{\partial x_i} + \frac{\partial m}{\partial t} = 0 \tag{8.25}$$

Substituting (8.21) and (8.22) into (8.25), we have

$$\kappa \nabla^2 p = \frac{3(\nu_u - \nu)}{2\mu B(1+\nu)(1+\nu_u)} \frac{\partial}{\partial t}(\sigma_{kk} + \frac{3}{B} p) \tag{8.26}$$

Finally, (8.26) can further be simplified in view of (8.24) to

$$c\nabla^2 (\sigma_{kk} + \frac{3}{B} p) = \frac{\partial}{\partial t}(\sigma_{kk} + \frac{3}{B} p) \tag{8.27}$$

where

$$c = \kappa [\frac{2\mu(1-\nu)}{(1-2\nu)}][\frac{B^2 (1+\nu_u)^2 (1-2\nu)}{9(1-\nu_u)(\nu_u - \nu)}] \tag{8.28}$$

This is the coefficient of diffusivity or the coefficient of consolidation. It is obvious from (8.21) that this mass conservation can equally be written as

$$c\nabla^2 m = \frac{\partial m}{\partial t} \tag{8.29}$$

It is clear that Rice and Cleary (1976) favored the stress formulation, instead of Biot's (1941) displacement formulation. Since both B and ν_u can be found by conventional triaxial test results, Rice–Cleary (1976) formulation is more attractive than Biot's original formulation.

8.2.3. Rudnicki's (1986) Constitutive Relation

In his derivation of the fluid mass source and point force solutions for linear elastic diffusive solids, Rudnicki (1986) proposed a slightly different form of the Rice–Cleary formulation by favoring undrained Lamé's constants and the pore pressure coefficient. In particular, long-term drained deformation is determined by the drained constitutive law as

$$\sigma_{ij} = \mu(u_{i,j} + u_{j,i}) + \lambda u_{k,k}\delta_{ij} \tag{8.30}$$

where μ and λ are the drained Lamé's constants. For deformations at other times, we have

$$\sigma_{ij} = \mu(u_{i,j} + u_{j,i}) + \lambda u_{k,k}\delta_{ij} - \zeta p \delta_{ij} \tag{8.31}$$

where

$$\zeta = 1 - \frac{K}{K_s'} \tag{8.32}$$

with K_s' defined in the last section by Rice and Cleary (1976) and K being the drained bulk modulus. The second constitutive law in terms of mass fluid alternation becomes

$$m - m_0 = \zeta \rho_0 [u_{k,k} + \frac{\zeta p}{\lambda_u - \lambda}] \tag{8.33}$$

This mathematical form looks more compact than that of Rice-Cleary (1976). It is straightforward to see that $\lambda < \lambda_u < \infty$. Applying the undrained condition ($m = m_0$) in (8.33) and using (8.31) to eliminate $u_{k,k}$, we have

$$p = -\frac{1}{3} B\sigma_{kk} \tag{8.34}$$

where

$$B = \frac{3(\lambda_u - \lambda)}{\zeta(3\lambda_u + 2\mu)} \tag{8.35}$$

The constitutive model is completed by Darcy's law:

$$q_i = -\rho_0 \kappa \frac{\partial p}{\partial x_i} \tag{8.36}$$

With (8.31), the equilibrium equation with body force becomes:

$$(\lambda + \mu)u_{i,ij} + \mu\nabla^2 u_j - \zeta p_{,j} + F_j = 0 \tag{8.37}$$

This can be rewritten in terms of fluid mass m by virtue of (8.33):

$$(\lambda_u + \mu)u_{i,ij} + \mu\nabla^2 u_j - \frac{\lambda_u - \lambda}{\zeta\rho_0} m_{,j} + F_j = 0 \tag{8.38}$$

Taking the divergence of (8.38) gives

$$\nabla^2[(\lambda_u + 2\mu)u_{i,i} - \frac{\lambda_u - \lambda}{\zeta\rho_0} m] + F_{k,k} = 0 \tag{8.39}$$

For the cases with nonzero fluid mass source, the equation of fluid mass conservation is

$$\boldsymbol{\nabla} \cdot \boldsymbol{q} + \frac{\partial m}{\partial t} = Q(\boldsymbol{x}, t) \tag{8.40}$$

where $Q(\boldsymbol{x},t)$ is the fluid mass source. The conservation equation (8.40) can be written as follows by combining it with Darcy's law (8.36), (8.33), and (8.39):

$$\frac{\partial m}{\partial t} = c\nabla^2 m + \{Q(\boldsymbol{x},t) + [\frac{\rho_0 \kappa(\lambda_u - \lambda)}{\zeta(\lambda_u + 2\mu)}]F_{k,k}\} \tag{8.41}$$

where the coefficient of diffusivity is

$$c = \frac{\kappa(\lambda_u - \lambda)(\lambda + 2\mu)}{\zeta^2(\lambda_u + 2\mu)} \tag{8.42}$$

For the case of no fluid source and negligible body force, (8.29) obtained by Rice and Cleary (1976) is recovered from (8.41). Instead of using B and ν_u , Rudnicki (1986) preferred ζ and λ_u. In addition, displacement formulation is used by Rudnicki (1986) instead of stress formulation.

8.2.4. Rudnicki's (1985) Anisotropic Diffusive Solids

Rudnicki (1985) presented the following tensor form of Biot's (1941) theory for general anisotropic diffusive solids:

$$\sigma_{ij} = L_{ijkl}\varepsilon_{kl} - M_{ij} p \tag{8.43}$$

$$m - m_0 = R_{ij}\varepsilon_{ij} + Qp \tag{8.44}$$

where Q is a scalar. The usual symmetries for L_{ijkl}, M_{ij}, and R_{ij} are observed in view of the symmetric properties of stress and strain. Actually, a somewhat similar form is also found in the Ph.D. thesis of Michael Cleary (1977), but we will stick to Rudnicki's (1985) form here. Because the solid response is elastic, the work increment in the Helmholtz function has the following differential form at constant temperature:

$$d\phi = \sigma_{ij} d\varepsilon_{ij} + \frac{1}{\rho} p dm = \frac{\partial \phi}{\partial \varepsilon_{ij}} d\varepsilon_{ij} + \frac{\partial \phi}{\partial m} dm \tag{8.45}$$

We can change the variable from m to p by using the following Legendre transformation:

$$\Phi_1 = \phi - \frac{1}{\rho} pm \tag{8.46}$$

Taking the differential of (8.46) gives

$$\begin{aligned} d\Phi_1 &= d\phi - \frac{1}{\rho} m dp - \frac{1}{\rho} p dm \\ &= \sigma_{ij} d\varepsilon_{ij} + \frac{1}{\rho} p dm - \frac{1}{\rho} m dp - \frac{1}{\rho} p dm \\ &= \sigma_{ij} d\varepsilon_{ij} - \frac{1}{\rho} m dp = \frac{\partial \Phi_1}{\partial \varepsilon_{ij}} d\varepsilon_{ij} + \frac{\partial \Phi_1}{\partial p} dp \end{aligned} \tag{8.47}$$

The last of (8.47) can be used to yield

$$\sigma_{ij} = \frac{\partial \Phi_1}{\partial \varepsilon_{ij}}, \quad -\frac{1}{\rho} m = \frac{\partial \Phi_1}{\partial p} \tag{8.48}$$

Taking the derivative of the first with respect to p and the second with respect to ε_{ij} we have the following constraint on the constitutive parameters:

$$\frac{\partial^2 \Phi_1}{\partial \varepsilon_{ij} \partial p} = \frac{\partial \sigma_{ij}}{\partial p} = -\frac{1}{\rho} \frac{\partial m}{\partial \varepsilon_{ij}} \tag{8.49}$$

Note that there is a typo in Eq. (15.4) of Rudnicki (1985) which omitted the minus sign as shown in (8.49). Mathematically, there are four different types of differential form for the work increment, depending what are the controlled variables. These choices and the corresponding Legendre transformation are given in Appendix C.

Substitution of (8.44) and (8.43) into (8.49) leads to

$$R_{ij} = \rho M_{ij} \tag{8.50}$$

For the drained condition, we can substitute $p = 0$ into (8.43) such that L_{ijkl} can be identified as the elastic tensor for the drained response. For the drained condition, we can substitute $m_0 = m$ into (8.44) in conjunction with (8.50) to obtain

$$p = -\frac{\rho}{Q} M_{ij} \varepsilon_{ij} \tag{8.51}$$

The undrained response can then be rewritten as

$$\sigma_{ij} = L^{u}_{ijkl}\,\varepsilon_{kl} \tag{8.52}$$

$$L^{u}_{ijkl} = L_{ijkl} + \frac{\rho}{Q} M_{ij} M_{kl} \tag{8.53}$$

In the absence of body force, Darcy's law can be expressed as

$$q_i = -\rho\kappa_{ij}\frac{\partial p}{\partial x_j} \tag{8.54}$$

This completes the general constitutive form for general anisotropic diffusive solids.

For the case of thc isotropic response, we have

$$M_{ij} = \zeta\delta_{ij}, \quad \kappa_{ij} = \kappa\delta_{ij} \tag{8.55}$$

$$L_{ijkl} = \mu(\delta_{ik}\delta_{jl} + \delta_{il}\delta_{jk}) + (K - \frac{2\mu}{3})\delta_{ij}\delta_{kl} \tag{8.56}$$

$$Q = \frac{\rho\zeta^2}{K_u - K} \tag{8.57}$$

where K_u is the undrained bulk modulus.

The constitutive model together with the field equations is summarized as:

$$\sigma_{ij} = (K - 2\mu/3)\varepsilon_{kk}\delta_{ij} + 2\mu\varepsilon_{ij} - \zeta p\delta_{ij} \tag{8.58}$$

$$m - m_0 = \zeta\rho[\varepsilon_{kk} + \frac{\zeta p}{K_u - K}] \tag{8.59}$$

$$q_i = -\rho\kappa\frac{\partial p}{\partial x_i} \tag{8.60}$$

$$(K + \mu/3)u_{i,ij} + \mu\nabla^2 u_j - \zeta p_{,j} = 0 \tag{8.61}$$

$$\nabla^2 m = \frac{1}{c}\frac{\partial m}{\partial t} \tag{8.62}$$

$$c = \frac{\kappa(K_u - K)(K + 4\mu/3)}{\zeta^2(K_u + 4\mu/3)} \tag{8.63}$$

Note however that there is a typo in the relation between B and ζ, given in (15.19) of Rudnicki (1985). The corrected one is:

$$B = \frac{(K_u - K)}{\zeta K_u} \tag{8.64}$$

Comparing to Rudnicki (1986), we find this constitutive form favoring K_u and ζ, instead of B and ν_u used by Rice and Cleary (1976), ζ and λ_u used by Rudnicki (1986), or R and H used by Biot (1941).

8.3 BIOT–VERRUIJT DISPLACEMENT FUNCTION

In 1956, Biot applied the Papkovitch–Neuber displacement functions in elasticity to solve problems in coupled thermoelasticity, and its completeness was considered by Verruijt (1969). This method was later extended to include fluid compressibility by Verruijt (1971). This Biot's (1956a) function would be extended to include solid constituent compressibility here. First, we recall from (4.53) of Chapter 4 that the Papkovitch–Neuber displacement function is first rewritten as (Verruijt, 1971)

$$\boldsymbol{u} = \boldsymbol{\nabla}(\varphi + \boldsymbol{r}\cdot\boldsymbol{\psi}) - 2\alpha\boldsymbol{\psi} \tag{8.65}$$

where α is a constant to be determined. The Laplacian of displacement becomes

$$\nabla^2\boldsymbol{u} = \nabla^2\boldsymbol{\nabla}(\varphi + \boldsymbol{r}\cdot\boldsymbol{\psi}) - 2\alpha\nabla^2\boldsymbol{\psi} \tag{8.66}$$

For Cartesian coordinates, we can show that

$$\nabla^2(\boldsymbol{r}\cdot\boldsymbol{\psi}) = 2\boldsymbol{\nabla}\cdot\boldsymbol{\psi} + \boldsymbol{r}\cdot(\nabla^2\boldsymbol{\psi}) \tag{8.67}$$

Following the same definition from the Papkovitch–Neuber displacement function, we first assume $\boldsymbol{\psi}$ is a harmonic function or

$$\nabla^2\boldsymbol{\psi} = 0 \tag{8.68}$$

With the mathematical form of (8.65) and (8.68), the volumetric strain is

$$\boldsymbol{\nabla}\cdot u = \nabla^2\varphi + 2(1-\alpha)\boldsymbol{\nabla}\cdot\psi \tag{8.69}$$

The equilibrium given in (8.37) in terms of displacement can be rewritten as

$$(\lambda + 2\mu)[\nabla^2\varphi + 2\boldsymbol{\nabla}\cdot\boldsymbol{\psi}] - 2\alpha(\lambda + \mu)\boldsymbol{\nabla}\cdot\boldsymbol{\psi} - \zeta p = 0) \tag{8.70}$$

Rearranging (8.70), the pore water pressure becomes

$$p = \frac{1}{\zeta}\{(\lambda + 2\mu)[\nabla^2\varphi + 2\boldsymbol{\nabla}\cdot\boldsymbol{\psi}] - 2\alpha(\lambda + \mu)\boldsymbol{\nabla}\cdot\boldsymbol{\psi}\} \tag{8.71}$$

When $\zeta = 1$ and $p \to -p$ we can recover (4) of Verruijt (1971) for the case of incompressible solid constituents. Substitution of (8.69) and (8.71) into (8.33) gives

$$m - m_0 = \frac{\zeta\rho_0(\lambda_u + 2\mu)}{(\lambda_u - \lambda)}(\nabla^2\varphi + 2\boldsymbol{\nabla}\cdot\boldsymbol{\psi}) - \frac{2\alpha\zeta\rho_0(\lambda_u + \mu)}{(\lambda_u - \lambda)}\boldsymbol{\nabla}\cdot\boldsymbol{\psi} \tag{8.72}$$

If we set

$$\alpha = \frac{\lambda_u + 2\mu}{\lambda_u + \mu}, \tag{8.73}$$

(8.72) is further simplified to

$$m - m_0 = \frac{\zeta\rho_0(\lambda_u + 2\mu)}{(\lambda_u - \lambda)}\nabla^2\varphi \tag{8.74}$$

Finally, substitution of (8.74) into (8.29) yields

$$\frac{\partial}{\partial t}(\nabla^2\varphi) - c\nabla^4\varphi = 0 \tag{8.75}$$

Equivalently, by using (2.29) we can express (8.73) as

$$\alpha = 2(1 - \nu_u) \tag{8.76}$$

This α is the same as Eq. (5.111) given by Detournay and Cheng (1993). When α = 1, we have $\lambda_u >> \mu$ or the case of incompressible fluid.

In summary, we have

$$\boldsymbol{u} = \boldsymbol{\nabla}(\varphi + \boldsymbol{r}\cdot\boldsymbol{\psi}) - 2\left(\frac{\lambda_u + 2\mu}{\lambda_u + \mu}\right)\boldsymbol{\psi} \tag{8.77}$$

$$\nabla^2 \boldsymbol{\psi} = 0\,, \quad \frac{\partial}{\partial t}(\nabla^2 \varphi) - c\nabla^4 \varphi = 0 \tag{8.78}$$

$$m - m_0 = \frac{\zeta \rho_0 (\lambda_u + 2\mu)}{(\lambda_u - \lambda)} \nabla^2 \varphi \tag{8.79}$$

$$p = \frac{1}{\zeta}[(\lambda + 2\mu)\nabla^2 \varphi + 2\mu(\frac{\lambda_u - \lambda}{\lambda_u + \mu})\nabla \cdot \boldsymbol{\psi}] \tag{8.80}$$

In this study, we called this particular form of displacement function the Biot–Verruijt displacement function. Actually, the form given here is more general than the form given by Verruijt (1971) to include solid constituent compressibility.

8. 4 MCNAMEE–GIBSON–VERRUIJT DISPLACEMENT FUNCTION

Verruijt (1971) identified that a generalized form of the displacement function of McNamee and Gibson (1960a) can be related to the Biot–Verruijt displacement function as

$$\varphi = -E(r,z,t)\,, \quad \psi_z = S(r,z,t)\,, \quad \psi_r = \psi_\theta = 0 \tag{8.81}$$

Here we called this McNamee–Gibson–Verruijt displacement as it has been generalized to include fluid compressibility (Verruijt, 1971). For axisymmetric cases, the following form was recorded by Detourney and Cheng (1993):

$$u_r = -\frac{\partial E}{\partial r} + z\frac{\partial S}{\partial r} \tag{8.82}$$

$$u_z = -\frac{\partial E}{\partial z} + z\frac{\partial S}{\partial z} - (\frac{\lambda_u + 3\mu}{\lambda_u + \mu})S \tag{8.83}$$

$$m - m_0 = -\frac{\zeta \rho_0 (\lambda_u + 2\mu)}{(\lambda_u - \lambda)} \nabla^2 E \tag{8.84}$$

$$\nabla^2 S = 0 \tag{8.85}$$

$$\frac{\partial}{\partial t}(\nabla^2 E) - c\nabla^4 E = 0 \tag{8.86}$$

$$\sigma_{rr} = 2\mu\left[\nabla^2 E - \frac{\partial^2 E}{\partial r^2} + z\frac{\partial^2 S}{\partial r^2} - (\frac{\lambda_u}{\lambda_u + \mu})\frac{\partial S}{\partial z}\right] \tag{8.87}$$

$$\sigma_{zz} = 2\mu\left[\nabla^2 E - \frac{\partial^2 E}{\partial z^2} + z\frac{\partial^2 S}{\partial z^2} - (\frac{\lambda_u + 2\mu}{\lambda_u + \mu})\frac{\partial S}{\partial z}\right] \tag{8.88}$$

$$\sigma_{rz} = 2\mu\left[-\frac{\partial^2 E}{\partial r \partial z} + z\frac{\partial^2 S}{\partial r \partial z} - (\frac{\mu}{\lambda_u + \mu})\frac{\partial S}{\partial r}\right] \tag{8.89}$$

$$p = -\frac{(\lambda + 2\mu)}{\zeta}\left[\nabla^2 E - (\frac{2\mu}{\lambda + 2\mu})(\frac{\lambda_u - \lambda}{\lambda_u + \mu})\frac{\partial S}{\partial z}\right] \tag{8.90}$$

The formulation by McNamee and Gibson (1960a) is recovered if we set $\lambda_u << \mu$ and $\zeta = 1$. Note that there is a typing error in the last term of the last equation in (18) of Verruijt (1971). Specifically, the $\partial S/\partial z$ should be $\partial S/\partial x$ in the last term in the x-z component of stress. The last term of (5.116g) of Detournay and Cheng (1993) should be $\partial S/\partial z$ instead of $\partial^2 S/\partial z^2$.

8.5 SCHIFFMAN–FUNGAROLI–VERRUIJT DISPLACEMENT FUNCTION

For nonaxisymmetric deformations, Schiffman and Fungaroli (1965) proposed three displacement functions to uncouple the equations of equilibrium. Verruijt (1971) has shown that the displacement functions proposed by Schiffman and Fungaroli (1965) are ad hoc modifications of McNamee and Gibson (1960a) by adding to the displacement vector a part accounting for the rotations in the horizontal plane. Because of this, the Schiffman–Fungaroli (1965) displacement function cannot be recovered as a special case of the Papkovitch–Neuber displacement function. In addition, Verruijt (1971) had modified the Schiffman–Fungaroli displacement function E (with one less derivative with respect to z and an extra minus sign for all other displacement functions) and had extended the analysis to include fluid compressibility.

The corresponding forms of Verruijt's (1971) displacement functions for Rudnicki's (1985) model given in Section 8.2.4 are reported here in Cartesian coordinates:

$$u_x = -\frac{\partial E}{\partial x} + z\frac{\partial S}{\partial x} + 2\frac{\partial Q}{\partial y} \tag{8.91}$$

$$u_y = -\frac{\partial E}{\partial y} + z\frac{\partial S}{\partial y} - 2\frac{\partial Q}{\partial x} \tag{8.92}$$

$$u_z = -\frac{\partial E}{\partial z} + z\frac{\partial S}{\partial z} - \left(\frac{\lambda_u + 3\mu}{\lambda_u + \mu}\right)S \tag{8.93}$$

$$p = -\frac{(\lambda + 2\mu)}{\zeta}\left[\nabla^2 E - \left(\frac{2\mu}{\lambda + 2\mu}\right)\left(\frac{\lambda_u - \lambda}{\lambda_u + \mu}\right)\frac{\partial S}{\partial z}\right] \tag{8.94}$$

The equilibrium equations for $\boldsymbol{u}$ lead to the following governing equations for E, S, and Q:

$$c\nabla^4 E = \nabla^2 \frac{\partial E}{\partial t}, \quad \nabla^2 S = 0, \quad \nabla^2 Q = 0 \tag{8.95}$$

where

$$\nabla^2 = \frac{\partial^2}{\partial x^2} + \frac{\partial^2}{\partial y^2} + \frac{\partial^2}{\partial z^2} \tag{8.96}$$

The total stresses are given in terms of the displacement functions as

$$\frac{\sigma_{xx}}{2\mu} = \nabla^2 E - \frac{\partial^2 E}{\partial x^2} + z\frac{\partial^2 S}{\partial x^2} - \left(\frac{\lambda_u}{\lambda_u + \mu}\right)\frac{\partial S}{\partial z} + 2\frac{\partial^2 Q}{\partial x \partial y} \tag{8.97}$$

$$\frac{\sigma_{yy}}{2\mu} = \nabla^2 E - \frac{\partial^2 E}{\partial y^2} + z\frac{\partial^2 S}{\partial y^2} - \left(\frac{\lambda_u}{\lambda_u + \mu}\right)\frac{\partial S}{\partial z} - 2\frac{\partial^2 Q}{\partial x \partial y} \tag{8.98}$$

$$\frac{\sigma_{zz}}{2\mu} = \nabla^2 E - \frac{\partial^2 E}{\partial z^2} + z\frac{\partial^2 S}{\partial z^2} - \left(\frac{\lambda_u + 2\mu}{\lambda_u + \mu}\right)\frac{\partial S}{\partial z} \tag{8.99}$$

$$p = -\frac{(\lambda + 2\mu)}{\zeta}\left[\nabla^2 E - \left(\frac{2\mu}{\lambda + 2\mu}\right)\left(\frac{\lambda_u - \lambda}{\lambda_u + \mu}\right)\frac{\partial S}{\partial z}\right] \tag{8.100}$$

$$\frac{\sigma_{xy}}{2\mu} = -\frac{\partial^2 E}{\partial x \partial y} + z\frac{\partial^2 S}{\partial x \partial y} - \frac{\partial^2 Q}{\partial x^2} + \frac{\partial^2 Q}{\partial y^2} \tag{8.101}$$

$$\frac{\sigma_{yz}}{2\mu} = -\frac{\partial^2 E}{\partial y \partial z} + z\frac{\partial^2 S}{\partial y \partial z} - (\frac{\mu}{\lambda_u + \mu})\frac{\partial S}{\partial y} - \frac{\partial^2 Q}{\partial x \partial z} \tag{8.102}$$

$$\frac{\sigma_{xz}}{2\mu} = -\frac{\partial^2 E}{\partial x \partial z} + z\frac{\partial^2 S}{\partial x \partial z} - (\frac{\mu}{\lambda_u + \mu})\frac{\partial S}{\partial x} - \frac{\partial^2 Q}{\partial y \partial z} \tag{8.103}$$

Note that the last equation of (29) in Verruijt (1971) has been mistakenly written as the *yz*-component of stress instead of the *xz*-component. When $\lambda_u >> \mu$, the expressions for incompressible fluid are obtained.

8.6. SCHIFFMAN–FUNGAROLI DISPLACEMENT FUNCTION

In this section, the three displacement functions of Schiffman and Fungaroli (1965) are given in cylindrical coordinates as

$$u_r = \frac{\partial^2 E}{\partial r \partial z} - \frac{2}{r}\frac{\partial Q}{\partial \theta} - z\frac{\partial S}{\partial r} \tag{8.104}$$

$$u_\theta = \frac{1}{r}\frac{\partial^2 E}{\partial \theta \partial z} + 2\frac{\partial Q}{\partial r} - \frac{z}{r}\frac{\partial S}{\partial \theta} \tag{8.105}$$

$$u_z = \frac{\partial^2 E}{\partial z^2} - z\frac{\partial S}{\partial z} + S \tag{8.106}$$

$$p = 2\mu\eta\frac{\partial}{\partial z}\nabla^2 E - 2\mu\frac{\partial S}{\partial z} \tag{8.107}$$

The corresponding governing equations for E, S, and Q are:

$$c\nabla^4 E = \nabla^2 \frac{\partial E}{\partial t}, \quad \nabla^2 S = 0, \quad \nabla^2 Q = 0 \tag{8.108}$$

where

$$\nabla^2 = \frac{\partial^2}{\partial r^2} + \frac{1}{r}\frac{\partial}{\partial r} + \frac{1}{r^2}\frac{\partial^2}{\partial \theta^2} + \frac{\partial^2}{\partial z^2} \tag{8.109}$$

Note that the problem is transformed to solving three *uncoupled* partial differential equations for displacement functions instead of four *coupled* partial differential equations for displacements and pore pressure. As a trade-off, the governing partial differential equations for E, S, and Q are of higher order. The total stress components and excess pore water pressure can then be written in terms of the displacement potentials. For axisymmetric deformation, the displacement functions, E^* and S^*, introduced by McNamee and Gibson (1960a) can be recovered by the following identifications:

$$-\frac{\partial E}{\partial z} = E^*, \quad Q = 0, \quad S = -S^* \tag{8.110}$$

where E and S are not functions of θ. Furthermore, it is straightforward to show that the governing equations for E^* and S^* are same as those for E and S. The expressions for the stresses are (Schiffman and Fungaroli, 1965):

$$\frac{\sigma_{rr}}{2\mu}=\frac{\partial}{\partial z}(\frac{\partial^2 E}{\partial r^2}-\nabla^2 E)-z\frac{\partial^2 S}{\partial r^2}+\frac{\partial S}{\partial z}-\frac{2}{r}\frac{\partial}{\partial \theta}(\frac{\partial Q}{\partial r}-\frac{Q}{r}) \tag{8.111}$$

$$\frac{\sigma_{zz}}{2\mu}=\frac{\partial}{\partial z}(\frac{\partial^2 E}{\partial z^2}-\nabla^2 E)-z\frac{\partial^2 S}{\partial z^2}+\frac{\partial S}{\partial z} \tag{8.112}$$

$$\frac{\sigma_{r\theta}}{2\mu}=\frac{1}{r}\frac{\partial^2}{\partial \theta \partial z}(\frac{\partial E}{\partial r}+\frac{E}{r^2})+\frac{z}{r}\frac{\partial}{\partial \theta}(\frac{S}{r}-\frac{\partial S}{\partial r})+\frac{\partial^2 Q}{\partial r^2}-\frac{1}{r}\frac{\partial Q}{\partial r}-\frac{1}{r^2}\frac{\partial^2 Q}{\partial \theta^2} \tag{8.113}$$

$$\frac{\sigma_{z\theta}}{2\mu}=\frac{1}{r}\frac{\partial}{\partial \theta}(\frac{\partial^2 E}{\partial z^2}-z\frac{\partial S}{\partial z})+\frac{\partial^2 Q}{\partial r \partial z} \tag{8.114}$$

$$\frac{\sigma_{rz}}{2\mu}=\frac{\partial}{\partial r}(\frac{\partial^2 E}{\partial z^2}-z\frac{\partial S}{\partial z})-\frac{1}{r}\frac{\partial^2 Q}{\partial \theta \partial z} \tag{8.115}$$

Note that the typo in (7d) of Schiffman and Fungaroli (1965) has been corrected as shown in (8.113).

8.7. LAPLACE–HANKEL TRANSFORM TECHNIQUE

For general loadings, we can expand the displacement functions in Fourier series as (e.g., Muki, 1960 or see Section 4.9)

$$E(r,\theta,z,t)=\sum_{n=0}^{\infty} E_n(r,z,t)\cos n\theta \tag{8.116}$$

$$S(r,\theta,z,t)=\sum_{n=0}^{\infty} S_n(r,z,t)\cos n\theta \tag{8.117}$$

$$Q(r,\theta,z,t)=\sum_{n=0}^{\infty} Q_n(r,z,t)\sin n\theta \tag{8.118}$$

Their government equations become

$$c\nabla_n^4 E_n=\nabla_n^2\frac{\partial E}{\partial t},\quad \nabla_n^2 S_n=0,\quad \nabla_n^2 Q_n=0 \tag{8.119}$$

where

$$\nabla_n^2=\frac{\partial^2}{\partial r^2}+\frac{1}{r}\frac{\partial}{\partial r}-\frac{n^2}{r^2}+\frac{\partial^2}{\partial z^2} \tag{8.120}$$

A standard solution technique for solving (8.119) is to apply the Hankel transform and the Laplace transform to r and t, respectively. This procedure has been applied successfully in several problems in fluid-saturated porous medium; for examples, see Apirathvorakij and Karasudhi (1980), Niumpradit and Karasudhi (1981), Kanok-Nukulchai and Chau (1990), Chau (1996), and Puswewala and Rajapakse (1988). More specifically, we assume that the variables r (radial coordinate) and t (time) of any function $\varphi(r,z,t)$ can be transformed to ξ and s as (Chau, 1996)

$$\varphi(r,z,t)=\frac{1}{2\pi i}\int_0^{\infty}\int_{\alpha-i\infty}^{\alpha+i\infty}\xi e^{st}J_n(r\xi)\overline{\varphi}(\xi,z,s)ds d\xi \tag{8.121}$$

$$\bar{\varphi}(\xi,z,s)=\int_0^{\infty}\int_0^{\infty} e^{-st} r\, J_n(r\xi)\varphi(r,z,t)drdt \qquad (8.122)$$

where $t > 0$ and α is chosen such that all singularities of the integrand lie on the left of it but it is otherwise an arbitrary real constant (Carrier et al., 1966; Spiegel, 1964). A brief introduction to the Laplace transform is given in Appendix B. By applying this double transform in time and radial coordinates, the (8.119) equations are reduced to the following ordinary differential equations:

$$(\frac{d^2}{dz^2}-\xi^2)(\frac{d^2}{dz^2}-\xi^2-\frac{s}{c})\bar{E}_n=0 \qquad (8.123)$$

$$(\frac{d^2}{dz^2}-\xi^2)\bar{Q}_n=0 \qquad (8.124)$$

$$(\frac{d^2}{dz^2}-\xi^2)\bar{S}_n=0 \qquad (8.125)$$

Their general solutions are

$$\bar{E}_n=A_m e^{-z\xi}+B_m e^{-z\gamma}+C_m e^{z\xi}+D_m e^{z\gamma} \qquad (8.126)$$

$$\bar{S}_n=E_m e^{-z\xi}+F_m e^{z\xi} \qquad (8.127)$$

$$\bar{Q}_n=G_m e^{-z\xi}+H_m e^{z\xi} \qquad (8.128)$$

where $\gamma=(\xi^2+s/c)^{1/2}$ and A_m, B_m, ... H_m are unknown constants to be determined by the boundary conditions. The subscript m describes the domain number for the case of the layered solid. For axisymmetric problems, the transformed McNamee–Gibson (1960a) displacement functions (S^* and E^*) have exactly the same solution form as S_n and E_n in (8.127) and (8.126). The Green's functions for poroelastic half-space are considered next, that is, a linear elastic fluid-infiltrated half-space subjected to interior vertical and horizontal point forces and fluid point source. This problem is the equivalent Mindlin's problem in linear elastic fluid-infiltrated half-space. All stresses and displacements can then be written in terms of these unknown constants.

These equations for stresses and displacements are given in Problems 8.16 and 8.17 and left as problems for the readers to solve.

8.8. POINT FORCES AND POINT FLUID SOURCE IN HALF-SPACE

As an example of how to solve time-dependent of poroelastic problems, we will consider point forces shown in Fig. 8.4(a) and the point source shown in Fig. 8.4(b) for linear elastic fluid-infiltrated halfspace using Biot's theory and the McNamee–Gibson function or the Schiffman–Fungaroli displacement function. The following presentation follows from Chau (1996). The analysis can easily be extended to compressible fluid and solid phases.

8.8.1. Vertical Point Force Solution

When a vertical point force of magnitude of F_z shown in Fig. 8.4 is suddenly applied (as a Heaviside step function) at a depth h from the free surface, we divide the half-space, occupying $z > 0$, into two domains by a fictitious horizontal plane at depth h. The upper domain and lower domain are denoted as domains 1 and 2, respectively. The displacement and stress fields are axisymmetric in nature if we choose the z axis passing through the applied vertical point force. Therefore, the McNamee–Gibson (1960a) displacement functions are sufficient for this case. There are 12 unknown constants for domains 1 and 2. However, the solutions for domain 2 should vanish as $z \to \infty$, and thus only the exponential terms with negative power remain (i.e., $C_2 = D_2 = F_2 = 0$). The remaining nine unknown constants have to be determined by nine conditions. In particular, the permeable and traction-free conditions on $z = 0$ are:

$$\sigma_{zz}^{(1)}(r,0,t) = 0, \quad \sigma_{zr}^{(1)}(r,0,t) = 0, \quad p^{(1)}(r,0,t) = 0 \tag{8.129}$$

where the superscript (1) denotes the components for domain 1. The continuities in displacements, shear stress, pore pressure, and fluid flow between domains 1 and 2 require

$$u_z^{(1)} = u_z^{(2)}, \quad u_r^{(1)} = u_r^{(2)} \tag{8.130}$$

$$\sigma_{zr}^{(1)} = \sigma_{zr}^{(2)} \tag{8.131}$$

$$p^{(1)} = p^{(2)}, \quad \frac{\partial p^{(1)}}{\partial z} = \frac{\partial p^{(2)}}{\partial z} \tag{8.132}$$

on $z = h$ and for all r and t. The only discontinuity between domains 1 and 2 appears in σ_{zz} at $z = h$ and $r = 0$; that is,

$$\sigma_{zz}^{(1)}(r,h,t) - \sigma_{zz}^{(2)}(r,h,t) = \frac{F_z}{2\pi r}\delta(r)H(t) \tag{8.133}$$

where F_z is the magnitude of the vertical force, $\delta(r)$ is the Dirac delta function with dimension as the inverse of length, and $H(t)$ is the Heaviside step function. Note that the right-hand side of (8.133) can be rewritten in terms of the discontinuous integral as (Chan et al., 1974; Kanok-Nukulchai and Chau, 1990)

$$\sigma_{zz}^{(1)}(r,h,t) - \sigma_{zz}^{(2)}(r,h,t) = \frac{1}{2\pi i}\int_0^{\infty}\int_{\alpha - i\infty}^{\alpha + i\infty} \frac{F_z}{2\pi}\xi J_0(\xi r)\frac{e^{st}}{s}\, ds\, d\xi \tag{8.134}$$

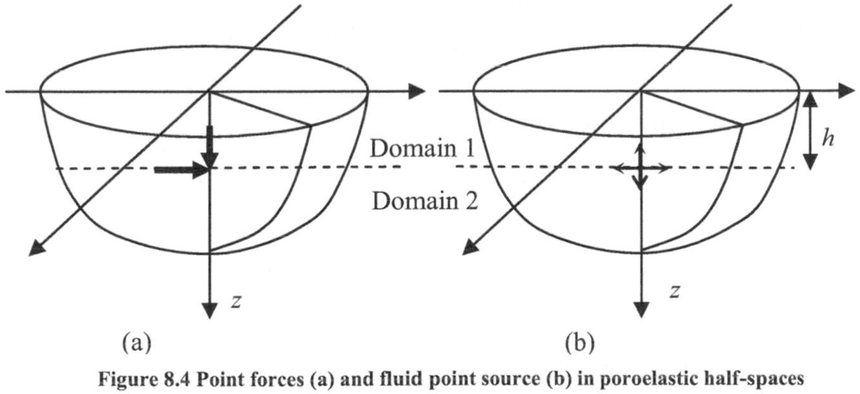

Figure 8.4 Point forces (a) and fluid point source (b) in poroelastic half-spaces

Thus, (8.129)–(8.132) and (8.134) provide a system of nine equations for nine unknown constants. The solutions for these constants require a straightforward, although tedious, algebraic manipulation to obtain. A symbolic manipulation program (such as Mathematica) can be used to assist the analysis.
The solutions for these constants are

$$A_1 = \Omega[-(\frac{\eta s}{c} + \xi^2 + \gamma\xi)(\xi - \frac{\eta s h}{c})e^{-h\xi} + 2\gamma(\frac{\eta s}{c} + \xi^2)e^{-h\gamma}] \quad (8.135)$$

$$F_1 = \Omega \frac{\eta s}{c}(\frac{\eta s}{c} + \xi^2 - \gamma\xi)e^{-h\xi} \quad (8.136)$$

$$B_1 = -\Omega[-2\xi^2(\xi - \frac{\eta s h}{c})e^{-h\xi} + \xi(\frac{\eta s}{c} + \xi^2 + \xi\gamma)e^{-h\gamma}] \quad (8.137)$$

$$D_1 = \Omega\xi(\frac{\eta s}{c} + \xi^2 - \gamma\xi)e^{-h\gamma} \quad (8.138)$$

$$C_1 = -\Omega(\frac{\eta s}{c} + \xi^2 - \gamma\xi)(\xi - \frac{\eta s h}{c})e^{-h\xi} \quad (8.139)$$

$$A_2 = A_1 + \Omega(\frac{\eta s}{c} + \xi^2 - \gamma\xi)(\xi + \frac{\eta s h}{c})e^{h\xi} \quad (8.140)$$

$$E_1 = \Omega[\frac{\eta s}{c}(\frac{\eta s}{c} - \xi^2 - \gamma\xi + \frac{2\xi\eta s h}{c})e^{-h\xi} + 2\gamma\frac{\eta s\xi}{c}e^{-h\gamma}] \quad (8.141)$$

$$B_2 = B_1 - \Omega\xi(\frac{\eta s}{c} + \xi^2 - \gamma\xi)e^{h\gamma} \quad (8.142)$$

$$E_2 = E_1 + \Omega\frac{\eta s}{c}(\frac{\eta s}{c} + \xi^2 - \gamma\xi)e^{h\xi} \quad (8.143)$$

where

$$\Omega = -\frac{F_z c^2}{8\pi\mu\xi\eta s^2(\xi^2 - \xi\gamma - \eta s / c)} \quad (8.144)$$

To get the numerical results, we have to take the inverse Laplace and Hankel transform. Closed form analytical solutions are in general not feasible. We refer to Chau (1996) on the issue of numerical Hankel-Laplace inversion.

8.8.2 Horizontal Point Force Solution

As for the case of vertical point force, the half-space is divided into two domains, one above the horizontal plane at depth h and the other below that plane. For this asymmetric case, the 16 unknown constants are reduced to 12 by the decaying condition at $z \to \infty$. The boundary conditions on the surface of the half-space are same as those given in (8.129) plus

$$\sigma_{z\theta}^{(1)}(r,0,t) = 0 \quad (8.145)$$

The continuity conditions for displacements, vertical normal stress, pore pressure, and fluid flow between domains 1 and 2 are, respectively,

$$u_z^{(1)} = u_z^{(2)}, \quad u_r^{(1)} = u_r^{(2)}, \quad u_\theta^{(1)} = u_\theta^{(2)} \tag{8.146}$$

$$\sigma_{zz}^{(1)} = \sigma_{zz}^{(2)} \tag{8.147}$$

$$p^{(1)} = p^{(2)}, \quad \frac{\partial p^{(1)}}{\partial z} = \frac{\partial p^{(2)}}{\partial z} \tag{8.148}$$

on $z = h$ and for all r and t. The applied horizontal point force can be modeled by the following discontinuities in shear stresses between domains 1 and 2 at $z = h$:

$$\sigma_{\theta z}^{(1)}(r,\theta,h,t) - \sigma_{\theta z}^{(2)}(r,\theta,h,t) = -\frac{F_r}{2\pi r}\delta(r)H(t)\sin\theta \tag{8.149}$$

$$\sigma_{zr}^{(1)}(r,\theta,h,t) - \sigma_{zr}^{(2)}(r,\theta,h,t) = \frac{F_r}{2\pi r}\delta(r)H(t)\cos\theta \tag{8.150}$$

where F_r is the magnitude of the horizontal point force, and $\delta(r)$ and $H(t)$ have the same definitions as those given in (8.133). Similar to the case of vertical point force, the Heaviside step function and the Dirac delta function can be represented by the following discontinuous integral:

$$\sigma_{\theta z}^{(1)}(r,\theta,h,t) - \sigma_{\theta z}^{(2)}(r,\theta,h,t) = -\frac{1}{2\pi i}\int_0^\infty \int_{\alpha-i\infty}^{\alpha+i\infty} \left(\frac{F_r \sin\theta}{2\pi}\right)\xi J_0(\xi r)\frac{e^{st}}{s}\,ds\,d\xi \tag{8.151}$$

$$\sigma_{rz}^{(1)}(r,\theta,h,t) - \sigma_{rz}^{(2)}(r,\theta,h,t) = \frac{1}{2\pi i}\int_0^\infty \int_{\alpha-i\infty}^{\alpha+i\infty} \left(\frac{F_r \cos\theta}{2\pi}\right)\xi J_0(\xi r)\frac{e^{st}}{s}\,ds\,d\xi \tag{8.152}$$

Thus, we have 12 conditions for evaluating all the unknown constants, and the solutions are

$$A_1 = -\Gamma\frac{\gamma^2}{\xi^2}\left[\left(\frac{\eta s}{c} + \xi^2 + \gamma\xi\right)\left(\frac{\eta s}{c} + \xi^2 - \frac{\eta s \xi h}{c}\right)e^{-h\xi} - 2\xi^2\left(\frac{\eta s}{c} + \xi^2\right)e^{-h\gamma}\right] \tag{8.153}$$

$$B_1 = \Gamma\left[2\xi\gamma\left(\frac{\eta s}{c} + \xi^2 - \frac{\eta s \xi h}{c}\right)e^{-h\xi} - \xi^2\left(\frac{\eta s}{c} + \xi^2 + \xi\gamma\right)e^{-h\gamma}\right] \tag{8.154}$$

$$C_1 = \Gamma\frac{\gamma^2}{\xi^2}\left(\frac{\eta s}{c} + \xi^2 - \frac{\eta s \xi h}{c}\right)\left(\frac{\eta sh}{c} + \xi^2 - \gamma\xi\right)e^{-h\xi} \tag{8.155}$$

$$D_1 = -\Gamma\,\xi^2\left(\frac{\eta s}{c} + \xi^2 - \gamma\xi\right)e^{-h\gamma} \tag{8.156}$$

$$E_1 = \Gamma\frac{\eta s \gamma^2}{c}\left[\left(\frac{\eta s}{c} + \xi^2 + \gamma\xi - \frac{2\xi\eta sh}{c}\right)e^{-h\xi} - 2\xi^2 e^{-h\gamma}\right] \tag{8.157}$$

$$B_2 = B_1 + \Gamma\xi^2\left(\frac{\eta s}{c} + \xi^2 - \gamma\xi\right)e^{h\gamma} \tag{8.158}$$

$$F_1 = G_1\xi = H_1\xi = -\Gamma\frac{\eta s \gamma^2}{c}\left(\frac{\eta s}{c} + \xi^2 - \gamma\xi\right)e^{-h\xi} \tag{8.159}$$

$$E_2 = E_1 + \Gamma\frac{\eta s \gamma^2}{c}\left(\frac{\eta s}{c} + \xi^2 - \gamma\xi\right)e^{h\xi} \tag{8.160}$$

$$A_2 = A_1 - \Gamma\frac{\gamma^2}{\xi^2}\left(\frac{\eta s}{c} + \xi^2 + \frac{\eta s \xi h}{c}\right)\left(\xi^2 - \gamma\xi + \frac{\eta s}{c}\right)e^{h\xi} \tag{8.161}$$

$$G_2=-\Gamma\frac{\eta s\gamma^2}{c\xi}(\frac{\eta s}{c}+\xi^2-\gamma\xi)(e^{-h\xi}+e^{h\xi}), \tag{8.162}$$

where

$$\Gamma=\frac{F_r c^2}{8\pi\mu\xi\eta s^2\gamma^2(\xi^2-\xi\gamma+\eta s/c)} \tag{8.163}$$

8.8.3 Fluid Point Source Solution

Similar to the cases of point forces, the half-space is divided into two domains. Since the problem is axisymmetric in nature, we again have $Q = 0$. Similar to the case for vertical point force, only nine unknown constants remain. The boundary conditions at $z = 0$ are exactly same as those for vertical point force. The continuity conditions at $z = h$ are same as those for vertical point force except that the normal stress along z-axis should be continuous (i.e., $\sigma_{zz}^{(1)} = \sigma_{zz}^{(2)}$ at $z = h$) and $\partial p/\partial z$ possesses a jump at $z = h$ as

$$\frac{\partial p^{(2)}}{\partial z}-\frac{\partial p^{(1)}}{\partial z}=\frac{q\gamma_w}{k}\frac{\delta(r)}{2\pi r}H(t) \tag{8.164}$$

or equivalently in integral form as

$$\frac{\partial p^{(2)}}{\partial z}-\frac{\partial p^{(1)}}{\partial z}=\frac{1}{2\pi i}\int_0^\infty\int_{\alpha-i\infty}^{\alpha+i\infty}(\frac{q\gamma_w}{2\pi k})\xi J_0(r\xi)\frac{e^{st}}{s}\,ds\,d\xi \tag{8.165}$$

where q is the strength of the point source (volume per unit time), k is the coefficient of permeability of the porous solid (length per time), and γ_w is the unit weight of the fluid. For this case, the nine unknown constants are:

$$A_1=\frac{\bar{Q}}{\xi}[(\frac{\eta s}{c}+\xi^2+\gamma\xi)e^{-h\xi}-2(\frac{\eta s}{c}+\xi^2)e^{-h\gamma}] \tag{8.166}$$

$$B_1=\frac{\bar{Q}}{\gamma}[-2\gamma\xi e^{-h\xi}+(\frac{\eta s}{c}+\xi^2+\xi\gamma)e^{-h\gamma}] \tag{8.167}$$

$$C_1=\frac{\bar{Q}}{\xi}(\frac{\eta s}{c}+\xi^2-\gamma\xi)e^{-h\xi} \tag{8.168}$$

$$D_1=-\frac{\bar{Q}}{\gamma}(\frac{\eta s}{c}+\xi^2-\gamma\xi)e^{-h\gamma} \tag{8.169}$$

$$E_1=E_2=\bar{Q}\frac{2\eta s}{c}(e^{-h\xi}-e^{-h\gamma}) \tag{8.170}$$

$$A_2=A_1+\frac{\bar{Q}}{\xi}(\frac{\eta s}{c}+\xi^2-\gamma\xi)e^{h\xi} \tag{8.171}$$

$$B_2=B_1-\frac{\bar{Q}}{\gamma}(\frac{\eta s}{c}+\xi^2-\gamma\xi)e^{h\gamma} \tag{8.172}$$

$$F_1=0 \tag{8.173}$$

where

$$\bar{Q} = \frac{q\gamma_w c^2}{4\pi\eta s^2 k(\xi^2 - \xi\gamma + \eta s / c)} \tag{8.174}$$

If we reverse the sign of the point source solution, we obtain the so-called point sink source solution. Point sink solution has been found useful in modeling subsidence due to ground water withdrawal (e.g., Booker and Carter, 1986a,b; Kanok-Nukulchai and Chau, 1990). When permeability of the half-space is anisotropic, the point sink solution has been obtained by Booker and Carter (1987a,b). For anisotropic poroelastic half-spaces, the point sink solution was considered by Tarn and Lu (1991).

8.9 CLEARY'S FUNDAMENTAL SOLUTION OF POINT FORCES IN FULL SPACE

One of the disadvantages of using the McNamee–Gibson or Schiffman–Fungaroli displacement functions in conjunction with the Hankel–Laplace transform is that no closed form solution is obtained. The final solutions are given in the transform space and transform inversion needs to be conducted numerically. In this section, we will present the elegant approach of Cleary (1977) for obtaining the fundamental solutions in fluid diffusive full-space. However, there is an algebraic error in Cleary's (1977) fluid source solution as pointed out and corrected by Rudnicki (1981). Therefore, we will only present the three-dimensional point force solution by Cleary (1977) (with typos corrected), and the fluid sources solution would be deferred to the next section and follow Rudnicki's (1986) approach.

The complete set of fundamental solutions provided in this and the next sections is very useful for generating other analytical solutions for poroelastic solids (e.g., Carvalho and Curran, 1998) as well as for boundary element formulation in solving practical problems numerically. In addition, the analytical methods used in obtaining these solutions merit detailed discussions.

8.9.1 Canonical Representation of Point Force Solution

The point force problem considered by Cleary (1976, 1977) is the equivalent Kelvin fundamental problem in elastic full space. The force term is

$$f_k = P_k H(t)\delta(x_1)\delta(x_2)\delta(x_3) \tag{8.175}$$

Cleary (1976, 1977) started by adopting the so-called canonical representation theorem (Wineman and Pipkin, 1964) that the pore pressure is a scalar function and that it must be linearly proportional to the applied force P_k and with a dimension of stress. There is no characteristic length scale in the problem, so the only length scale is $(ct)^{1/2}$. Therefore, there is only one possible canonical form for the pressure p:

$$p = \frac{P_k x_k}{r^3} f_1(\xi), \quad \xi \equiv \frac{r}{\sqrt{ct}} \tag{8.176}$$

This kind of scaling idea was discussed in detail by Barenblatt (1996). Using a similar argument, Cleary (1976, 1977) postulated that the displacement must have the following form:

$$\mu u_i = \frac{P_k x_k}{r^3} x_i f_2(\xi) + \frac{P_i}{r} f_3(\xi) \tag{8.177}$$

Substituting (8.176) and (8.177) into the stress-strain relation given in (8.20) and in view of the following formulas,

$$\frac{\partial \xi}{\partial x_i} = \frac{\xi x_i}{r^2}, \quad \frac{\partial r}{\partial x_i} = \frac{x_i}{r}, \quad \xi^4 \frac{d}{d\xi}[\frac{1}{\xi^3} f_2(\xi)] = -3 f_2 + \xi f_2' \tag{8.178}$$

$$\xi^2 \frac{d}{d\xi}[\frac{1}{\xi} f_3(\xi)] = -f_3 + \xi f_3' \tag{8.179}$$

we have

$$2\mu\varepsilon_{ij} = \frac{P_k x_k}{r^5} x_i x_j F_1 + \frac{(P_i x_j + P_j x_i)}{r^3} F_2 + \frac{2P_k x_k}{r^3} f_2 \delta_{ij} \tag{8.180}$$

$$\sigma_{ij} = \frac{x_i x_j}{r^2}[\frac{P_k x_k}{r^3} F_1(\xi)] + \frac{(P_i x_j + P_j x_i)}{r^3} F_2(\xi) + \frac{P_k x_k}{r^3} \delta_{ij} F_3(\xi) \tag{8.181}$$

where

$$F_1(\xi) = 2\xi^4 \frac{d}{d\xi}[\frac{1}{\xi^3} f_2(\xi)], \quad F_2(\xi) = f_2 + \xi^2 \frac{d}{d\xi}[\frac{1}{\xi} f_3(\xi)] \tag{8.182}$$

$$F_3(\xi) = \frac{2}{1-2\nu}[(1-\nu) f_2 - \nu f_3 + \nu\xi(f_2' + f_3') - \frac{3(\nu_u - \nu)}{2B(1+\nu_u)} f_1] \tag{8.183}$$

8.9.2 Determination of Evolution Functions

Substitution of (8.181) into the equilibrium equation yields

$$[F_2 + F_3 + \xi F_2'] r^2 P_i + [\xi F_1' - 3(F_2 + F_3) + \xi(F_2' + F_3')] P_k x_k x_i = 0 \tag{8.184}$$

Therefore, we must have independently

$$[F_2 + F_3 + \xi F_2'] = 0 \tag{8.185}$$

$$\xi \frac{d}{d\xi}[F_1 + 4F_2 + F_3] = 0 \tag{8.186}$$

Actually, the integration of the force on the sphere S_ρ shown in Fig. 8.5 can be used to find the integration of (8.186). More specifically, we can integrate the following force equilibrium with body force

$$\sigma_{kl,l} + f_k = 0 \tag{8.187}$$

over the sphere of S_ρ with body force given in (8.175):

$$\begin{aligned} -\int_{r\le\rho} f_i dV &= -P_i \int_{r\le\rho} H(t)\delta(x_1)\delta(x_2)\delta(x_3) dV = -P_i \\ &= \int_{r\le\rho} \sigma_{ij,j} dV = \int_{r=\rho} \sigma_{ij} n_j dS \end{aligned} \tag{8.188}$$

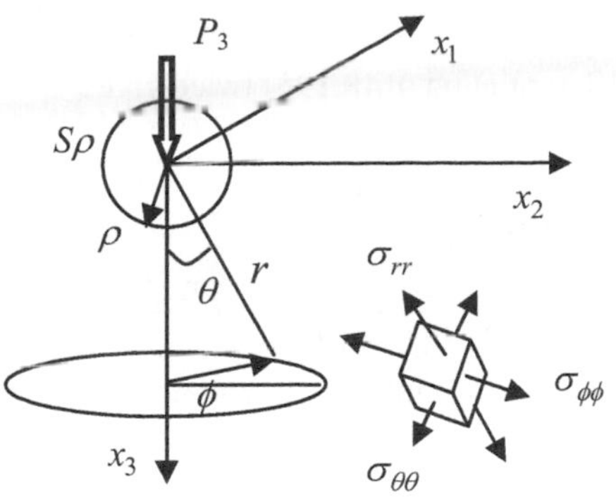

Figure 8.5 Point force applied within diffusive full space

The last equation is the result of applying the divergence theorem given in Section 1.6.1. Now let us consider the case of vertical force P_3 shown in Fig. 8.5 (i.e., $P_1 = P_2 = 0$) . In view of (8.181), (8.188) becomes

$$-P_3 = \int_0^\rho \frac{x_3^2 P_3}{r^4}(F_1 + F_2 + F_3)dS + \int_0^\rho \frac{P_3}{r^2} F_2 dS \tag{8.189}$$

Now we use spherical coordinates as

$$x_3 = \rho\cos\theta, \quad dS = \rho^2 \sin\theta d\theta d\phi \tag{8.190}$$

The integration can be carried out as

$$\begin{aligned} -P_3 &= \int_0^\pi \int_0^{2\pi} \frac{\rho^2 \cos^2\theta P_3}{\rho^4}(F_1 + F_2 + F_3)\rho^2 \sin\theta d\theta d\phi \\ &\quad + \int_0^\pi \int_0^{2\pi} \frac{P_3 F_2}{\rho^2}\rho^2 \sin\theta d\theta d\phi \\ &= 2\pi(F_1 + F_2 + F_3)P_3 \int_0^\pi \cos^2\theta \sin\theta d\theta + 2\pi P_3 F_2 \int_0^\pi \sin\theta d\theta \\ &= \frac{4\pi}{3}(F_1 + F_2 + F_3)P_3 + 4\pi P_3 F_2 \end{aligned} \tag{8.191}$$

Integration leads to

$$F_1 + 4F_2 + F_3 = -\frac{3}{4\pi} \tag{8.192}$$

Taking the trace of (8.181) gives

$$\sigma_{kk} = (F_1 + 2F_2 + 3F_3)\frac{x_k P_k}{r^3} \tag{8.193}$$

Combining this equation with (8.176), we have

$$\sigma_{kk} + \frac{3}{B}p = (F_1 + 2F_2 + 3F_3 + 3\frac{f_1}{B})\frac{x_k P_k}{r^3} = \frac{x_k P_k}{r^3}F(\xi) \tag{8.194}$$

Substituting (8.194) into (8.21), we obtain

$$m-m_0=\frac{3\rho_0(\nu_u-\nu)}{2\mu B(1+\nu)(1+\nu_u)}\frac{x_kP_k}{r^3}F(\xi) \tag{8.195}$$

Taking the spatial differentiation of (8.195) twice gives

$$\nabla^2 m=\frac{3\rho_0(\nu_u-\nu)}{2\mu B(1+\nu)(1+\nu_u)}\frac{x_kP_k}{r^5}\xi^2(F''-\frac{2}{\xi}F') \tag{8.196}$$

The time derivative of (8.195) is

$$\frac{\partial m}{\partial t}=-\frac{3\rho_0(\nu_u-\nu)}{2\mu B(1+\nu)(1+\nu_u)}(\frac{x_kP_k}{r^3})(\frac{\xi}{2t})F'(\xi) \tag{8.197}$$

Substituting (8.196) and (8.197) into (8.29) and noting the second part of (8.176)

$$\frac{1}{t}=\frac{c\xi^2}{r^2} \tag{8.198}$$

we arrive at an ordinary differential equation for $F(\xi)$:

$$F''(\xi)-(\frac{2}{\xi}-\frac{\xi}{2})F'(\xi)=0 \tag{8.199}$$

By observation, we find that the solution is

$$F(\xi)=\frac{F_\infty}{2\sqrt{\pi}}\int_\xi^\infty \eta^2\exp(-\eta^2/4)d\eta \tag{8.200}$$

where F_∞ is the value of $F(\xi)$ as $\to\infty$ and we have used the undrained condition that $t\to 0$ implies $\Delta m=0$ or $F=0$.

To verify that (8.200) is a solution of (8.199) we can use Leibniz's rule (Spiegel, 1963) to find the differentiation of the integral given in (8.200):

$$\frac{\partial}{\partial t}\int_{f(t)}^{g(t)}h(t,\zeta)d\zeta=g'(t)h[t,g(t)]-f'(t)h[t,f(t)]+\int_{f(t)}^{g(t)}\frac{\partial h(t,\zeta)}{\partial t}d\zeta \tag{8.201}$$

Using (8.201), we find that

$$F'(\xi)=\frac{F_\infty}{2\sqrt{\pi}}[-\xi^2\exp(-\xi^2/4)] \tag{8.202}$$

$$F''(\xi)=\frac{F_\infty\xi^2}{2\sqrt{\pi}}[-\frac{2}{\xi}+\frac{\xi}{2}]\exp(-\xi^2/4) \tag{8.203}$$

and it is obvious that (8.200) is indeed a solution of (8.199).

Finally, we rewrite (8.24) in terms of $F(\xi)$ as

$$\begin{aligned}\sigma_{kk}+\frac{6(\nu_u-\nu)}{B(1-\nu)(1+\nu_u)}p&=(F_1+2F_2+3F_3)\frac{P_kx_k}{r^3}+\frac{6(\nu_u-\nu)}{B(1-\nu)(1+\nu_u)}\frac{P_kx_k}{r^3}f_1(\xi)\\&=\{F(\xi)-\frac{3(1+\nu)(1-\nu_u)}{B(1-\nu)(1+\nu_u)}f_1(\xi)\}\frac{P_kx_k}{r^3}=G(\xi)\frac{P_kx_k}{r^3}\end{aligned} \tag{8.204}$$

With this new definition for $G(\xi)$, (8.24) gives the following ordinary differential equation:

$$G''(\xi)-\frac{2}{\xi}G'(\xi)=0 \tag{8.205}$$

The solution of $G(\xi)$ is

$$G(\xi) = k_1\xi^3 + k_2 \tag{8.206}$$

For long term $t \to \infty$, we have $G = F_\infty - 0 = F_\infty$ or $k_2 = F_\infty$. When $t \to 0$, for boundedness we must have $k_1 = 0$. Therefore, we have

$$G(\xi) \equiv F_\infty \tag{8.207}$$

With this solution for G, f_1 can be solved from the last part of (8.204) as

$$f_1(\xi) = [F(\xi) - F_\infty]\frac{B(1-\nu)(1+\nu_u)}{3(1+\nu)(1-\nu_u)} \tag{8.208}$$

Note that $F(\xi)$ can be simplified as

$$\begin{aligned} F(\xi) &= \frac{F_\infty}{2\sqrt{\pi}}\int_\xi^\infty \eta^2 \exp(-\eta^2/4)d\eta \\ &= \frac{F_\infty}{2\sqrt{\pi}}\left[\int_0^\infty \eta^2 \exp(-\eta^2/4)d\eta - \int_0^\xi \eta^2 \exp(-\eta^2/4)d\eta\right] \end{aligned} \tag{8.209}$$

The first integral can be evaluated by the following change of variables

$$\eta^2 = x, \quad 2\eta d\eta = dx \tag{8.210}$$

to give

$$\begin{aligned} \int_0^\infty \eta^2 \exp(-\eta^2/4)d\eta &= \frac{1}{2}\int_0^\infty x^{1/2}\exp(-x/4)dx \\ &= (\frac{1}{2})\frac{1}{(1/4)^{3/2}}\mathbf{\Gamma}(\frac{3}{2}) = 4\mathbf{\Gamma}(\frac{3}{2}) \end{aligned} \tag{8.211}$$

The result of this definite integral can be found in standard handbook (e.g., Formula 3.381.4 of Gradshteyn and Ryzhik, 1980). From Formulas 6.1.12 and 6.1.8 of Abramowitz and Stegun (1964), we have

$$\Gamma(\frac{3}{2}) = \Gamma(1+\frac{1}{2}) = \frac{1}{2}\Gamma(\frac{1}{2}) = \frac{\sqrt{\pi}}{2} \tag{8.212}$$

We finally have

$$\begin{aligned} F(\xi) &= \frac{F_\infty}{2\sqrt{\pi}}\int_\xi^\infty \eta^2 \exp(-\eta^2/4)d\eta \\ &= F_\infty\left[1 - \frac{1}{2\sqrt{\pi}}\int_0^\xi \eta^2 \exp(-\eta^2/4)d\eta\right] \end{aligned} \tag{8.213}$$

Substitution of (8.213) into (8.208) gives

$$f_1(\xi) = -\frac{2B(1-\nu)(1+\nu_u)}{3(\nu_u-\nu)}\frac{\omega}{2\sqrt{\pi}}\int_0^\xi \eta^2 \exp(-\eta^2/4)d\eta \tag{8.214}$$

where

$$\omega = \frac{F_\infty(\nu_u - \nu)}{2(1+\nu)(1-\nu_u)} \tag{8.215}$$

Thus, the pore water pressure is obtained by substituting (8.214) into (8.176).

Substitution of (8.207) and (8.214) into (8.204) gives

$$F_1 + 2F_2 + 3F_3 = \frac{F_\infty}{(1+\nu)(1-\nu_u)}\left\{(1-\nu)(1+\nu_u) - \frac{2(\nu_u-\nu)}{2\sqrt{\pi}}\int_\xi^\infty \eta^2 e^{-\eta^2/4}d\eta\right\} \tag{8.216}$$

Equation (8.185) can be rearranged in the following form:

$$F_2 - F_3 = \frac{1}{\xi}\frac{d}{d\xi}(\xi^2 F_2) \tag{8.217}$$

Therefore, (8.216), (8.217), and (8.192) provide three equations for three unknowns F_1, F_2, and F_3.

Subtraction of (8.216) from (8.192) gives

$$-2F_2 + 2F_3 = \frac{3}{4\pi} + F_\infty \frac{(1-\nu)(1+\nu_u)}{(1+\nu)(1-\nu_u)} - \frac{4\omega}{2\sqrt{\pi}} \int_\xi^\infty \eta^2 e^{-\eta^2/4} d\eta \tag{8.218}$$

Combining (8.217) and (8.218) results in

$$-\frac{2}{\xi}\frac{d}{d\xi}(\xi^2 F_2) = \frac{3}{4\pi} + F_\infty \frac{(1-\nu)(1+\nu_u)}{(1+\nu)(1-\nu_u)} - \frac{4\omega}{2\sqrt{\pi}} \int_\xi^\infty \eta^2 e^{-\eta^2/4} d\eta \tag{8.219}$$

Integrating (8.219) once, we obtain

$$-\xi^2 F_2 = \frac{3\xi^2}{16\pi} + \frac{F_\infty \xi^2}{4}\frac{(1-\nu)(1+\nu_u)}{(1+\nu)(1-\nu_u)} - \frac{\omega}{\sqrt{\pi}} \int_0^\xi s \int_s^\infty \eta^2 e^{-\eta^2/4} d\eta ds \tag{8.220}$$

The last integral can be integrated by part as

$$\begin{aligned} I_0 &= \int_0^\xi s \int_s^\infty \eta^2 e^{-\eta^2/4} d\eta ds = \frac{1}{2}\int_0^\xi \int_s^\infty \eta^2 e^{-\eta^2/4} d\eta d(s^2) \\ &= \frac{1}{2}[\xi^2 \int_\xi^\infty \eta^2 e^{-\eta^2/4} d\eta + \int_0^\xi \eta^4 e^{-\eta^2/4} d\eta] \\ &= \frac{1}{2}[\xi^2 \int_0^\infty \eta^2 e^{-\eta^2/4} d\eta - \xi^2 \int_0^\xi \eta^2 e^{-\eta^2/4} d\eta + \int_0^\xi \eta^4 e^{-\eta^2/4} d\eta] \\ &= \frac{1}{2}[\xi^2 2\sqrt{\pi} + \xi^2 \int_0^\xi (\frac{\eta^2}{\xi^2} - 1)\eta^2 e^{-\eta^2/4} d\eta] \end{aligned} \tag{8.221}$$

In the above derivation, we have used the result of (8.211). Back-substitution of (8.221) into (8.220) gives

$$F_2(\xi) = -\frac{1}{4}[\frac{3}{4\pi} + F_\infty] + \frac{\omega}{2\sqrt{\pi}} \int_0^\xi (\frac{\eta^2}{\xi^2} - 1)\eta^2 e^{-\eta^2/4} d\eta \tag{8.222}$$

Note that this equation differs from Eq. (28) on p. 81 of Cleary (1976). There is a typo in his equation (i.e., a power of 2 is missing in ξ within the integral). Substitution of (8.222) into (8.218) results in the solution of F_3:

$$F_3(\xi) = \frac{1}{4}[\frac{3}{4\pi} + F_\infty] + \frac{\omega}{2\sqrt{\pi}} \int_0^\xi (\frac{\eta^2}{\xi^2} + 1)\eta^2 e^{-\eta^2/4} d\eta \tag{8.223}$$

With these results, (8.192) gives the solution for F_1

$$F_1(\xi) = -\frac{3}{4}[\frac{1}{4\pi} - F_\infty] + \frac{\omega}{2\sqrt{\pi}} \int_0^\xi (3 - 5\frac{\eta^2}{\xi^2})\eta^2 e^{-\eta^2/4} d\eta \tag{8.224}$$

Substituting (8.224) into the first part of (8.182) and integrating the resulting equation once gives

$$f_2(\xi) = \frac{1}{8}[\frac{1}{4\pi} - F_\infty] + \frac{\omega\xi^3}{4\sqrt{\pi}} \int_\xi^\infty \int_0^s (\frac{3}{s^4} - 5\frac{\eta^2}{s^6})\eta^2 e^{-\eta^2/4} d\eta ds \tag{8.225}$$

By applying integration by part and by Leibniz's rule of differentiation for the integral given in (8.225), we obtain

$$I_{11} = \int_0^\xi \int_0^s \frac{3}{s^4}\eta^2 e^{-\eta^2/4} d\eta ds = -[\frac{1}{\xi^3}\int_0^\xi \eta^2 e^{-\eta^2/4} d\eta - \int_0^\xi \frac{1}{\eta} e^{-\eta^2/4} d\eta] \quad (8.226)$$

$$I_{12} = \int_0^\xi \int_0^s \frac{5}{s^6}\eta^4 e^{-\eta^2/4} d\eta ds = -[\frac{1}{\xi^5}\int_0^\xi \eta^4 e^{-\eta^2/4} d\eta - \int_0^\xi \frac{1}{\eta} e^{-\eta^2/4} d\eta] \quad (8.227)$$

Substitution of (8.226) and (8.227) into (8.225) yields

$$f_2(\xi) = \frac{1}{8}[\frac{1}{4\pi} - F_\infty] - \frac{\omega}{4\sqrt{\pi}}\int_0^\xi (1-\frac{\eta^2}{\xi^2})\eta^2 e^{-\eta^2/4} d\eta \quad (8.228)$$

Finally, the second of (8.182) can be used to evaluate f_3:

$$f_3(\xi) = \frac{1}{8}[\frac{7}{4\pi} + F_\infty] - \frac{\omega\xi}{4\sqrt{\pi}}\int_0^\xi \int_0^\zeta \frac{1}{\zeta^2}(1-\frac{\eta^2}{\zeta^2})\eta^2 e^{-\eta^2/4} d\eta d\zeta \quad (8.229)$$

Similar to the above procedure, integration by parts and using Leibniz's rule leads to:

$$\int_0^\xi \int_0^\zeta \frac{1}{\zeta^2}(1-\frac{\eta^2}{\zeta^2})\eta^2 e^{-\eta^2/4} d\eta d\zeta = -\frac{1}{\xi}\int_0^\xi (1-\frac{\eta^2}{3\xi^2})\eta^2 e^{-\eta^2/4} d\eta + \frac{2}{3}\int_0^\xi \eta e^{-\eta^2/4} d\eta \quad (8.230)$$

The last integral can be evaluated by integration by parts as

$$\frac{2}{3}\int_0^\xi \eta e^{-\eta^2/4} d\eta = -\frac{4}{3} e^{-\xi^2/4} \quad (8.231)$$

Finally, substitution of (8.230) and (8.231) into (8.229) yields

$$f_3(\xi) = \frac{1}{8}[\frac{7}{4\pi} + F_\infty] + \frac{\omega}{4\sqrt{\pi}}[\int_0^\xi (1-\frac{\eta^2}{3\xi^2})\eta^2 e^{-\eta^2/4} d\eta + \frac{4\xi}{3} e^{-\xi^2/4}] \quad (8.232)$$

Note that f_3 given in Eq. (29) on p. 81 of Cleary (1976) is incorrect. The last term in (8.232) is missing from his equation. This typo has been corrected in Cleary (1977).

By now all evolution functions are determined. There is only one unknown F_∞ that remains to be determined.

8.9.3 Determination of Unknown Constant F_∞

Finally, we have to evaluate the constant F_∞. To do so, we can substitute these evolution functions obtained in the last section into (8.183). In particular, the differentiations of f_2 and f_3 are

$$f_2'(\xi) = -\frac{2\omega}{4\sqrt{\pi}\xi^3}\int_0^\xi \eta^4 e^{-\eta^2/4} d\eta \quad (8.233)$$

$$f_3'(\xi) = \frac{\omega}{4\sqrt{\pi}}[\frac{2}{3\xi^3}\int_0^\xi \eta^4 e^{-\eta^2/4} d\eta + \frac{4}{3} e^{-\xi^2/4}] \quad (8.234)$$

$$f_2' + f_3' = \frac{\omega}{12\sqrt{\pi}} e^{-\xi^2/4} - \frac{\omega}{3\sqrt{\pi}} \frac{1}{\xi^3} \int_0^{\xi} \eta^4 e^{-\eta^2/4} d\eta \tag{8.235}$$

Collection of the constant terms in (8.183) gives

$$F_{\infty} = -\frac{(1+\nu)}{4\pi(1-\nu)} \tag{8.236}$$

It is straightforward, although somewhat lengthy, to show that both the coefficients for the integral and exponential terms are identically zero. With this result, ω can now be evaluated by substituting (8.236) into (8.215):

$$\omega = -\frac{(\nu_u - \nu)}{8\pi(1-\nu)(1-\nu_u)} \tag{8.237}$$

8.9.4 Final Solutions

In summary, the pore water pressure, displacements, and stresses are

$$p = \frac{P_k x_k}{r^3} f_1(\xi), \quad \xi \equiv \frac{r}{\sqrt{ct}} \tag{8.238}$$

$$\mu u_i = \frac{P_k x_k}{r^3} x_i f_2(\xi) + \frac{P_i}{r} f_3(\xi) \tag{8.239}$$

$$\sigma_{ij} = \frac{x_i x_j}{r^2}[\frac{P_k x_k}{r^3} F_1(\xi)] + \frac{(P_i x_j + P_j x_i)}{r^3} F_2(\xi) + \frac{P_k x_k}{r^3} \delta_{ij} F_3(\xi) \tag{8.240}$$

where the time-dependent functions are summarized as:

$$8\pi(1-\nu)\begin{Bmatrix} -F_1 \\ -F_2 \\ F_3 \\ f_1 \\ 2f_2 \\ 2f_3 \end{Bmatrix} = \begin{Bmatrix} 3 \\ (1-2\nu) \\ (1-2\nu) \\ 0 \\ 1 \\ (3-4\nu) \\ -\frac{4\xi}{3}(\frac{\nu_u-\nu}{1-\nu_u})\frac{e^{-\xi^2/4}}{2\sqrt{\pi}} \end{Bmatrix} + \frac{\nu_u-\nu}{2\sqrt{\pi}(1-\nu_u)} \int_0^{\xi} d\eta \begin{Bmatrix} 3-5\eta^2/\xi^2 \\ \eta^2/\xi^2-1 \\ -\eta^2/\xi^2-1 \\ \frac{2B(1+\nu_u)(1-\nu)}{3(\nu_u-\nu)} \\ 1-\eta^2/\xi^2 \\ -1+\eta^2/(3\xi^2) \end{Bmatrix} \eta^2 e^{-\eta^2/4}$$

(8.241)

Note that there is a typo in the last row in this vector in Cleary (1976), and the result is corrected here. The typo in Cleary (1976) is partially corrected in Cleary (1977) but the constant $2\pi^{1/2}$ is still missing in last row of the first vector on the right-hand side of (8.241).

Although the derivation given in this section is lengthy, the approach is elegant and allows a closed-form analytical solution to be obtained. Since the fluid sources solution is incorrect as shown by Rudnicki (1981), we will consider the fluid source fundamental solution following Rudnicki's (1986) approach in the next section. It will be demonstrated that the point force solution given in

(8.238)–(8.241) can be derived in a more elegant approach proposed by Rudnicki (1986).

8.10 RUDNICKI'S FUNDAMENTAL SOLUTIONS IN FULL SPACE

8.10.1 Impulsive Fluid Source

The complete set of fundamental solution for diffusive half-spaces consists of the point force solution considered above and the fluid mass point source solution to be considered in this section. The constitutive formulation of Biot's theory follows that given in Section 8.2.3 proposed by Rudnicki (1986). The fluid mass source is given by

$$Q(\boldsymbol{x},t) = Q_0 \delta(t)\delta(\boldsymbol{x}) \tag{8.242}$$

where δ is the Dirac delta function. The solution of (8.41) with the source given in (8.242) is well known (Carslaw and Jaeger, 1959):

$$m(\boldsymbol{x},t) = \frac{Q_0}{(4\pi ct)^{3/2}} \exp(-\frac{r^2}{4ct}) \tag{8.243}$$

It is straightforward to show that (8.243) satisfies (8.41) for the case of zero body force.

8.10.2 Canonical Form of Displacement Solution

By following the argument of Cleary (1976) for dimensional analysis, linearity, and spherical symmetry and isotropy, Rudnicki (1986) proposed the displacement solution as

$$u_i(\boldsymbol{x},t) = \frac{Q_0}{\rho_0}(\frac{x_i}{r^3})U(\xi) \tag{8.244}$$

Taking the spatial differentiation of (8.244) once and twice gives:

$$\frac{\partial u_i(\boldsymbol{x},t)}{\partial x_j} = \frac{Q_0}{\rho_0}[(\frac{U}{r^3})\delta_{ij} + \frac{x_i x_j}{r^5}(\xi U' - 3U)] \tag{8.245}$$

$$\frac{\partial u_i(\boldsymbol{x},t)}{\partial x_j \partial x_k} = \frac{Q_0}{\rho_0}\{[\frac{x_k \delta_{ij}}{r^5} + \frac{\delta_{ik}x_j + \delta_{kj}x_i}{r^5} - \frac{5x_i x_j x_k}{r^7}](\xi U' - 3U) + \frac{x_i x_j x_k}{r^7}\xi(\xi U'' - 2U')\} \tag{8.246}$$

Equation (8.246) can be specialized to give

$$\frac{\partial u_i(\boldsymbol{x},t)}{\partial x_i \partial x_j} = \frac{\partial u_j(\boldsymbol{x},t)}{\partial x_k \partial x_k} = \frac{Q_0}{\rho_0}\frac{x_j \xi}{r^5}(\xi U'' - 2U') \tag{8.247}$$

Differentiation of (8.243) results in

$$\frac{\partial m(\boldsymbol{x},t)}{\partial x_j} = -\frac{Q_0}{(4\pi ct)^{3/2}}(\frac{x_j}{2ct})\exp(-\xi^2/4) \tag{8.248}$$

where ξ is defined (8.176). Substitution of (8.247) and (8.248) into (8.38) with $F_j = 0$ gives

$$(\lambda_u + 2\mu)(\xi U'' - 2U')\frac{x_j \xi}{r^5} + \frac{(\lambda_u - \lambda)x_j}{16\pi\zeta\sqrt{\pi}}(\frac{\xi}{r})^5 e^{-\xi^2/4} = 0 \tag{8.249}$$

Multiplying (8.249) by x_j, we obtain the following ordinary differential equation for U:

$$\xi U'' - 2U' + \frac{(\lambda_u - \lambda)}{8\pi(\lambda_u + 2\mu)\zeta}(\frac{1}{2\sqrt{\pi}})\xi^4 e^{-\xi^2/4} = 0 \tag{8.250}$$

Note that both r and t only enter this differential equation through ξ. Therefore, (8.250) is an ordinary differential equation in ξ. It is straightforward to show that the solution of U is:

$$U(\xi) = [\frac{(\lambda_u - \lambda)}{4\pi(\lambda_u + 2\mu)\zeta}]g(\xi) \tag{8.251}$$

where

$$g(\xi) = \frac{1}{2\sqrt{\pi}}\int_0^{\xi} \eta^2 e^{-\eta^2/4} d\eta \tag{8.252}$$

In the next, we will show that function g given in (8.252) can be expressed in terms of error function which is defined in Abramowitz and Stegun (1964).

8.10.3 Error Function Representation

The integral in (8.252) can be expressed in terms of the error function commonly encountered in thermoelasticty (see Carslaw and Jaeger, 1959). In particular, we apply the following change of variables:

$$x = \eta^2 / 4, \quad dx = \frac{\eta}{2} d\eta \tag{8.253}$$

and the integral becomes

$$\int_0^{\xi} \eta^2 e^{-\eta^2/4} d\eta = 4\int_0^{\xi^2/4} \sqrt{x} e^{-x} dx = 4\gamma(\frac{3}{2}, \frac{\xi^2}{4}) \tag{8.254}$$

where $\gamma(x,y)$ is the incomplete gamma function (Abramowitz and Stegun, 1964). The last part of (8.254) is obtained by using Formula 3.381 of Gradshteyn and Ryzhik (1980). By applying Formula 8.356.1 of Gradshteyn and Ryzhik (1980)

$$\gamma(\alpha + 1, x) = \alpha\gamma(\alpha, x) - x^{\alpha} e^{-x} \tag{8.255}$$

and the Formula 6.5.16 of Abramowitz and Stegun (1964)

$$\gamma(1/2, x^2) = 2\int_0^x e^{-t^2} dt = \sqrt{\pi}\, \mathrm{erf}(x) \tag{8.256}$$

to (8.254), we find the following relation

$$g(\xi) = \frac{1}{2\sqrt{\pi}}\int_0^{\xi} \eta^2 e^{-\eta^2/4} d\eta = \mathrm{erf}(\frac{\xi}{2}) - \frac{\xi}{\sqrt{\pi}} e^{-\xi^2/4} \tag{8.257}$$

where the error function erf is defined in the last part of (8.256) (Abramowitz and Stegun, 1964). Note that $t \to \infty$ gives $\xi \to 0$ and with erf(0) = 0, we have g(0) = 0. When $t \to 0$ gives $\xi \to \infty$ and with erf(∞) = 1, we have by virtue of L'Hôpital's rule for the indeterminate limit (e.g., Speigel, 1963):

$$\lim_{\xi\to\infty}\frac{\xi}{e^{\xi^2/4}}=\lim_{\xi\to\infty}\frac{2}{\xi e^{\xi^2/4}}=0 \tag{8.258}$$

L'Hôpital's rule was published in the first-ever book on calculus written by L'Hôpital in 1696, according to Maor (1994) it was actually discovered by Johann Bernoulli (1667–1748). Thus, we have $g(\infty) = 1$. At initial time the displacement field given in (8.244) has the mathematical form of a center of dilatation (see Section 132 of Love, 1944). This observation agrees with our common sense that the fluid point force is indeed a kind of center dilatation.

Note that the differentiation of U gives

$$U'(\xi)=[\frac{(\lambda_u-\lambda)}{4\pi(\lambda_u+2\mu)\zeta}]g'(\xi)=[\frac{(\lambda_u-\lambda)}{4\pi(\lambda_u+2\mu)\zeta}]\frac{1}{2\sqrt{\pi}}(\xi^2e^{-\xi^2/4}) \tag{8.259}$$

The last equation is resulted from applying Leibniz's rule of differentiation given in (8.201) to the integral form of g given in (8.257). Substitution of (8.243) and (8.244) into (8.33) with $m_0 = 0$ gives

$$p=\frac{Q_0}{\rho_0}[\frac{(\lambda+2\mu)(\lambda_u-\lambda)}{(\lambda_u+2\mu)\zeta^2}](4\pi ct)^{-3/2}e^{-\xi^2/4} \tag{8.260}$$

Then, the stress tensor can be computed from (8.31) as

$$\sigma_{ij}=\frac{Q_0}{2\pi\rho_0 r^3}[\frac{\mu(\lambda_u-\lambda)}{(\lambda_u+2\mu)\zeta}]\left\{\delta_{ij}(g-\xi g')+\frac{x_ix_j}{r^2}(\xi g'-3g)\right\} \tag{8.261}$$

8.10.4 Suddenly Applied Fluid Mass Source

We now proceed to integrate the solution for impulsive fluid mass obtained in the last section to get the solution for continuous injection of fluid mass. In particular, we replace Q_0 by $qd\tau$, t by $t-\tau$ and integrate τ from 0 to t:

$$Q(\boldsymbol{x},t)=\int_0^t q\delta(t-\tau)\delta(\boldsymbol{x})d\tau=qH(t)\delta(\boldsymbol{x}) \tag{8.262}$$

where $H(t)$ is the Heaviside step function. Likewise, (8.243) can be integrated to get

$$m(\boldsymbol{x},t)=\int_0^t\frac{q}{[4\pi c(t-\tau)]^{3/2}}\exp[-\frac{r^2}{4c(t-\tau)}]d\tau \tag{8.263}$$

We apply the following change of variable

$$t-\tau=s,\quad -d\tau=ds \tag{8.264}$$

to get

$$m(\boldsymbol{x},t)=-\int_t^0\frac{qe^{-r^2/(4cs)}}{[4\pi cs]^{3/2}}ds=\frac{q}{8\pi\sqrt{\pi}}\int_0^t\frac{e^{-r^2/(4cs)}}{(cs)^{3/2}}ds \tag{8.265}$$

Applying another round of change of variable

$$\eta^2=\frac{r^2}{4cs},\quad 2\eta d\eta=-\frac{r^2}{4cs^2}ds \tag{8.266}$$

we obtain

$$m(x,t)=\frac{q}{\pi cr}\frac{1}{2\sqrt{\pi}}\int_{r/(2\sqrt{ct})}^{\infty}\exp(-\eta^2)d\eta=\frac{q}{4\pi cr}\mathrm{erfc}(\frac{\xi}{2}) \tag{8.267}$$

where erfc(x) = 1 − erf(x), which is the complementary error function (Abramowitz and Stegun, 1964). Comparison of (8.267) and (8.265) yields

$$\int_0^t \frac{e^{-r^2/(4cs)}}{(cs)^{3/2}}ds=\frac{2\sqrt{\pi}}{cr}\mathrm{erfc}(\frac{\xi}{2}) \tag{8.268}$$

Replacing Q_0 by $qd\tau$, t by $t-\tau$, and integrating τ from 0 to t, we can also integrate (8.244) with (8.251) and (8.252) to give the displacement as

$$u_i(x,t)=\frac{q}{\rho_0}(\frac{x_i}{r^3})[\frac{(\lambda_u-\lambda)}{4\pi(\lambda_u+2\mu)\zeta}]\frac{1}{2\sqrt{\pi}}\int_0^t\int_0^{r/\sqrt{c(t-\tau)}}\eta^2 e^{-\eta^2/4}d\eta d\tau \tag{8.269}$$

We apply the following change of variable

$$t-\tau=s,\quad -d\tau=ds \tag{8.270}$$

to get

$$u_i(x,t)=\frac{q}{\rho_0}(\frac{x_i}{r^3})[\frac{(\lambda_u-\lambda)}{4\pi(\lambda_u+2\mu)\zeta}]\frac{1}{2\sqrt{\pi}}\int_0^t\int_0^{r/\sqrt{cs}}\eta^2 e^{-\eta^2/4}d\eta ds \tag{8.271}$$

By applying integration by parts and Leibniz's rule, the integral can be reduced to

$$\begin{aligned}\int_0^t\int_0^{r/\sqrt{cs}}\eta^2 e^{-\eta^2/4}d\eta ds&=\frac{r^2}{c\xi^2}\int_0^{\xi}\eta^2 e^{-\eta^2/4}d\eta+\frac{1}{2}\int_0^t\frac{r^3}{(cs)^{3/2}}e^{-r^2/(4cs)}ds\\&=\frac{2\sqrt{\pi}r^2}{c\xi^2}g(\xi)+\frac{r^2\sqrt{\pi}}{c}\mathrm{erfc}(\frac{\xi}{2})\end{aligned} \tag{8.272}$$

The last expression is the result of applying (8.257) and (8.268). With the result of (8.272) we have the displacement as

$$u_i(x,t)=\frac{q}{\rho_0 c}(\frac{x_i}{r})[\frac{(\lambda_u-\lambda)}{8\pi(\lambda_u+2\mu)\zeta}]u(\xi) \tag{8.273}$$

where

$$u(\xi)=\mathrm{erfc}(\frac{\xi}{2})+\frac{2}{\xi^2}g(\xi) \tag{8.274}$$

The limiting value of u for $t\to 0$ and $\xi\to\infty$, we have $u(\infty)=0$. For the long term (or $t\to\infty$ and $\xi\to 0$), we have

$$\lim_{\xi\to 0}\frac{g(\xi)}{\xi^2}=\lim_{\xi\to 0}\frac{g'(\xi)}{2\xi}=\lim_{\xi\to 0}\frac{\xi^2 e^{-\xi^2/4}}{(2\sqrt{\pi})2\xi}=0 \tag{8.275}$$

That is, we have $u(0)=1$. Finally, stress and pore pressure can be calculated from (8.31) and (8.33). In particular, by noting the following identities

$$g'(\xi)=\frac{1}{2\sqrt{\pi}}(\xi^2 e^{-\xi^2/4}) \tag{8.276}$$

$$u'(\xi)=-\frac{4}{\xi^3}g(\xi) \tag{8.277}$$

$$\frac{d}{d\xi}\mathrm{erf}(\frac{\xi}{2}) = \frac{1}{\sqrt{\pi}} e^{-\xi^2/4} \tag{8.278}$$

we have

$$u_{k,k} = \frac{q}{\rho_0 c r}[\frac{\lambda_u - \lambda}{4\pi\zeta(\lambda_u + 2\mu)}]\mathrm{erfc}(\frac{\xi}{2}) \tag{8.279}$$

$$u_{i,j} = \frac{q}{\rho_0 c}[\frac{\lambda_u - \lambda}{8\pi\zeta(\lambda_u + 2\mu)}]\left\{[\mathrm{erfc}(\frac{\xi}{2}) + \frac{2}{\xi^2} g]\frac{\delta_{ij}}{r} - \frac{x_i x_j}{r^3}[\mathrm{erfc}(\frac{\xi}{2}) + \frac{6}{\xi^2} g]\right\} \tag{8.280}$$

Substitution of (8.279) and (8.267) into (8.33) with $m_0 = 0$ leads to

$$p = \frac{q}{\rho_0 \zeta^2}(\frac{1}{4\pi c r})[\frac{(\lambda_u - \lambda)(\lambda + 2\mu)}{(\lambda_u + 2\mu)}]\mathrm{erfc}(\frac{\xi}{2}) \tag{8.281}$$

Substitution of (8.279)–(8.281) into (8.31) yields

$$\sigma_{ij} = -\frac{q}{\rho_0 c r}[\frac{\mu(\lambda_u - \lambda)}{4\pi\zeta(\lambda_u + 2\mu)}]\left\{\delta_{ij}[\mathrm{erfc}(\frac{\xi}{2}) - \frac{2}{\xi^2} g] + \frac{x_i x_j}{r^2}[\mathrm{erfc}(\frac{\xi}{2}) + \frac{6}{\xi^2} g]\right\} \tag{8.282}$$

which is the same as Eq. (26) of Rudnicki (1986).

8.10.5 Equivalence of Fluid Mass Dipole and Body Force

It was recognized by Cleary (1977) that the body force term in (8.41) contributes to the solution with an equal effect of a fluid mass dipole. For example, we can consider a fluid source term of

$$Q(\boldsymbol{x},t) = Q_0 f(\boldsymbol{x},t) \tag{8.283}$$

The corresponding fluid mass dipole can be found by superimposing a source of strength Q_0/ε at $\boldsymbol{x} - \varepsilon\boldsymbol{\xi}$ and a sink of strength $-Q_0/\varepsilon$ at $\boldsymbol{x}$ and taking the limit of $\varepsilon \to 0$. The resulting dipole solution is (Rudnicki, 1986)

$$Q_{dipole}(\boldsymbol{x},t) = -h_k \frac{\partial f(\boldsymbol{x},t)}{\partial x_k} = \lim_{\varepsilon \to 0} \frac{Q_0}{\varepsilon}[f(\boldsymbol{x} - \varepsilon\boldsymbol{\xi}, t) - f(\boldsymbol{x},t)] \tag{8.284}$$

where

$$h_k = \xi_k Q_0 \tag{8.285}$$

We can see from the second term of (8.41) that this term has the same mathematical contribution of a body force of

$$F_j(\boldsymbol{x},t) = P_j f(\boldsymbol{x},t) \tag{8.286}$$

provided that we set the dipole strength as

$$h_j = -\frac{\rho_0 \kappa P_j (\lambda_u - \lambda)}{\zeta(\lambda_u + 2\mu)} \tag{8.287}$$

Therefore, the solution for a body force distribution of $P_j f(\boldsymbol{x},t)$ is the same for a dipole source distribution of strength given by (8.287). The beauty of Rudnicki's (1986) approach lies in this observation of equivalence between fluid mass dipole and body force. This solution will be employed in the next section to determine the point force solution.

8.10.6 Fluid Mass Dipoles

Using the fluid source solution obtained in Section 8.10.4, if a particular field quantity is given by $qF(\boldsymbol{x}, t)$ for a source of strength q, the corresponding quantity for a dipole of strength q and direction λ_k is given by

$$F_{dipole}(\boldsymbol{x},t) = -q\lambda_k \frac{\partial F(\boldsymbol{x},t)}{\partial x_k} \tag{8.288}$$

Applying (8.288) to (8.267), we have

$$m(\boldsymbol{x},t) = -\frac{h_k}{4\pi c}\frac{\partial}{\partial x_k}\left\{\frac{1}{r}\operatorname{erfc}(\frac{\xi}{2})\right\} \tag{8.289}$$

By noting that

$$\frac{\partial}{\partial x_k}\left\{\operatorname{erfc}(\frac{\xi}{2})\right\} = \frac{\partial}{\partial \xi}\left\{\operatorname{erfc}(\frac{\xi}{2})\right\}\frac{\partial \xi}{\partial x_k} = -(\frac{\xi x_k}{r^2})\frac{1}{\sqrt{\pi}}e^{-\xi^2/4} \tag{8.290}$$

we obtain

$$m_{dipole}(\boldsymbol{x},t) = \frac{h_k}{4\pi c}(\frac{x_k}{r^3})[1-g(\xi)] \tag{8.291}$$

Similarly, we can apply the same procedure shown in (8.288) to water pressure given in (8.281) and the result is

$$p_{dipole} = \frac{h_k}{\rho_0 \kappa}(\frac{x_k}{4\pi r^3})[1-g(\xi)] \tag{8.292}$$

Similarly, the displacement and stress for fluid dipole is

$$u_{i(dipole)} = -\frac{h_k \zeta}{\rho_0\kappa(\lambda+2\mu)}(\frac{1}{8\pi r})\left\{\delta_{ik}u(\xi) + \frac{x_i x_k}{r^2}[\xi u'(\xi) - u(\xi)]\right\} \tag{8.293}$$

$$\begin{aligned}\sigma_{ij(dipole)} = {} & \frac{h_k \zeta \mu}{\rho_0\kappa(\lambda+2\mu)}(\frac{1}{4\pi r^3})[\delta_{ij}(\xi\boldsymbol{\Sigma}_1' - \boldsymbol{\Sigma}_1)x_k + (\delta_{jk}x_i + \delta_{ik}x_j)\boldsymbol{\Sigma}_2 \\ & - \frac{3x_i x_j x_k}{r^2}(\boldsymbol{\Sigma}_2 - \frac{\xi}{3}\boldsymbol{\Sigma}_2')]\end{aligned} \tag{8.294}$$

where

$$\boldsymbol{\Sigma}_1(\xi) = u(\xi) + \xi u'(\xi) = \operatorname{erfc}(\frac{\xi}{2}) - \frac{2}{\xi^2}g(\xi) \tag{8.295}$$

$$\boldsymbol{\Sigma}_2(\xi) = u(\xi) - \xi u'(\xi) = \operatorname{erfc}(\frac{\xi}{2}) + \frac{6}{\xi^2}g(\xi) \tag{8.296}$$

These solutions will be employed in the next section to determine the point force solution.

8.10.7 Point Force Solution by Rudnicki (1986)

Although point force solution in poroelastic solids was derived by Cleary (1976, 1977), Rudnicki (1986) rederived this fundamental solution following a more elegant way that the physical meaning of the solution can easily be recognized. More specifically, the solution can be obtained by superimposing the Kelvin solution using the undrained moduli onto the fluid dipole solution with appropriate strength (as given in the last section).

In particular, for a suddenly applied load with components P_j at the origin we have

$$F_j = P_j \delta(\boldsymbol{x}) H(t) \tag{8.297}$$

The corresponding Kelvin solution for an undrained solid is (Eq. (12) on p.185 of Love, 1944)

$$u_i = \frac{P_j}{8\pi r \mu(\lambda_u + 2\mu)} \left\{ (\lambda_u + 3\mu)\delta_{ij} + (\lambda_u + \mu)\frac{x_i x_j}{r^2} \right\} \tag{8.298}$$

where the undrained Lamé's constant has been used. The corresponding stress components are

$$\sigma_{ij} = \frac{P_k}{4\pi r^2 (\lambda_u + 2\mu)} \{ \frac{\mu}{r} [x_k \delta_{ij} - (x_i \delta_{jk} + x_j \delta_{ik})] - (\lambda_u + \mu)\frac{3x_i x_j x_k}{r^3} \} \tag{8.299}$$

The point force solution is now obtained by adding (8.298) to (8.293) with (8.287) as

$$\mu u_i = \frac{P_j}{8\pi r(\lambda_u + 2\mu)} \left\{ (\lambda_u + 3\mu)\delta_{ij} + (\lambda_u + \mu)\frac{x_i x_j}{r^2} \right\}$$
$$+ \frac{P_j}{8\pi r} \left\{ \left[\frac{\lambda + 3\mu}{\lambda + 2\mu} - \frac{\lambda_u + 3\mu}{\lambda_u + 2\mu} \right] u(\xi)\delta_{ij} + \left[\frac{\lambda + \mu}{\lambda + 2\mu} - \frac{\lambda_u + \mu}{\lambda_u + 2\mu} \right] \frac{x_i x_j}{r^2} [u(\xi) - \xi u'(\xi)] \right\}$$
$$\tag{8.300}$$

Note that in deriving (8.300) we have used the following identity:

$$\frac{(\lambda_u - \lambda)}{(\lambda + 2\mu)(\lambda_u + 2\mu)} = \frac{1}{\mu}\left[\frac{\lambda + 3\mu}{\lambda + 2\mu} - \frac{\lambda_u + 3\mu}{\lambda_u + 2\mu} \right] = -\frac{1}{\mu}\left[\frac{\lambda + \mu}{\lambda + 2\mu} - \frac{\lambda_u + \mu}{\lambda_u + 2\mu} \right] \tag{8.301}$$

These alternative forms proposed by Rudnicki (1986) made the physical meaning of these time-dependent terms apparent. That is, the magnitudes of time-dependent terms of (8.300) are the differences between the drained and undrained responses. Substitution of (8.287) into (8.291) gives

$$m(\boldsymbol{x},t) = -\frac{\rho_0 \zeta P_k}{4\pi(\lambda + 2\mu)} (\frac{x_k}{r^3})[1 - g(\xi)] \tag{8.302}$$

We can substitute (8.299) into (8.34) and adding the pressure from (8.292) after substituting (8.287) to get

$$p(\boldsymbol{x},t) = \frac{(\lambda_u - \lambda)}{\zeta(\lambda_u + 2\mu)} (\frac{P_k x_k}{4\pi r^3}) g(\xi) \tag{8.303}$$

Finally, substituting (8.287) into (8.294) and adding (8.299) we have

$$4\pi r^3\sigma_{ij} = P_k\{\frac{\mu}{\lambda_u+2\mu}[x_k\delta_{ij}-(x_i\delta_{jk}+x_j\delta_{ik})]-(\frac{\lambda_u+\mu}{\lambda_u+2\mu})\frac{3x_ix_jx_k}{r^2}\}$$

$$+P_k\{\left[\frac{\mu}{\lambda+2\mu}-\frac{\mu}{\lambda_u+2\mu}\right][(\mathbf{\Sigma}_1-\xi\mathbf{\Sigma}_1^{'})\delta_{ij}x_k-(\delta_{jk}x_i+\delta_{ik}x_j)\mathbf{\Sigma}_2] \quad (8.304)$$

$$-\left[\frac{\lambda+\mu}{\lambda+2\mu}-\frac{\lambda_u+\mu}{\lambda_u+2\mu}\right]\frac{3x_ix_jx_k}{r^2}(\mathbf{\Sigma}_2-\frac{\xi}{3}\mathbf{\Sigma}_2^{'})\}$$

Note that there is a minor typo in Rudnicki (1986) in the last term of (8.304). In deriving (8.304) we have used (8.301) and the following identity:

$$\frac{(\lambda_u-\lambda)}{(\lambda+2\mu)(\lambda_u+2\mu)}=\frac{1}{\mu}\left[\frac{\mu}{\lambda+2\mu}-\frac{\mu}{\lambda_u+2\mu}\right] \quad (8.305)$$

Again the physical meaning of the time-dependent terms of (8.304) is clear because of the recognition of (8.301) and (8.305). This demonstrates that proper regrouping is of crucial importance in recognizing the physical meaning of the obtained solutions. By now, we have a complete set of fundamental solutions of point fluid source and point forces.

8.11 THERMOELASTICITY VS. POROELASTICITY

Mathematically, coupled thermoelasticity is equivalent to Biot's theory of poroelasticity. For example, coupled thermoelasticity can expressed as (Carslaw and Jaeger, 1959; Boley and Wiener, 1960)

$$\sigma_{ij}=\lambda\varepsilon_{kk}\delta_{ij}+2\mu\varepsilon_{ij}-\frac{E\alpha}{1-2\nu}\delta_{ij}T \quad (8.306)$$

where α is the coefficient of linear thermal expansion. Comparison of (8.306) with (8.31) or (8.58) gives the following identifications between thermoelasticity and poroelasticity:

$$\zeta \leftarrow \frac{E\alpha}{1-2\nu},\ p\leftarrow T \quad (8.307)$$

With these identifications, all solutions discussed here for poroelasticity apply equally to thermoelastic problems.

8.12 SUMMARY AND FURTHER READING

8.12.1 Summary

In this chapter, we summarize various generalized forms of Biot's theory given by McNamee and Gibson (1960a,b), Rice and Cleary (1976), Rudnicki (1986), and Rudnicki (1985). The use of displacement potentials are summarized, including the Biot–Verruijt potential, the McNamee–Gibson–Verruijt potential, and the Schiffman–Fungaroli–Verruijt potential. The Verruijt–Biot potential has been extended to include soil compressibility. The use of the Hankel–Laplace transform technique is discussed, and illustrated by applying it to the cases of point forces and point source in poroelastic fluid diffusive half-spaces. The full-space fundamental solution by Cleary (1977) is rederived step by step, and some typos by Cleary (1976, 1977) are corrected. The same solution obtained by Rudnicki (1986) by following a different approach is also summarized. The equivalence of Rudnicki (1986) and Cleary (1977) is left as a problem at the end of this chapter. A minor misprint in the paper by Rudnicki (1986) is also corrected. These fundamental solutions proposed by Cleary and Rudnicki are obtained in an elegant way and in closed form. These fundamental solutions are useful in boundary element formulation.

8.12.2 Further Reading

In this chapter, we have not covered any wave propagation issues in poroelastic solids. For such consideration, the readers are referred to Biot (1956b). We also have restricted our discussion to Biot's theory, and mixture theory for poroelastic solids is not covered. The readers can refer to de Boer (2000) for mixture theory for poroelastic solids. The fundamental solution by Cleary and Rudnicki discussed here is restricted to static situations. For the dynamic Green's function for poroelastic half-spaces, we refer to Rajapakse and Senjuntichai (1993), and Senjuntichai (1994a,b). For layered half-spaces, Green's function was considered by Pan (1999) and Vardoulakis and Harnpattanapanich (1986). The fundamental solution for transversely isotropic poroelastic solids was given by Taguchi and Kurashige (2002). Consolidation problems in inhomogeneous layers have been considered by Mahmoud and Deresiewicz (1980a,b).

Dislocation and crack problems in poroelastic solids have been considered by Rudnicki (1987, 1991, 1996), Rudnicki and Hsu (1988), Rudnicki et al. (1993), Simons (1977), and Rice and Cleary (1976). Many of these solutions were also compiled by Wang (2000). In addition, complex variable technique has been developed for 2-D poroelastic problems, and readers can refer to Rice and Cleary (1976). For the boundary element method for poroelastic solids, we refer to Detournay and Cheng (1993).

We have only dealt with saturated situations in this book, and for unsaturated soil mechanics we refer to Fredlund and Rahardjo (1993). Regarding fluid infiltrated solids, multi-phase coupling becomes an important topic in soil mechanics (Oka and Kimoto, 2012), but it is outside the scope of the present chapter.

8.13 PROBLEMS

Problem 8.1 Express ζ and λ_u of the Rudnicki (1986) model in terms of B and ν_u of Rice–Cleary (1976) model.

Problem 8.2 Express K_u and ζ, of Rudnicki (1985) model in terms of B and ν_u of Rice–Cleary (1976) model.

Problem 8.3 Show the validity of (8.180) and (8.181).

Problem 8.4 Prove the following identity between Rudnicki's (1986) model and Rice-Cleary's (1976) model:

$$\zeta = \frac{3(\nu_u - \nu)}{B(1-2\nu)(1+\nu_u)} \tag{8.308}$$

Problem 8.5 Prove the validity of (8.184).

Problem 8.6 Prove the validity of (8.186) from (8.184) and (8.185).

Problem 8.7 As shown in Fig. 8.6, consider the problem of a suddenly applied point source at a depth h from the free surface $z = 0$ of a half-space with an impervious surface (i.e., $\partial p/\partial z = 0$). First, we divide the half-space occupying $z > 0$ into two domains by a fictitious horizontal plane at the depth where the point source is applied. Solve this problem using the McNamee–Gibson displacement function given in Section 8.4.

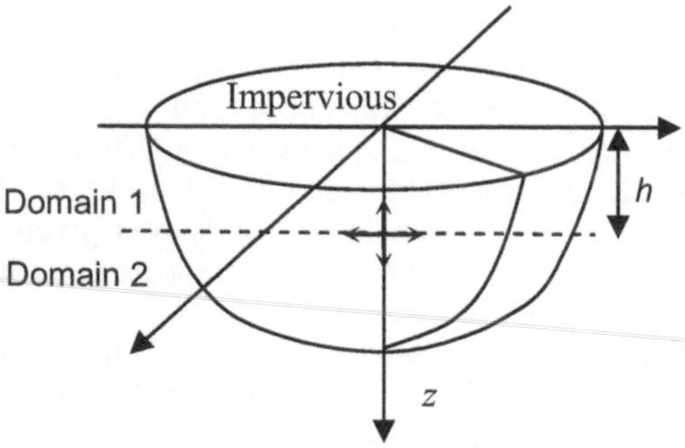

Figure 8.6 Point source in poroelastic half-space with impervious surface

Answer:

$$A_1 = \frac{\bar{Q}}{\xi s}[(\xi^3 + \xi^2\gamma - \eta\gamma\xi^2 + \eta\gamma^3)e^{-h\xi} - 2\xi^3 e^{-h\gamma}] \tag{8.309}$$

$$B_1 = \frac{\bar{Q}}{\gamma s}[-2\gamma\xi^2 e^{-h\xi} + (\xi^3 + \xi^2\gamma + \eta\xi^2\gamma - \eta\gamma^3)e^{-h\gamma}] \tag{8.310}$$

$$C_1 = \frac{q\gamma_w c}{4\eta\pi k\xi s}\frac{e^{-h\xi}}{(s - c\xi^2)} \tag{8.311}$$

$$D_1 = -\frac{q\gamma_w c}{4\eta\pi k\gamma s^2}e^{-h\gamma} \tag{8.312}$$

$$E_1 = E_2 = \bar{Q}\frac{2\eta s}{c^2}(\gamma e^{-h\xi} - \xi e^{-h\gamma}) \tag{8.313}$$

$$A_2 = A_1 + \frac{\bar{Q}}{\xi s}(\xi^3 - \xi^2\gamma - \eta\gamma\xi^2 + \eta\gamma^3)e^{h\xi} \tag{8.314}$$

$$B_2 = B_1 + \frac{\bar{Q}}{\gamma s}(-\xi^3 + \gamma\xi^2 + \eta\gamma\xi^2 - \eta\gamma^3)e^{h\gamma} \tag{8.315}$$

$$F_1 = 0 \tag{8.316}$$

where

$$\bar{Q} = \frac{q\gamma_w c}{4\pi\eta sk(\gamma - \xi)(\xi^2 - \xi\gamma + \eta s / c)} \tag{8.317}$$

Problem 8.8 In Rudnicki's (1986) appendix, all the evolution functions used in Cleary's (1976) solutions were expressed in terms of Rudnicki's (1986) solutions. They are

$$f_1(\xi) = [\frac{(\lambda_u - \lambda)}{4\pi\zeta(\lambda_u + 2\mu)}]g(\xi) \tag{8.318}$$

$$f_2(\xi) = \frac{(\lambda_u + \mu)}{8\pi(\lambda_u + 2\mu)}\left\{1 - \frac{\mu(\lambda_u - \lambda)}{(\lambda + 2\mu)(\lambda_u + \mu)}(u - \xi u')\right\} \tag{8.319}$$

$$f_3(\xi) = \frac{(\lambda_u + 3\mu)}{8\pi(\lambda_u + 2\mu)}\left\{1 + \frac{\mu(\lambda_u - \lambda)}{(\lambda + 2\mu)(\lambda_u + 3\mu)}u(\xi)\right\} \tag{8.320}$$

$$F_1(\xi) - \frac{-3(\lambda_u + \mu)}{4\pi(\lambda_u + 2\mu)}\{1 - \left[1 - \frac{(\lambda_u + 2\mu)(\lambda + \mu)}{(\lambda + 2\mu)(\lambda_u + \mu)}\right](\mathbf{\Sigma}_2 - \frac{\xi}{3}\mathbf{\Sigma}_2')\} \tag{8.321}$$

$$F_2(\xi) = \frac{-\mu}{4\pi(\lambda_u + 2\mu)}\{1 - \left[1 - \frac{(\lambda_u + 2\mu)}{(\lambda + 2\mu)}\right]\mathbf{\Sigma}_2(\xi)\} \tag{8.322}$$

$$F_3(\xi) = \frac{\mu}{4\pi(\lambda_u + 2\mu)}\{1 - \left[1 - \frac{(\lambda_u + 2\mu)}{(\lambda + 2\mu)}\right](\mathbf{\Sigma}_1 - \xi\mathbf{\Sigma}_1')\} \tag{8.323}$$

However, the equivalence of these expressions with those given by Cleary (1977) is not demonstrated explicitly in the paper of Rudnicki (1986). Show that these expressions are, indeed, equal to (8.241), in which Cleary's (1976, 1977) typos has been corrected.

Problem 8.9 Demonstrate that (8.251) and (8.252) are indeed the solution of (8.250).

Problem 8.10 For the case of compressible fluid and solid phase, consider the following cylindrical form of displacement potential given in (8.104) to (8.107). Prove the following identities:

$$\frac{\partial^2}{\partial r\partial z}(\nabla^2 E) = \nabla^2\frac{\partial^2 E}{\partial r\partial z} - \frac{2}{r^3}\frac{\partial^3 E}{\partial z\partial\theta^2} - \frac{1}{r^2}\frac{\partial^2 E}{\partial r\partial z} \tag{8.324}$$

$$-\frac{2}{r}\frac{\partial}{\partial\theta}(\nabla^2 Q) = -\nabla^2(\frac{2}{r}\frac{\partial Q}{\partial\theta}) - \frac{4}{r^2}\frac{\partial^2 Q}{\partial r\partial\theta} - \frac{2}{r^3}\frac{\partial Q}{\partial\theta} \tag{8.325}$$

$$-z\frac{\partial}{\partial r}(\nabla^2 S) = -\nabla^2 (z\frac{\partial S}{\partial r}) + \frac{2z}{r^3}\frac{\partial^2 S}{\partial \theta^2} + \frac{z}{r^2}\frac{\partial S}{\partial r} + 2\frac{\partial^2 S}{\partial r \partial z} \quad (8.326)$$

Problem 8.11 By virtue of (8.327)–(8.330), show that (8.1) is identically satisfied if

$$\nabla^2 S = 0\,, \quad \nabla^2 Q = 0 \quad (8.327)$$

Problem 8.12 By virtue of (8.327), show that (8.4) is identically satisfied if

$$c\nabla^4 E = \nabla^2 \frac{\partial E}{\partial t} \quad (8.328)$$

Problem 8.13 Prove the following identities:

$$\nabla^2 (\frac{1}{r}\frac{\partial^2 E}{\partial r \partial \theta \partial z}) = \frac{1}{r}\frac{\partial^2}{\partial r \partial \theta \partial z}(\nabla^2 E) + \frac{1}{r^3}\frac{\partial^2 E}{\partial z \partial \theta} - \frac{2}{r^2}\frac{\partial^3 E}{\partial r \partial \theta \partial z} \quad (8.329)$$

$$2\frac{\partial}{\partial r}(\nabla^2 Q) = 2\nabla^2 (\frac{\partial Q}{\partial r}) - \frac{2}{r^2}\frac{\partial Q}{\partial r} - \frac{4}{r^3}\frac{\partial^2 Q}{\partial \theta^2} \quad (8.330)$$

$$-\frac{z}{r}\frac{\partial}{\partial \theta}(\nabla^2 S) = -\nabla^2 (\frac{z}{r}\frac{\partial S}{\partial \theta}) + \frac{z}{r^3}\frac{\partial S}{\partial \theta} - \frac{2z}{r^2}\frac{\partial^2 S}{\partial r \partial \theta} + \frac{2}{r}\frac{\partial^2 S}{\partial z \partial \theta} \quad (8.331)$$

Problem 8.14 By virtue of (8.333)–(8.335), show that (8.2) is identically satisfied if

$$\nabla^2 S = 0\,, \; \nabla^2 Q = 0 \quad (8.332)$$

Problem 8.15 Prove the following identity:

$$z\frac{\partial}{\partial z}\nabla^2 S = \nabla^2 (z\frac{\partial S}{\partial z}) - 2\frac{\partial^2 S}{\partial z^2} \quad (8.333)$$

Problem 8.16 By virtue of (8.337), show that (8.3) is identically satisfied if

$$\nabla^2 S = 0 \quad (8.334)$$

Problem 8.17 By using (8.104)–(8.107), (8.116)–(8.118), and (8.126)–(8.128), show the validity of the following formulas for the m-th harmonics:

$$\frac{u_r}{\cos m\theta} + \frac{u_\theta}{\sin m\theta} = \frac{1}{2\pi i}\int_0^\infty \int_{\alpha - i\infty}^{\alpha + i\infty} \xi^2 J_{m+1}(r\xi) e^{st} [\xi e^{-z\xi} A_m + \gamma e^{-z\gamma} B_m - \xi e^{z\xi} C_m - \gamma e^{z\gamma} D_m + z e^{-z\xi} E_m + z e^{z\xi} F_m - 2e^{-z\xi} G_m - 2e^{z\xi} H_m] ds d\xi \quad (8.335)$$

$$\frac{u_r}{\cos m\theta} - \frac{u_\theta}{\sin m\theta} = \frac{1}{2\pi i}\int_0^\infty \int_{\alpha - i\infty}^{\alpha + i\infty} \xi^2 J_{m-1}(r\xi) e^{st} [-\xi e^{-z\xi} A_m - \gamma e^{-z\gamma} B_m + \xi e^{z\xi} C_m + \gamma e^{z\gamma} D_m - z e^{-z\xi} E_m - z e^{z\xi} F_m - 2e^{-z\xi} G_m - 2e^{z\xi} H_m] ds d\xi \quad (8.336)$$

$$u_z = \frac{1}{2\pi i}\int_0^\infty \int_{\alpha-i\infty}^{\alpha+i\infty} \xi J_m(r\xi)e^{st}[\xi^2 e^{-z\xi} A_m + \gamma^2 e^{-z\gamma} B_m$$
$$+\xi^2 e^{z\xi} C_m + \gamma^2 e^{z\gamma} D_m + (z\xi+1)e^{-z\xi} E_m + (1-z\xi)ze^{z\xi} F_m]dsd\xi \cos m\theta \tag{8.337}$$

$$\frac{p}{2\mu} = \frac{1}{2\pi i}\sum_{m=0}^{\infty} \int_0^\infty \int_{\alpha-i\infty}^{\alpha+i\infty} \xi J_1(r\xi)e^{st}[\eta\gamma(\gamma^2-\xi^2)e^{-z\gamma} B_m$$
$$+\eta\gamma(\xi^2-\gamma^2)e^{z\xi} D_m - \xi e^{-z\xi} E_m + \xi e^{z\xi} F_m]dsd\xi \cos m\theta \tag{8.338}$$

Problem 8.18 By using (8.111)–(8.115), (8.116)–(8.118), and (8.126)–(8.128), show the validity of the following formulas for the m-th harmonics:

$$\frac{1}{2\mu\cos m\theta}(\sigma_{\theta\theta}+\sigma_{rr}) = \frac{1}{2\pi i}\int_0^\infty \int_{\alpha-i\infty}^{\alpha+i\infty} \xi J_m(r\xi)e^{st}[\xi^3 e^{-z\xi} A_m + \gamma(2\gamma^2-\xi^2)e^{-z\gamma} B_m$$
$$+\xi^3 e^{z\xi} C_m + \gamma(\xi^2-2\gamma^2)e^{z\gamma} D_m + \xi(z\xi-2)e^{-z\xi} E_m + \xi(z\xi+2)e^{z\xi} F_m]dsd\xi \tag{8.339}$$

$$\frac{\sigma_{zz}}{2\mu} = \frac{1}{2\pi i}\int_0^\infty \int_{\alpha-i\infty}^{\alpha+i\infty} \xi^2 J_m(r\xi)e^{st}[-\xi^2 e^{-z\xi} A_m - \xi\gamma e^{-z\gamma} B_m$$
$$+\xi^2 e^{z\xi} C_m + \gamma\xi e^{z\gamma} D_m - (z\xi+1)e^{-z\xi} E_m + (1-z\xi)e^{z\xi} F_m]dsd\xi \cos m\theta \tag{8.340}$$

$$\frac{\sigma_{r\theta}}{2\mu\sin m\theta} + \frac{u_r}{r\cos m\theta} + \frac{u_\theta}{r\sin m\theta} =$$
$$\frac{-1}{2\pi i}\int_0^\infty \int_{\alpha-i\infty}^{\alpha+i\infty} \xi^2 J_m(r\xi)e^{st}[\xi e^{-z\xi} G_m + \xi e^{z\xi} H_m]dsd\xi \tag{8.341}$$

$$\frac{1}{2\mu}\left(\frac{\sigma_{rz}}{\cos m\theta} + \frac{\sigma_{\theta z}}{\sin m\theta}\right) = \frac{1}{2\pi i}\int_0^\infty \int_{\alpha-i\infty}^{\alpha+i\infty} \xi^2 J_{m+1}(r\xi)e^{st}[-\xi^2 e^{-z\xi} A_m - \gamma^2 e^{-z\gamma} B_m$$
$$-\xi^2 e^{z\xi} C_m - \gamma^2 \xi e^{z\gamma} D_m - z\xi e^{-z\xi} E_m + z\xi e^{z\xi} F_m + \xi e^{-z\xi} G_m - \xi e^{z\xi} H_m]dsd\xi \tag{8.342}$$

$$\frac{1}{2\mu}\left(\frac{\sigma_{\theta z}}{\sin m\theta} - \frac{\sigma_{rz}}{\cos m\theta}\right) = \frac{1}{2\pi i}\int_0^\infty \int_{\alpha-i\infty}^{\alpha+i\infty} \xi^2 J_{m-1}(r\xi)e^{st}[-\xi^2 e^{-z\xi} A_m - \gamma^2 e^{-z\gamma} B_m - \xi^2 e^{z\xi} C_m$$
$$-\gamma^2 \xi e^{z\gamma} D_m - z\xi e^{-z\xi} E_m + z\xi e^{z\xi} F_m - \xi e^{-z\xi} G_m + \xi e^{z\xi} H_m]dsd\xi \tag{8.343}$$

$$\frac{\sigma_{rr}}{2\mu\cos m\theta} + \frac{u_r}{r\cos m\theta} + \frac{u_\theta}{r\sin m\theta} = \frac{1}{2\pi i}\int_0^\infty \int_{\alpha-i\infty}^{\alpha+i\infty} \xi J_m(r\xi)e^{st}[\xi^3 e^{-z\xi} A_m + \gamma^3 e^{-z\gamma} B_m$$
$$-\xi^3 e^{z\xi} C_m - \gamma^3 e^{z\gamma} D_m + \xi(z\xi-1)e^{-z\xi} E_m + \xi(z\xi+1)e^{z\xi} F_m]dsd\xi \tag{8.344}$$

CHAPTER NINE

Waves Propagations and Dynamics Problems in Geomaterials

9.1 INTRODUCTION

The propagation of mechanical disturbances in solids is of profound importance in many disciplines including physical sciences and engineering. In these kinds of problems, the loading or disturbance is applied at such a "fast" rate that the effect of inertia cannot be ignored (as we have been doing in this book so far). These loadings are described as suddenly applied like a dropping mass sticking to the ground (modeled as Heaviside step function in time) or applied impulsively like in a dynamic impact (modeled by Dirac delta function in time). These loadings are referred to as dynamic loadings instead of quasi-static loadings. Energy of disturbance or waves may propagate in the solids at different wave speeds, depending on the nature of the disturbances and on whether they are dilatational or shear in nature. Both displacement and stress responses of the solids are functions of time (rather than in the sense of the viscoelastic type of creeping or relaxation discussed in Chapter 7). For isotropic solids of infinite extent, these waves are either dilatational or compressional waves (also called P-waves) or shear waves (also called S-waves). The phase velocities of the particle movements are parallel and perpendicular to the direction of wave propagation for dilatational and shear waves, respectively. Referring to a fixed coordinate system, we can further decompose shear waves as SH-waves or SV-waves, corresponding to the polarized components along the "horizontal" and "vertical" directions, respectively. For anisotropic solids, these waves would no longer be purely dilatational or purely transverse. The mathematical techniques used in solving dynamic problems are more tedious and lengthy; only some simple situations have been solved analytically.

In geomechanics applications, seismic wave propagation induced by earthquakes is of major concerns. Manmade structures are vulnerable to ground shakings. Both responses and failure mechanisms in soils or rocks can be highly sensitive to dynamic loadings.

Many important topics need to be included in a chapter on dynamics and wave propagations. In view of the size limitations of this book, only some selected topics in dynamics and wave propagations in geomaterials are included in this chapter. Although one-dimensional wave propagations and dynamic problems provide very useful insight into the problems of dynamics, in view of its limited usefulness in geomechanics we will start with 3-D formulations of wave equations in solids. Surface and interfacial waves, including Rayleigh waves, Love waves, and Stoneley waves, will be discussed because of their relevance to seismic wave propagations. The nature of elastic-plastic waves in geomaterials will be discussed.

Strain localization, a special case of acceleration waves in geomaterials, will be investigated. Dynamic problems in viscoelastic solids will be introduced, and essential results in dynamic fracture mechanics and soil dynamics will be summarized.

This chapter is intended to arouse the readers' interest in dynamic problems in geomaterials, and hopefully provide the first stepping stone in exploring fascinating phenomena in dynamics and wave propagations.

9.2 SEISMIC WAVES

Inside the Earth's surface, fault ruptures can release energy stored along a stick-slip fault segment and set the ground shaking by radiating energy through waves. The focal mechanism of faulting can be normal faulting, reverse faulting, strike-slip faulting, or a combination of these (oblique faulting). Data from far-field seismic stations can be used to identify the source mechanisms as well as the location and depth (i.e., hypocenter of the earthquake); but we should be careful that this focal mechanism is only the initial starting mechanism, which may change as a function of time as the rupture process continues along the fault plane. This is especially true for large earthquakes.

As we will see from the next few sections, P-waves travel with the greatest wave speed, following by S-waves, Love waves, and Rayleigh waves. Figure 9.1 illustrates the relatively speeds and their associated particle motions of these waves. However, since P-waves and S-waves are body waves they spread three dimensionally, while surface waves (like Love and Rayleigh waves) are plane waves spread two dimensionally. Body waves decay as $1/R^2$ whereas surface waves decay as $1/\sqrt{R}$, where R is the distance measured from the sources. Surface waves attenuate much more slowly and normally carry more energy than body waves in the far-field. Therefore, earthquake damages in the far-field are normally associated with surface waves. Thus, they are more important than body waves. Shear waves can further be decomposed vertically and horizontally, leading to SV- and SH-waves as shown in Fig. 9.1. In seismology, when the surface waves are the results of multiple reflections and refractions of waves in a layered stratum, they are also called coda waves.

9.3 WAVES IN INFINITE ELASTIC ISOTROPIC SOLIDS

Wave propagations in infinite or unbounded solids are of great simplicity because two different types of waves exist, and they propagate independently. Wave propagation in unbounded solids were first considered by George Green in 1838 and G.G. Stokes in 1849 (Lamb, 1904).

In this section, we start with the equation of motion in terms of the displacement field for isotropic solids (compare (2.72)):

$$(\lambda + \mu)\nabla(\nabla \cdot \boldsymbol{u}) + \mu\nabla^2 \boldsymbol{u} + \rho \boldsymbol{f} = \rho \ddot{\boldsymbol{u}} \tag{9.1}$$

where ρ is the density of the solid. By using the following vector identity (see (1.50)),

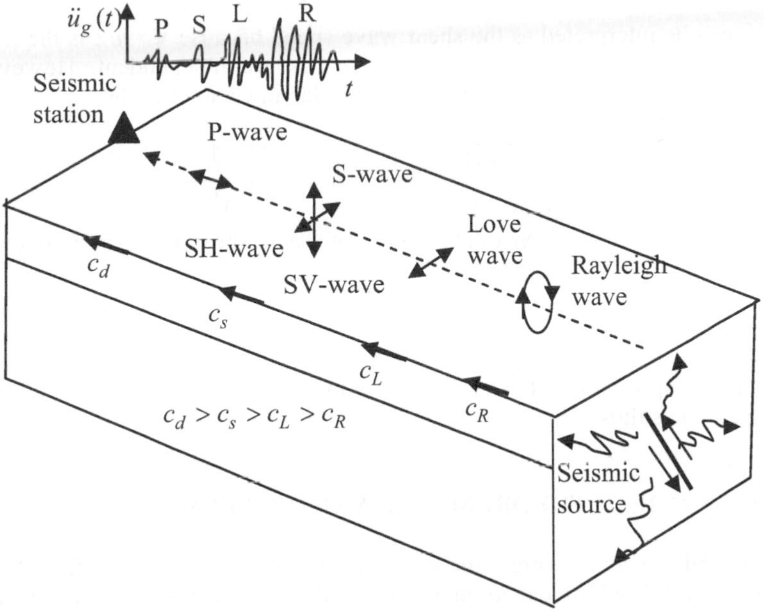

Figure 9.1 Illustrations of P-wave, S-wave, Love wave, and Rayleigh wave from a seismic source and their observations at a seismic station

$$\nabla^2 \boldsymbol{u} = \nabla(\nabla \bullet \boldsymbol{u}) - \nabla \times (\nabla \times \boldsymbol{u}) \tag{9.2}$$

We can rewrite (9.1) as

$$c_d^2 \nabla(\nabla \bullet \boldsymbol{u}) - c_s^2 \nabla \times (\nabla \times \boldsymbol{u}) + \boldsymbol{f} = \ddot{\boldsymbol{u}} \tag{9.3}$$

where

$$c_d = \sqrt{\frac{\lambda + 2\mu}{\rho}}, \quad c_s = \sqrt{\frac{\mu}{\rho}} \tag{9.4}$$

To see the physical meaning of these parameters, we first neglect body force and take the divergence of (9.3) to get

$$c_d^2 \nabla^2 (\nabla \bullet \boldsymbol{u}) = \frac{\partial^2 \nabla \bullet \boldsymbol{u}}{\partial t^2} \tag{9.5}$$

In deriving (9.5), we have used the vector identity that the divergence of the curl of a vector is identically zero. Since $\nabla \cdot \boldsymbol{u}$ is the dilatation, c_d is the dilatational wave speed. Similarly, taking the curl of (9.3) gives

$$c_d^2 \nabla \times \nabla(\nabla \bullet \boldsymbol{u}) - c_s^2 \nabla \times \nabla \times (\nabla \times \boldsymbol{u}) = \frac{\partial^2 (\nabla \times \boldsymbol{u})}{\partial t^2} \tag{9.6}$$

Applying the following vector identities

$$\nabla \times [\nabla \times (\nabla \times \boldsymbol{u})] = \nabla[\nabla \bullet (\nabla \times \boldsymbol{u})] - \nabla^2 (\nabla \times \boldsymbol{u}), \quad \nabla \times (\nabla g) = 0 \tag{9.7}$$

we find

$$c_s^2\nabla^2(\boldsymbol{\nabla}\times\boldsymbol{u}) = \frac{\partial^2}{\partial t^2}(\boldsymbol{\nabla}\times\boldsymbol{u}) \tag{9.8}$$

Thus, c_s can be interpreted as the shear wave speed because $\nabla\times\boldsymbol{u}/2$ is the rotation tensor. If the solid is unbounded, these waves are independent. However, at boundaries they do interact. By applying formulas in Chapter 2, these wave speeds can be rewritten as

$$c_d = \sqrt{\frac{E(1-\nu)}{\rho(1+\nu)(1-2\nu)}}, \quad c_s = \sqrt{\frac{E}{2\rho(1+\nu)}} \tag{9.9}$$

Therefore, the ratio between dilatational wave and shear wave is a function of Poisson's ratio:

$$\frac{c_d}{c_s} = \sqrt{\frac{2(1-\nu)}{(1-2\nu)}} \tag{9.10}$$

For highly compressible solids ($\nu = 0$), we have $c_d/c_s = 1.4142$, and $c_d/c_s \to \infty$ for incompressible solids.

9.4 HELMHOLTZ THEOREM AND WAVE SPEEDS

There is another way to express the wave equations in solids. Recall from Chapter 4 that Helmholtz's additive theorem can be applied to decompose the displacement vector:

$$\boldsymbol{u} = \boldsymbol{\nabla}\phi + \boldsymbol{\nabla}\times\boldsymbol{\psi} \tag{9.11}$$

We have

$$\boldsymbol{\nabla}\boldsymbol{\cdot}\boldsymbol{u} = \nabla^2\phi + \boldsymbol{\nabla}\boldsymbol{\cdot}(\boldsymbol{\nabla}\times\boldsymbol{\psi}) = \nabla^2\phi \tag{9.12}$$

Applying (9.12) to (9.5), we arrive at a commonly used form of wave equation for the scalar function:

$$\nabla^2[c_d^2\nabla^2\phi - \frac{\partial^2\phi}{\partial t^2}] = 0 \tag{9.13}$$

Strictly speaking, we should have

$$c_d^2\nabla^2\phi - \frac{\partial^2\phi}{\partial t^2} = f \tag{9.14}$$

where f is any arbitrary harmonic function. However, most researchers have been restricted to the special case of f being zero. That is,

$$c_d^2\nabla^2\phi = \frac{\partial^2\phi}{\partial t^2} \tag{9.15}$$

On the other hand, the curl of (9.11) gives

$$\begin{aligned}\boldsymbol{\nabla}\times\boldsymbol{u} &= \boldsymbol{\nabla}\times\boldsymbol{\nabla}\phi + \boldsymbol{\nabla}\times\boldsymbol{\nabla}\times\boldsymbol{\psi} = \boldsymbol{\nabla}(\boldsymbol{\nabla}\boldsymbol{\cdot}\boldsymbol{\psi}) - \nabla^2\boldsymbol{\psi} \\ &= -\nabla^2\boldsymbol{\psi}\end{aligned} \tag{9.16}$$

In obtaining (9.16), we have applied the vector identity given in Chapter 1 and have used the commonly adopted constraint $\boldsymbol{\nabla}\cdot\boldsymbol{\psi} = 0$, which is also known as

Lamé's representation. Substitution of (9.16) into (9.8) and dropping the arbitrary harmonic function gives

$$c_s^2 \nabla^2 \boldsymbol{\psi} = \frac{\partial^2 \boldsymbol{\psi}}{\partial t^2} \tag{9.17}$$

For shallow earthquakes, typical values of these wave speeds are $c_d \approx 5.5$–6.2 km/s whereas $c_s \approx 3.2$–3.6 km/s, depending on the crustal properties.

9.5 RAYLEIGH WAVES

9.5.1 Characteristics Equation for Rayleigh Wave Speed

As illustrated in Fig. 9.1, Rayleigh waves propagate on the surface of the Earth and travel more slowly than body waves but their magnitude attenuate much slower than body waves. In mathematical terms, we are seeking a decaying SV-wave component propagating along the surface of a half-space. The mathematical formulation of this wave is first done by Lord Rayleigh in 1887, who was a Nobel Prize winner in physics in 1904, and this surface wave was named after him. As we will see in Section 9.10 that it is of fundamental importance in dynamic fracture mechanics. In terms of scalar and vector, we have

$$c_d^2 \nabla^2 \phi = \frac{\partial^2 \phi}{\partial t^2}, \quad c_s^2 \nabla^2 \psi_2 = \frac{\partial^2 \psi_2}{\partial t^2} \tag{9.18}$$

with $\psi_1 = \psi_3 = 0$. Specializing (9.11) gives the displacement components in terms of these Helmholtz scalar and vector fields as:

$$u_3 = \frac{\partial \phi}{\partial x_3} + \frac{\partial \psi_2}{\partial x_1}, \quad u_1 = \frac{\partial \phi}{\partial x_1} - \frac{\partial \psi_2}{\partial x_3} \tag{9.19}$$

with

$$\phi = \phi(x_1, x_3, t), \quad \psi_2 = \psi_2(x_1, x_3, t) \tag{9.20}$$

It can be shown that $u_2 = 0$. As shown in Fig. 9.2, the wave is expressed to propagate on the surface and its magnitude drops rapidly with depth.

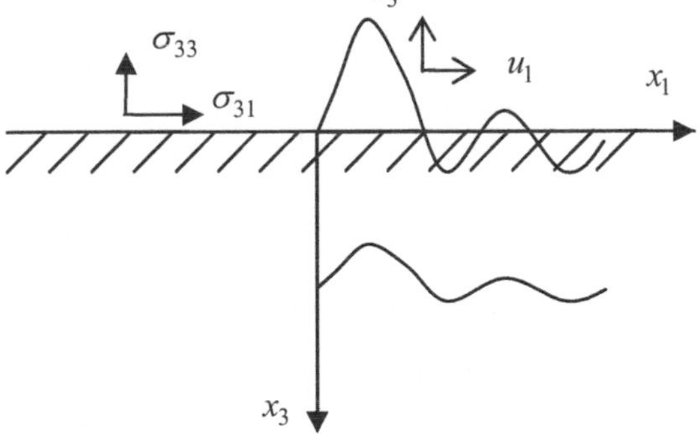

Figure 9.2 Schematic diagram of Rayleigh waves

Mathematically, the following form is expected:

$$\phi(x_1, x_3, t) = F(x_3)\exp i(\omega t - kx_1) \tag{9.21}$$

$$\psi_2(x_1, x_3, t) = G(x_3)\exp i(\omega t - kx_1) \tag{9.22}$$

Substitution of (9.21) and (9.22) into (9.18) gives

$$F'' - r^2 F = 0, \quad G'' - s^2 G = 0 \tag{9.23}$$

where

$$r = k(1 - c_R^2 / c_d^2)^{1/2}, \quad s = k(1 - c_R^2 / c_s^2)^{1/2}, \quad c_R = \omega / k \tag{9.24}$$

The Rayleigh wave speed is denoted by c_R. Thus, in view of the decay condition at $x_3 \to \infty$, we have

$$\phi(x_1, x_3, t) = Ae^{-rx_3}\exp i(\omega t - kx_1) \tag{9.25}$$

$$\psi_2(x_1, x_3, t) = Be^{-rx_3}\exp i(\omega t - kx_1) \tag{9.26}$$

The traction-free conditions are

$$\sigma_{33} = (\lambda + 2\mu)\frac{\partial^2\phi}{\partial x_3^2} + \lambda\frac{\partial^2\phi}{\partial x_1^2} + 2\mu\frac{\partial^2\psi_2}{\partial x_1 \partial x_3} = 0, \tag{9.27}$$

$$\sigma_{31} = 2\mu\frac{\partial^2\phi}{\partial x_1 \partial x_3} + \mu\left(\frac{\partial^2\psi_2}{\partial x_1^2} - \frac{\partial^2\psi_2}{\partial x_3^2}\right) = 0 \tag{9.28}$$

on $x_3 = 0$. Substitution of (9.25) and (9.26) into (9.27) and (9.28) yields

$$\begin{bmatrix} (\lambda + 2\mu)r^2 - \lambda k^2 & 2i\mu sk \\ 2irk & -(k^2 + s^2) \end{bmatrix} \begin{Bmatrix} A \\ B \end{Bmatrix} = \begin{Bmatrix} 0 \\ 0 \end{Bmatrix} \tag{9.29}$$

In terms of the dilatational and shear wave speeds, (9.29) can be expressed as:

$$\begin{bmatrix} \dfrac{c_d^2}{c_s^2} r^2 - \left(\dfrac{c_d^2}{c_s^2} - 2\right)k^2 & 2isk \\ 2irk & -(k^2 + s^2) \end{bmatrix} \begin{Bmatrix} A \\ B \end{Bmatrix} = \begin{Bmatrix} 0 \\ 0 \end{Bmatrix} \tag{9.30}$$

For nontrivial solutions, we must have the determinant of (9.30) be zero. That is,

$$\left(2 - \frac{c_R^2}{c_s^2}\right)^2 - 4\left(1 - \frac{c_R^2}{c_d^2}\right)^{1/2}\left(1 - \frac{c_R^2}{c_s^2}\right)^{1/2} = 0 \tag{9.31}$$

The determination of the roots of c_R is not straightforward and a more robust and routine approach proposed by Segel (1987) will be introduced later in Section 9.5.3. Some authors present the characteristic equation (9.31) in terms of wave slowness (or the inverse wave speed), which was introduced by Sir W.R. Hamilton in the context of optics (Lamb, 1904).

9.5.2 Rayleigh Wave in Solids Satisfying Poisson Condition

First, let us consider a special case. To determine the actual roots of the Rayleigh wave speed, we consider the Poisson condition (Love, 1944). That is, $\nu = ¼$ or $\lambda = \mu$. The characteristic equation for Rayleigh wave speed becomes

$$x^3 - 8x^2 + \frac{56}{3}x - \frac{32}{3} = 0 \tag{9.32}$$

where $x = (c_R/c_s)^2$. There are four roots for this equation:

$$x_1 = 0, \quad x_2 = 4, \quad x_3 = 2 + \frac{2}{\sqrt{3}}, \quad x_4 = 2 - \frac{2}{\sqrt{3}} \tag{9.33}$$

Some of these solutions are extraneous solutions that come from the squaring process, and the actual solution is $x_4 = 0.8453$. The Rayleigh wave speed is smaller than both body wave speeds:

$$c_R = 0.9194 c_s = 0.5308 c_d \tag{9.34}$$

Therefore, Rayleigh wave speed is about half of the dilatation wave speed and 90% of the shear wave speed.

More generally, the Rayleigh wave speed is a function of Poisson's ratio ν. Viktorov (1967) proposed a simple approximation to the Rayleigh wave speed:

$$c_R = \frac{0.862 + 1.14\nu}{1+\nu} c_s \tag{9.35}$$

Note that the Rayleigh wave speed is not a function of the frequency of the wave, and it is called a non-dispersive wave (Achenbach, 1973). Using the second part of (9.30) gives

$$B = \frac{2irk}{k^2 + s^2} A \tag{9.36}$$

Substitution of (9.36) into (9.25) and (9.26) and the results into (9.19) gives

$$u_1 = kA\left\{e^{-rx_3} - \frac{2rs}{k^2+s^2} e^{-sx_3}\right\}\sin(\omega t - kx_1) = \tilde{u}_1(x_3)\sin(\omega t - kx_1) \tag{9.37}$$

$$u_3 = -rA\left\{e^{-rx_3} - \frac{2k^2}{k^2+s^2} e^{-sx_3}\right\}\cos(\omega t - kx_1) = \tilde{u}_3(x_3)\cos(\omega t - kx_1) \tag{9.38}$$

The variations of these displacements with depth have been given in Fig. 5.13 of Achenbach (1973) for two different Poisson's ratios. The corresponding stresses are plotted in Fig. 5.14 of Achenbach (1973). In obtaining the above equations, we have taken the real parts of the displacement fields. Clearly, (9.37) and (9.38) can be combined to give an equation of an ellipse:

$$\left[\frac{u_3}{\tilde{u}_3(x_3)}\right]^2 + \left[\frac{u_1}{\tilde{u}_1(x_3)}\right]^2 = 1 \tag{9.39}$$

To consider the nature of particle movement for Rayleigh wave, we first specialize (9.39) to the ground surface and consider the Poisson condition of $\nu = 1/4$. Thus, we have

$$\left[\frac{u_3}{0.6204kA}\right]^2 + \left[\frac{u_1}{0.4226kA}\right]^2 = 1 \tag{9.40}$$

Figure 9.3 shows the particle on the ground surface. The major axis is along the vertical x_3-direction and the minor axis is along the horizontal x_1-direction with the ellipticity of 1.468. In addition, we can see that the horizontal displacement u_1 given in (9.37) changes sign at a depth h, which is given by

$$h = \frac{1}{s-r} \ln(\frac{2rs}{k^2 + s^2}) = \frac{1.2098}{k} = 0.1925\lambda_R \tag{9.41}$$

where the Rayleigh wavelength is defined by

$$\lambda_R = \frac{2\pi}{k} \tag{9.42}$$

Thus, the horizontal particle movement changes sign at a depth of about 20% of the wavelength. However, the vertical component will not change sign (see Problem 9.11). To consider the sense of direction of particle movement, we note the following forms on the ground surface:

$$u_1 = 0.4226kA \sin \omega\tau \tag{9.43}$$

$$u_3 = 0.6204kA \cos \omega\tau \tag{9.44}$$

where

$$\tau = (t - x_1 / c_R) \tag{9.45}$$

Now, let us consider the direction of movement at different time ranges within a period of $2\pi/\omega$:

(i) First quadrant: $0 < \tau < \dfrac{\pi}{2\omega}$: $u_1 > 0$, $u_3 > 0$

(ii) Second quadrant: $\dfrac{\pi}{2\omega} < \tau < \dfrac{\pi}{\omega}$: $u_1 > 0$, $u_3 < 0$

(iii)Third quadrant: $\dfrac{\pi}{\omega} < \tau < \dfrac{3\pi}{2\omega}$: $u_1 < 0$, $u_3 < 0$

(iv) Fourth quadrant: $\dfrac{3\pi}{2\omega} < \tau < \dfrac{2\pi}{\omega}$: $u_1 < 0$, $u_3 > 0$

As illustrated in Fig. 9.3, the particle movement is counter-clockwise in elliptical motion with a larger vertical component. The motion then turns to clockwise at a depth larger than 0.1925 of the wavelength. Therefore, near the surface, when the surface wave is traveling to the right, the particle will travel backward to the left. At a depth of about 1.6 wavelengths, the displacement will drop to about 5% of its value at the ground surface.

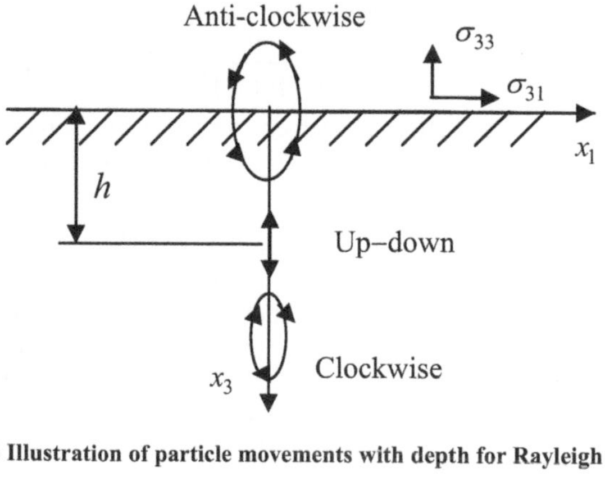

Figure 9.3 Illustration of particle movements with depth for Rayleigh waves

9.5.3 Segel (1987) Method for Arbitrary Poisson's Ratio

Although (9.35) provides a simple approximation of the Rayleigh wave speed with Poisson's ratio, a simple but general technique for determining the Rayleigh wave speed is still desirable. In this section, we will present a general technique from Segel (1987). In particular, we can rewrite (9.31) as

$$c_0^6 - 8c_0^4 + (24 - 16\xi^2)c_0^2 - 16(1-\xi^2) = 0, \tag{9.46}$$

where

$$c_0^2 = \frac{c_R^2}{c_s^2}, \quad \xi^2 = \frac{c_s^2}{c_d^2} \tag{9.47}$$

with c_0 in the following range

$$0 < c_0^2 < 1 \tag{9.48}$$

For $c_0 = 0$, the left-hand side of (9.46) equals $-16(1-\xi^2)$ and thus is negative. At the other extreme, for $c_0 = 1$ the left-hand side of (9.46) equals 1. Therefore, there must be a real root within the physical range.

To give a general method for finding c_0 in (9.46), we first note that the second part of (9.47) can be expressed in terms of Poisson's ratio as

$$\xi^2 = \frac{c_s^2}{c_d^2} = \frac{1-2\nu}{2(1-\nu)} \tag{9.49}$$

Inverting (9.49) gives

$$\nu = \frac{1-2\xi^2}{2(1-\xi^2)} \tag{9.50}$$

On the other hand, (9.46) can be used to solve for ξ^2 in terms of c_0 as

$$\xi^2 = \frac{c_0^6 - 8c_0^4 + 24c_0^2 - 16}{16(c_0^2 - 1)}, \tag{9.51}$$

Therefore, instead of solving for c_0 we know from (9.48) that the range of c_0 is from 0 to 1, so we can calculate backward. That is, for a given value of c_0 we can find ξ^2 from (9.51), then ν from (9.50). Thus, the Rayleigh wave speed c_R can be found as a function of Poisson's ratio. Of course, we have to reject the results and stop our calculations whenever $\xi > 1$ or $0 \le \nu \le 1/2$ is violated.

This is a genius way to solve this seemingly difficult problem. This particular technique illustrates that sometimes a difficult problem becomes a simple one if you look at the problem from another point of view (in this particular example, we look at the problem backward).

In the last section, we showed that for solids satisfying Poisson's condition (i.e., the $\nu = 1/4$ or $\lambda = \mu$), the particle movement changes from counter-clockwise to clockwise if a critical depth is exceeded. At this critical depth, there is only vertical particle movement. We can show that this is true for other values of Poisson's ratio.

Let us start with the magnitude of the displacement fields. From (9.37) and (9.38), we have

$$\tilde{u}_1(x_3) \propto e^{-sx_3}\left\{e^{-(r-s)x_3} - \frac{2rs}{k^2+s^2}\right\} \tag{9.52}$$

$$\tilde{u}_3(x_3) \propto e^{-sx_3}\left\{e^{-(r-s)x_3} - \frac{2k^2}{k^2+s^2}\right\} \tag{9.53}$$

Whether the horizontal and vertical displacements will change sign with the increasing depth (x_3) depends on the relative values of the two terms in the brackets in (9.52) and (9.53). From the definitions of (9.24) that we have $r > s$ because $c_d > c_s$. Therefore, the first term in these brackets is a decreasing function of x_3 and it attains a maximum value of one when $r \to s$ or $c_d \to c_s$. Consequently, the criteria for having the brackets in (9.52) and (9.53) to change sign for increasing x_3 are:

$$1 > \frac{2rs}{k^2+s^2} \quad \text{and} \quad 1 > \frac{2k^2}{k^2+s^2} \tag{9.54}$$

Substituting the definition of (9.24) into the second term inside the brackets of (9.52), we find

$$\frac{2rs}{k^2+s^2} = \frac{2k^2(1-\frac{c_R^2}{c_d^2})^{1/2}(1-\frac{c_R^2}{c_s^2})^{1/2}}{k^2+s^2} \tag{9.55}$$

However, c_R must satisfy the characteristic equation for Rayleigh wave speed (9.31), thus (9.55) becomes

$$\frac{2rs}{k^2+s^2} = \frac{\frac{k^2}{2}(2-\frac{c_R^2}{c_s^2})^2}{k^2+s^2} = \frac{1}{2}\left[\frac{(2-\frac{c_R^2}{c_s^2})^2}{1+(s/k)^2}\right] = \frac{1}{2}\left\{\frac{[1+(s/k)^2]^2}{1+(s/k)^2}\right\} = \frac{1}{2}(2-\frac{c_R^2}{c_s^2}) < 1 \tag{9.56}$$

Since $c_R < c_s$, therefore (9.56) must be smaller than 1. Therefore, it is always possible to find a depth h such that horizontal displacement vanishes. On the other hand, it is straightforward to see the following identity:

$$\frac{2k^2}{k^2+s^2} = \frac{2}{1+(s/k)^2} = \frac{2}{2-c_R^2/c_s^2} > 1 \tag{9.57}$$

Therefore, it is impossible for the vertical displacement to change sign. Therefore, the illustration given in Fig. 9.3 is true regardless of the value of Poisson's ratio.

9.6 LOVE WAVES

Note that a Rayleigh wave is a combination of P-waves and SV-waves near the surface. However, seismic surface waves recorded at seismic stations also show horizontal components of SH-waves. It is thus natural to investigate this possibility. This mathematical problem was first solved by Love (1911), and this surface wave is therefore named after him. In particular, these waves do not exist in homogeneous half-space, but are observed only when there is a low velocity or softer layer overlying a high velocity or firmer layer. They usually travel slightly faster than the Rayleigh waves, about 90% of the S-wave velocity. On the Earth's surface, geological layering is quite common and thus Love waves are also observed during earthquakes.

9.6.1 Non-existence of SH-Wave in Homogeneous Half-Space

For SH-waves, we have the so-called anti-plane motions, and both u_1 and u_3 are zero. The u_2 component is

$$u_2(x_1, x_3, t) = Ae^{-bx_3} \exp i(kx_1 - \omega t) \tag{9.58}$$

For a homogeneous half-space, the boundary conditions at ground surface are

$$\sigma_{33} = 0, \qquad \sigma_{32} = \mu \frac{\partial u_2}{\partial x_3} = 0 \tag{9.59}$$

on $x_3 = 0$. The wave equation for anti-plane motions is (e.g., Freund, 1998):

$$c_s^2 \nabla^2 u_2 = \frac{\partial^2 u_2}{\partial t^2} \tag{9.60}$$

Substitution of (9.58) into (9.60) gives

$$b^2 = k^2[1-(c_L / c_s)^2] \tag{9.61}$$

where $c_L = \omega/k$. It is obvious that the first part of (9.59) is identically satisfied since $u_3 = 0$ whereas the second part of (9.59) gives

$$Ab = 0 \tag{9.62}$$

Therefore, the SH-type surface wave does not exist on homogeneous half-spaces.

9.6.2 Love Waves in an Elastic Layer on a Half-Space

We now turn to the problem of an elastic layer of thickness H on the top of a homogeneous half-space, as shown in Fig. 9.4. With the origin located at the interface, we have the boundary conditions at ground surface as

$$\sigma_{33}^I = 0, \qquad \sigma_{32}^I = 0 \tag{9.63}$$

on $x_3 = -H$ where I indicates the top layer. The interface continuity gives

$$\sigma_{33}^I = \sigma_{33}^{II}, \quad \sigma_{32}^I = \sigma_{32}^{II}, \quad u_2^I = u_2^{II} \tag{9.64}$$

on $x_3 = 0$. The wave equations of the top layer I and the underlying half-space II are

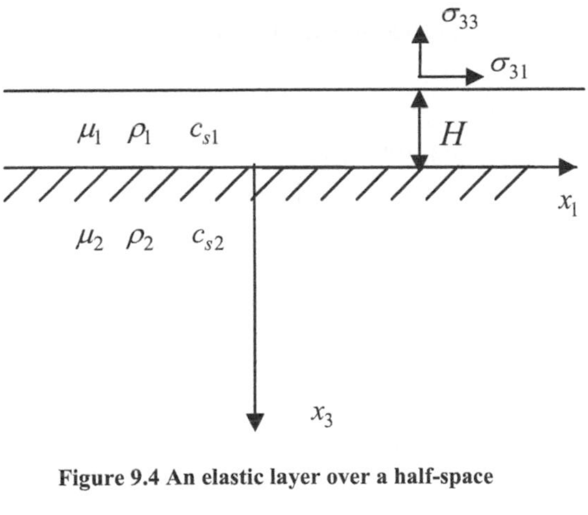

Figure 9.4 An elastic layer over a half-space

$$c_{s1}^2 \nabla^2 u_2^{(I)} = \frac{\partial^2 u_2^{(I)}}{\partial t^2} \qquad c_{s2}^2 \nabla^2 u_2^{(II)} = \frac{\partial^2 u_2^{(II)}}{\partial t^2} \tag{9.65}$$

We obtain the solutions of u_2 for layer I and half-space II as

$$u_2^I(x_1, x_3, t) = f_1(x_3) \exp ik(x_1 - c_L t) \tag{9.66}$$

$$u_2^{II}(x_1, x_3, t) = f_2(x_3) \exp ik(x_1 - c_L t) \tag{9.67}$$

where c_L is the Love wave speed:

$$c_L = \omega / k \tag{9.68}$$

Substitution of (9.66) and (9.67) into (9.65) gives

$$\frac{d^2 f_1}{dx_3^2} - (k^2 - \frac{\omega^2}{c_{s1}^2}) f_1 = 0, \qquad \frac{d^2 f_2}{dx_3^2} - (k^2 - \frac{\omega^2}{c_{s2}^2}) f_2 = 0 \tag{9.69}$$

Since the solution in the half-space must decay to zero as $x_3 \to \infty$, the solution form of f_2 must be

$$f_2 = A \exp(-bx_3) \tag{9.70}$$

where

$$b^2 = k^2 [1 - (c_L / c_{s2})^2] \tag{9.71}$$

However, for layer one there are two possibilities for the solution form:

$$\text{Case 1: } k^2 > \frac{\omega^2}{c_{s1}^2}, \qquad \text{Case 2: } k^2 < \frac{\omega^2}{c_{s1}^2} \tag{9.72}$$

For Case 1, the solution form is

$$u_2^I(x_1, x_3, t) = [D_1 \exp(q_2 x_3) + D_2 \exp(-q_2 x_3)] \exp[i(x_1 - c_L t)] \tag{9.73}$$

$$u_2^{II}(x_1, x_3, t) = A \exp(-bx_3) \exp ik(x_1 - c_L t) \tag{9.74}$$

where

$$q_2 = (k^2 - \frac{\omega^2}{c_{s1}^2})^{1/2} \tag{9.75}$$

The first part of (9.63) is satisfied identically since $u_3 \equiv 0$. The second part of (9.63) gives

$$D_1 = D_2 = D \tag{9.76}$$

Then the solution for the layer becomes

$$u_2^I(x_1, x_3, t) = D \cosh(q_2 x_3) \exp ik(x_1 - c_L t) \tag{9.77}$$

The second and third continuity conditions given in (9.64) can be expressed as

$$\mu_1 \frac{\partial u_2^I}{\partial x_3} = \mu_2 \frac{\partial u_2^{II}}{\partial x_3}, \qquad u_2^I = u_2^{II} \tag{9.78}$$

on $x_3 = 0$. Substitution of (9.73) and (9.74) into (9.78) gives two equations for the unknown constants:

$$\mu_1 q_2 D \sinh(q_2 H) = -\mu_2 b A \exp(-bH) \tag{9.79}$$

$$D \cosh(q_2 H) = A \exp(-bH) \tag{9.80}$$

The condition for the nontrivial solution gives

$$\tanh(q_2 H) = -\frac{b\mu_2}{q_2\mu_1} \tag{9.81}$$

Since $q_2 > 0$, it is obvious that (9.81) cannot be satisfied. That is, the Love wave does not exist for Case 1.

We now turn to Case 2, where the solution forms are

$$u_2^I(x_1, x_3, t) = [B_1 \sin(q_1 x_3) + B_2 \sin(q_1 x_3)] \exp ik(x_1 - c_L t) \tag{9.82}$$

$$u_2^{II}(x_1, x_3, t) = A \exp(-bx_3) \exp ik(x_1 - c_L t) \tag{9.83}$$

where

$$q_1 = (\frac{\omega^2}{c_{s1}^2} - k^2)^{1/2} \tag{9.84}$$

The boundary condition on ground surface ($x_3 = 0$) requires

$$\frac{\partial u_2^I}{\partial x_3} = 0 \tag{9.85}$$

This condition leads to

$$B_1 = 0 \tag{9.86}$$

Continuity conditions of (9.78) give two equations for the unknown constants:

$$\mu_1 q_1 B_2 \sin(q_1 H) = \mu_2 b A \exp(-bH) \tag{9.87}$$

$$B_2 \cos(q_1 H) = A \exp(-bH) \tag{9.88}$$

Nontrivial solution gives the following characteristic equation for the Love wave speed:

$$\tan\left\{kH[(c_L / c_{s1})^2 - 1]^{1/2}\right\} = \frac{\mu_2}{\mu_1}\left[\frac{1-(c_L / c_{s2})^2}{(c_L / c_{s1})^2 - 1}\right]^{1/2} \tag{9.89}$$

where

$$c_{s2} = \sqrt{\frac{\mu_2}{\rho_2}}, \text{ and } \qquad c_{s1} = \sqrt{\frac{\mu_1}{\rho_1}} \tag{9.90}$$

Since $q_1 > 0$ and $b > 0$, we must have from (9.71) and (9.84) that

$$c_{s1} \le c_L \le c_{s2} \tag{9.91}$$

That is, the half-space must be stiffer than the elastic layer in order to have a Love wave exist.

We now consider two special cases. If $c_L \to c_{s2}$, we have the right-hand side of (9.89) approaching zero. This corresponds to $k \to 0$ or $\lambda \to \infty$, and thus it is the long wavelength limit. If $c_L \to c_{s1}$, we have the right hand side of (9.89) approaching infinity. The only possibility is that the argument in the tangent function on the left must be a multiple of $\pi/2$:

$$\frac{2\pi H}{\lambda}[(c_L / c_{s1})^2 - 1]^{1/2} \to (2n+1)\frac{\pi}{2} \tag{9.92}$$

However, the bracket term on the left side of (9.92) approaches zero, and thus the only possibility is for $\lambda \to 0$. Therefore, $c_L \to c_{s1}$ corresponds to short wavelength limit.

That is, the longer the Love wave wavelength, the faster is its wave speed. Therefore, Love waves are dispersive.

For an anisotropic solid, the characteristic equation of the Love wave was given by Stoneley (1949).

9.6.3 Dispersion Characteristics of Love Waves

To examine the dispersion characteristics of Love waves, we first rewrite (9.84) as

$$q_1 = \frac{2\pi}{\lambda}\left(\frac{c_L^2}{c_{s1}^2} - 1\right)^{1/2} \tag{9.93}$$

Equivalently, we have

$$c_L^2 = c_{s1}^2\left(\frac{\lambda^2 q_1^2}{4\pi^2} + 1\right) \tag{9.94}$$

Using (9.94), we can eliminate c_L from (9.89) to get

$$\tan(q_1 H) = \frac{\mu_2}{\mu_1}\left[\frac{4\pi^2}{\lambda^2 q_1^2}\left(1 - \frac{c_{s1}^2}{c_{s2}^2}\right) - \frac{c_{s1}^2}{c_{s2}^2}\right]^{1/2} \tag{9.95}$$

Alternatively, (9.95) can be rewritten as

$$\frac{2\pi}{\lambda q_1} = \left[\frac{c_{s1}^2}{c_{s2}^2 - c_{s1}^2} + \frac{c_{s2}^2}{c_{s2}^2 - c_{s1}^2}\left(\frac{\mu_1}{\mu_2}\right)^2 \tan^2(q_1 H)\right]^{1/2} \tag{9.96}$$

We can take the derivative of (9.94) to get

$$\frac{dc_L}{d\lambda} = \frac{c_{s1}}{4\pi^2}\left\{\frac{\lambda q_1}{[\lambda^2 q_1^2/(4\pi^2) + 1]^{1/2}}\right\}\frac{d\lambda q_1}{d\lambda} \tag{9.97}$$

Taking the derivative of (9.96) with respect to λ gives

$$\left(\frac{KH}{\lambda} + \frac{4\pi^2}{q_1^3 \lambda^3}\right)\frac{d\lambda q_1}{d\lambda} = \frac{KH}{\lambda^2}(\lambda q_1) \tag{9.98}$$

where

$$K = \frac{c_{s2}^2}{c_{s2}^2 - c_{s1}^2}\left(\frac{\mu_1}{\mu_2}\right)^2 \tan\left(\lambda q_1 \frac{H}{\lambda}\right)\sec^2\left(\lambda q_1 \frac{H}{\lambda}\right) \tag{9.99}$$

In deriving (9.98), we use the following identity

$$\frac{d(q_1 H)}{d\lambda} = \frac{d}{d\lambda}\left(\lambda q_1 \frac{H}{\lambda}\right) = \frac{H}{\lambda}\frac{d}{d\lambda}(\lambda q_1) - \frac{q_1 H}{\lambda} \tag{9.100}$$

We now note that since $c_{s2} > c_{s1}$ (see (9.91)), we must have $K > 0$ and, thus, from (9.98) $d(\lambda q_1)/d\lambda > 0$. Therefore, (9.97) shows that $dc_L/d\lambda > 0$. Therefore, the dispersion is normal (Achenbach, 1973). The group velocity can be defined as (Achenbach, 1973)

$$c_g = c_L - \lambda \frac{dc_L}{d\lambda} \tag{9.101}$$

and thus, the group velocity of the Love wave is always smaller than the phase velocity.

We can also consider the penetration depth of the Love wave by considering the following ratio:

$$\frac{u_2^{II}(x_3)}{u_2^{II}(0)} = \exp[-\frac{2\pi x_3}{\lambda}(1-\frac{c_L^2}{c_{s2}^2})^{1/2}] \tag{9.102}$$

For a depth of one wavelength, we have

$$\frac{u_2^{II}(\lambda)}{u_2^{II}(0)} = \exp[-2\pi(1-\frac{c_L^2}{c_{s2}^2})^{1/2}] \tag{9.103}$$

The decay ratio is smaller for larger c_L, and from (9.94) we see that a longer wavelength gives a larger c_L. Thus, long wavelength Love waves travel faster and penetrate farther into the ground.

One major observation is that dispersion of waves only occurs in problems with a length scale. For Love waves, the length scale is the layer thickness *H*. For Rayleigh waves, there is no length scale, and thus the wave is nondispersive. That is, the wave speed is independent of the frequency of the waves. Naturally, if there is no length scale in the problem, whether wave are "long" or "short" is not meaningful and how fast a wave travels is always meaningless without a length scale. Therefore, Rayleigh waves must be nondispersive.

9.7 STONELEY WAVES

Rayleigh waves were considered in Section 9.5, and they propagate along a free surface. Stoneley has extended the analysis to the case that waves travel along an interface between two half-spaces (see Fig. 9.5).

The traction continuity on the interface at $x_3 = 0$ is

$$\sigma_{33}^{(1)} = \sigma_{33}^{(2)}, \quad \sigma_{31}^{(1)} = \sigma_{31}^{(2)} \tag{9.104}$$

and the displacement continuity on the interface at $x_3 = 0$ is

$$u_3^{(1)} = u_3^{(2)}, \quad u_2^{(1)} = u_2^{(2)} \tag{9.105}$$

The following displacement fields in the half-space (1) are sought:

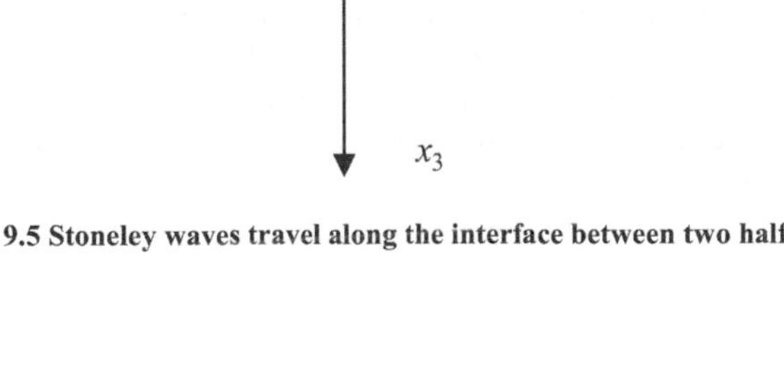

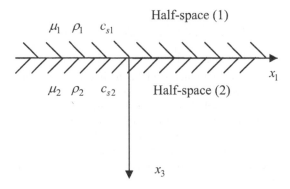

Figure 9.5 Stoneley waves travel along the interface between two half-spaces

$$u_1^{(1)}(x_1,x_3,t)=[A_3e^{b_3x_3}+A_4e^{b_4x_3}]\exp ik(x_1-ct) \tag{9.106}$$

$$u_3^{(1)}(x_1,x_3,t)=[\frac{b_3}{ik}A_3e^{b_3x_3}-\frac{ik}{b_4}A_4e^{b_4x_3}]\exp ik(x_1-ct) \tag{9.107}$$

and these displacements decay to zero for $x_3 \to -\infty$. Substitution of (9.106) and (9.107) into the wave equations gives

$$b_3=k[1-(c/c_{L1})^2]^{1/2}, \qquad b_4=k[1-(c/c_{s1})^2]^{1/2} \tag{9.108}$$

In half-space (2), the displacement fields are

$$u_1^{(2)}(x_1,x_3,t)=[A_1e^{-b_1x_3}+A_2e^{-b_2x_3}]\exp ik(x_1-ct) \tag{9.109}$$

$$u_3^{(2)}(x_1,x_3,t)=[-\frac{b_1}{ik}A_1e^{-b_1x_3}+\frac{ik}{b_2}A_2e^{-b_2x_3}]\exp ik(x_1-ct) \tag{9.110}$$

Substitution of (9.109) and (9.110) into the wave equations gives

$$b_1=k[1-(c/c_{L2})^2]^{1/2}, \qquad b_2=k[1-(c/c_{s2})^2]^{1/2} \tag{9.111}$$

Interface continuity leads to four homogenous equations for four unknowns. For nontrivial solutions, the determinant of the system of equations must vanish. This leads to

$$\det\begin{vmatrix} 1 & 1 & -1 & -1 \\ b_1/k & k/b_2 & b_3/k & k/b_4 \\ 2b_1/k & (2-\frac{c^2}{c_{s2}^2})\frac{k}{b_2} & 2\frac{\mu_1}{\mu_2}(\frac{b_3}{k}) & \frac{\mu_1}{\mu_2}(2-\frac{c^2}{c_{s2}^2})\frac{k}{b_4} \\ (2-\frac{c^2}{c_{s2}^2}) & 2 & -\frac{\mu_1}{\mu_2}(2-\frac{c^2}{c_{s2}^2}) & -2\frac{\mu_1}{\mu_2} \end{vmatrix}=0 \tag{9.112}$$

Written out (9.112) explicitly gives (Ewing et al., 1957)

$$\{(\rho_1-\rho_2)^2-(\rho_1A_2+\rho_2A_1)(\rho_1B_2+\rho_2B_1)\}c^4$$
$$+4(\mu_1-\mu_2)\{\rho_1A_2B_2-\rho_2A_1B_1-\rho_1+\rho_2\}c^2+4(\mu_1-\mu_2)^2(1-A_1B_1)(1-A_2B_2)=0 \tag{9.113}$$

where

$$A_1=[1-(c/c_{L1})^2]^{1/2}, \qquad B_1=[1-(c/c_{s1})^2]^{1/2} \tag{9.114}$$

$$A_2=[1-(c/c_{L2})^2]^{1/2}, \qquad B_2=[1-(c/c_{s2})^2]^{1/2} \tag{9.115}$$

It is straightforward to show that when we take the limiting cases of $\rho_1=0$ and $\mu_1=0$, the Rayleigh wave speed equation given in (9.31) is recovered as a special case.

Ewing et al. (1957) concluded that the Stoneley wave speed is slightly larger than the Rayleigh wave speed in the stiffer half-space. Only for the cases where the density ratio and shear wave speed ratio between the half-spaces are close to unity can the Stoneley wave exist. This condition is normally called Wiecherts' condition (Scholte, 1947). The following condition for the existence of Stoneley waves is simplified from Cagniard (1962):

$$\alpha^2\{4[(1-\beta_1)(1-\beta_3)]^{1/2}-(2-\beta_3)^2\}$$
$$+\alpha\beta_3\{4-2\beta_3+\beta_3[(1-\beta_2)(1-\beta_3)]^{1/2}-4[(1-\beta_1)(1-\beta_3)]^{1/2}\} \quad (9.116)$$
$$-\beta_3^2\{1-[(1-\beta_1)(1-\beta_3)]^{1/2}\}<0$$

where

$$\alpha=\rho_1/\rho_2\,,\quad \beta_1=c_{s2}^2/c_{d1}^2\,,\quad \beta_2=c_{s2}^2/c_{d2}^2\,,\quad \beta_3=c_{s2}^2/c_{s1}^2 \quad (9.117)$$

9.8 ELASTIC-PLASTIC WAVES

9.8.1 Acceleration Waves in Solids

We will present the Hadamard compatibility for acceleration waves in elastic-plastic solids in this section. The classification of waves in elastic-plastic solids was considered by Mandel in 1962. The presentation mainly follows that of Lubliner (1990). Consider an acceleration wave front in a solid with unit normal $\boldsymbol{n}$ and speed c. A time-dependent field ϕ is assumed to be continuous across a wave front but with discontinuous derivative of ϕ across it, as shown in Fig. 9.6. The total differential of the field ϕ is

$$d\phi=\frac{\partial\phi}{\partial\boldsymbol{x}}\bullet d\boldsymbol{x}+\frac{\partial\phi}{\partial t}dt=\frac{\partial\phi}{\partial\boldsymbol{x}}\bullet\boldsymbol{n}cdt+\frac{\partial\phi}{\partial t}dt \quad (9.118)$$

The jump across the wave front is denoted by a square bracket [...], and the jump in total differential of ϕ is $[d\phi]=0$. Thus, we have

$$[\frac{\partial\phi}{\partial\boldsymbol{x}}]\bullet\boldsymbol{n}c+[\frac{\partial\phi}{\partial t}]=0 \quad (9.119)$$

However, since only the component of ϕ along $\boldsymbol{n}$ undergoes a jump, we must have the spatial derivative proportional n_i. Consequently, we have

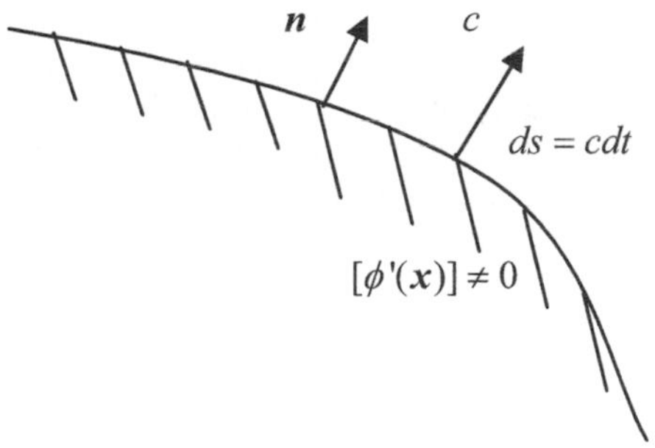

Figure 9.6 Jump of derivative of ϕ across the acceleration wave front

$$[\frac{\partial \phi}{\partial x_i}]c = -[\frac{\partial \phi}{\partial t}]n_i \tag{9.120}$$

Now consider the velocity field, and the jump in velocity gradient becomes

$$c[v_{i,j}] = -[\dot{v}_i]n_j \tag{9.121}$$

This is called Hadamard's compatibility condition. In addition, the momentum balance across the wave front is

$$[\dot{\sigma}_{ij}]n_j = -\rho c[\dot{v}_i] \tag{9.122}$$

In terms of rate formulation, we can write the stress rate in terms of the tangent modulus tensor as

$$[\dot{\sigma}_{ij}] = \bar{C}_{ijkl}[\dot{\varepsilon}_{kl}] \tag{9.123}$$

By recognizing the following definition of strain rate for small deformation,

$$[\dot{\varepsilon}_{kl}] = \frac{1}{2}([v_{k,l}] + [v_{l,k}]) \tag{9.124}$$

Combining Hadamard's compatibility condition, momentum balance, and constitutive law, we have

$$[\dot{\sigma}_{ij}]n_j = -\frac{1}{c}\bar{C}_{ijkl}n_j n_l[\dot{v}_k] = -\rho c[\dot{v}_i] \tag{9.125}$$

Rearrangement of (9.125) gives

$$(\bar{C}_{ijkl}n_j n_l - \rho c^2 \delta_{ik})[\dot{v}_k] = 0 \tag{9.126}$$

For a nontrivial solution for the speed of acceleration waves, we must have

$$\det(\bar{C}_{ijkl}n_j n_l - \rho c^2 \delta_{ik}) = \det(A_{ik} - \rho c^2 \delta_{ik}) = 0 \tag{9.127}$$

The tensor A_{ik} is called the acoustic tensor (Love, 1944).

9.8.2 Shear Banding as Stationary Acceleration Wave

Rice (1976) interpreted ρc^2 as the eigenvalue of A_{ik}. When a solid is subject to a perturbation, if ρc^2 is real and larger than zero, we have stability, whereas if ρc^2 is real and smaller than zero, we have divergent growth of the deformation caused by the perturbation. When ρc^2 is complex, the solid may undergo fluttering instability. However, it is normally argued that fluttering instability is impossible in solids because it needs continuous supply of energy to sustain this kind of instability. Therefore, from stable to divergent growth response, we pass through $c = 0$. A stationary acceleration wave (or $c = 0$) in a solid corresponds to the so-called strain localization or shear banding (e.g., Rudnicki and Rice, 1975; Rudnicki, 1977):

$$\det(\bar{C}_{ijkl}n_j n_l) = 0 \tag{9.128}$$

Recently, Osinov and Wu (2009) showed that the correspondence between stationary acceleration wave and shear band analysis is not true for the case of incrementally nonlinear solids. Earlier strain localization works in Russia can be found in the review by Guz (1985). The result of (9.128) should be interpreted as the upper bound solution of a linear comparison solid (Vardoulakis, 1994). For the case of lower bound bifurcation stress, we refer to the work by Raniecki and Bruhns (1981).

9.8.3 Acoustic Tensor for Geomaterials

We now recall the result in Chapter 5 that the deformation theory of a plastically deforming solid satisfying the non-normality rule gives the tangent modulus tensor as:

$$\bar{C}_{ijkl} = C^e_{ijkl} - \frac{C^e_{ijmn} P_{mn} Q_{pq} C^e_{pqkl}}{h + Q_{rs} C^e_{rsmn} P_{mn}} \tag{9.129}$$

where $\boldsymbol{P}$ and $\boldsymbol{Q}$ can be related to the yield function f and plastic potential g as

$$Q_{ij} = \frac{\partial f}{\partial \sigma_{ij}}, \quad P_{ij} = \frac{\partial g}{\partial \sigma_{ij}} \tag{9.130}$$

The scalar function h is a tangent hardening modulus. The yielding direction in the stress space is given by $\boldsymbol{Q}$ wheras the plastic flow direction is controlled by $\boldsymbol{P}$. If they are the same, the solids are said to obey the normality rule. But for geomaterials, such a normality rule usually does not hold, or deformation is governed by the non-normality rule. When bifurcation analysis for strain localization given in (9.128) is applied to solids obeying (9.129) with $\boldsymbol{P} \neq \boldsymbol{Q}$, the solid is normally called a linear comparison solid because no unloading is considered explicitly.

If the solid deforms elastically, we have the acoustic tensor as

$$A^e_{ik} = C^e_{ijkl} n_j n_l \tag{9.131}$$

This is also called Christoffel stiffness. The wave speed in elastic anisotropic solids, in general, is not purely transverse or purely dilatational like in isotropic solids as discussed in Section 9.3. For the special case of isotropic solids, we have the acoustic tensor as

$$A^e_{ik} = (\lambda + \mu) n_i n_k + \mu \delta_{ik} \tag{9.132}$$

Substitution of (9.132) into (9.127) gives two wave speeds:

$$c_d = \sqrt{\frac{\lambda + 2\mu}{\rho}}, \quad c_s = \sqrt{\frac{\mu}{\rho}} \tag{9.133}$$

which describe wave speed normal to the wave front and parallel to the wave front, respectively. These dilatational and shear wave speeds were obtained in (9.4) in Section 9.3 following a different approach.

For general elastic-plastic or elasto-plastic solids, (9.127) can be expressed as

$$\dot{\sigma}_{ij} = C^e_{ijkl} \dot{\varepsilon}_{kl} - \frac{C^e_{ijmn} P_{mn}}{h + Q_{rs} C^e_{rsmn} P_{mn}} < g > \tag{9.134}$$

where

$$< g > = Q_{pq} C^e_{pqkl} \dot{\varepsilon}_{kl} \tag{9.135}$$

The Macauley bracket is defined as

$$\begin{aligned} \langle g \rangle &= 0 \qquad & g \le 0 \\ &= g \qquad & g > 0 \end{aligned} \tag{9.136}$$

The inclusion of this Macauley bracket in (9.134) makes the second term on the right of (9.134) only appear when continuous plastic loading is applied. For unloading, the strain rate becomes negative, and this leads to zero contribution from the Macauley bracket.

We now consider the jump of the stress rate across the wave front:

$$[\dot{\sigma}_{ij}] = C^e_{ijkl}[\dot{\varepsilon}_{kl}] - \frac{C^e_{ijmn}P_{mn}}{h + Q_{rs}C^e_{rsmn}P_{mn}}[< g >] \tag{9.137}$$

with

$$[\langle g \rangle] = \eta[g] \tag{9.138}$$

Four different cases of wave propagation in elastic-plastic solids can be identified (Lubliner, 1990):

Case 1: plastic wave $\quad g^+ > 0, \quad g^- > 0; \quad \eta = 1 \qquad (9.139)$

Case 2: unloading wave $\quad g^+ > 0, \quad g^- \le 0, \quad 0 < \eta < 1 \qquad (9.140)$

Case 3: reloading wave $\quad g^+ \le 0, \quad g^- > 0, \quad 0 < \eta < 1 \qquad (9.141)$

Case 4: elastic wave $\quad g^+ \le 0, \quad g^- \le 0, \quad \eta = 0 \qquad (9.142)$

where $g+$ and $g-$ are the values of g ahead of and behind the wave front, respectively. In Case 1, the solid is plastically deforming on both sides of the wave front; in Case 2, the solid is plastically deforming in front of the wave front but is elastically unloading behind the wave front; Case 3 is called the reloading wave because the material is elastically deforming ahead of the wave front but plastically deforming behind the wave front; and Case 4 represents that the body is deforming elastically when the wave passes through. Since $\eta = 0$ for case 4, the second term in (9.137) vanishes and thus Case 4 represents an elastic wave.

Lubliner (1990) considered two more cases of the material ahead of the wave front elastic but about to yield (.i.e., $g^+ > 0$), and η is defined as:

$$\eta = \frac{g^-}{g^- - g^+} \tag{9.143}$$

Then, two more cases occur:

Case 5: strong loading wave $\quad g^- > g^+ > 0; \quad \eta > 1 \qquad (9.144)$

Case 6: weak loading wave $\quad g^+ > g^- > 0; \quad \eta < 0 \qquad (9.145)$

The effective tangent modulus can now be expressed as:

$$\bar{C}_{ijkl} = C^e_{ijkl} - \eta N_{ij} M_{kl} \tag{9.146}$$

where

$$N_{ij} = C^e_{ijkl}P_{kl} / \sqrt{h + Q_{mn}C^e_{mnkl}P_{kl}} \tag{9.147}$$

$$M_{ij} = C^e_{ijkl}Q_{kl} / \sqrt{h + Q_{mn}C^e_{mnkl}P_{kl}} \tag{9.148}$$

Consequently, the acoustic tensor becomes

$$A_{ik} = A^e_{ik} - \eta d_i \tilde{d}_k \tag{9.149}$$

where the elastic acoustic tensor is defined in (9.131) and

$$d_i = N_{ij}n_j, \quad \tilde{d}_i = M_{ij}n_j \tag{9.150}$$

9.8.4 Wave Speed Analysis

Note from (9.149) that for geomaterials the acoustic tensor is not symmetric (i.e., $A_{ij} \neq A_{ji}$). However, when there is no plastic deformation, the elastic acoustic tensor is symmetric. Therefore, the eigenvalues of elastic acoustic tensor $\boldsymbol{A}^e$ and $\boldsymbol{A}$ are assumed as:

$$A_1^e \geq A_2^e \geq A_3^e\,, \quad A_1 \geq A_2 \geq A_3 \tag{9.151}$$

Note in general that A_α ($\alpha = 1,2,3$) may not be real because $\boldsymbol{A}$ is unsymmetric. For the time being, we assume that real eigenvalue exists for $\boldsymbol{A}$ such that its value is A. The characteristic equation for A is:

$$\det(\boldsymbol{A} - \boldsymbol{I}A) = \begin{vmatrix} A_1^e - \eta d_1\tilde{d}_1 - A & -\eta d_1\tilde{d}_2 & -\eta d_1\tilde{d}_3 \\ -\eta d_2\tilde{d}_1 & A_2^e - \eta d_2\tilde{d}_2 - A & -\eta d_2\tilde{d}_3 \\ -\eta d_3\tilde{d}_1 & -\eta d_3\tilde{d}_2 & A_3^e - \eta d_3\tilde{d}_3 - A \end{vmatrix} = 0 \tag{9.152}$$

or

$$\begin{aligned} F(A) &= (A_1^e - A)(A_2^e - A)(A_3^e - A) \\ &-\eta[d_1\tilde{d}_1(A_2^e - A)(A_3^e - A) + d_2\tilde{d}_2(A_1^e - A)(A_3^e - A) + d_3\tilde{d}_3(A_1^e - A)(A_2^e - A)] = 0 \end{aligned} \tag{9.153}$$

For $\eta = 0$, we have only elastic wave speeds. For $\eta > 0$, it is straightforward to see that

$$F(A_1^e) \leq 0\,, \quad F(A_2^e) \geq 0\,, \quad F(A_3^e) \leq 0 \tag{9.154}$$

Therefore, we must have:

$$A_1^e \geq A_1 \geq A_2^e \geq A_2 \geq A_3^e \geq A_3 \tag{9.155}$$

Note also that A_α decreases monotonically with η. For $\eta = 1$, we have the plastic wave speed, and for $0 < \eta < 1$ we have the unloading wave speed. Thus, the elastic wave speed is larger than the unloading wave speed and even larger than the plastic wave speed:

$$A_\alpha^p \leq A_\alpha^u \leq A_\alpha^e \qquad \alpha = ,1,2,3 \tag{9.156}$$

9.9 WAVES IN VISCOELASTIC SOLIDS

9.9.1 Complex Moduli

In Chapter 7, we considered quasi-static problems in viscoelastic solids. In this section, we will consider wave speeds in viscoelastic solids. For dynamic viscoelastic problems, it is more advisable to use the Fourier transform instead of the Laplace transform for the time variables. For example, we can again start with the force equilibrium of the three-parameter standard linear solids or Model A given in Fig. 7.3(a):

$$(1 + \frac{\eta_2}{\mu_1 + \mu_2}\frac{\partial}{\partial t})\sigma = \frac{\mu_1}{\mu_1 + \mu_2}(\mu_2 + \eta_2\frac{\partial}{\partial t})\gamma \tag{9.157}$$

Let us consider a loading of harmonic type. The corresponding harmonic response is:

$$\sigma(t) = \sigma^* e^{i\omega t}, \quad \gamma(t) = \gamma^* e^{i\omega t} \tag{9.158}$$

$$(\mu_1 + \mu_2 + \eta_2 i\omega)\sigma^* = \mu_1(\mu_2 + \eta_2 i\omega)\gamma^* \tag{9.159}$$

Therefore, a complex shear modulus can be defined as

$$G_S^*(i\omega) = \frac{\sigma^*}{\gamma^*} = \frac{\mu_1(\mu_2 + \eta_2 i\omega)}{\mu_1 + \mu_2 + \eta_2 i\omega} = G'(\omega) + iG''(\omega) \tag{9.160}$$

It is straightforward to show that

$$G'(\omega) = \frac{(\mu_1 + \mu_2)\mu_1\mu_2 + \mu_1\omega^2\eta_2^2}{(\mu_1 + \mu_2)^2 + \omega^2\eta_2^2} \tag{9.161}$$

$$G''(\omega) = -\frac{[\mu_1\mu_2 + (\mu_1 + \mu_2)\mu_1]\omega\eta_2}{(\mu_1 + \mu_2)^2 + \omega^2\eta_2^2} \tag{9.162}$$

Note that the superscript prime in (9.161) and (9.162) has nothing to do with differentiation. For general time harmonic motions, we have the constitutive law in transformed space as

$$s_{ij}^*(\omega) = 2G_s^*(\omega)e_{ij}^*, \quad \sigma_{kk}^*(\omega) = 2G_B^*(\omega)\varepsilon_{kk}^* \tag{9.163}$$

The spring-dashpot model of standard linear solid gives the complex shear modulus but not the complex bulk modulus. As discussed in Chapter 7, the complex bulk modulus is normally assumed to be elastic. For viscoelastic responses, the stress can be expressed in terms of the relaxation functions in shear and in bulk deformations:

$$\sigma_{ij} = \delta_{ij}\int_0^t [G_B(t-s) - \frac{2}{3}G_S(t-s)]\frac{d\varepsilon_{kk}}{ds}ds + 2\int_0^t G_S(t-s)\frac{d\varepsilon_{ij}}{ds}ds \tag{9.164}$$

Alternatively, in Fourier transform space (9.164) can be expressed as

$$\sigma_{ij}^*(\omega) = \delta_{ij}[G_B^*(\omega) - \frac{2}{3}G_s^*(\omega)]\varepsilon_{kk}^* + 2G_s^*(\omega)\varepsilon_{ij}^* \tag{9.165}$$

where

$$\sigma_{ij} = \sigma_{ij}^* e^{i\omega t}, \quad \varepsilon_{ij} = \varepsilon_{ij}^* e^{i\omega t} \tag{9.166}$$

9.9.2 Longitudinal and Transverse Waves Speeds

We consider a time harmonic wave of the form:

$$\boldsymbol{u} = A\boldsymbol{m}\exp[i(k\boldsymbol{x}\bullet\boldsymbol{n} - \omega t)], \tag{9.167}$$

where $\boldsymbol{n}$ and $\boldsymbol{m}$ are the unit vectors along the direction of propagation and the direction of particle motion, respectively. Substitution of (9.167) into the displacement-strain relation gives

$$\boldsymbol{\varepsilon} = \frac{ikA}{2}(\boldsymbol{mn} + \boldsymbol{nm})\exp[i(k\boldsymbol{x}\bullet\boldsymbol{n} - \omega t)], \tag{9.168}$$

Substitution of (9.165), (9.167), and (9.168) into the Fourier transform of the following equation of motion

$$\nabla\bullet\boldsymbol{\sigma} = \rho\ddot{\boldsymbol{u}} \tag{9.169}$$

gives

$$[k^2 G_s^*(\omega) - \rho\omega^2]\boldsymbol{m} + k^2[G_B^*(\omega) + \frac{1}{3}G_s^*(\omega)](\boldsymbol{m}\bullet\boldsymbol{n})\boldsymbol{n} = 0 \tag{9.170}$$

If the particle motion is along the direction of propagation, we have

$$\boldsymbol{m} = \pm\boldsymbol{n}, \quad \boldsymbol{m}\bullet\boldsymbol{n} = 1 \tag{9.171}$$

Thus, we have for this longitudinal wave

$$k = [\frac{\rho}{G_B^*(\omega) + \frac{4}{3}G_S^*(\omega)}]^{1/2}\omega, \text{ or } \quad c_d = \frac{\omega}{k} = [\frac{G_B^*(\omega) + \frac{4}{3}G_S^*(\omega)}{\rho}]^{1/2} \tag{9.172}$$

If the motion is perpendicular to the direction of propagation, we have

$$\boldsymbol{m}\bullet\boldsymbol{n} = 0 \tag{9.173}$$

For this shear or transverse wave, we have

$$k = [\frac{\rho}{G_S^*(\omega)}]^{1/2}\omega, \text{ or } \quad c_s = \frac{\omega}{k} = [\frac{G_S^*(\omega)}{\rho}]^{1/2} \tag{9.174}$$

From (9.160), (9.172), and (9.174) that wave number k must be complex for both longitudinal and shear waves. That is, waves in viscoelastic solids are dispersive and thus energy dissipating. Substitution of a complex k into (9.167) illustrates that the wave motion is a decreasing function of x. Therefore, viscoelastic behavior induces both dispersion and attenuation. The shape of a propagating pulse will change with distance in viscoelastic solids.

In general, we have discussed the correspondence principle between static and viscoelastic problems. Actually, the same principle can also be extended to solve dynamic problems (Achenbach, 1973). Details will not be given here. Full coverage of wave propagation in viscoelastic layered media is given by Borcherdt (2009). Because of the complex nature of (9.172) and (9.174), Borcherdt (2009) showed that the waves in homogenous isotropic linear viscoelastic solids can be classified as P-wave (elliptical particle motion), Type I S-wave (elliptical particle motion), and Type II S-wave (linear particle motion). In addition, Borcherdt (2009) showed that Rayleigh-type and Love-type waves also exist in homogenous isotropic linear viscoelastic half-spaces or layered half-spaces. For example, the P-wave motion is not parallel to the free surface whereas Type I S-wave motion is not perpendicular to the free surface. Because of attenuation, the reciprocal quality factor or the Q factor also enters the solution. These solutions can be applied to seismology and geophysics problems.

9.10 DYNAMIC FRACTURE MECHANICS

Introduction to fracture mechanics was covered in Chapter 6, but it was restricted to static loading and quasi-static crack propagation. However, under dynamic loadings, the stress intensity factor as well as the fracture propagation process is a function of time. In general, four cases can be classified (Achenbach, 1974):

Case 1: Quasi-static loading and quasi-static fracture
Case 2: Quasi-static loading and dynamic fracture
Case 3: Dynamic loading and dynamic fracture
Case 4: Dynamic loading and quasi-static fracture

For case 1, the inertia effect can be completely neglected but the fracture can still propagate nonlinearly. Development of dynamic fracture mechanics started in the 1950s. Some fundamental problems in dynamic fracture mechanics were considered by Yoffe in 1951, Craggs in 1960, Broberg in 1960, Baker in 1962, Freund in the early 1970s, and many others (Rice, 1968b; Ravi-Chandar, 2004). For some historical developments on this topic, we refer to Rice (1968b), Freund (1998), Broberg (1999), and Ravi-Chandar (2004). In this section, we will not go into the details of the analysis of dynamic fracture; instead, some essential results of dynamic fracture mechanics reported in Freund (1998) and Ravi-Chandar (2004) will be summarized as an introduction to dynamic fracture mechanics.

9.10.1 Dynamic Solutions for a Stationary Crack

A special situation of case 4 is a dynamic stress intensity factor for a stationary crack (see Fig. 9.7). A suddenly applied traction on the crack face leads to the following elastodynamic solution near the tip of a stationary crack (Freund, 1998):

$$\sigma_{yy}(x,0,t) \approx \frac{K_I(t)}{\sqrt{2\pi x}}, \quad \sigma_{xy}(x,0,t) \approx \frac{K_{II}(t)}{\sqrt{2\pi x}}, \quad \sigma_{zy}(x,0,t) \approx \frac{K_{III}(t)}{\sqrt{2\pi x}} \tag{9.175}$$

These solutions are for modes I, II, and III, respectively. The corresponding stress intensity factors are:

$$K_I(t) = 2\sigma^* \sqrt{\frac{c_d t(1-2\nu)}{\pi(1-\nu)}}, \quad K_{II}(t) = 2\tau^* \sqrt{\frac{2c_s t}{\pi(1-\nu)}}, \quad K_{III}(t) = 2\tau^* \sqrt{\frac{2c_s t}{\pi}} \tag{9.176}$$

These solutions can be obtained by the Wiener–Hopf method based on the Laplace transform (Freund, 1998).

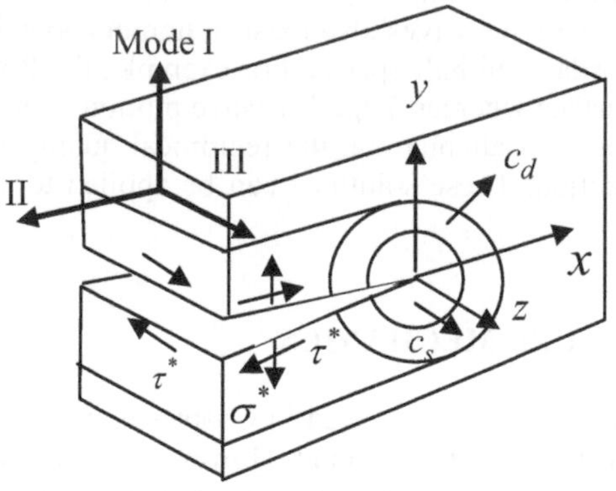

Figure 9.7 Dynamic loading on stationary crack

9.10.2 Asymptotic Fields near a Moving Crack Tip

The asymptotic expansion of the stress field around a moving crack tip, shown in Fig. 9.8 can be written in a universal form:

$$\sigma_{ij} \approx \frac{K_I}{\sqrt{2\pi r}} \Sigma_{ij}^I(\theta, v) + \frac{K_{II}}{\sqrt{2\pi r}} \Sigma_{ij}^{II}(\theta, v) + \frac{K_{III}}{\sqrt{2\pi r}} \Sigma_{ij}^{III}(\theta, v) \tag{9.177}$$

For mode I, the characteristic angular variations for the case of subsonic crack speed are (i.e., $v \leq c_s$) (Ravi-Chandar, 2004; Freund, 1998)

$$\Sigma_{11}^I(\theta, v) = \frac{1}{D}\left\{(1+\alpha_s^2)(1+2\alpha_d^2-\alpha_s^2)\frac{1}{\sqrt{\gamma_d}}\cos\frac{\theta_d}{2} - 4\alpha_s\alpha_d\frac{1}{\sqrt{\gamma_s}}\cos\frac{\theta_s}{2}\right\} \tag{9.178}$$

$$\Sigma_{12}^I(\theta, v) = \frac{2\alpha_d(1+\alpha_s^2)}{D}\left\{\frac{1}{\sqrt{\gamma_d}}\sin\frac{\theta_d}{2} - \frac{1}{\sqrt{\gamma_s}}\sin\frac{\theta_s}{2}\right\} \tag{9.179}$$

$$\Sigma_{22}^I(\theta, v) = -\frac{1}{D}\left\{(1+\alpha_s^2)^2\frac{1}{\sqrt{\gamma_d}}\cos\frac{\theta_d}{2} - 4\alpha_s\alpha_d\frac{1}{\sqrt{\gamma_s}}\cos\frac{\theta_s}{2}\right\} \tag{9.180}$$

where

$$\alpha_s = \sqrt{1 - v^2/c_s^2}, \quad \alpha_d = \sqrt{1 - v^2/c_d^2}, \quad \gamma_d = \sqrt{1 - (v\sin\theta/c_d)^2},$$

$$\gamma_s = \sqrt{1 - (v\sin\theta/c_s)^2}, \quad \tan\theta_d = \alpha_d\tan\theta, \quad \tan\theta_s = \alpha_s\tan\theta,$$

$$\theta = \tan^{-1}(\xi_2/\xi_1), \quad D = 4\alpha_s\alpha_d - (1+\alpha_s^2)^2 \tag{9.181}$$

Note that D is the Rayleigh function given on the left-hand side of (9.31). For mode II, the characteristic angular variations for the case of subsonic speed are (i.e., $v \leq c_s$) (Ravi-Chandar, 2004; Freund, 1998):

$$\Sigma_{11}^{II}(\theta, v) = -\frac{2\alpha_s}{D}\left\{(1+2\alpha_d^2-\alpha_s^2)\frac{1}{\sqrt{\gamma_d}}\sin\frac{\theta_d}{2} - (1+\alpha_s^2)\frac{1}{\sqrt{\gamma_s}}\sin\frac{\theta_s}{2}\right\} \tag{9.182}$$

$$\Sigma_{12}^{II}(\theta, v) = \frac{1}{D}\left\{4\alpha_d\alpha_s\frac{1}{\sqrt{\gamma_d}}\cos\frac{\theta_d}{2} - (1+\alpha_s^2)^2\frac{1}{\sqrt{\gamma_s}}\cos\frac{\theta_s}{2}\right\} \tag{9.183}$$

$$\Sigma_{22}^{II}(\theta, v) = \frac{2\alpha_s(1+\alpha_s^2)}{D}\left\{\frac{1}{\sqrt{\gamma_d}}\sin\frac{\theta_d}{2} - \frac{1}{\sqrt{\gamma_s}}\sin\frac{\theta_s}{2}\right\} \tag{9.184}$$

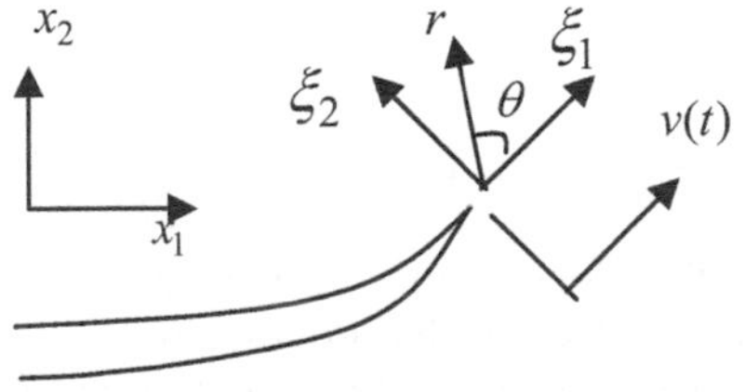

Figure 9.8 Asymptotic field near a moving crack tip

For mode III, the characteristic angular variations for the case of subsonic speed are (i.e., $v \leq c_s$) (Ravi-Chandar, 2004; Freund, 1998)

$$\Sigma_{13}^{III}(\theta,v) = -\frac{1}{\alpha_s\sqrt{\gamma_s}}\sin\frac{\theta_s}{2}, \qquad \Sigma_{23}^{III}(\theta,v) = \frac{1}{\sqrt{\gamma_s}}\cos\frac{\theta_s}{2} \tag{9.185}$$

where

$$\theta_s = \tan^{-1}(\alpha_s\xi_2 / \xi_1) \tag{9.186}$$

The angular variation differs little from the corresponding quasi-static results for $v/c_s < 0.4$, but the velocity effect cannot be neglected for larger v. In addition, the terminal crack speed is normally much smaller than the Rayleigh wave speed of the solids. A concise derivation of the universal forms given in (9.177) can be found in Ravi-Chandar (2004). More generally, the crack propagation speed can in general be classified into three regimes:

Subsonic: $$v < c_s \tag{9.187}$$

Transonic: $$c_s < v < c_d \tag{9.188}$$

Supersonic: $$v > c_d \tag{9.189}$$

Photographs of isochromatic fringes around a moving crack tip for the transonic case can be found in Figs. 3.10–3.11 of Ravi-Chandar (2004). The crack tip fields given in (9.177) remains valid for crack moving with nonuniform speed, with the exception that $K_I = K_I\,(t,v)$ (Ravi-Chandar, 2004). There were also different terminologies for crack speed classification used in the literature. For example, Slepyan (2002) defined (9.188) as intersonic, and the subsonic region can further be divided into "sub-Rayleigh" and "super-Rayleigh" regions. From the energy release point of view, the latter case corresponds to an energy source propagating at the crack tip and it was argued that it is possible as a result of the action of residual stress (Slepyan, 2002). Another term commonly used for transonic defined in (9.188) is "supershear" (Lu et al., 2010), and there is observational evidence of the 2001 Kokoxili (Tibet) earthquake rupturing at supershear speed (Vallée et al., 2008).

9.10.3 Dynamic Energy Release Rate

The energy release rate introduced by George Irwin can be extended to dynamic cases. A universal form can be obtained which is independent of applied loading and body configuration. In particular, the energy release rate is (Freund, 1998)

$$G = \frac{1-\nu^2}{E}[A_I(v)K_I^2 + A_{II}(v)K_{II}^2] + \frac{1}{2\mu}A_{III}(v)K_{III}^2 \tag{9.190}$$

where

$$A_I(v) = \frac{v^2\alpha_d}{(1-\nu)c_s^2 D}, \qquad A_{II}(v) = \frac{v^2\alpha_s}{(1-\nu)c_s^2 D}, \qquad A_{III}(v) = \frac{1}{\alpha_s} \tag{9.191}$$

For $v \to 0^+$ we have all functions $A_I(v)$, $A_{II}(v)$ and $A_{III}(v) \to 1$. The definitions for other parameters have been given in the previous sections. The limiting speed for modes I and II is c_R (i.e., $D(c_R) = 0$), whereas the limiting speed for mode III is c_s.

9.10.4 Dynamic Fracture Toughness

In Chapter 6, we discussed the concept that once the stress intensity factor at a crack tip equals the critical stress intensity factor or fracture toughness. However, the fracture toughness for dynamic crack propagation is more complicated than static crack growth. The following discussion is adopted from Ravi-Chandar (2004). In particular, similar to the quasi-static case we can assume that dynamic crack propagation occurs when the dynamic stress intensity factor attains a critical value:

$$K_I^{dyn} = K_{Id}(T, \dot{K}_I^{dyn}) \qquad (9.192)$$

where K_{Id} is the dynamic initiation toughness and is in general a function of temperature T and loading rate. Once crack growth starts the propagation speed will jump to a large finite value instantaneously, as shown by the arrow in Fig. 9.9. The subsequent growth must be determined by another growth criterion called dynamic crack growth toughness K_{ID}:

$$K_I^{dyn} = K_{ID}(v; T, \dot{K}_I^{dyn}) \qquad (9.193)$$

Experiments showed that this dynamic crack growth toughness is also a function of crack speed v (e.g., see Figure 1.28 of Kannien and Popelar, 1985). Note that the dynamic initiation toughness K_{Id} is not on the K_{ID} curve shown in Fig. 9.9. That is,

$$K_{ID}(v \to 0; T, \dot{K}_I^{dyn}) \neq K_{Id} \qquad (9.194)$$

However, the measurement of crack growth at low speed v is very difficult because of the jump character of crack speed during initiation (Ravi-Chandar, 2004). Therefore, the lower part K_{ID} curve is shown by a dotted line with a "question mark." Crack propagation will stop if

$$K_I^{dyn} < K_{Ia}(T) \qquad (9.195)$$

For more detailed discussion see Ravi-Chandar (2004). The situation of dynamic fracture toughness closely resembles the case of static and the kinetic frictional coefficient of a sliding block on a rough surface. That is, the kinetic frictional coefficient is smaller than the static friction coefficient once the sliding motion starts.

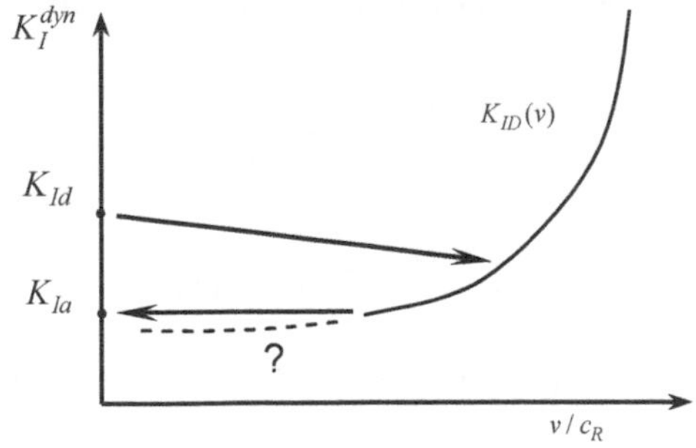

Figure 9.9 Dynamic crack growth criteria (after Ravi-Chandar (2004) with permission from Elsevier)

9.11 VIBRATIONS AND SOIL DYNAMICS

One of the main problems in wave propagation in soil is the amplification of seismic waves during an earthquake. In this section, following Das (1993), we will consider a simple case of soil of thickness H overlying a rock stratum and subject to a seismic shear motion of u_g(t). The equation of motion for a soil column of unit cross-section area is

$$\rho\frac{\partial^2 u}{\partial t^2}+c\frac{\partial u}{\partial t}-G\frac{\partial^2 u}{\partial y^2}=-\rho\frac{\partial^2 u_g}{\partial t^2} \tag{9.196}$$

where ρ, G, and c are the density, shear modulus, and damping coefficient of the soil, respectively. We look for a solution of the following form:

$$u(y,t)=\sum_{n=1}^{\infty}Y_n(y)X_n(t) \tag{9.197}$$

With the traction boundary on the free surface, we have

$$Y_n(y)=\cos[\frac{y}{2H}(2n-1)] \tag{9.198}$$

Substitution of (9.198) into (9.196) gives

$$\ddot{X}_n(y)+2D_n\omega_n\dot{X}_n+\omega_n^2X_n=(-1)^n[\frac{4}{(2n-1)\pi}]\ddot{u}_g \tag{9.199}$$

where

$$\omega_n=[\frac{(2n-1)\pi}{2H}]\sqrt{\frac{G}{\rho}},\quad D_n=\frac{c}{2\rho\omega_n} \tag{9.200}$$

Once u_g(t) is given, the displacement solution can be evaluated by solving (9.199). Actually, this formulation can easily be extended to a multi-layered system. For details see Idriss and Seed (1968). For real soil, both the shear modulus G and damping coefficient D_n is a function shear strain γ. Therefore, we have to update the modulus and damping. Once we find a solution, we can find an equivalent shear strain for the whole soil layer. This value can be substituted into experimental curves of $G(\gamma)$ and $D_n(\gamma)$ or an empirical model, to yield the updated modulus and damping. For example, Hardin and Drnevich (1972) made the following predictions:

$$G=\frac{G_{\max}}{1+\gamma_1},\quad D_n=D_{\max}(\frac{\gamma_2}{1+\gamma_2}) \tag{9.201}$$

where

$$G_{\max}=\frac{1230(2.97-e)^2}{1+e}(OCR)^K\bar{\sigma}_0^{1/2} \tag{9.202}$$

$$D_{\max}(\%)=31-(3+0.03f)\bar{\sigma}_0^{1/2}+1.5f^{1/2}-1.5\log N \tag{9.203}$$

$$\gamma_1=(\frac{\gamma}{\gamma_r})[1+a_1e^{-b_1(\gamma/\gamma_r)}],\quad \gamma_2=(\frac{\gamma}{\gamma_r})[1+a_2e^{-b_2(\gamma/\gamma_r)}] \tag{9.204}$$

$$a_1=1+0.25\log N,\quad a_2=1+0.2f^{1/2} \tag{9.205}$$

$$b_1 = 1.3, \quad b_2 = 0.2f(e^{-\bar{\sigma}_0}) + 2.25\bar{\sigma}_0 + 0.3\log N \tag{9.206}$$

$$\bar{\sigma}_0 = \frac{1}{3}(\bar{\sigma}_v + 2K_0\bar{\sigma}_v) \tag{9.207}$$

where e is the void ratio and OCR is the over-consolidation ratio. The number of cycles of loading is denoted by N, the frequency is f, K is a function of the plasticity index ranging from 0 to 0.5, and K_0 is the lateral earth pressure coefficient at rest.

Once the updated G and D_n are found, they can be back-substituted into (9.199) to give an updated solution. The newly calculated γ can be used to find another updated G and D_n. This iterative procedure will be continued until an converged solution is obtained. This is sometimes called an equivalent linear wave analysis for nonlinear constitutive soil response.

For the complicated cases of soil–pile–structure interactions, an analytical solution has been obtained by Koo et al. (2003). For soil–pile interaction, the nonlinear dynamic stiffness of the soil–pile system has been calculated using the equivalent linear analysis discussed in this section (Chau and Yang, 2005).

9.12 SUMMARY AND FURTHER READING

9.12.1 Summary

In this chapter, we have summarized some essential results for seismic waves, including P-waves, S-waves (body waves), Rayleigh waves (surface waves), Love waves (SH surface waves), and Stoneley waves (interface waves). The wave equations in elastic isotropic solids can be obtained by Helmholtz theorem for vector decomposition, which has been introduced in Section 4.2.1 of Chapter 4 for "Method of Solution for 3-D Elasticity." The wave speed and wave motion for Rayleigh waves are discussed in full detail when considering the special solution for solids satisfying Poisson's conditions in Section 9.5.1 and 9.5.2. The solution of the Rayleigh wave speed together with the particle motion for the arbitrary Poisson's ratio is discussed in Section 9.5.3. It was shown that the solution for Rayleigh wave characteristic equation can be evaluated easily if we look at the problem from a different point of view. Love waves and their dispersion characteristics are introduced in Section 9.6, and Stoneley waves are discussed in Section 9.7. The general wave equation for elastic-plastic waves in geomaterials is discussed in Section 9.8, including Hadamard's jump condition for acceleration waves, and the interpretation of strain localization as a stationary acceleration wave. Wave speed, dispersion, and attenuation in viscoelastic solids are summarized in Section 9.9. Some essential results of dynamic fracture mechanics are introduced in Section 9.10, including the time-dependent stress intensity factors for modes I, II, and III of a stationary crack subject to a suddenly induced traction on the crack faces. The general asymptotic stress field near the tip of a moving crack is given in Section 9.10.2 and the universal form of speed dependence of the dynamic energy release rate is summarized in Section 9.10.3. Finally, soil amplification of ground motion is discussed in Section 9.11.

9.12.2 Further Reading

Dynamic problems are in general more complicated than static problems because of the inertia effect and time dependence. It is impossible to give a comprehensive review in a chapter of this size. Here are some recommendations for further reading:

9.12.2.1 Waves in Solids and Elastodynamics

For detailed but more elementary introduction to elastodynamic problems, we highly recommend Chapter 8 of Mal and Singh (1991). Many important dynamic problems are presented in a simplified manner that most readers with basic training in engineering mathematics should be able to follow. For wave propagation in inelastic solids, readers can refer to Nowacki (1978). For general wave propagations in elastic solids, we recommend the book by Achenbach (1973). Important solution techniques included Green's function method, integral representation of wave solutions in terms of body forces, Kirchhoff's formula for inhomogeneous wave equation, the Laplace transform, the Fourier transform, the Mellin transform, the Hankel transform, the method of deepest descent and stationary phase, the Wiener–Hopf technique, and the Cagniard–de Hoop method. Lamb's problem of impulsive point force and line loads on half-space, which is importance to many applications, is also covered. More advanced treatment on elastodynamics is given by Eringen and Suhubi (1975).

9.12.2.2 Seismic Waves on Earth

For more advanced treatment on seismic waves, we refer the readers to the comprehensive book by Ben-Menahem and Singh (1981), Ewing et al. (1957), and Cagniard (1962), and Aki and Richards (1980). Surface waves traveling along the sea bottom can be considered a special case of Stoneley waves, and this is also called Scholte waves (see Cagniard, 1962). For seismic waves traveling in layered half-space, we refer to Ewing et al. (1957) and Ben-Menahem and Singh (2000). The problem of a point force being applied impulsively on the surface of a half-space (also known as Lamb's problem) is of particular importance in dynamics of geomaterials (Lamb, 1904), and it is the dynamic equivalent problem of the Boussinesq problem for elastic half-space. This paper was considered by some the start of theoretical seismology. Another related problem is the free vibrations of the Earth. After the 1960 Chile earthquake, the Earth was set into vibrations for days. This provided an opportunity to study the interior structures of the Earth. The most fundamental paper is by Lamb (1882) who showed that vibrations of an elastic sphere can be divided into two classes: the first-class vibrations or the toroidal modes (zero dilatation and radial displacement), and the second-class vibrations or the spheroidal modes (zero radial component of the curl of displacement). Some related papers were reviewed by Chau (1998b) when he considered the toroidal vibrations of anisotropic elastic spheres with spherical isotropy.

9.12.2.3 Waves in Porous Media

Wave propagation in poro-elastic solids was considered by Biot (1956b). Surface waves in porous media are covered by Deresiewicz (1961, 1962), and Stoneley wave in poroelastic solids by Markov (2009).

9.12.2.4 Dynamic Fracture Mechanics

A good introduction to dynamic fracture mechanics is given in a review by Achenbach (1974). The most comprehensive coverage on dynamic fracture mechanics is the book by Freund (1998), and experimental results are compiled in the book by Ravi-Chandar (2004), which covers some of the most fundamental problems, including Yoffe's problem (a crack of fixed length propagating at a steady speed in a solid under plane strain far-field tension) and Broberg's problem (self-similar problem of crack growth from zero initial length under far-field tension). The book by Broberg (1999) provides an excellent introduction and a lot of early references on dynamic fracture mechanics and detailed coverage of the topic. Slepyan (2002) also provided some fundamental solutions (like the mode II Yoffe's problem) in dynamic fracture mechanics.

9.12.2.5 Dynamic Fragmentation

Section 8.7 of Freund (1998) gives a brief review on microcracking and fragmentation under dynamic loading. This topic has become more important recently in geomechanics, and it relates to mining, blasting, and explosion problems. For example, the size of fragments can be estimated as

$$d = 2.9\left\{\frac{K_{cr}}{\rho c \dot{\varepsilon}_0}\right\}^{2/3} \tag{9.208}$$

where K_{cr}, ε_0 , c, and ρ are the representative fracture toughness for high-speed crack growth, the initial applied strain, wave speed of energy propagation, and density of the solid, respectively. Although the power of 2/3 appears to fit experimental data, the fragment size is overestimated by 30%. Chau et al. (2000) and Wu et al. (2004) conducted some experiments on the fragmentation of spheres under diametral impacts. The fragmentation process is also related to the stress focusing phenomenon of induced waves in spheres (Wu and Chau, 2006). For current research on particle breakage, we refer to a *Powder Technology Handbook* volume, *Particle Breakage*, compiled by Salman et al. (2007).

9.13 PROBLEMS

Problem 9.1. Use equations (9.50) and (9.51) to plot the Rayleigh wave speed versus Poisson's ratio (rejected unphysical range of the results).

Problem 9.2. Use the results of Problem 9.1 to plot the relative horizontal and vertical displacements against the normalized depth to 1.5 wavelengths of Poisson's ratio for 0.2 and 0.4 (Hint: Similar results have been plotted in Figure 5.13 in Achenbach (1973) for $\nu = 0.25$ and 0.35).

Problem 9.3. Use results of Problem 9.1 and plot the relative stresses against the normalized depth to 1.5 wavelengths for Poisson's ratio of 0.2 and 0.4 (Hint: Similar results have been plotted in Figure 5.14 in Achenbach (1973) for $\nu = 0.25$ and 0.35.)

Problem 9.4. Extend the analysis for Love waves to the problem of two elastic layers on the top of an elastic half-space, as shown in Fig. 9.10. More specifically, we are searching for the SH-component of surface waves, which decay in the half-space. Show that the characteristic equation for wave speed c is

$$\frac{\alpha}{q_2}(\frac{\mu_0}{\mu_2})[\tan(q_2 h_2) - \frac{q_2}{q_1}(\frac{\mu_2}{\mu_1})\cot(q_1 h_1)] + [1 + \frac{q_2}{q_1}(\frac{\mu_2}{\mu_1})\cot(q_1 h_1)\tan(q_2 h_2)] = 0 \tag{9.209}$$

where

$$\alpha = k\sqrt{1-(c/c_0)^2}\,, \quad q_1 = k\sqrt{(c/c_1)^2 - 1}\,, \quad q_2 = k\sqrt{(c/c_2)^2 - 1} \tag{9.210}$$

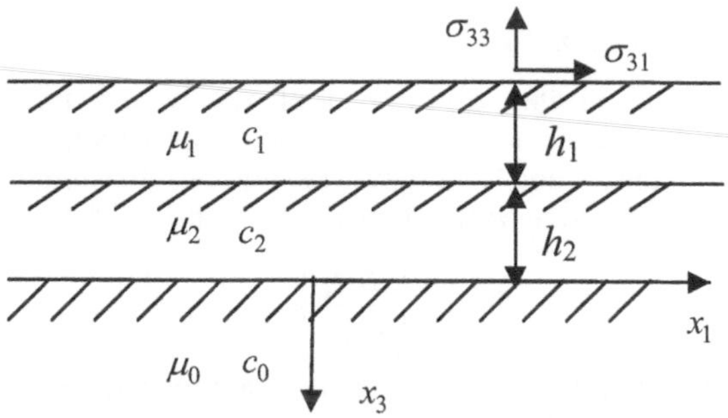

Figure 9.10 Two finite layers over an elastic half-space

Problem 9.5. Demonstrate that by taking the appropriate limit the result given in Problem 9.4 converges to the characteristic equation for Love waves.

Problem 9.6. Show that the SH-type of interface wave (analogous to the Stoneley wave) does not exist between the interface between two elastic half-spaces.

Problem 9.7. Derive the following characteristic equation for wave speed for the SH-type of interface wave for an elastic layer of thickness h between two elastic half-spaces (see Fig. 9.11):

$$\tan(q_2 h)[1 + \frac{\mu_2}{\mu_1}(\frac{\mu_2}{\mu_0})\frac{q_2^2}{\alpha_1 \alpha_0}] - q_2 \mu_2 (\frac{1}{\alpha_0 \mu_0} + \frac{1}{\alpha_1 \mu_1}) = 0 \tag{9.211}$$

where

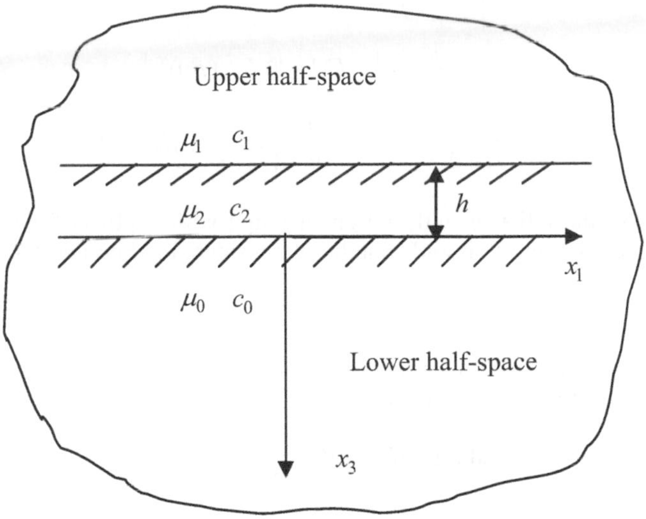

Figure 9.11 An elastic layer between two half-spaces

$$\alpha_1 = k\sqrt{1-(c/c_1)^2}\,, \quad \alpha_0 = k\sqrt{1-(c/c_0)^2}\,, \quad q_2 = k\sqrt{(c/c_2)^2-1} \tag{9.212}$$

Problem 9.8 Demonstrate that by taking the appropriate limit the result given in Problem 9.7 converges to the characteristic equation for Love waves.

Problem 9.9. Write a simple computer program to solve (9.116) and plot $\alpha = \rho_1/\rho_2$ with α as the vertical axis and $(c_{s1}/c_{s2})^2$ as the horizontal axis. Show the region in the parameter space where the Stoneley wave exists (Hints: similar plot is available in Cagniard (1962), and some useful Fortran subroutines can be used directly from *Numerical Recipes* by Press et al. (1992).)

Problem 9.10. Rederive the formulas in Section 9.9 for the case of nonhomogeneous soil layer of thickness H governed by the following equation:

$$\rho\frac{\partial^2 u}{\partial t^2} + c\frac{\partial u}{\partial t} - \frac{\partial}{\partial y}[G_0 y^n \frac{\partial u}{\partial y}] = -\rho\frac{\partial^2 u_g}{\partial t^2} \tag{9.213}$$

Hint: The answer can be found in Idriss and Seed (1968):

$$Y_n(y) = (\frac{1}{2}\beta_n)^b \Gamma(1-b)(\frac{y}{H})^{b/\theta} J_{-b}[\beta_n(\frac{y}{H})^{1/\theta}] \tag{9.214}$$

$$\ddot{X}_n(y) + 2D_n\omega_n\dot{X}_n + \omega_n^2 X_n = -R_n\ddot{u}_g \tag{9.215}$$

$$\omega_n = \frac{\beta n \sqrt{G_0 / \rho}}{\theta H^{1/\theta}}, \quad D_n = \frac{c}{2\rho\omega_n} \tag{9.216}$$

$$R_n = [(\frac{1}{2}\beta n)^{1+b} \Gamma(1-b) J_{1-b}(\beta n)]^{-1} \tag{9.217}$$

$$\theta = -\frac{2}{n-2}, \qquad b = \frac{n-1}{n-2}, \quad J_{-b}(\beta_n) = 0 \tag{9.218}$$

Problem 9.11. Show that u_3 will not change sign with depth x_3 for the Rayleigh wave in solids satisfying Poisson's condition considered in Section 9.5.2.

Problem 9.12. Take the differentiation of (9.96) and derive (9.98).

Problem 9.13. Show that by setting $\rho_1 = 0$ and $\mu_1 = 0$ in (9.113), the Rayleigh wave speed given in (9.31) is recovered.

Problem 9.14. Show the validity of (9.170).

APPENDICES

Appendix A: Nanson Formula

A.1 NANSON FORMULA

The Nanson formula has been used in the derivation of the first Piola–Kirchhoff stress and the second Piola–Kirchhoff stress in Chapter 2. The proof of the Nanson formula is given here. As shown in Fig. A.1, consider a surface dS_0 with normal $\boldsymbol{N}$ in the reference body V_0 which is mapped to the deformed surface dS with normal $\boldsymbol{n}$ in the current body. The sizes of these reference and deformed elements can be related by deformation gradient as

$$d\overline{\boldsymbol{x}} = \boldsymbol{F}\cdot d\boldsymbol{x}\,, \quad \delta\overline{\boldsymbol{x}} = \boldsymbol{F}\cdot\delta\boldsymbol{x} \tag{A.1}$$

The areas of these elements are

$$\boldsymbol{N}dS_0 = d\boldsymbol{x}\times\delta\boldsymbol{x}\,, \quad \boldsymbol{n}dS = d\overline{\boldsymbol{x}}\times\delta\overline{\boldsymbol{x}} \tag{A.2}$$

In component form, we have

$$N_i dS_0 = e_{ijk}dx_j\delta x_k\,, \quad n_i dS = e_{ijk}d\overline{x}_j\delta\overline{x}_k \tag{A.3}$$

Recall from (2.10) that the Jacobian of the deformation is defined as

$$J = \frac{dV}{dV_0} = \frac{\rho_0}{\rho} \tag{A.4}$$

The Jacobian can be related to the inverse of the deformation gradient $\boldsymbol{F}$ as

$$e_{rst}\frac{\rho}{\rho_0} = e_{rst}J^{-1} = e_{ijk}F_{ir}^{-1}F_{js}^{-1}F_{kt}^{-1} \tag{A.5}$$

Using (A.1), the first part of (A.3) becomes

$$N_i dS_0 = e_{ijk}F_{jm}^{-1}d\overline{x}_m F_{kn}^{-1}\delta\overline{x}_n\,, \tag{A.6}$$

Multiplying another inverse of deformation gradient gives

$$\begin{aligned} N_i F_{il}^{-1}dS_0 &= e_{ijk}F_{jm}^{-1}F_{kn}^{-1}F_{il}^{-1}d\overline{x}_m\delta\overline{x}_n \\ &= e_{lmn}\frac{\rho}{\rho_0}d\overline{x}_m\delta\overline{x}_n \\ &= \frac{\rho}{\rho_0}n_l dS \end{aligned} \tag{A.7}$$

The second of (A.7) is a consequence of (A.5), and the last of (A.7) is obtained by applying the second of (A.3). Rearranging (A.7) gives

$$\boldsymbol{n}dS = \frac{\rho_0}{\rho}\boldsymbol{N}\bullet\boldsymbol{F}^{-1}dS_0 \tag{A.8}$$

which is the Nanson formula used in Chapter 2.

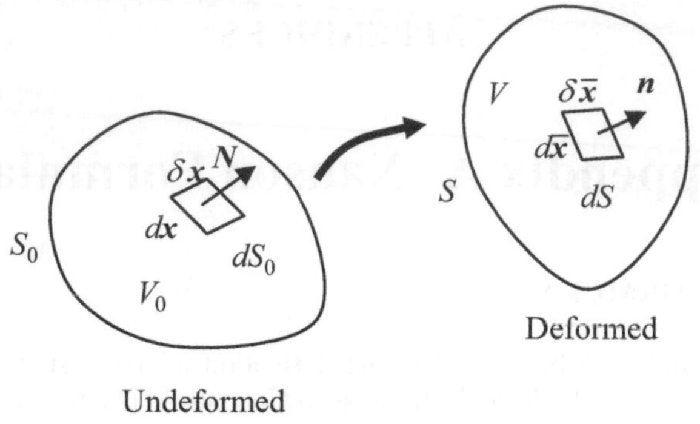

Figure A.1 Deformed area for deriving Nanson formula

Appendix B: The Laplace Transform

B.1 LAPLACE TRANSFORM

The Laplace transform was used in Chapters 7 on viscoelasticity and 8 on poroelasticity to handle the time-dependent effect of viscoelasticity on relaxation and creeping and time-dependent pore pressure dissipation. One should note that the Laplace transform is very useful for removing time dependence and replacing it by algebraic dependence on the transformed parameter space. However, it cannot be applied to a number of problems, such as problems with a moving boundary or a boundary condition that depends on the solution of the problem (such as the Stefan problem of moving interface of solidification discussed in Davis, 2001), problems with traction boundary changing to displacement boundary with time, and vice versa, and problems governed by differential equations between stress and strain having time-dependent coefficients. Hetnarski and Ignaczak (2011) remarked that the integral involved in the Laplace transform was first considered by Euler in 1744, but it was Laplace's work in 1782 that introduced the method in a way that now is so useful. The application of the Laplace transform to differential equations was largely due to Heaviside in the 1890s when he worked on electric circuit problems in what was then called operational calculus. Later contributors who put the Laplace transform on a rigorous mathematical basis include Bromwich, Carson, and van der Pol.

By assuming the existence of the Laplace transform of a function $f(x)$, the function has to be continuous for $x \geq 0$, and $f(0) = 0$, and more importantly, $f(x)$ does not grow faster than exponential form $e^{\gamma x}$ for $x \geq 0$, with the real part of γ larger than zero (note from the definition of the inverse Laplace transform that the function does not depend on γ).

B.2 FALTUNG OR CONVOLUTION THEOREM OF LAPLACE TRANSFORM

To prove the Faltung theorem used in Chapter 7, we first assume the existence of two time functions $g(t)$ and $f(t)$ and their Laplace transforms:

$$\hat{g}(s) = \mathfrak{L}[g(t)] = \int_0^\infty g(t) e^{-st} dt \tag{B.1}$$

$$\hat{f}(s) = \mathfrak{L}[f(t)] = \int_0^\infty f(t) e^{-st} dt \tag{B.2}$$

Their inverse transforms can be written as

$$f(t) = \mathfrak{L}^{-1}[\hat{f}(s)] = \frac{1}{2\pi i} \int_{\gamma - i\infty}^{\gamma + i\infty} e^{st} \hat{f}(s) ds \tag{B.3}$$

$$g(t) = \mathfrak{L}^{-1}[\hat{g}(s)] = \frac{1}{2\pi i} \int_{\gamma - i\infty}^{\gamma + i\infty} e^{st} \hat{g}(s) ds \tag{B.4}$$

where γ is a real number such that it lies on the right of all singularities, including poles and branch points. With these definitions, we consider the following inverse of the Laplace transform:

$$\begin{aligned}\frac{1}{2\pi i}\int_{\gamma-i\infty}^{\gamma+i\infty} e^{st}\hat{f}(s)\hat{g}(s)ds &= \frac{1}{2\pi i}\int_{\gamma-i\infty}^{\gamma+i\infty} e^{st}\hat{f}(s)\int_0^{\infty} g(\tau)e^{-s\tau}d\tau ds \\ &= \int_0^{\infty} g(\tau)\frac{1}{2\pi i}\int_{\gamma-i\infty}^{\gamma+i\infty}\hat{f}(s)e^{s(t-\tau)}dsd\tau \\ &= \int_0^{\infty} g(\tau)f(t-\tau)d\tau\end{aligned} \tag{B.5}$$

Since $f(t-\tau) = 0$ if $t-\tau < 0$, the integrand in the last part of (B.5) can be rewritten as

$$\frac{1}{2\pi i}\int_{\gamma-i\infty}^{\gamma+i\infty} e^{st}\hat{f}(s)\hat{g}(s)ds = \int_0^{t} g(\tau)f(t-\tau)d\tau \tag{B.6}$$

Now we can take the Laplace transform for both sides of (B.6), and we have

$$\hat{f}(s)\hat{g}(s) = \mathfrak{L}[\int_0^{t} g(\tau)f(t-\tau)d\tau] \tag{B.7}$$

This is the Faltung or convolution theorem of the Laplace transform. For example, to consider the Laplace transform of (7.31) we can make the following identifications:

$$f(t-\tau) \leftarrow G_1(t-\tau), \quad g(t) \leftarrow \frac{de_{ij}}{dt} \tag{B.8}$$

Then, we have

$$\hat{f}(s) = \hat{G}_1(s), \quad g(s) = s\hat{e}_{ij}(s) \tag{B.9}$$

The Faltung or convolution theorem of the Laplace transform can now be applied to give

$$\mathfrak{L}[s_{ij}(t)] = \hat{s}_{ij}(s) = \mathfrak{L}[\int_0^{t} G_1(\tau-t)\frac{de_{ij}(\tau)}{d\tau}d\tau] = s\hat{G}_1(s)\hat{e}_{ij}(s) \tag{B.10}$$

This gives the Laplace transform of (7.31) given in (7.59). Similarly, (7.60)–(7.62) can be obtained in the same way.

B.3 LAPLACE TRANSFORM OF DERIVATIVES

One of the main advantages of using an integral transform is to remove differential operators by algebraic operators in any differential equations. Algebraic equations are, of course, much easier to solve, but the price to pay is to deal with the subsequent inverse Laplace transform. More importantly, we have the following Laplace transform of derivatives:

$$\mathfrak{L}[\frac{ds_{ij}(t)}{dt}] = \int_0^{\infty}\frac{ds_{ij}(t)}{dt}e^{-st}dt = -s_{ij}(0) + s\hat{s}_{ij}(s) \tag{B.11}$$

$$\mathfrak{L}[\frac{d^2 s_{ij}(t)}{dt^2}] = \int_0^\infty \frac{d^2 s_{ij}(t)}{dt^2} e^{-st} dt = -\frac{ds_{ij}(0)}{dt} - ss_{ij}(0) + s^2 \hat{s}_{ij}(s) \quad \text{(B.12)}$$

$$\mathfrak{L}[\frac{d^n s_{ij}(t)}{dt^n}] = s^n \hat{s}_{ij}(s) - s^{n-1} s_{ij}(0) - s^{n-2} \frac{ds_{ij}(0)}{dt} - \ldots - s\frac{d^{n-2} s_{ij}(0)}{dt^{n-2}} - \frac{d^{n-1} s_{ij}(0)}{dt^{n-1}} \quad \text{(B.13)}$$

If the initial deviatoric stress and its derivatives are zeros at t = 0, we have the simple form of

$$\mathfrak{L}[\frac{d^n s_{ij}(t)}{dt^n}] = s^n \hat{s}_{ij}(s) \quad \text{(B.14)}$$

This is the formula that we used to obtain the transforms of the hereditary integrals in Chapter 7.

B.4 LAPLACE TRANSFORM OF UNIT STEP AND DELTA FUNCTION

Two of the most popular situations that we encounter in solving viscoelastic problems involve time functions of Heaviside unit step and Dirac delta functions. Their proofs are given here:

$$\hat{H}(s) = \int_{0^-}^\infty H(t) e^{-st} dt = -\frac{1}{s}[e^{-st}]_{0^-}^\infty = \frac{1}{s} \quad \text{(B.15)}$$

$$\hat{\delta}(s) = \int_{0^-}^\infty \delta(t) e^{-st} dt = \int_{0^-}^\infty \frac{dH(t)}{dt} e^{-st} dt = -H(0^-) + s\hat{H}(s) = 0 + \frac{s}{s} = 1 \quad \text{(B.16)}$$

Note from Fig. 7.8 and (7.168) that the Dirac delta function $\delta(t)$ can be defined as the derivative of the Heaviside step function $H(t)$. The derivative in (B.16) can be evaluated by using (B.11). These formulas are also compiled in Table B.1.

B.5 TABLE OF LAPLACE TRANSFORM

For the inversion of the Laplace transform, a number of mathematical handbooks should be consulted (Erdelyi, 1954; Abramowitz and Stegun, 1964; Gradshteyn and Ryzhik, 1980; Spiegel, 1965; Nixon, 1965). There are 166 formulas in Spiegel (1968), 55 formulas in Gradshteyn and Ryzhik (1980), 129 formulas in Abramowitz and Stegun (1964), and 33 formulas in Carslaw and Jaeger (1959).

If F(t) is the inverse Laplace transform of $f(s)$, we have the following general properties:

$$\mathfrak{L}^{-1}[f'(s)] = -tF(t) \quad \text{(B.17)}$$

$$\mathfrak{L}^{-1}[\int_s^\infty f(x)dx] = \frac{1}{t} F(t) \quad \text{(B.18)}$$

$$\mathfrak{L}^{-1}[f(cs-b)]=\frac{1}{c}e^{(b/c)t}F(\frac{t}{c}) \quad \text{for } c>0 \tag{B.19}$$

$$\mathfrak{L}^{-1}[\frac{p(s)}{(s-a)^r}]=e^{at}\sum_{n=1}^{r}\frac{p^{(r-n)}(a)}{(r-n)!}\frac{t^{n-1}}{(n-1)!} \quad \text{for } a>0 \tag{B.20}$$

Table B.1 Table of Laplace transform

$f(t)$	$\hat{f}(s)$
$H(t)$	$1/s$
$H(t-\alpha)$	$e^{-\alpha s}/s$
$H(t-\alpha)-H(t-\alpha-\varepsilon)$	$e^{-\alpha s}(1-e^{-\varepsilon s})/s$
$t^{n-1}/(n-1)!$	$1/s^n \qquad (n>0)$
$\delta(t)$	1
$e^{-\alpha t}$	$1/(\alpha+s)$
$\frac{t^{n-1}e^{-\alpha t}}{(n-1)!}$	$1/(\alpha+s)^n$
$\frac{1}{\alpha}(1-e^{-\alpha t})$	$1/[s(\alpha+s)]$
$\frac{t}{\alpha}-\frac{1}{\alpha^2}(1-e^{-\alpha t})$	$1/[s^2(\alpha+s)]$
$\int_0^t f(\tau)g(t-\tau)d\tau$	$\hat{f}(s)\hat{g}(s)$
$\sum_{k=1}^{n}\frac{P(\alpha_k)}{Q'(\alpha_k)}e^{\alpha_k t}$	$\frac{P(s)}{Q(s)}$, $P(s)=$ polynomial of degree less than n $Q(s)=(s-\alpha_1)(s-\alpha_2)...(s-\alpha_n)$ where all α_1, α_2,..., α_n are all distinct
$\int_0^t f(\tau)d\tau$	$\hat{f}(s)/s$
$\frac{e^{-bt}\sinh at}{a}$	$\frac{1}{(s+b)^2-a^2}$
$e^{-bt}\cosh at$	$(s+b)/[(s+b)^2-a^2]$
$\frac{e^{-bt}\sin at}{a}$	$\frac{1}{(s+b)^2+a^2}$
$\frac{e^{-bt}\cos at}{a}$	$(s+b)/[(s+b)^2+a^2]$

$$\mathfrak{L}^{-1}[f^{(n)}(s)] = (-1)^n t^n F(t) \tag{B.21}$$

Some formulas are more useful in solving viscoelasticity problems presented in Chapter 7. They are compiled in Table B.1 for easy reference.

B.6 APPROXIMATE METHODS FOR INVERSION OF THE LAPLACE TRANSFORM

Here we summarize essential ideas about approximate methods for the inverse Laplace transform. The standard approach is to use Bromwich's integral formula to evaluate the inverse:

$$s_{ij}(t) = \mathfrak{L}^{-1}[\hat{s}_{ij}(s)] = \frac{1}{2\pi i}\int_{\gamma-i\infty}^{\gamma+i\infty} e^{st}\hat{s}_{ij}(s)ds \tag{B.22}$$

The integration is to be performed along a line $s = \gamma$ in the complex plane where $s = x + iy$. The real number γ is chosen so that $s = \gamma$ lies to the right of all the singularities (poles, branch points, or essential singularities) but is otherwise arbitrary. The Cagniard–de Hoop method deals exclusively with this inverse transform (Achenbach, 1973).

B.6.1 Widder's General Inversion Formula

Widder's (1946) inversion formula is based on the property of Dirac delta function, which is defined as

$$\int_0^\infty h(t)\delta(t-\tau)dt = h(\tau) \tag{B.23}$$

where $\tau > 0$ and the Dirac delta function is defined as

$$\delta(t-\tau) = \begin{cases} 0 & t \neq \tau \\ \lim\limits_{\Delta t\to 0}\dfrac{1}{\Delta t} & t = \tau \end{cases} \tag{B.24}$$

$$\int_0^\infty \delta(t-\tau)dt = 1 \tag{B.25}$$

with $\tau > 0$. Equation (B.23) is called the shifting property of the Dirac delta function (see Fig. B.1). We are going to find an approximation for the Dirac delta function. To do so, we first take differentiation of a Laplace transform n times under the integration sign (compare (B.21)):

$$\frac{d^n \hat{f}(s)}{ds^n} = (-1)^n \int_0^\infty f(t)[t^n e^{-st}]dt \tag{B.26}$$

The key is that the term in the brackets [..] on the right-hand side can be approximately regarded as a delta function after normalization. We note from formula 3.351.3 of Gradshteyn and Ryzhik (1980) that

$$\int_0^\infty [t^n e^{-st}]dt = \frac{n!}{s^{n+1}} \tag{B.27}$$

Thus, the following approximation is made:

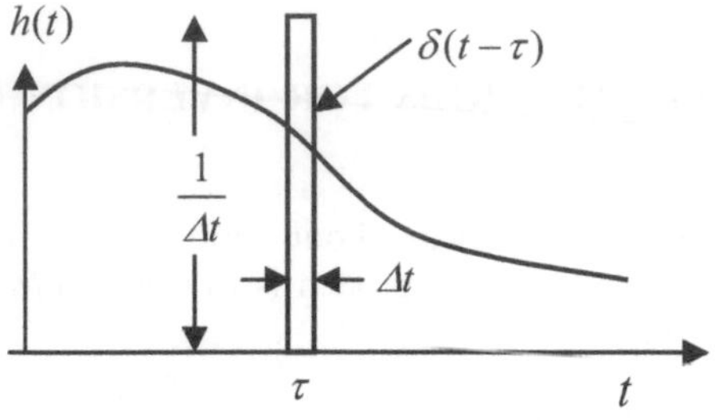

Figure B.1 Shifting property of Dirac delta function

$$\frac{s^{n+1}}{n!} t^n e^{-st} \approx \delta(t - t_0) \tag{B.28}$$

To find t_0, we can find the maximum of the function on the left of (B.28)

$$\frac{d}{dt}\{\frac{s^{n+1}}{n!} t^n e^{-st}\} = \frac{s^{n+1}}{n!}\{nt^{n-1}e^{-st} - st^n e^{-st}\} = 0 \tag{B.29}$$

Therefore, we find that this function has a maximum at $t = t_0$, where

$$t_0 = \frac{n}{s} \tag{B.30}$$

Thus, the Dirac delta function can be replaced by a distribution functions

$$\delta(t - \frac{n}{s}) \approx \frac{s^{n+1}}{n!} t^n e^{-st} \tag{B.31}$$

This distribution sense of delta function approaches the Dirac function as $n \to \infty$. We can now multiply (B.31) by $f(t)$ and integrating from 0 to ∞, we have

$$\begin{aligned} \frac{s^{n+1}}{n!} \int_0^\infty f(t) t^n e^{-st} dt &\approx \int_0^\infty f(t) \delta(t - \frac{n}{s}) dt \\ (-1)^n \frac{s^{n+1}}{n!} \frac{d\hat{f}(s)}{ds^n} &= f(t)\Big|_{t=n/s} \end{aligned} \tag{B.32}$$

The second part of (B.32) is a result of the substitution of (B.31) and (B.27) into the first part of (B.32). Finally, rewriting (B.19) gives the approximate Widder's (1946) formula:

$$f(t) = \lim_{n \to \infty} \left[(-1)^n \frac{s^{n+1}}{n!} \frac{d^n \hat{f}(s)}{ds^n} \right]_{s=n/t} \tag{B.33}$$

B.6.2 Alfrey's and ter Haars' Approximation

If we set $n = 1$, Alfrey's formula is obtained:

$$f(t) = \left[-s^2 \frac{d\hat{f}(s)}{ds} \right]_{s=1/t} \tag{B.34}$$

Clearly, Alfrey's formula is the first approximation of Widder's formula. Sometimes taking the differentiation of the transformed space can be tedious. An alternate way of alleviating this problem is to rewrite (B.26) by setting $n = 0$:

$$\hat{f}(p) = \int_0^\infty \frac{f(t)}{t} [te^{-st}] dt \tag{B.35}$$

Note that this formula is exact. We now approximate the delta function by the first order term in (B.31) we have the following approximate for delta function

$$te^{-st} \approx \frac{1}{s^2} \delta(t - t_0) \tag{B.36}$$

Substitution of (B.36) into (B.35) gives

$$s^2 \hat{f}(s) = \int_0^\infty \frac{f(t)}{t} \delta(t - t_0) dt = \left[\frac{f(t)}{t} \right]_{t=t_0} \tag{B.37}$$

It is similar to the first-order Widder's formula with $n = 1$, but no differentiation with respect to s is needed. The location of the peak of the distribution function is $t_0 = 1/s$, and with this result we get

$$s^2 \hat{f}(s) = [sf(t)]_{t=1/s} \tag{B.38}$$

Alternatively (B.38) can be rewritten as

$$f(t) = \left[s\hat{f}(s) \right]_{s=1/t} \tag{B.39}$$

This is ter Haars' formula.

B.6.3 Schapery's Direct Method of Approximation

Schapery's (1962) direct method of approximation gives the exact result as ter Haar's formula except for the relation between t and s. The key assumption of Schapery's method is that

$$s\hat{f}(s) \propto \log_{10} s \tag{B.40}$$

This assumption implies that

$$\frac{d[s\hat{f}(s)]}{d(\log s)} \tag{B.41}$$

is a slowly varying function. Now, let us write

$$\hat{f}(u) = s\hat{\psi}(s), \quad f(v) = \psi(t) \tag{B.42}$$

where

$$u = \log s, \quad s = 10^u \tag{B.43}$$

$$v = \log t, \quad t = 10^v \tag{B.44}$$

If we define

$$w = v + u = \log s + \log t = \log(st) \tag{B.45}$$

we have

$$\tilde{\psi}(s) = \int_0^\infty \psi(t) e^{-st} dt = \int_0^\infty f(v) e^{-10^w} dt \tag{B.46}$$

Multiplying both sides by *p*, we obtain

$$s\tilde{\psi}(s) = \int_0^\infty f(w-u) e^{-10^w} sdt \tag{B.47}$$

Using the following change of variables

$$st = 10^w = e^{w \ln 10} \tag{B.48}$$

we have

$$sdt = \ln(10) 10^w dw \tag{B.49}$$

With this result (B.42) is now rewritten as

$$\hat{f}(u) = \ln 10 \int_{-\infty}^\infty f(w-u) 10^w e^{-10^w} dw \tag{B.50}$$

The weighting function is now treated as the delta function

$$\ln(10) 10^w e^{-10^w} \approx \delta(w - w_0) \tag{B.51}$$

which is illustrated in Fig. B.2. Note that (B.51) is the key of the whole approximate method. Then, we have

$$\hat{f}(u) = \int_{-\infty}^\infty f(w-u) \delta(w - w_0) dw \approx f(w_0 - u) \tag{B.52}$$

The choice of w_0 is somewhat arbitrary. One way is to expand $f(v)$ in a Taylor series about the point $v_0 = w_0 - u$:

$$f(v) = f(v_0) + f'(v_0)(v - v_0) + \frac{1}{2} f''(v_0)(v - v_0)^2 + \ldots \tag{B.53}$$

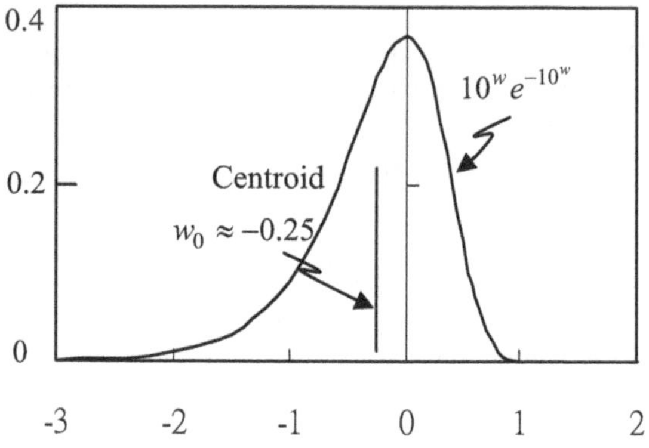

Figure B.2 Weighting function given in (B.38) used as the Dirac delta function

Substitution of (B.53) into (B.52) gives

$$\begin{aligned}\hat{f}(u) &= \ln 10 \int_{-\infty}^{\infty} \{f(v_0) + f'(v_0)(v - v_0)\} 10^w e^{-10^w} dw \\ &= \ln 10 \int_{-\infty}^{\infty} f(v_0) 10^w e^{-10^w} dw + \ln 10 \int_{-\infty}^{\infty} f'(v_0)(v - v_0) 10^w e^{-10^w} dw\end{aligned} \tag{B.54}$$

Substitution of (B.52) into the first of (B.54) gives

$$\hat{f}(u) = f(v_0) + \ln 10 f'(v_0) \int_{-\infty}^{\infty} (w - w_0) 10^w e^{-10^w} dw \tag{B.55}$$

Actually, (B.52) implies that the integral in (B.55) should be zero in order to have the same order of approximation. Thus, we have

$$\ln 10 f'(v_0) \int_{-\infty}^{\infty} (w - w_0) 10^w e^{-10^w} dw = 0 \tag{B.56}$$

Schapery (1962) gave the following result for w_0 as

$$w_0 = \frac{\int_0^{\infty} (\ln t) e^{-t} dt}{\ln(10)} \approx \frac{-0.58}{\ln(10)} \tag{B.57}$$

The last part of (B.57) is a consequence of (B.56). This value of w_0 can now be substituted into (B.52), and we recall (B.42):

$$s\hat{\psi}(s) = f(w_0 - u) = f(v_0) = \psi(t) \tag{B.58}$$

Note from (B.44) that

$$\log t = v_0 = w_0 - u = -\frac{C}{\ln(10)} - u \tag{B.59}$$

Substitution of (B.45) into (B.58) gives

$$-\frac{C}{\ln(10)} = \log t + \log s = \log(st) \tag{B.60}$$

This gives

$$st = e^{-C} = 0.56 \tag{B.61}$$

Back substitution of (B.61) into (B.58) gives

$$\psi(t) = [s\hat{\psi}(s)]_{s=0.5/t} \tag{B.62}$$

Schapery (1962) suggested that 0.5 should be used instead of 0.56 in (B.61). This is the Schapery's approximate formula used in Chapter 7.

B.7 INITIAL-VALUE AND FINAL-VALUE THEOREMS

The short-term solution for $t \to 0$ can be obtained exactly by using

$$\lim_{t \to 0} \psi(t) = \lim_{s \to \infty} [s\hat{\psi}(s)] \tag{B.63}$$

The long-term or final-value solution can be found by

$$\lim_{t \to \infty} \psi(t) = \lim_{s \to 0} [s\hat{\psi}(s)] \tag{B.64}$$

It is obvious from (B.62) that Schapery's (1962) approximate formula is exact for both initial-value and final-value solutions, for $t \to 0$ and $t \to \infty$, respectively.

Appendix C: Legendre Transform and Work Increments

The Legendre transform or more precisely the Legendre differential transform was originally proposed by Legendre for transforming variables in differential equations. For example, consider a function $f(x,y)$:

$$df = udx + vdy = \frac{\partial f}{\partial x}dx + \frac{\partial f}{\partial y}dy \tag{C.1}$$

There exists another but related function $g_1(u,y)$ defined as

$$g_1(u,y) = f - ux \tag{C.2}$$

The differential form is

$$dg_1 = df - xdu - udx = -xdu + vdy = \frac{\partial g_1}{\partial u}du + \frac{\partial g_1}{\partial y}dy \tag{C.3}$$

We can actually also define other related functions as

$$g_2(x,v) = f - vy\,,\quad g_3(u,v) = f - vy - ux \tag{C.4}$$

This transformation is simple but a powerful technique to write a differential equation in different forms. Its application to thermodynamics was first proposed by Gibbs for transforming between internal energy, enthalpy, the Helmholtz function, and the Gibbs free energy. In analytic mechanics, the Legendre transform links Lagrangian to Hamiltonian (Arnold, 1989). In elasticity, it links strain energy to complementary energy. In variational principles, it links Hu-Washizu principle to Hellinger-Reissner principle. In essence, the Legendre transform can be used to rewrite the problems in different variables (u versus x or v versus y in the example above).

In this appendix, we will express all possibilities for the case of poroelasticity. The work increment for a poroelasticity solid under constant temperature can be written if strain evolves under stress and fluid content evolves under pore-water pressure:

$$d\Phi_0 = \sigma_{ij}d\varepsilon_{ij} + \frac{1}{\rho}pdm = \frac{\partial \Phi_0}{\partial \varepsilon_{ij}}d\varepsilon_{ij} + \frac{\partial \Phi_0}{\partial m}dm \tag{C.5}$$

In this differential form, the variables are ε_{ij} and m. Alternatively, we can also use σ_{ij} and p as variables. There are altogether four possible choices of two combinations of variables.

Introducing the following potential, we can change the variable from m to p by using the following Legendre transformation:

$$\Phi_1 = \Phi_0 - \frac{1}{\rho}pm \tag{C.6}$$

$$
\begin{aligned}
d\Phi_1 &= d\Phi_0 - \frac{1}{\rho} mdp - \frac{1}{\rho} pdm \\
&= \sigma_{ij} d\varepsilon_{ij} + \frac{1}{\rho} pdm - \frac{1}{\rho} mdp - \frac{1}{\rho} pdm \qquad \text{(C.7)} \\
&= \sigma_{ij} d\varepsilon_{ij} - \frac{1}{\rho} mdp = \frac{\partial \Phi_1}{\partial \varepsilon_{ij}} d\varepsilon_{ij} + \frac{\partial \Phi_1}{\partial p} dp
\end{aligned}
$$

Alternatively, we can change the variable from ε_{ij} to σ_{ij} by using the following Legendre transformation:

$$\Phi_2 = \Phi_0 - \varepsilon_{ij}\sigma_{ij} \qquad \text{(C.8)}$$

$$
\begin{aligned}
d\Phi_2 &= d\Phi_0 - \sigma_{ij} d\varepsilon_{ij} - \varepsilon_{ij} d\sigma_{ij} \\
&= \sigma_{ij} d\varepsilon_{ij} + \frac{1}{\rho} pdm - \sigma_{ij} d\varepsilon_{ij} - \varepsilon_{ij} d\sigma_{ij} \qquad \text{(C.9)} \\
&= -\varepsilon_{ij} d\sigma_{ij} + \frac{1}{\rho} pdm = \frac{\partial \Phi_2}{\partial \sigma_{ij}} d\sigma_{ij} + \frac{\partial \Phi_2}{\partial m} dm
\end{aligned}
$$

The final choice of Legendre transformation is

$$\Phi_3 = \Phi_0 - \frac{1}{\rho} pm - \varepsilon_{ij}\sigma_{ij} \qquad \text{(C.10)}$$

$$
\begin{aligned}
d\Phi_3 &= d\Phi_0 - \sigma_{ij} d\varepsilon_{ij} - \varepsilon_{ij} d\sigma_{ij} - \frac{1}{\rho} pdm - \frac{1}{\rho} mdp \\
&= \sigma_{ij} d\varepsilon_{ij} + \frac{1}{\rho} pdm - \sigma_{ij} d\varepsilon_{ij} - \varepsilon_{ij} d\sigma_{ij} - \frac{1}{\rho} pdm - \frac{1}{\rho} mdp \qquad \text{(C.11)} \\
&= -\varepsilon_{ij} d\sigma_{ij} - \frac{1}{\rho} mdp = \frac{\partial \Phi_3}{\partial \sigma_{ij}} d\sigma_{ij} + \frac{\partial \Phi_3}{\partial p} dp
\end{aligned}
$$

The results shown in (C.5)–(C.11) for these differential forms can be summarized in Table C.1.

Table C.1 Table of Legendre transform

No.	Potentials	Work Increment	Variables
1	Φ_0	$d\Phi_0 = \boldsymbol{\sigma} : d\boldsymbol{\varepsilon} + (1/\rho) pdm$	$\boldsymbol{\varepsilon}, m$
2	$\Phi_1 = \Phi_0 - pm/\rho$	$d\Phi_1 = \boldsymbol{\sigma} : d\boldsymbol{\varepsilon} - (1/\rho) mdp$	$\boldsymbol{\varepsilon}, p$
3	$\Phi_2 = \Phi_0 - \varepsilon_{ij}\sigma_{ij}$	$d\Phi_2 = -\boldsymbol{\varepsilon} : d\boldsymbol{\sigma} + (1/\rho) pdm$	$\boldsymbol{\sigma}, m$
4	$\Phi_3 = \Phi_0 - \varepsilon_{ij}\sigma_{ij} - pm/\rho$	$d\Phi_3 = -\boldsymbol{\varepsilon} : d\boldsymbol{\sigma} - (1/\rho) mdp$	$\boldsymbol{\sigma}, p$

The constitutive constraint can be expressed as four different forms and the one used in (8.49) is only one of them. The four are

$$\frac{\partial^2 \Phi_0}{\partial \varepsilon_{ij} \partial m} = \frac{\partial \sigma_{ij}}{\partial m} = \frac{1}{\rho} \frac{\partial p}{\partial \varepsilon_{ij}} \qquad \text{(C.12)}$$

$$\frac{\partial^2 \Phi_1}{\partial \varepsilon_{ij} \partial p} = \frac{\partial \sigma_{ij}}{\partial p} = -\frac{1}{\rho} \frac{\partial m}{\partial \varepsilon_{ij}} \tag{C.13}$$

$$\frac{\partial^2 \Phi_2}{\partial \sigma_{ij} \partial m} = -\frac{\partial \varepsilon_{ij}}{\partial m} = \frac{1}{\rho} \frac{\partial p}{\partial \sigma_{ij}} \tag{C.14}$$

$$\frac{\partial^2 \Phi_3}{\partial \sigma_{ij} \partial p} = -\frac{\partial \varepsilon_{ij}}{\partial p} = -\frac{1}{\rho} \frac{\partial m}{\partial \sigma_{ij}} \tag{C.15}$$

The physical meaning of these potentials is illustrated in Fig. C.1. More specifically, these potentials are

$$\Phi_0 = \int \sigma : d\varepsilon + \frac{1}{\rho} \int p dm \tag{C.16}$$

$$\Phi_1 = \int \sigma : d\varepsilon - \frac{1}{\rho} \int m dp \tag{C.17}$$

$$\Phi_2 = -\int \varepsilon : d\sigma + \frac{1}{\rho} \int p dm \tag{C.18}$$

$$\Phi_3 = -\int \varepsilon : d\sigma - \frac{1}{\rho} \int m dp \tag{C.19}$$

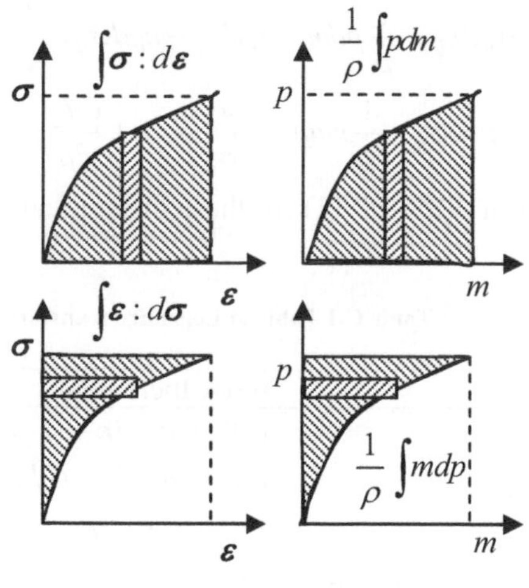

Figure C.1 Strain energy vs. complementary energy

SELECTED BIOGRAPHIES

Knowing the background of scientists can lead to a better appreciation on the part of the reader of the significance of their work. Many of the individuals described here have made great discoveries in diverse areas that have greatly impacted our daily lives. Some of these stories are inspiring. Biographies of a number of individuals whose works are covered or mentioned in this book are included here. The main references for this section are Jenkins-Jones (1996) and Millar et al. (2002). This section will hopefully form a mini-Who's Who in Mechanics and Geomechanics that will motivate readers to develop an interest in mechanics.

Achenbach, J.D. (1935–) is a Netherlands-born American mechanician who has made significant contributions to wave motions, acoustics, and fracture mechanics, including major contributions to the inspection of fatigue cracks in airplanes. His book *Wave Propagation in Elastic Solids*, published in 1973, remains a classic. He is a member of the National Academy of Engineering and the National Academy of Sciences (USA). He received the Timoshenko Medal in 1992, the National Medal of Science in 2005, and the von Karman Medal of the ASCE in 2010.

Airy, G.B. (1801–1892) was a British mathematician, astronomer, and geophysicist. The Airy stress function has been of great importance to the analysis of 2-D elasticity problems. He considered the bending of beams and published in 1862 the use of stress function on rectangular beam. However, he did not consider compatibility condition at the time and thus his formulation was incomplete. Nevertheless, it was the first time that a stress function was used. His mathematical skills were used to establish the border between Canada and the USA. He also was involved in laying the transatlantic telegraph cable, and the construction of the clock of Big Ben. Airy was, however, better known for serving as the Astronomer Royal for 46 years and for measuring Greenwich mean time by stars crossing the meridian observed through his telescope. In geophysics, Airy proposed a floating mountain theory to explain gravitational anomalies. Airy was arrogant and perhaps best known for his failure to exploit Adam's prediction of a new planet, Neptune. While still an undergraduate, Adam sent his prediction to Airy, but Airy was sceptical and ignored it. Nine months later Leverrier made the same prediction, which led to the discovery of Neptune.

Bazant, Z.P. (1937-) is a Czech-born American mechanician who has made significant contributions to size and scale effect in strength, creeping, stability, fractures, and damages of solids or structures. He is the fifth-generation civil engineer in his family. He is a member of the National Academy of Engineering and the National Academy of Sciences (USA). He received the Timoshenko Medal in 2009. He is also a downhill-skiing enthusiast who has patented a safety ski binding called ZPB binding.

Biot, M. (1905–1985) was a Belgian-born American physicist and engineer, who proposed a theory of poroelasticity which is now known as Biot's theory. This theory is the basis for Chapter 8. Biot also developed the response spectrum method for earthquake engineering. Biot contributed to irreversible thermodynamics, viscoelasticity, and thermoelasticity. He received the Timoshenko Medal in 1962.

Beltrami, E. (1835–1900) was an Italian mathematician who made notable contributions to differential geometry, non-Euclidean geometry, and mathematical physics. He developed the singular decomposition theory for matrices. In 1892, Beltrami derived the compatibility equation and the stress function, which were discussed in Chapters 2 and 4, respectively.

Bessel, F.W. (1784–1846) was a German astronomer and mathematician. He was the first to measure a star's distance by parallax. He studied the perturbation of planetary and stellar motions and he developed a mathematical function, now called the Bessel function. This function had wide applications in many other areas of mechanics. The Hankel transform for cylindrical coordinates is based on the Bessel function. Bessel made fundamental contributions to positional astronomy, geodesy, and calculating the sizes of stars, galaxies, and clusters of galaxies. Based on irregularities of Uranus' orbit, he predicted the existence of Neptune in 1840, but died a few months before its discovery.

Boltzmann, L.E. (1844–1906) was an Austrian physicist who is the founder of statistical mechanics in physics. He studied the kinetic theory of gases and derived a formula giving the number of molecules with a given energy at a specific temperature, the so-called Boltzmann constant. His theories about atomistic structures were opposed by others and he was so depressed that he committed suicide. His integral on viscoelastic formulation is classic, and viscoelastic solids are also called Boltzmann solids. The Boltzman lattice model has been widely used in solid mechanics recently.

Boussinesq, J.V. (1842–1929) was a French mathematician and physicist who made significant contributions to hydrodynamics, vibration, light, and heat. He also laid down the mathematical theory of solitons, which was observed by Russell in channels. He also derived the Boussinesq equation in fluid mechanics and made significant contributions on turbulent flows in fluids. Boussinesq's solution for point force on half-space is considered by some the start of geomechanics research.

Brace, W.F. (1926–) is an American rock mechanician and geophysicist who has made major contributions to the development of theoretical and experimental fracture mechanics for rocks. Together with J.B. Walsh, Brace contributed to experimental rock fractures, including fracture energy, rock fabric, and microcracks. Brace was elected to the National Academy of Sciences (USA) in 1971. He received the Bucher Medal of the American Geophysical Society and the Distinguished Achievement Award from the U.S. National Committee on Rock Mechanics in 1987.

Budiansky, B. (1925–1999) was an American mechanician who made significant contributions to solid and structural mechanics, microcracked rocks, the slip theory of plasticity, and yield vertex theory in plasticity. His Ph.D. advisor at Brown was William Prager. Budiansky's research interests included elasticity, fracture mechanics, plasticity, buckling and post-buckling behavior, biomechanics, and aeroelasticity. He received the von Karman medal in 1982, the Eringen medal in 1985, and the Timoshenko Medal in 1989. He was a member of the National Academy of Sciences and the National Academy of Engineering (USA).

Burgers, J.M. (1895–1981) was a Dutch physicist who made significant contributions to fluid dynamics, turbulence, dislocation theory, and viscoelasticity. In hydrodynamics, he derived the Burgers equation, which is used extensively in condensed matter and cosmology. Burgers vector in dislocation theory is named after him. In viscoelasticity, a Maxwell model in series with the Kelvin–Voigt model is called Burgers material. He was one of the co-founders of the International Union of Theoretical and Applied Mechanics (IUTAM) in 1946.

Cauchy, A.L. (1789–1857) was a French civil engineer, mathematician, and mechanician who founded complex analysis and contour integration (on which the inverse Laplace transform is based). He also made major contributions in continuum mechanics and elasticity. He published 7 books and over 700 papers, on such topics as calculus, definite integrals, limits, probability, convergence of infinite series, mechanics, astronomy, geometry, wave modulation, and complex functions. There are 16 concepts and theorems named after him, the most of any mathematician. The story is told that when Cauchy presented his theory of convergence of series, Laplace rushed home and checked those series that he used in his books on celestial mechanics (luckily they all converged). He was a devoted teacher, was the most carefully in citing other people's works, and candidly admitted errors in his publications.

Cerruti, V. (1850–1909) was an Italian mathematician and civil engineer who derived the horizontal point force solution on the surface of a half-space. The problem is now called the Cerruti problem and is of fundamental importance to geomechanics. Cerruti also made contributions to rational mechanics and structural analysis. Already as a student, he published a paper on analytical geometry. He extended Betti's reciprocal theorem from statics to dynamics. In 1873, Cerruti submitted his thesis on elastic truss analysis simultaneously with C.A. Castigliano's similar thesis to the Royal technical school of application to engineers. They were awarded the same marks by the committee, but somehow Cerruti was awarded first and Castigliano second. The rest is history as we know Castigliano's work as the "Castigliano principle" whereas Cerruti's work on elastic trusses has almost been forgotten (Capecchi and Ruta, 2011).

Clapeyron, B.P.E. (1799–1864) was a French engineer and physicist who made contributions to thermodynamics, phase transition, perfect gas, bridge engineering, and elasticity. The Clapeyron theorem in elasticity gives the form of strain energy in elastic bodies. The Clausius–Clapeyron relation in thermodynamics and Clapeyron theorem in beam bending bear his name.

Coulomb, C.A. de (1736–1806) was a French scientist, military engineer, and physicist who discovered the inverse square law of electric and magnetic attraction. The force between two electric charges is called Coulomb's law of force. He extended Amontons' friction law (which is now believed first discovered by Leonardo da Vinci in the 15th century but Leonardo's result was never published) to kinetic friction law. He invented torsion balance to measure electric force, and thus the unit of electric charge, the Coulomb, is named for him.

Dirac, P.A.M. (1902–1984) was a British theoretical physicist, a main contributor to quantum mechanics who predicted the existence of the positron and other anti-particles, which is the theoretical basis for the popular TV program "Fringe." He worked out the relativistic theory of electrons. The Fermi–Dirac statistics for determining the distribution of electrons at different energy levels is named in part for him. He also worked on large number hypothesis, and how it relates to the ratio of different kinds of forces, such as the electrical force of electrons to gravitational force of protons as 10^{39}. He predicted the meeting of an electron and a positron can result in mutual annihilation, giving out energy as photons or light. All these predictions were observed experimentally by Anderson in 1932. The same argument leads to the existence of antiparticles of all particles. Dirac shared the 1933 Nobel prize in physics with Schrödinger. Dirac delta function is commonly used in modeling dynamic problems.

Drucker, D.C. (1918–2001) was an American mechanician and engineer who made significant contributions to photoelasticity and plasticity. His Drucker's postulate provided the framework for metal plasticity. He was the first Ph.D. student of Mindlin. His work with Prager formed the basis of plasticity theory for geomaterials. His leadership and service appear unparalleled. He was the president of American Academy of Mechanics, the American Society of Mechanical Engineering, and the International Union of Theoretical and Applied Mechanics. He is remembered as a brilliant scholar, a leader in education, a spokesman for engineering, and a thoughtful, kind, generous gentleman. He received the Timoshenko Medal in 1983 and the National Medal of Science in 1988. The ASME Drucker Medal was named in his honor in 1998.

Dundurs, J. (1922–) is a Latvia-born American engineer and mechanician who made major contributions in classical elasticity, singular stress field, cracks and dislocations. The Dundurs constants for bimaterial or composite material are recognized internationally. He is known for his lucid lectures and as a pre-eminent educator. He was a recipient of the Theodore von Karman Medal from ASCE in 1990.

Einstein, A. (1879–1955) was a Germany-born American theoretical physicist who developed the theory of general relativity, resulting in a revolution in physics. The Einstein notation has been widely adopted in tensor analysis as discussed in Chapter 1. For this achievement, Einstein is often regarded as the father of modern physics. He received the 1921 Nobel Prize in Physics "for his discovery of the law of the photoelectric effect." Einstein published more than 300 scientific papers.

Eshelby, J.D. (1916–1981) was a British scientist who made fundamental contributions to micromechanics, plastic deformation, fractures, and dislocations. After a serious illness at age 13, Eshelby did not go to school. He was mostly self-educated. His seminal paper "The force on an elastic singularity" was not widely known before Rice derived his *J*-integral. Ellipsoidal inclusion is sometimes referred as "Eshelby inclusion." Eshelby was elected a Fellow of the Royal Society in 1974, and he was awarded the Timoshenko Medal in 1977.

Euler, L. (1707–1783) was a Swiss mathematician, physicist, and astronomer. He is recognized as the greatest mathematician genius of all time. He wrote almost 900 papers, memoirs, books, and other works, and is one of the most prolific mathematicians ever (his 900 papers are second only to Paul Erdös' 1,500 papers). In terms of mechanics, he contributed to the principle of superposition, the principle of virtual work, the free-body and section principle, the Euler multiplier (also called the Lagrangian multiplier), tidal theory, and the Laplace equation in potential flow. His investigation of the seven bridge problem of Konigsberg marked the beginning of graph theory. Euler made major contributions to all areas in mathematics, engineering, and science, including calculus, differential equations, analytic and differential geometry of curves and surfaces, number theory, infinite series (such as Euler's constant in infinite series), calculus of variations, optics, acoustics, light, and hydrodynamics. It was estimated that three-quarters of analytical mechanics consists of Euler's contributions. He also contributed to the design of telescopes, microscopes, and ships. His solution of the three-body problem of Earth, Moon, and Sun improved navigational tables. He developed much of classical perturbation theory. In geometry, the beautiful Euler's formula for polyhedron relates numbers of vertices, edges, and faces. Euler's formula of $e^{\pi i}+1 = 0$ is considered by many to be the most famous and beautiful formula in all of mathematics. In structural mechanics, Euler's buckling formula for columns remains a classical result today. He investigated the base of the natural logarithm e (Euler's number). Eulerian formulation for large deformations is named in honor of him. We define Eulerian strain in Chapter 2. Most of our modern mathematical notations are those of Euler. After Euler lost his eye in Russia, he said "now I have less distraction and can focus more." Euler processed prodigious memory and could perform complex calculations in his head when he became blind in his old age.

Filon, L.N.G. (1875–1938) was a French-born British applied mathematician and elastician who contributed to the theory of elasticity. He coined the term anti-plane. He also independently proposed the use of complex variable technique in elasticity in 1903. He was a follower of Karl Pearson and derived the 3-D solution for solid cylinders subject to axisymmetric loads, and non-uniform stress of cylinders under compression with end constraint (Meleshko and Selvadurai, 2003). Although this solution has been revised by many subsequent mechanicians, it is one of the classical solutions for a practical problem. He was elected to Fellowship of the Royal Society in 1910. After World War I, Filon served as vice-chancellor of the University of London (Jeffery, 1938).

Freund, L.B. (1942–) is a American mechanician who has made major contributions to dynamic fracture mechanics. His Ph.D. advisor at Northwestern

University is J.D. Achenbach. Freund worked in many areas, including stress waves in solids, fracture mechanics, seismology, computational mechanics, dislocation theory, thin films, microstructure evolution in films, and engineering education. He is a member of the National Academy of Engineering (1994) and the National Academy of Sciences (1997). He received the Timoshenko Medal in 2003.

Fung, Y.C. (1919–) is a Chinese-born American who has made major contributions to bioengineering and is recognized as the father of biomechanics. He is a member of the National Academy of Engineering and the National Academy of Sciences (USA). He received the Timoshenko Medal in 1991 and National Medal of Science in 2000. In 1957, Fung went on sabbatical to Germany and got into biomechanics by accident. At that time his mother was suffering from glaucoma (eye disease caused by high intraocular pressure) in China, and he translated all newly published papers on glaucoma and sent them to his mother's surgeon. He then quit his job at Caltech and started a new career at the University of California at San Diego in physiology and bioengineering. Later in his career, he also conducted research on blood pressure because of his wife's strategy of taking medicine as needed, instead of finishing a dosage as advised by doctor. It turns out that his wife is right, Based on continuum mechanics prediction, he concluded that she was correct. His book *Foundations of Solid Mechanics* has been quoted extensively in this book.

Galerkin, B.G. (1871–1945) was a Russian/Soviet structural engineer, mathematician, elastician, and engineer who made significant contributions to numerical methods for solving differential equations and to the theory of three-dimensional elasticity by extending Love's potential to 3-D cases. He grew up in a poor family and went to work in the Russian Court as a calligrapher at age of 12. At college years, he had to work as a private tutor and draftsman to support himself. His involvement in political activities when he worked as a railway engineer resulted in a 1.5 year jail sentence. It was the turning point in his life. He lost interest in politics and devoted himself to science and engineering. He wrote his first paper (130 pages) while in prison. In 1915, Galerkin published a paper on the approximate solution of differential equations applied to plate bending problems. This method is now known as Galerkin method. This method forms the basis of the finite element method. He was a member of the Academy of Sciences (USSR).

Gauss, K.F. (1777–1855) was a German mathematician considered by many to be one of the greatest of all mathematicians. He contributed to all areas of mathematics, especially number theory, statistics, and topology. In statistics, normal distribution is called Gaussian distribution. Gauss also originated the method of least squares for best fit curves among data points. In science, Gauss made contributions in geodesy, electric telegraph, crystallography, optics, mechanics, electricity, magnetism, and capillarity. His book on arithmetic is the basis of modern number theory. The Gauss theorem, introduced in Chapter 1, is of great impact in mechanics. Note its use in the proof of the *J*-integral in Chapter 6.

Gibbs, J.W. (1839–1903) was an American physical chemist and theoretical physicist. He developed the mathematical approach to thermodynamics and founded vector methods in mechanics and physics. He was awarded the first American doctorate in engineering from Yale in 1863. He introduced the concept of enthalpy (or heat content) and entropy (a measure of the disorder of a chemical system) in thermodynamics. He was not a good teacher and few understood his work. Some of his works were later rediscovered by Planck and Einstein, but even Poincaré, the leading French mathematician of his time, found Gibbs' papers difficult to read (Millar et al., 2002).

Gibson, R.E. (1928–2008) was a British geotechnical engineer and mathematician who proposed the NcNamee–Gibson displacement potential for Biot's theory. A class of nonlinear soil originated by him was called Gibson soil. He proposed analytical solutions for a model to describe the consolidation behavior of very soft soils.

Green, G. (1793–1841) was a British mathematical physicist who introduced Green's theorem and Green's function method for partial differential equations. These methods had huge impacts in applied mathematics and mechanics. His work on potential theory ran parallel to that of Gauss. He introduced the Green strain tensor and Green-elastic (or hyperelastic) discussed in Chapter 2, and wave propagation in infinite solid discussed in Chapter 9. Green's story is remarkable in that he was almost entirely self-taught; he only had one year of formal education at the age of eight. The son of a baker, he worked his childhood years in a bakery, except for one year of formal schooling at Robert Goodacre Academy. He published his famous Green's theorem "An Essay on the Application of Mathematical Analysis to the Theories of Electricity and Magnetism" at his own expense at the age of 35 in 1828. This work was considered by some to be one of the most significant mathematical works of all time. The way that he acquired his mathematical skill remains a mystery. He was encouraged by Sir Bromhead to enroll as an undergraduate at Cambridge University at the age of 40. But he died before his work were discovered and publicized by Lord Kelvin (see biography of Lord Kelvin). His works were further developed by James Maxwell to formulate the electromagnetic theory. The Cauchy–Green strain tensor in elasticity bears his name. Green was the first one formulated elastic constitutive law using 21 constants (Timoshenko, 1953). To commemorate the 200th anniversary of his birth in 1993, a plaque bearing Greens's name was placed in Westminster Abbey near Isaac Newton's grave. Similar honors have been given to Michael Faraday, William Thomson (Lord Kelvin), and James Clerk Maxwell.

Griffith, A.A. (1893–1963) was an English aircraft engineer who worked on fracture. His 1920 paper on fractures triggered the subsequent development of fracture mechanics.

Gurtin, M.E. (1934–) is an American mathematician, engineer, and mathematical physicist who has made significant contributions in nonlinear continuum mechanics and thermodynamics, and dynamical phase transitions. His Ph.D. advisor was E. Sternberg, and their work on viscoelasticity is of fundamental importance. Gurtin developed nonclassical theories for phase transitions, fracture dynamics, atomic diffusion, and crystalline plasticity. He received the Timoshenko Medal in 2004.

Hadamard, J. (1865–1963) was a French mathematician who founded the area functional analysis and calculus of variations. Hadamard was one of the most influential mathematicians of his time. He published over 300 papers containing novel and highly creative works. He made contributions to logic, complex analytic functions, number theory, geodeics, and hydrodynamics. He proved the prime number theorem (proposed by Gauss and Riemann) independently with Poussin that the number of prime numbers less than x approach $x/\ln x$ as $x \to \infty$. This remains perhaps the most important result in number theory. He published book on psychology of mathematical minds and initiated the concept of "well posed" in differential equations. He was an acclaimed and inspiring lecturer. Hadamard's compatibility condition is of fundamental importance to acceleration waves and localization analysis, as discussed in Chapter 9.

Hankel, H. (1839–1873) was a German mathematician who made significant contributions to complex and hypercomplex numbers, and the theory of function. The Hankel functions provided a solution to Bessel equation. The Hankel transform used in Chapter 8 bears his name. He originated the "measure" theory of point sets which are useful in probability, cybernetics, and electronic.

Heaviside, O. (1850–1925) was a British physicist and mathematician. Lacking a university education, he worked initially as a telegraph operator until deafness forced him to stop. Working alone, he developed much of the mathematics behind telegraphy and electric circuits. The Heaviside step function was adopted as a standard in mechanics. Together with Gibbs in the USA, he was the founder of vector calculus and analysis (Struik, 1987). He was one of the developers of the rigor of the Laplace transform in solving electromagnetic theory. Heaviside predicted independently and almost simultaneously with Kennelly the existence of an ionized layer in the upper atmosphere, which was known as the Kennelly–Heaviside layer or the E layer of the ionosphere. Radio signals would not be able to transmit around the world without reflecting off the layer of the ionosphere. Most of his earlier works was ignored, but eventually he was recognized by being electing a Fellow of the Royal Society. He never held an academic position and died in poverty.

Helmholtz, H. von (1821–1894) was a German physicist, mathematician, and physiologist. He discovered the law of conservation of energy, developed a theory on the nature of harmony and musical sound (he was a skillful musician), and invented the ophthalmoscope for viewing human retina. Boltzmann was one of his students. Helmholtz was considered the most versatile scientist of his century. He has been called the last scholar whose work covered science, physiology, and arts. Helmholtz believed that his diversified interests helped him adopt novel ideas in research. Together with Kirchhoff, he was one of the main contributors to mathematical physics in Germany in the 19th century. His work on Riemann's quadratic measures led to the "Lie–Helmholtz space problem" which is important to Einstein's relativity, group theory, and physiology (Struik, 1987). Helmholtz's decomposition theorem is of fundamental importance to both dynamic and static elastic problems, as discussed in this book.

Hill, R. (1921–2011) was a British applied mathematician who made significant contributions to mathematical theory of plasticity and to the uniqueness and stability in nonlinear continuum mechanics. In 2008, the International Union of Theoretical and Applied Mechanics established the Rodney Hill Prize in his honor.

Hu, H.C. (1928–2011) was a Chinese mechanician who made significant contributions to the variational principle and the general solutions of transversely isotopic elastic solids. He was in charge of theoretical development for China's first satellite. He was a member of the China Academy of Sciences. The Hu–Washizu variational principle used all 15 components of displacements, strains, and stresses as variables, and is a powerful tool in finite element formulation. Although Hu's theory on transversely isotropic elastic solids is less known in Western countries, his paper was published in English and was widely cited in the Russian literature.

Irwin, G.R. (1907–1998) was an American scientist who made significant contributions to fracture mechanics. The stress intensity factor and fracture toughness and energy release rate were first defined by Irwin. He was a member of the National Academy of Engineering in the U.S. He received the Timoshenko Medal in 1986.

Jaeger, J.C. (1907–1979) was an Australian mathematician, physicist, and engineer. At Sydney University, he changed from engineering to science under the influence of Prof. H.S. Carslaw, with whom he co-authored the classic book *Conduction of Heat in Solids*. After graduating with first class honors in both physics and mathematics, Jaeger continued his graduate studies at Cambridge University. Despite his Tripos success in applied mathematics, Jaeger lost out on a research fellowship to S. Chandrasekhar (the India-born U.S. physicist studying white dwarf stars who obtained the 1983 Nobel Prize in physics) in theoretical physics and never enrolled in Ph.D. studies. After some ups and downs at Cambridge while working on quantum mechanics, he returned to Australia and began to work in applied mathematics, including the Laplace transform, the Bessel function and hypergeometric functions. In addition to working on heat conduction with Carslaw, he also worked on rock mechanics, fractures, charcoal production, radio waves, meteorology, heating of retina by solar radiation, palaeomagnetism, seismology, in-situ measurement of rock stress, geothermal flux, rock joints, rock friction, and prediction of dust thickness on the Moon's surface through solar eclipse. His book *Fundamentals of Rock Mechanics* co-authored with N.G.W. Cook is a classic in rock mechanics. He was elected a fellow of the Royal Society (UK) and a fellow of the Australian Academy of Sciences.

Karman, von T. (1881–1963) was a Hungarian-born American mathematician, physicist, and aerospace engineer who made significant contributions to aerodynamics and is often referred as the father of rocket science. His name is associated with at least 16 equations, theories, phenomena, constant, and models, from aerodynamics, turbulence, crystallography, transportation, and aerofoil theory to vortex theory. He has been called a "genius's genius." He invented the triaxial test for testing marbles and sandstones, which is now widely adopted for soil testing as well. He was awarded the Timoshenko Medal in 1958, and was first recipient of the National Medal of Science (USA) in 1963. Since 1960 the American Society of Civil Engineers (ASCE) has awarded the Theodore von Karman medal in his honor.

Keer, L.M. (1934–) is an American mechanician, who made major contributions to tribology, contact mechanics, integral equations for elasticity, fracture mechanics, earthquake mechanics, and wave propagation in solids. His contact mechanics results led to more reliable design of gears, bearings, and railway rails. He was elected to the National Academy of Engineering in 1997. He was awarded the Drucker Medal in 2003 and the Mindlin Medal in 2011. He is known as an excellent teacher who has influenced and trained generations of students with his teaching.

Kelvin, Lord (Thomson, William) (1824–1907) was an Irish mathematician, physicist, and mechanician. Kelvin is probably best known for his introduction of the absolute temperature scale Kelvin. As a young man, he discovered Green's work, then little known, and publicized it. Since then Green's method has become a powerful tools in mathematical physics. His work on the conservation of energy led to the second law of thermodynamics. He was an unusual scientist with unparalleled enthusiasm, energy, and talent. He invented the tide gauge, an improved compass, and simpler method for fixing a ship's position at sea. He investigated many different areas of science. He published 661 papers and many books and was the author of several patents. He coined the term "turbulence" in fluid mechanics. The Kelvin solution in elasticity remains one of the most fundamental contributions to applied mechanics. He always shared ideas and gave credit to others. For his role in the Kelvin–Stokes Theorem see the biography of G.G. Stokes. He directed the first successful project for a transatlantic cable telegraph, which became operational in 1866, and brought him considerable wealth. The *Cambridge Dictionary of Scientists* says he was "probably the first scientist to become wealthy through science" (Millar et al., 2002).

Kirchhoff, G.R. (1824–1887) was a German physicist and a pioneer in spectroscopy. He also made major contributions to plate theory and elasticity. An early accident made him a wheelchair user but did not alter his cheerful character or hinder his scientific curiosity. He formulated Kirchhoff's law for electrical networks. Kirchhoff and his lifelong friend and colleague Bunsen established spectroscopy as an analytical technique in chemical analysis. Using spectroscopy, they discovered the elements caesium and rubidium, and were able to analyze the chemical element present in the Sun's atmosphere (see, however, the biography of Stokes for his role in the development of spectroscopy). The spectrometer, telescope, and microscope are the most dominant scientific instruments of our time. The first and second Piola–Kirchhoff stresses are named for him.

Kolosov, G.V. (1867–1936) was a Russia/ Soviet mathematician and engineer who proposed the complex variable method for plane elasticity in his doctoral thesis in 1908. The method was formalized by his student N.I. Muskhelishvili. He was a corresponding member of the Russian Academy of Sciences.

Lamb, H. (1849–1934) was a British applied mathematician who made major contributions to elasticity, hydrodynamics, and mechanics. His book *Hydrodynamics* remains a classic. His 1882 paper on vibrations of spheres identified two kinds of vibrations; the first class is now known as toroidal modes with zero dilatation and radial displacement and the second one is now known as spheroidal modes with zero radial component of the curl of the displacement. His predicted vibration period was observed after the 1960 Chile earthquake set the Earth in free vibrations. His 1904 paper on a suddenly applied point force on a elastic half-space was seminal contribution in theoretical seismology. This is now known as Lamb's problem. Lamb was also the first to use the integral transform for elasticity problems. He was an excellent lecturer.

Lamé, G. (1795–1870) was a French engineer and elastician who made significant contributions to the development of the theory of elasticity. The Lamé strain potential is discussed in Chapter 4. Lamé published the first book on elasticity in French. He was the first to formulate elastic problems in cylindrical and spherical coordinates. For isotropic elastic solids, two constants were found sufficient, and these constants are now known as Lamé's constants in the theory of elasticity.

Laplace, P.-S. (1749–1827) was a French mathematician, astronomer, and mathematical physicist. The story has often been told of how D'Alembert gave him difficult mathematical problems to test his ability, and found that Laplace was able to solve them overnight. Much impressed, D'Alembert helped secure Laplace a teaching job at the École Militaire in Paris. He is one of the founders of probability, and he made his name in celestial mechanics by publishing a five-volume survey of celestial mechanics. He theorized that the solar system originated from a cloud of gas (called nebular hypothesis). Laplace developed the concept of 'potential' and the study of the Laplace equation. He was from a poor family, but he was appointed minister and later senator by Napoleon. The Laplace transform, that bears his name, is of fundamental importance for solving differential equations. Many considered Laplace being the most illustrious scientist of France's golden age, and one of the most influential scientists of all time. Our current unit of length, the meter, was proposed by Laplace in 1790. Laplace is considered only second to Newton in scientific talent. He was known for his arrogance, and he frequently neglected to acknowledge the sources of his results. He was notorious for overusing the term "it is obvious" in mathematical derivations when it was far from obvious (James, 2002).

Lee, E.H. (1916–2006) was a British-born American mechanician who made significant contributions to plasticity, viscoelasticity, and inelastic wave propagation. He recognized the correspondence between elastic and viscoelastic problems in 1955. He became a member of the National Academy of Engineering in 1975 and was a recipient of the Timoshenko Medal in 1976.

Legendre, A.M. (1752–1833) was a French mathematician who contributed to number theory, celestial mechanics, and elliptical functions. In celestial mechanics, he derived the Legendre equations and the Legendre polynomials. As shown in Chapter 4, it is closely related to spherical harmonics. Legendre also invented the theory of least squares. The Legendre transformation, a powerful tool in mechanics, is discussed in Appendix C.

Love, A.E.H. (1863–1940) was a British elastician who made significant contributions to elasticity and geophysics. His book *A Treatise on the Mathematical Theory of Elasticity* is a classic work on elasticity even by today's standards. He also shown the possibility of a surface seismic wave in a layer half-space, which is now known as Love's wave. The Love's strain potential was developed for solving axisymmetric elastic problems.

Maxwell, J.C. (1831–1879) was a Scottish physicist who derived the Maxwell equations coupling electricity, magnetism, and light. Maxwell was the most able theoretician of the 19th century. He showed all colors are derived from three primary colors: red, green, and blue. This led him to produce the first color photograph in 1861. He and Boltzmann each independently developed kinetic gas theory. He theorized the rings of Saturn must consist of many small objects. Many believe his theory of electromagnetism can only be matched by that of Newton and Einstein. In structural engineering, he formulated virtual force and reciprocal theorem for the analysis of statically indeterminate frames (Kurrer, 2008). The Maxwell stress function in 3-D elasticity discussed in Chapter 4 demonstrates his diversified interests.

Michell, J.H. (1863–1940) was an Australian mathematician who made significant contributions to elasticity and hydrodynamics. He was elected a Fellow of the Royal Society in 1902. The compatibility condition equation bears his name.

Mindlin, R.D. (1906–1987) was an American mechanician who made seminal contributions to applied mechanics. His Ph.D. thesis at Columbia University was on the fundamental problem of a point applied within the interior of a half-space, and the result was published in 1936, the year he obtained his Ph.D. (Deresiewicz, 1987). It is amazing to note that he completed the work without any guidance at Columbia. His other contributions include dynamics of package cushioning, wave propagation, contact problems, micropolar elasticity, photoelasticity, plate theory, and piezoelectric crystals. He made major contributions to the development of proximity fuse during the wartime, for which he received the Presidential Medal for Merit. He was a member of the National Academy of Engineering and the National Academy of Sciences (USA), and received the Theodore von Karman medal in 1961, and the Timoshenko Medal in 1964, and the National Medal of Science in 1979. To honor his contributions to engineering mechanics, ASCE founded and awarded the Mindlin Medal in 2006 in his name.

Mohr, O. (1835–1918) was a German civil and structural engineer. He published the first paper on influence lines on structural analysis. Further developing Culman's work, Mohr completed the analysis of stress at a point, what is now called Mohr's circle. The failure of solid based on shear is known as the

Mohr–Coulomb failure condition, which is still used widely for concrete, soil, and rocks, and is presented in Chapter 5.

Morera, G. (1856–1909) was an Italian engineer and mathematician who derived the Morera theorem in complex variable theory and the Morera stress function for 3-D elasticity. His teachers are among some of the most famous mathematicians and engineers of his time, including E. Beltrami, E. Betti, F. Klein, H. von Helmholtz, G. Kirchhoff, L. Kronecker, and K. Weierstrass. Morera also contributed to the fields of differential equations, rational mechanics, the potential theory of ellipsoidal harmonics, and differential geometry.

Muki, R. (1929–2004) was a Japanese-born American civil engineer and elastician, whose seminal work "Asymmetric Problems of the Theory of Elasticity for a Semi-infinite Solid and a Thick Plate" is the basis of the Laplace-Hankel transform method used in Chapter 8 on poroelasticity. He obtained post-doctoral training at Brown University under E. Sternberg.

Mura, T. (1925–2009) was a Japanese-born American applied mathematician who made significant contributions to micromechanics, fractures, dislocation, and fatigue. His book *Micromechanics of Defects in Solids* is a bible in the field. He coined the term "eigenstrain" for the "transformation strain" of Eshelby and developed a powerful and unified method to solve micromechanics problems. In dislocation theory, he derived a line integral for curved dislocation loops in anisotropic solids which is called the Mura formula. In 1956, he and Kinoshita derived the boundary integral equation for elasticity, which is the basis of the boundary element method used today. Mura was a member of the National Academy of Engineering. He was a highly respected professor. He was open-minded and generous in sharing research ideas.

Muskhelishvili, N. (1891–1976) was a Georgian and Soviet mathematician, who continued the work of Kolosov and formalized the use of complex variable technique in 2-D elasticity.

Nemat-Nasser, S. (1937–) is a Iranian-born American mechanician who made significant contributions to a wide spectrum of problems, including structural instability, granular materials, brittle rocks, compressive fractures, high strain-rate response, composites, polymers, micromechanics, finite plasticity, liquefaction, crack kinking, curving and branching, integrated sensors for structural health monitoring, shape memory alloy, dislocation, and viscoplasticity. A series of theoretical and experimental papers co-authored with H. Horii formed the foundation for compressive fracturing in brittle solids. He is known for being energetic, enthusiatic, humorous, and rigorous. He received the Timoshenko Medal and the Theodore von Karman Medal in 2008.

Neuber, H. (1906–1989) was a German mechanical engineer. He showed that a three-dimensional solution for elasticity theory can be expressed in terms of four harmonic functions, which is now known as the Papkovitch–Neuber potential. He was a student of L. Foppl. He helped standardize the aircraft contruction method in Germany (Kuhn, 2006). He also published books on stress concentrations at holes.

Papkovitch, P.F. (1887–1946) was a Soviet scientist and structural engineer who helped design and build warships and passenger vessels in Russia. He mainly worked on the structural mechanics of ships. Papkovitch showed that elasticity problems can be solved by introducing four harmonic functions in 1932. The general solution derived independently by Neuber in 1934 and Grodski in 1935 (Goodman, 1974), is now called the Papkovitch–Neuber displacement function.

Piola, G. (1794–1850) was an Italian physicist who made contributions to pendulums, the calculus of variations, analytical mechanics of Lagrange, and large deformation of bodies. The first and second Piola–Kirchhoff stresses bear his name in part.

Poisson, S.D. (1781–1840) was a French mathematician and physicist who made contributions to probability theory, elasticity, electricity, magnetism, heat, and sound. In probability, we have the Poisson distribution (the basis for modern hazard analysis) and in elasticity we have Poisson ratio. In complex analysis, he was the first to carry out path or contour integration of complex functions (called contour integration). He was the first to recognize that there are compressional waves and shear waves in isotropic elastic solids.

Prager, W. (1903–1980) was a German-born American applied mathematician. Together with Synge, he developed the hypercircle method for solving the equilibrium of an elastic body. Prager made contributions to plasticity, continuum mechanics, and structural optimization. Drucker and Prager's (1952) work formed the basis of plasticity applied to geomaterials. He was a member of the National Academy of Engineering and the National Academy of Sciences (USA). He received the Timoshenko Medal in 1966. Since 1983 the Society of Engineering Society has awarded the Prager Medal in his honor.

Rayleigh, Lord (Strutt, J.W.) (1842–1919) was a British mathematician and physicist and Nobel Prize winner for his work on gas density and on argon. His work on Rayleigh waves is only a small part of his wide range of interests in physical problems. Rayleigh made major contributions to sound, light, and electricity. He wrote his classic book *Theory of Sound* partly on boathouse on the Nile. He inherited the title Lord Rayleigh from his father, and succeeded Maxwell in Cambridge. Rayleigh explained the blue color of sky from the scattering of light by dust particles in the air. His enthusiasm on precise measurement led him to the standardization of electrical units in 1884: the ohm, ampere, and volt. The inconsistency of the Rayleigh–Jeans equation (published by Rayleigh in 1900), which describes the distribution of wavelengths in black-body radiation, led Planck to the formulation of quantum theory. In numerical analysis, the Rayleigh–Ritz method is a powerful approximate method that bears his name.

Rice, J.R. (1940–) is an American mechanician and geophysicist who has made significant contributions to many aspects of solid mechanics. His undergraduate project with George Sih led to an influential paper in bimaterial crack problem. His B.S., M.S., and Ph.D. advisor at Lehigh University was Ferdinand Beer. He finished the three degrees in 6 years, which remains a Lehigh record (Chuang and

Rudnicki, 2001). Rice went to Brown University for post-doctoral work under D.C. Drucker and continued his academic career there before moving to Harvard University in 1981. The *J*-integral in fracture mechanics that he derived has played a great role in the development of fracture mechanics because of its path independence. He made major contributions to shear band and localization, dislocation theory, fracture mechanics, material science, plasticity, thermodynamics and state variables, earthquake mechanics and dynamics, nonlinear finite element analysis of solids, poroelasticity, slope stability, and state dependent friction law. Rice is a member of the National Academy of Sciences and the National Academy of Engineering (USA). Rice received the ASME Timoshenko Medal in 1994 and the ASCE Biot Medal in 2008. In addition to his academic accomplishments, Rice is known for his intellectual honesty and integrity, and his generosity in sharing ideas and insights.

Riemann, G.F.B. (1826–1866) was a German mathematician, who originated Riemann geometry which was used by Einstein in the theory of general relativity. He also made breakthrough in conceptual understanding of theory of functions, vector analysis, differential geometry, and topology. He was a student of Gauss. Riemann took into account the possible interaction between space and the bodies placed in it. In complex variable theory, he developed the concept of Riemann surface which separates multi-connected surfaces by branch cuts. The differentiable condition for complex variable is now known as Cauchy-Riemann relation. He defined the Riemann zeta function and formulated a Riemann hypothesis of this function. It remains one of the most important unsolved problems of number theory and analysis. The Clay Mathematical Institute of Cambridge offered 1 million US dollars for its proof (Sabbagh, 2003). Riemann died at the age of 39 because of tuberculosis.

Roscoe, K.H. (1914–1970) was a British geotechnical engineer who was one of the founders of the Cam-clay model. Roscoe was a prominent athlete at Cambridge University where he excelled at ruby and cricket. He was unable to take his examination because of arm injury and thus was awarded an aegrotat degree, but he requested to stay on for the fourth year and obtained the First Class Honors. Shortly thereafter he served in the British army despite his arm injury, and was sent to France in 1939. He was captured and spent five years in a prison camp. In the prison, he helped organize a university course on mathematics and science without the use of textbooks. After the war, he went back to Cambridge and motivated by Hvorslev's thesis, he became one of the founders of the Cam-clay model in 1958.

Rudnicki, J.W. (1952–) is an American mechanician who has made significant contributions to theory of localizations of rocks, poroelasticity, and earthquake mechanics. His yield vertex model is widely adopted. His 1975 paper with Rice has become a classic (this was actually his master's thesis). His publications has been recognized as precise, rigorous, and elegant. He derived a series of solutions for crack propagation in poroelastic solid, and applied them to predict well water fluctuation before earthquake. By correcting Cleary's fundamental solutions, Rudnicki was the first to derive the point force and fluid point source solutions for poroelastic full space. He is recognized as a dedicated teacher. He received the ASCE Biot Medal in 2006 and the ASME Drucker Medal in 2011.

Schiffman, R.L. (1923–1997) was an American geotechnical engineer who proposed the Schiffman–Fungaroli displacement function for solving the three-dimensional Biot theory. He was probably influenced by the stimulating lectures of Mindlin while a student at Colombia University. He also made contributions to finite strain application to consolidation problems, tailing dams, sedimentation, waste disposal, viscoelasticity application to secondary consolidation, numerical analysis in geomechanics, and offshore engineering problems.

Seed, H.B. (1922–1989) was a British-born American civil engineer. He made major contributions to geotechnical earthquake engineering, and is recognized as the father of this discipline. He worked on seismic site responses, potential of liquefaction, earthquake-resistant design of earth dams, and soil-pile interactions. He supervised 50 Ph.D. students and published over 300 papers and reports. He was a member of the National Academy of Sciences (USA) and was awarded the National Medal of Science by President Reagan. He is remembered as a generous, compassionate, and witty scholar and educator with incisive insights (NAE, 1992a).

Segel, L.A. (1932–2005) was an American applied mathematician, who made significant contributions to mathematical biology and hydrodynamics. His Ph.D. advisor at MIT was C.C. Lin, who was the Timoshenko Medal recipient in 1975. He studied nonlinear convection of fluid heated from below (Rayleigh–Benard convection), and had explained the spontaneous appearance of rolling pattern. The amplitude equation for this phenomenon is now known as Newell–Whitehead–Segel equation. The Society of Mathematical Biology established the Lee Segel Prize in his honor. His classic book *Mathematics Applied to Continuum Mechanics* has been referenced in this book.

Sternberg, E. (1917–1988) was an Austrian-born American mechanician. He made major contributions to applied mechanics, worked on stress concentrations around holes and cavities, static and dynamic thermoelasticty, viscoelasticity, finite deformation effect on stress singularities, spheres under diametral point loads, and completeness of general solutions for three-dimensional elasticity. Sternberg was a member of the National Academy of Engineering and the National Academy of Sciences (USA). When he accepted the Timoshenko Medal in 1985, he said "As you know, medals—much like arthritis—are a common symptom of advancing years" (NAE, 1992b). He was described as humble, charming, and full of humor and warmth. He was beloved by his students and colleagues.

Stokes, G.G. (1819–1903) was an Irish mathematician and physicist who made fundamental contributions to fluid dynamics. The most general governing equations for fluid dynamics are called the Navier–Stokes equations. Stokes described the phenomenon of fluorescence in 1852. His Stokes law for a sphere settling in a fluid also bears his name. He was the first to explain the fundamentals of spectroscopy. When Kirchhoff published his work on spectroscopy, Stokes modestly disclaimed "any part of Kirchhoff's admirable discovery." However, the

Stokes theorem discussed in Chapter 1 was in fact discovered by Lord Kelvin and communicated to Stokes in 1850, and Stokes set the theorem as a question for the 1854 Smith's prize exam, which led to the result bearing his name. Therefore, some mathematicians called it the Kelvin-Stokes theorem. Stokes served as president of the Royal Society.

Stoneley, R. (1894–1976) was a British mathematician, geologist, seismologist, and geophysicist. He was a fellow of the Royal Society and the president of the International Seismological Association. Stoneley developed a passion for chemistry at the age of 12, and obtained a chemistry scholarship at Cambridge. He then moved to mathematics under the influence of his mathematics professors. One of his first papers was on the interface wave trapped between two half-spaces, and this wave is now known as the Stoneley wave. He was interested in music and an amateur pianist (Jeffrey, 1977).

Timoshenko, S.P. (1878–1972) was a Ukrainian-born American engineer and elastician who is considered the modern father of engineering mechanics. His books on elasticity, elastic stability, plate and shells, and strength of materials were adopted as textbook worldwide. His textbooks have been published in 36 languages. He made significant contributions to buckling, stability, beam deflection, Rayleigh method, railway contacts, and plate and shells. In 1957, the ASME established the Timoshenko Medal in his honor, and he was the first recipient. Some consider the Timoshenko Medal as the equivalent of the Noble Prize in applied mechanics.

Vardoulakis, I. (1949–2009) was a Greek scientist who contributed to the modeling of geomaterials, geohazards, and geotechnical processes. He was one of the pioneers of the bifurcation theory of geomechanics, and his book *Bifurcation Analysis in Geomechanics* became the standard textbook for such analysis. He also made constitutive modeling of shear bands and worked extensively in experimental geomechanics.

Verruijt, A. (1940–) is a Dutch geotechnical engineer who extended McNamee-Gibson and Schiffman–Fungaroli displacement functions to allow for fluid compressibility. These functions are called the McNamee–Gibson–Verruijt and the Schiffman–Fungaroli–Verruijt displacement functions in this book. Many wonderful teaching tools for soil mechanics can be found on his website: http://geo.verruijt.net/.

Voigt, W. (1850–1919) was a German physicist who made contributions to thermodynamics, crystal physics, and electro-optics. The current usage of the term "tensor" was introduced by him. He demonstrated that isotropic solids must be described by two constants and eneral anisotropic solids by 21 constants. One of the mechanical models of viscoelasticity discussed in Chapter 7 also bears his name. He also formulated a framework of Voigt transformation (which is the coordinate transformation between a rest reference frame and a moving reference frame) which is similar to the Lorentz transformation. Both transformations formed the basis for the theory of relativity.

Volterra, V. (1860–1940) was an Italian mathematician who contributed to functional theory, nonlinear integro-differential equations, biological and population growth, and dislocation theory. Volterra integral equations were named after him. His contribution to dislocation was introduced in Chapter 2. During the World War I, he was involved in designing armaments, and he was also the first to propose the use of helium to replace hydrogen in airships.

Westergaard, H.M. (1888–1950) was a Danish-born American mechanician, whose Westergaard stress function was a standard tool in obtaining the stress intensity factor in 2-D crack analysis. He also made contributions to plate bending, concrete pavement analysis, structural buckling, and seismic dam analysis (Newmark, 1974). Together with Casagrande, Westergaard was in charge of the seismic resistance and atomic-bond resistance of the Panama Canal in 1947. In 1938, Westergaard extended the Bousinesq half-space solution to a half-space being laterally constrained, simulating a half-space of alternative layers of stiff and soft materials. This solution has been widely adopted in foundation engineering manuals.

Wong, T.F. (1952–) is a Macau-born American applied mathematician, mechanician, and geophysicist who has made significant contributions to both experimental and theoretical geomechanics, seismology, earthquake mechanics, fluid flow in rocks, micromechanics and grain crushing. Wong obtained his undergraduate degree from Brown University, his master's degree in applied mathematics from Harvard University, and his Ph.D. in Geophysics from Massachusetts Institute of Technology (MIT). His advisors were B. Budiansky at Harvard and Bill Brace at MIT. With this diversified training, he is probably the most renowned geomechanician in the world, excelling in both experiments and theories. He received the Basic Research Award from the U.S. National Committee for Rock Mechanics in 1986, and the Louis Néel Medal from the European Geosciences Union in 2010 for his outstanding contributions to rock mechanics.

REFERENCES

Abramowitz, M. and Stegun, I.A., 1964, *Handbook of Mathematical Functions* (New York: Dover).

Achenbach, J.D., 1973, *Wave Propagation in Elastic Solids* (Amsterdam: North-Holland).

Achenbach, J.D., 1974, Dynamic effects in brittle fracture. In *Mechanics Today*, Vol. 1, edited by S. Nemat-Nasser (Elmsford: Pergamon), pp. 1–57.

Adachi, T. and Oka, F., 1982, Constitutive equations for normally consolidated clay based on elasto-viscoplasticity. *Soil and Foundations*, **22**(4), pp. 55–70.

Aki, K. and Richards, P.G., 1980, *Quantitative Seismology—Theory and Methods* (San Francisco: W.H. Freeman).

Apirathvorakij, V. and Karasudhi, P., 1980, Quasi-static bending of a cylindrical elastic bar partially embedded in a saturated elastic half-space. *International Journal of Solids and Structures,* **16**, pp. 625–644.

Arnold, V.I., 1989, *Mathematical Methods of Classical Mechanics* (New York: Springer-Verlag).

Ashby, M.F. and Hallam, S.D., 1986, The failure of brittle solids containing small cracks under compressive stress states. *Acta Metallurgy*, **34**(3), pp. 497–510.

Atkinson, B.K., 1984, Subcritical crack growth in geological materials. *Journal of Geophysical Research,* **89**(B6), pp. 4077–4114.

Atkinson, B.K., (Ed.), 1987, *Rock Fracture Mechanics* (New York: Academic Press).

Atkinson, J.H. and Bransby, P.L., 1978, *The Mechanics of Soils: An Introduction to Critical State Soil Mechanics* (Maidenhead: McGraw-Hill).

Barber, J.R., 2002, *Elasticity*, 2nd ed. (Dordrecht: Kluwer).

Bardet, J.P. and Iai, S., 2002, Axisymmetric instability of fluid saturated pervious cylinders. *Journal of Applied Mechanics ASME*, **69**, pp. 717-723.

Barenblatt, G.I., 1996, *Scaling, Self-Similarity, and Intermediate Asymptotics* (Cambridge: Cambridge University Press).

Bažant, Z.P. and Cedolin, L., 1991, *Stability of Structures: Elastic, Inelastic, Fracture, and Damage Theories* (New York: Oxford University Press).

Bažant, Z.P., Kim, J.J., Daniel, I.M., Becq-Giraudon, E., and Zi, G., 1999, Size effect on compression strength of fiber composites failing by kink band propagation. *International Journal of Fracture,* **95**, pp. 103–141.

Bažant, Z.P. and Planas, J., 1998, *Fracture and Size Effect in Concrete and Other Quasibrittle Materials* (Boca Raton: CRC Press).

Bažant, Z.P. and Rajapakse, Y.D.S., 1991, *Fracture Scaling* (Dordrecht: Kluwer).

Bellman, R., Kalaba, R.E., and Lockett, J.A., 1966, *Numerical Inversion of the Laplace Transform: Applications to Biology, Economics, Engineering, and Physics* (New York: Elsevier).

Ben-Menahem, A. and Singh, S.J., 2000, *Seismic Waves and Sources* (New York: Dover).

Bert, C.W., 1968, Comments on "A note on the general solution of the two-dimensional linear elasticity problem in polar coordinates." *AIAA Journal,* **6**, p. 568.

Bertsch, P.K. and Findley, W.N., 1962, An experimental study of subsequent yield surfaces—corners, normality, Bauschinger and allied effects. *Proceedings of the 4th U.S. National Congress of Applied Mechanics,* pp. 893–907.

Besuelle, P. and Rudnicki, J.W., 2004, Localization: Shear bands and compaction bands. In *Mechanics of Fluid-Saturated Rocks,* Chapter 5, edited by Y. Guéguen and M. Boutéca (New York: Academic) pp. 219–321.

Bigoni, D., 2012, *Nonlinear Solid Mechanics: Bifurcation Theory and Material Instability* (New York: Cambridge University Press).

Biot, M.A., 1941, General theory of three dimensional consolidation. *Journal of Applied Physics,* **12**, pp. 155–164.

Biot, M.A., 1956a, General solutions of the equations of elasticity and consolidation for a porous material. *Journal of Applied Physics,* **78**, pp. 91–95.

Biot, M.A., 1956b, Theory of propagation of elastic waves in a fluid—saturated porous solid. *Journal of the Acoustic Society of America,* **28**(2), pp. 168–191.

Bland, D.R., 1960. *The Theory of Linear Viscoelasticity* (New York: Pergamon Press).

Bodner, S.R. and Symonds, P.S., 1960. Plastic deformations in impact and impulsive loading of beams. In *Plasticity*, edited by E.H. Leard and P.S. Symonds (New York: Pergamon Press), pp. 488–500.

Boley, B.A. and Weiner, J.H., 1960, *Theory of Thermal Stresses* (New York: Wiley).

Booker, J.R., 1991, Analytic methods in geomechanics. *Computer Methods and Advances in Geomechanics*, Beer, G., Booker, J., and Carter, J. (Eds.) (Rotterdam: Balkema), pp. 3–14.

Booker, J.R. and Carter, J.P., 1986a, Analysis of a point sink embedded in a half space. *International Journal of Numerical and Analytical Methods in Geomechanics,* **10**, pp. 137–150.

Booker, J.R. and Carter, J.P., 1986b, Long term subsidence due to fluid extraction from a saturated, anisotropic, elastic soil mass. *Quarterly Journal of Mechanics and Applied Mathematics,* **39**, pp. 85–97.

Booker, J.R. and Carter, J.P., 1987a, Elastic consolidation around a point sink embedded in a half-space with anisotropic permeability. *International Journal of Numerical and Analytical Methods in Geomechanics,* **11**, pp. 61–77.

Booker, J.R. and Carter, J.P., 1987b, Withdrawal of a compressible pore fluid from a point sink in an isotropic elastic half space with anisotropic permeability. *International Journal of Solids and Structures,* **23**(3), pp. 369–385.

Borcherdt, R.D., 2009, *Viscoelastic Waves in Layered Media* (Cambridge: Cambridge University Press).

Boresi, A.P., Chong, K.P., and Lee, J.D., 2011, *Elasticity in Engineering Mechanics,* 3rd ed. (Hoboken, N.J.: Wiley).

Brace, W.F., 1960, An extension of the Griffith theory of fracture to rocks. *Journal of Geophysical Research,* **65**, pp. 3477–3480.

Britton, J.R., Kriegh, R.B. and Rutland, L.W., 1966, *Calculus and Analytic Geometry* (San Francisco: Freeman and Company).

Broberg, K.B., 1999, *Cracks and Fracture* (San Diego: Academic Press).

Budiansky, B., 1959, A reassessment of deformation theories of plasticity. *Journal of Applied Mechanics ASME,* **26**, pp. 259–264.

Budiansky, B. and O'Connell, R.J., 1976, Elastic moduli of a cracked solid. *International Journal of Solids and Structures,* **12**, pp. 81–97.

Cagniard, L., 1962, *Reflection and Refraction of Progressive Seismic Waves,* translated and revised by E.A. Flinn and C.H. Dix (New York: McGraw-Hill).

Campbell, H.G., 1980, *Linear Algebra with Applications* (Englewood Cliffs: Prentice-Hall).

Capecchi, D., and Ruta, G., 2011, Cerruti's treatment of linear elastic trusses. *Meccanica,* **46**, pp. 1283–1298.

Carrier, G.F., Krook, M., and Pearson, C.E., 1966, *Functions of a Complex Variable: Theory and Technique* (New York: McGraw Hill).

Carslaw, H.S. and Jaeger, J.C., 1959. *Conduction of Heat in Solids*, 2nd ed. (Oxford: Oxford University Press).

Carvalho, J.L. and Curran, J.H., 1992, Two-dimensional Green's functions for elastic bi-materials. *Journal of Applied Mechanics ASME*, **59**, pp. 321–327.

Carvalho, J.L. and Curran, J.H., 1998, Three-dimensional displacement discontinuity solutions for fluid-saturated porous media. *International Journal of Solids and Structures,* **35**, pp. 4887–4893.

Chan, K. S., Karasudhi, P., and Lee, S.L., 1974, Force at a point in the interior of a layered elastic half space. *International Journal of Solids and Structures,* **10**, pp. 1179–1199.

Chatterjee, A.K., Mal, A.K., and Knopoff, L., 1978, Elastic moduli of two-component systems. *Journal of Geophysical Research,* **83**, pp. 1785–1792.

Chau, K.T., 1991, Localization of Deformation and Inelastic Deformation of Pressure-Sensitive Dilatant Materials, Ph.D. thesis submitted to Northwestern University.

Chau, K.T., 1992, Non-normality and bifurcation in a compressible pressure-sensitive circular cylinder under axisymmetric tension and compression. *International Journal of Solids and Structures*, **29**, pp. 801–824.

Chau, K.T., 1993, Anti-symmetric bifurcation in a compressible pressure-sensitive circular cylinder under axisymmetric tension and compression. *Journal of Applied Mechanics ASME*, **60**, pp. 282–289.

Chau, K.T., 1994a, Half-space instabilities and short wavelength bifurcations in cylinders and rectangular blocks. *Journal of Applied Mechanics ASME*, **61,** pp. 742–744.

Chau, K.T. 1994b, Vibrations of transversely isotropic finite circular cylinders. *Journal of Applied Mechanics ASME,* **61,** pp. 964–970.

Chau, K.T., 1995a, Buckling, barrelling, and surface instabilities of a finite, transversely isotropic circular cylinder. *Quarterly of Applied Mathematics*, **53**, pp. 225–244.

Chau, K.T., 1995b, Bifurcations at a spherical cavity in a compressible solid with spherical isotropy. *International Journal of Numerical and Analytical Methods in Geomechanics,* **19,** pp.381–398.

Chau, K.T., 1995c, Landslides modeled as bifurcations of creeping slopes with nonlinear friction law. *International Journal of Solids and Structures,* **32**(23), pp. 3451–3464 .

Chau, K.T., 1996, Fluid point source and point forces in linear elastic diffusive half-spaces. *Mechanics of Materials*, **23**, pp. 241–253.

Chau, K.T., 1997, Young's modulus interpreted from compression tests with end friction. *Journal of Engineering Mechanics ASCE*, **123**, pp. 1–7.

Chau, K.T., 1998a, A semi-analytic approach for bifurcation analysis of thick-walled cylinders under different internal and external pressures. *4th*

International Workshop on Localization and Bifurcation Theory for Soils and Rocks, Gifu, Japan, September 28–October 2, 1997, edited by T. Adachi, F. Oka, and A. Yashima (Dordrecht: Balkema), pp.11–20.

Chau, K.T., 1998b, Toroidal vibrations of anisotropic spheres with spherical isotropy. *Journal of Applied Mechanics ASME*, **65**, pp. 59–65.

Chau, K.T., 1998c, Analytic solutions for diametral point load strength tests. *Journal of Engineering Mechanics ASCE*, **124**(8), pp. 875–883.

Chau, K.T., 1999, Young's modulus interpreted from plane compressions of geomaterials between rough end blocks. *International Journal of Solids and Structures*, **36**, pp. 4963–4974.

Chau, K.T., 1999b, Onset of natural terrain landslides modeled by linear stability analysis of creeping slopes with a two state variable friction law. *International Journal of Numerical and Analytical Methods in Geomechanics*, **23**, pp. 1835–1855.

Chau, K.T. and Choi, S.K., 1998, Bifurcations of thick-walled hollow cylinders of geomaterials under axisymmetric compression. *International Journal of Numerical and Analytical Methods in Geomechanics*, **22**, pp. 903–919.

Chau, K.T. and Rudnicki, J.W., 1990, Bifurcations of compressible pressure-sensitive materials in plane strain tension and compression. *Journal of the Mechanics and Physics of Solids*, **38**, pp. 875–898.

Chau, K.T. and Shao, J.F., 2006, Subcritical crack growth of edge and center cracks in façade rock panels subject to periodic surface temperature variations. *International Journal of Solids and Structures*, **43**, pp. 807–827.

Chau, K.T. and Wang, G.S., 2001, Condition for the onset of bifurcation in en echelon crack arrays. *International Journal of Numerical and Analytical Methods in Geomechanics*, **25**(3), pp. 289–306.

Chau, K.T. and Wang, Y.B., 1998a, A new boundary integral formulation for plane elastic bodies containing cracks and holes. *International Journal of solids and Structures*, **36**, pp. 251–274.

Chau, K.T. and Wang, Y.B., 1998b, Singularity analysis and boundary integral equation method for frictional crack problems in two-dimensional elasticity. *International Journal of Fracture*, **90**, pp. 251–274.

Chau, K.T. and Wei, X.X., 1999, Spherically isotropic, elastic spheres subject to diametral point load strength test. *International Journal of Solids and Structures*, **36**, pp. 4473–4496.

Chau, K.T. and Wei, X.X., 2000, Finite solid circular cylinders subjected to arbitrary surface load: part I. Analytic solution. *International Journal of Solids and Structures* **37**, pp. 5707–5732.

Chau, K.T. and Wei, X.X., 2001a, A new analytic solution for the diametral point load strength test on finite solid circular cylinders. *International Journal of Solids and Structures*, **38**, pp. 1459–1481.

Chau, K.T. and Wei, X.X., 2001b, Stress concentration reduction at a reinforced hole loaded by a bonded circular inclusion. *Journal of Applied Mechanics ASME*, **68**, pp. 405–411.

Chau, K.T., Wei X.X., Wong R.H.C., and Yu, T.X., 2000, Fragmentation of brittle spheres under static and dynamic compressions: experiments and analyses. *Mechanics of Materials*, **32**, pp. 543–554.

Chau, K.T. and Wong, R.C.K., 2009, Interactions of a center of dilatation and an interface crack in a two-dimensional viscoelastic biomaterial. *Mechanics of Materials,* **41**, pp. 1072–1082.

Chau, K.T. and Wong R.H.C., 1996, Uniaxial compressive strength and point load strength of rocks. *International Journal of Rock Mechanics and Mining Science & Geomechanics Abstract,* **33**(2), pp.183–188.

Chau, K.T. and Wong, R.H.C., 1997, Effective moduli of microcracked-rocks: Theories and Experiments. *International Journal of Damage Mechanics*, **6**, pp. 258–277.

Chau, K.T. and Yang, X., 2005, Nonlinear interaction of soil-pile in horizontal vibration. *Journal of Engineering Mechanics ASCE,* **131**(8), pp. 847–858.

Chau, K.T., Yang, X., and Wong, R.C.K., 2000, Interactions of a penny-shaped crack with a center of dilatation in an elastic half-space. *Mechanics of Materials,* **32**, pp. 645–662.

Chau, K.T., Yang, X., and Wong, R.C.K., 2002, Interactions of a sub-surface crack with a center of dilatation in an elastic half-plane. *Engineering Fracture Mechanics,* **69**, pp. 1827–1844.

Chen, W.F., 1975, *Limit Analysis and Soil Plasticity* (Amsterdam: Elsevier).

Chen, W.F. and Baladi, G.Y., 1985, *Soil Plasticity: Theory and Implementation* (Amsterdam: Elsevier).

Chen, W.F. and Liu, X., 1990, *Limit Analysis in Soil Mechanics* (Amsterdam: Elsevier).

Chen, W.F. and Mizuno E., 1990, *Nonlinear Analysis in Soil Mechanics* (Amsterdam: Elsevier).

Chen, W.F. and Saleeb, A.F., 1988, *Constitutive equations for Engineering Materials*, Vols. 1 and 2 (New York: John Wiley).

Cherepanov, G.P., 1969, On crack propagation in solids. *International Journal of Solids and Structures,* **5**, pp. 863–871.

Chinnery, M.A., 1961, The deformation of the ground around surface faults. *Bulletin of the Seismological Society of America*, 51, pp. 355–372.

Chinnery, M.A., 1963, The stress changes that accompany strike-slip faulting. *Bulletin of the Seismological Society of America*, 53, pp. 921–932.

Chou, P.C. and Pagano, N.J., 1967, *Elasticity: Tensors, Dyadic, and Engineering Approaches* (Princeton: Van Nostrand).

Christensen, R.M., 1971, *Theory of Viscoelasticity: An Introduction* (New York: Academic Press).

Christoffersen, J. and Hutchinson, J.W., 1979, A class of phenomenological corner theories of plasticity. *Journal of the Mechanics and Physics of Solids*, **27**, pp. 465–487.

Chuang, T.-J. and Rudnicki, J.W., 2001, *Multiscale Deformation and Fracture in Materials and Structures: The James R. Rice 60th Anniversary Volume*, pp. xv–xxvi (New York: Springer).

Cleary, M.P., 1976, Fundamental Solutions for Fluid-Saturated Porous Media and Application to Localized Rupture Phenomena, Ph.D. thesis submitted to Brown University.

Cleary, M.P., 1977, Fundamental solutions for a fluid-saturated porous solid. *International Journal of Solids and Structures*, **13**, pp. 785–806.

Cleary, M.P. and Rudnicki, J.W., 1976, Initiation and propagation of dilatant rupture zones in geological materials. In *The Effects of Voids on Material*

Deformation, edited by S. C. Cowin, ASME Applied Mechanics Division, Vol. 16, pp. 13–30.
Costin, L.A., 1983, Microcrack model for deformation of brittle rock. *Journal of Geophysical Research*, **88**(1), pp. 9485–9492.
Costin, L.A., 1985, Damage mechanics in the post-failure regime. *Mechanics of Materials*, **4**(2), pp. 149–160.
Cristescu, N., 1994, Viscoplasticity of geomaterials. In *Visco-Plastic Behavior of Geomaterials* (New York: Springer-Verlag), pp. 103–207.
Das, B.M., 1993, *Principles of Soil Dynamics* (Boston: PWS-KENT).
Davis, R.O. and Selvadurai, A.P.S., 1996, *Elasticity and Geomechanics* (Cambridge: Cambridge University Press).
Davis, R.O. and Selvadurai, A.P.S., 2002, *Plasticity and Geomechanics* (Cambridge: Cambridge University Press).
Davis, S.H., 2001, *Theory of Solidification* (Cambridge: Cambridge University Press).
De Boer, R., 2000, *Theory of Porous Media. Highlights in Historical Development and Current State* (Berlin: Springer-Verlag).
Deresiewicz, H., 1961, The effect of boundaries on wave propagation in a liquid-filled porous solid: Love waves in a porous layer. *Bulletin of the Seismological Society of America,* **51**, pp. 51–59.
Deresiewicz, H., 1962, The effect of boundaries on wave propagation in a liquid-filled porous solid: surface waves in a half-space. *Bulletin of the Seismological Society of America,* **52**(3), pp. 627–638.
Deresiewicz, H., 1987, Raymond D. Mindlin—A Bio/Bibliographical Sketch. http://www.olemiss.edu/sciencenet/mindlin/mindlin-bio.pdf.
Derski, W., 1964, A method of solving of the system of equations of consolidation Theory. *Bulletin of the Polish Academy of Sciences-Technical Sciences*, **12**, pp. 489–493.
Derski, W., 1965, Equations of the consolidation theory for the case of a source of fluid. *Bulletin of the Polish Academy of Sciences-Technical Sciences*, **13**, pp. 37–43.
Desai, C.S. and Zhang, D., 1987, Viscoplastic models for geologic materials with generalized flow rule. *International Journal of Numerical and Analytical Methods in Geomechanics,* **11**, pp. 603–620.
Detournay, E. and Cheng, A.H.-D., 1993, Fundamentals of poroelasticity. In *Comprehensive Rock Engineering: Principles, Practice & Projects*, edited by Hudson, J.A. (London: Pergamon), Vol. 2, Chapter 5, pp. 113–171.
Dmowska, R. and Rice, J.R., 1986, Fracture theory and its seismological applications. In *Continuum Theories in Solid Earth Physics,* edited by R. Teisseyre (Amsterdam: Elsevier), pp. 187–255.
Drucker, D.C. and Prager, W., 1952, Soil mechanics and plastic analysis or limit design. *Quarterly of Applied Mathematics*, **10**(2), pp. 157–175.
Drucker, D.C., Gibson, R.E., and Henkel, D.J., 1957, Soil mechanics and work-hardening theories of plasticity. *Transactions of ASCE*, **122**, pp. 338–346.
Dundurs, J., 1969, Discussion. *Journal of Applied Mechanics ASME*, **91**, pp. 650–652.
Dundurs, J., 1988, Class Notes of Elasticity II, Department of Civil Engineering, Spring Quarter, 1988, Northwestern University.

Dundurs, J., 2008, Dislocations as Green's functions in plane elasticity. In *The Mechanics of Solids: History and Evolution,* edited by M.H. Santare and M.J. Chajes (Newark: University of Delaware Press).

Dundurs, J. and Hetenyi, M., 1965, Transmission of force between two semi-infinite solids. *Journal of Applied Mechanics ASME*, **32**, pp. 671–674.

Dusseault, M.B., Bruno, M.S., and Barrera, J., 2001, Casing shear: causes, cases, cures. *SPE Drill Completion*, **16**(2), pp. 98–107.

England, A.H., 1971, *Complex Variable Methods in Elasticity* (New York: Wiley).

Erdelyi, A. (Ed.), 1954, *Bateman Manuscript Project, Tables of Integral Transforms* (New York: McGraw-Hill).

Eringen, A.C. and Suhubi, E.S., 1975, *Elastodynamics*, Vols. I and II. (New York: Academic Press).

Eshelby, J.D., 1957, The determination of the elastic field of an ellipsoidal inclusion, and related problems. *Proceedings of the Royal Society of London Series A*, **241**, pp. 376–396.

Ewing, W.M., Jardetzky, W.S., and Press, F., 1957, *Elastic Waves in Layered Media* (New York: McGraw-Hill).

Fairbairn, E.M.R. and Ulm F.-J., 2002, A tribute to Fernando L. L. B. Carneiro (1913–2001) engineer and scientist who invented the Brazilian test. *Materials and Structures,* **35**, pp. 195–196

Filon, L.N.G., 1902, On the equilibrium of circular cylinders under certain practical systems of load. *Philosophical Transactions of Royal Society of London Series A,* **198**, pp. 147–233.

Filon, L.N.G., 1903, On the approximate solution for the bending of a beam of rectangular cross-section under any system of load, with special reference to points of concentrated or discontinuous loading. *Transactions of the Royal Society of London Series A,* **201**, pp. 63–155.

Filonenko-Borodich, M., 1965, *Theory of Elasticity*, translated from the Russian by M. Konayeva (New York: Dover).

Flugge, W., 1967, *Viscoelasticity* (Waltham: Blaisdell Publishing Company).

Fredlund, D.G. and Rahardjo, H., 1993, *Soil Mechanics for Unsaturated Soils* (New York: Wiley).

Freund, L.B., 1998, *Dynamic Fracture Mechanics* (Cambridge: Cambridge University Press).

Fung, Y.C., 1965, *Foundations of Solid Mechanics* (Englewood Cliffs: Prentice-Hall).

Gharpuray, V.M., Keer, L.M., and Lewis, J.L., 1990, Cracks emanating from circular voids or elastic inclusions in PMMA near a bone-implant interface. *Journal of Biomechanical Engineering ASME,* **112**, pp. 22–28.

Gladwell, G.M.L., 1980, *Contact Problems in the Classical Theory of Elasticity* (Alphen aan den Rijn: Sijthoff & Noordhoff).

Goodman, L.E., 1974, Development of the three-dimensional theory of elasticity. In *R.D. Mindlin and Applied Mechanics*, edited by G. Herrmann (New York: Pergamon), pp. 25–65.

Goodman, R.E., 1989, *Introduction to Rock Mechanics,* 2nd ed. (New York: John Wiley & Sons).

Gradshteyn, I.S. and Ryzhik, I.M., 1980, *Table of Integrals, Series, and Products* (New York: Academic Press).

Green, A.E. and Zerna, W., 1968, *Theoretical Elasticity,* 2nd ed. (London: Oxford University Press).

Griffith, A.A., 1920, The phenomena of rupture and flow in solids. *Philosophical Transactions of Royal Society of London,* **221**, pp. 163–198.

Gurtin, M.E. and Herrera, I., 1965, On dissipation inequalities and linear viscoelasticity. *Quarterly of Applied Mathematics,* **23**, pp. 235–245.

Gurtin, M.E. and Sternberg, E., 1962, On the linear theory of viscoelasticity. *Archive of Rational Mechanics and Analysis*, **11**, pp. 291–356.

Guz, A.N., 1985, Three-dimensional stability theory of deformed bodies. Internal instability. *Soviet Applied Mechanics,* **21**(11), pp. 1023–1034 (English version).

Haddad, Y.M., 1995, *Viscoelasticity of Engineering Materials* (London: Chapman & Hall).

Hardin, B.O. and Drnevich, V.P., 1972, Shear modulus and damping in soils: Design equations and curves. *Journal of the Soil Mechanics and Foundation Division,* ASCE, **98**(SM7), pp. 667–692.

Hashin, Z., 1988, The differential scheme and its application to cracked materials. *Journal of the Mechanics and Physics of Solids*, **36**, pp. 719–734.

Hearmon R.F.S., 1961, *An Introduction to Applied Anisotropic Elasticity* (London: Oxford University Press).

Heeres, O.M., Suiker, A.S.J., and de Borst R., 2002, A comparison between the Perzyna viscoplastic model and the consistency viscoplastic model. *European Journal of Mechanics A/Solids*, **21**, pp. 1–12.

Hellan, K., 1984, *Introduction to Fracture Mechanics* (New York: McGraw-Hill).

Hetnarski, R.B. and Ignaczak, J., 2011, *The Mathematical Theory of Elasticity* (Boca Raton: CRC Press).

Henyey, F.S. and Pomphery, N., 1982, Self-consistent elastic moduli of a cracked solid. *Geophysical Research Letters,* **9**, pp. 903–906.

Hill, D.A., Kelly, P.A., Dai, D.N., and Korsunsky, A.M., 1996, *Solution of Crack Problems: The Distributed Dislocation Technique* (Dordrecht: Kluwer).

Hill, R., 1950, *The Mathematical Theory of Plasticity* (London: Oxford University Press).

Hill, R., 1967, The essential structure of constitutive laws for metal composites and polycrystals. *Journal of the Mechanics and Physics of Solids*, **15**, pp. 79–95.

Hill, R., 1973, Elastic potentials and the structure of inelastic constitutive laws. *SIAM Journal of Applied Mathematics*, **25**, pp. 448–461.

Hill, R., 1978, Aspects of invariance in solid mechanics. In *Advances in Applied Mechanics,* edited by C.S. Yih, Vol. 48, pp. 1–75.

Hill, R. and Rice, J.R., 1973, Elastic potentials and the structure of inelastic constitutive laws. *SIAM Journal of Applied Mathematics*, 25, 1973, pp. 448–461.

Hirth, J.P. and Lothe, J., 1982, *Theory of Dislocations.* 2nd ed. (New York: John Wiley).

Ho, K.C. and Chau, K.T., 1997, An infinite plane loaded by a rivet of a different material. *International Journal of Solids and Structures,* **34**, pp. 2477–2496.

Ho, K.C. and Chau, K.T., 1999, A finite strip loaded by a bonded-rivet of a different material. *Computers and Structures*, **70**, pp. 203–218.

Hobson, E.W., 1955, *The Theory of Spherical Harmonics* (New York: Chelsea).

Hoek, E. and Bieniawski, Z.T., 1965, Brittle rock fracture propagation in rock under compression. *International Journal of Fracture Mechanics,* **1**(3), pp. 137(155.

Hoek, E., 1986, Practical rock mechanics(developments over the past 25 years. Conference on Rock Engineering and Excavation in an Urban Environment (Hong Kong: Institution of Mining and Metallurgy), pp. iv(xvi.

Hohenemser, K. and Prager, W., 1932, Fundamental equations and definitions concerning the mechanics of isotropic continua. Journal of Rheology, 3, pp. 16(22.

Holcomb, D.J. and Costin, L.S., 1986, Detecting damage surfaces in brittle materials using acoustic emissions. *Journal of Applied Mechanics ASME,* **53**(3), pp. 536–544.

Holcomb, D.J. and Rudnicki, J.W., 2001, Inelastic constitutive properties and shear localization in Tennessee marble. *International Journal for Numerical and Analytical Methods in Geomechanics*, **25**, pp. 109–129.

Hondros, G., 1959, The evaluation of Poisson's ratio and the modulus of materials of a low tensile resistance by the Brazilian (indirect tensile) test with particular reference to concrete. *Australian Journal of Applied Science*, **10**, pp. 243–268.

Horii, H. and Nemat-Nasser, S., 1985, Compression-induced micro-crack growth in brittle solids: Axial splitting and shear failure. *Journal of Geophysical Research*, **90**(B4), pp. 3105–3125.

Horii, H. and Sahasakmontri, K., 1989, Mechanical properties of cracked solids: Validity of the self-consistent method. In *Micromechanics and Inhomogeneity: The Toshio Mura 65th Anniversary Volume,* edited by G. J. Weng, M. Taya, and H. Abe (New York: Springer), pp. 137–159.

Hu, H.C., 1954, On the three-dimensional problems of the theory of elasticity of a transversely isotropic body. *Acta Scientia Sinica,* **2**, pp. 141–151.

Hudson, J.A., 1980, Overall properties of a cracked solid. *Mathematical Proceedings of Cambridge Philosophical Society,* **88**, pp. 371–384.

Hughes, W.F. and Gaylord, E.W., 1964, *Schaum's Outline Series: Basic Equations of Engineering Science* (New York: McGraw Hill).

Hyman, B.I., 1968, Comments on "A note on the general solution of the two-dimensional linear elasticity problem in polar coordinates." *AIAA Journal,* **6**, pp. 568–569.

Idriss, I.M. and Seed, H.B., 1968, Seismic response of horizontal soil layers. *Journal of the Soil Mechanics and Foundation Division* ASCE, **94**(SM4), pp. 1003–10331.

Inglis, C.E., 1913, Stresses in a plate due to the presence of cracks and sharp comers. *Transactions of the Royal Institute of Naval Architects,* **55**, pp. 219–230.

Irwin, G.R., 1956, Onset of fast crack propagation in high strength steel and aluminum alloys. 1955 *Sagamore Conference Proceedings,* Vol. II**,** Syracuse University Press, pp. 289–305.

ISRM, 1988, Suggested methods for determining the fracture toughness of rock. *International Journal of Rock Mechanics and Mining Science & Geomechanics Abstract,* **25**(2), pp. 71–96.

Issen, K.A. and Rudnicki, J.W., 2000, Conditions for compaction bands in porous rock. *Journal of Geophysical Research,* **105**(B9), pp. 21529–21536.

Jaeger, J.C., 1967, Failure of rocks under tensile conditions. *International Journal of Rock Mechanics,* **4**, pp. 219–227.

Jaeger, J.C. and Cook, N.G.W., 1976, *Fundamentals of Rock Mechanics* (London: Chapman & Hall).

James, I., 2002, *Remarkable Mathematicians: From Euler to von Neumann* (Cambridge: Cambridge University Press).

Jenkins-Jones, S., 1996, *The Hutchinson Dictionary of Scientists* (Bath: Helicon Publishing Ltd.).

Jeyakumaran, M. and Rudnicki, J.W., 1995, The sliding wing crack - Again! *Geophysical Research Letters*, **22**, pp. 2901–2904.

Jeyakumaran, M., Rudnicki, J.W., and Keer, L.M., 1992, Modeling slip zones with triangular dislocation elements. *Bulletin of the Seismological Society of America*, **82**(5), pp. 2153–2169.

Jeffery, G.B., 1938, Louis Napoleon George Filon. *Journal of London Mathematical Society,* **13,** pp. 310–318.

Jeffrey, H., 1977, Robert Stoneley. *Quarterly Journal of the Royal Astronomical Society*, **18**, pp. 302–309.

Kachanov, L.M., 1986, *Introduction to Continuum Damage Mechanics* (Dordrecht: Nijhoff).

Kalandiya, A.I., 1975, *Mathematical Methods of Two-dimensional Elasticity* (translated from the Russian by M. Konyaeva) (Moscow: Mir Publishers).

Kannien, M.F. and Popelar, C.H., 1985, *Advanced Fracture Mechanics* (New York: Oxford University Press).

Kanok-Nukulchai, W. and Chau, K.T., 1990, Point sink fundamental solutions for subsidence prediction. *Journal of Engineering Mechanics ASCE,* **116**, pp. 1176–1182.

Karman, von T., and Edison, L., 1967, *The Wind and Beyond—Theodore von Karman: The Pioneer in Aviation and Pathfinder in Space* (Little Brown and Co.).

Karasudhi, P., 1991, *Foundations of Solid Mechanics* (Dordrecht: Kluwer).

Keer, L.M., 1967, Mixed boundary value problems for an elastic half-space. *Proceedings of the Cambridge Philosophical Society*, **63**, pp. 1379–1386.

Kimoto, S., Oka, F., and Higo, Y., 2004, Strain localization analysis of elasto-viscoplastic soil considering structural degradation. *Computer Methods in Applied Mechanics and Engineering*, **193**, pp. 2845–2866.

Koehler, J.S., 1941, On the dislocation theory of plastic deformation. *Physical Review II*, **60**, pp. 397–410.

Krajcinovic, D., 1996, *Damage Mechanics* (Amsterdam: Elsevier).

Kranz, R.L., 1979, Crack-crack and crack-pore interactions in stressed granite. *International Journal of Rock Mechanics and Mining Science & Geomechanics Abstract,* **16**, pp. 37–47.

Kreyszig, E., 1996, *Advanced Engineering Mathematics*, 8th ed. (New York: Wiley).

Koo, K.K., Chau, K.T., Yang, X., Lam, S.S., and Wong, Y.L., 2003, Soil-pile-structure interactions under SH waves. *Earthquake Engineering and Structural Dynamics*, **32**(3), pp. 395–415.

Kuhn, G., 2006, Prof. Dr.-Ing. Dr.rer.nat. h.c. Heinz Neuber–100 years. *Archives of Applied Mechanics*, **76,** pp. 615–616.

Kupradze, V.D. (Ed.), 1979, *Three-Dimensional Problems of the Mathematical Theory of Elasticity and Thermoelasticity* (Amsterdam: North-Holland).

Kurrer, K.-E., 2008, *The History of the Theory of Structures: From Arch Analysis to Computational Mechanics* (Berlin: Ernst & Sohn).

Kuruppu, M.D. and Chong, K.P., 2012, Fracture toughness testing of brittle materials using semi-circular bend (SCB) specimen. *Engineering Fracture Mechanics*, **91**, pp. 133–150.

Lade, P.V. and Duncan, J.M., 1975, Elastoplastic stress-strain theory for cohesionless soil. *Journal of Geotechnical Engineering Division, ASCE*, **101** (GT10), pp. 1037–1053.

Lama, R.D. and Vutukuri, V.S., 1978, Time-dependent properties of rocks. In *Handbook on Mechanical Properties of Rocks-Testing Techniques and Results,* Vol. **3**, Chapter 9, pp. 209–323, Trans Tech Publications, Clausthal, Germany.

Lamb, H., 1882, On the vibrations of an elastic sphere. *Proceedings of London Mathematical Society Series 1,* **13**, pp. 189–212.

Lamb, H., 1902, On Boussinesq's problem. *Proceedings of London Mathematical Society*, **34**, pp. 276–284.

Lamb, H., 1904, On the propagation of tremors over the surface of an elastic solid. *Philosophical Transactions of the Royal Society of London,* **203A**, pp. 1–42.

Lee, E.H., 1955, Stress analysis in viscoelastic bodies. *Quarterly of Applied Mathematics,* **13**, pp. 183–190.

Lekhnitskii, S.G., 1963, *Theory of Elasticity of an Anisotropic Elastic Body,* translated by P. Fern (San Francisco: Holden-Day).

Lemaître, J., 1996, *A Course on Damage Mechanics* (Berlin: Springer).

Li, V.C., 1987, Mechanics of shear rupture applied to earthquakes zones. In *Rock Fracture Mechanics*, edited by B.K. Atkinson (London: Academic Press), pp. 351–458.

Lim, I.L., Johnston, I.W., Choi, S.K., and Boland, J.N., 1994, Fracture Testing of a soft rock with semi-circular specimens under three-point bending. Part 2–mixed-mode. *International Journal of Rock Mechanics and Mining Science*, **31**(3), pp.199–212.

Little, R.W., 1973, *Elasticity* (Englewood Cliffs: Prentice Hall).

Love, A.E.H., 1911, *Some Problems of Geodynamics* (Cambridge: Cambridge University Press) (republished by New York: Dover, 1967).

Love, A.E.H., 1944, *A Treatise on the Mathematical Theory of Elasticity,* 4th ed. (New York: Dover).

Lu, X., Rosakis, A.J., and Lapusta, N., 2010, Rupture modes in laboratory earthquakes: Effect of fault prestress and nucleation conditions. *Journal of Geophysical Research,* **115**B, B12302, doi:10.1029/2009JB006833, pp. 1–25.

Lubliner, J., 1990, *Plasticity Theory,* rev. ed. (New York: Macmillan Publishing).

Luré, A.I., 1964, *Three-Dimensional Problems of the Theory of Elasticity,* translated from the Russian by D.B. McVean (New York: Interscience).

Mahmoud, M.S. and Deresiewicz, H., 1980a, Settlement of inhomogeneous consolidating soils–I: The single-drained layer under confined compression. *International Journal of Rock Mechanics and Mining Science*, **4**, pp. 57–72.

Mahmoud, M.S. and Deresiewicz, H., 1980b, Settlement of inhomogeneous consolidating soils–II: The symmetrically loaded mass. *International Journal of Numerical and Analytical Methods in Geomechanics*, **4**, pp. 73–88.

Mal, A.K. and Singh, S.J., 1991, *Deformation of Elastic Solids* (Englewood Cliffs: Prentice-Hall).

Malvern, L.E., 1951, The propagation of longitudinal waves of plastic deformation in a bar of material exhibiting a strain-rate effect. *Journal of Applied Mechanics ASME*, **18**, pp. 203–208.

Malvern, L.E., 1969, *Introduction to the Mechanics of a Continuum Medium* (Englewood Cliffs: Prentice-Hall).

Mann, E.H., 1949, An elastic theory of dislocations. *Proceedings of the Royal Society of London, Series A,* **199**, pp. 376–394.

Maor, E., 1994, *e: The Story of a Number* (Oxford: Princeton University Press).

Markov, M.G., 2009, Low-frequency Stoneley wave propagation at the interface of two porous half-spaces. *Geophysical Journal International,* **177**, pp. 603–608.

Mase, G.E., 1964, *Schaum's Outline Series: Theory and Problems of Continuum Mechanics* (New York: McGraw-Hill).

Mase, G.T. and Mase G.E., 1999, *Continuum Mechanics for Engineers,* 2nd ed. (Boca Raton: CRC Press).

Matsuoka, T. and Nakai, T., 1974, Stress-deformation and strength characteristics of soil under three different principal stresses. *Proceedings of the Japan Society of Civil Engineers,* **232**, pp. 59–70.

McClintock, F.A. and Walsh, J.B., 1962, Friction of Griffith cracks in rock under pressure. *Proceedings of the Fourth U.S. National Congress of Applied Mechanics*, pp. 1015–1021.

McCollum, P.A. and Brown, B.F., 1965, *Laplace Transform Tables and Theories* (New York: Holt, Rinehart & Winston).

McNamee, J. and Gibson, R.E., 1960a, Displacement functions and linear transform applied to diffusion through porous elastic media. *Quarterly Journal of Mechanics and Applied Mathematics,* **13**, pp. 98–111.

McNamee, J. and Gibson, R.E., 1960b, Plane strain and axially symmetric problems of the consolidation of semi-infinite clay stratum. *Quarterly Journal of Mechanics and Applied Mathematics,* **13**, pp. 210–227.

Meleshko, V.V. and Selvadurai, A.P.S., 2003, Contributions to the theory of elasticity by Louis Napoleon George Filon as viewed in the light of subsequent developments in biharmonic problems in applied mechanics and engineering mathematics. *Journal of Engineering Mathematics,* **46**, pp. 191–212.

Meleshko, V.V., 2003, Selected topics in the history of the two-dimensional biharmonic problem. *Applied Mechanics Reviews*, **56**(1), pp. 33–85.

Michell, J.H., 1899, On the direct determination of stress in an elastic solid, with application to the theory of plates. *Proceedings of the London Mathematics Society,* **31**, pp. 100–121.

Michell, J.N., 1902, The inversion of plane stress. *Proceedings of the London Mathematics Society*, **34**, pp. 134–142.

Millar, D., Millar, I., Millar, J., and Millar, M., 2002, *The Cambridge Dictionary of Scientists,* 2nd ed. (Cambridge: Cambridge University Press).

Milne-Thomson, L.M., 1962, *Antiplane Elastic Systems* (New York: Academic Press).

Milne-Thomson, L.M., 1968, *Plane Elastic Systems,* 2nd ed. (Berlin: Springer-Verlag).

Mindlin, R.D., 1936a, Note on the Galerkin and Papkovitch stress functions. *Bulletin of the American Mathematics Society*, **42**, pp. 373–376.

Mindlin, R.D., 1936b, Force at a point in the interior of a semi-infinite solid. *Physics,* 7, pp. 195–202.

Mindlin, R.D., 1953, Force at a point in the interior of a semi-infinite solid. *Proceedings of the First Midwestern Conference*, pp. 56–59.

Muhlhaus, H.B., Chau, K.T. and Ord, A., 1996, Bifurcation of crack pattern in arrays of two-dimensional cracks. *International Journal of Fracture,* **77**(1), pp. 1–14.

Muhlhaus, H.B. and Vardoulakis, I., 1987, The thickness of shear bands in granular materials. *Geotechnique,* **37**(3), pp. 271–283.

Muki, R., 1960, Asymmetric problems of the theory for a semi-infinite solid and a thick plate. In *Progress in Solid Mechanics,* edited by I.N. Sneddon and R. Hill (Amsterdam: North Holland), Vol. 1, pp. 399–439.

Mura, T., 1963, Continuous distributions of moving dislocations. *Philosophical Magazine,* **8**, pp. 843–857.

Mura, T., 1987, *Micromechanics of Defects in Solids,* 2nd ed. (Dordrecht: Martinus Nijhoff Publishers).

Murakami, Y. (Ed.), 1987, *Stress Intensity Factors Handbook,* Vols. 1 and 2 (Oxford: Pergamon Press).

Muskhelishvili, N.I., 1953, *Singular Integral Equations,* 2nd ed., translated by J.R.M. Radok (Groningen: Noordhoff).

Muskhelishvili, N.I., 1975, *Some Basic Problems of the Mathematical Theory of Elasticity*, translated by J.R.M. Radok, 2nd English ed. (Groningen: Noordhoff).

Nakai, T., 2012, *Constitutive Modeling of Geomaterials* (Boca Raton: CRC Press).

National Academy of Engineering (NAE), 1992a, Harry Bolton Seed. In *Memorial Tributes: National Academy of Engineering*, Vol. 5 (Washington: National Academies Press), pp. 246-251.

National Academy of Engineering (NAE), 1992b, Eli Sternberg. In *Memorial Tributes: National Academy of Engineering*, Vol. 5 (Washington: National Academies Press), pp. 270-275.

Nayak, G.C. and Zienkiewicz, O.C., 1972, Convenient form of stress invariants for plasticity. *Journal of the Structural Division ASCE*, **25**, pp. 201–209.

Newmark, N.M., 1974, Westergaard, Harold Malcolm. In *Dictionary of American Biography*, Suppl. 4, pp. 873–874 (New York: Scribner).

Nemat-Nasser, S., 1983, On finite plastic flow of crystalline solids and geomaterials. *Journal of Applied Mechanics ASME*, **50**, pp. 1114–1126.

Nemat-Nasser, S., 2004, *Plasticity: A Treatise on Finite Deformation of Heterogeneous Inelastic Materials* (Cambridge: Cambridge University Press).

Nemat-Nasser, S. and Hori, M., 1993, *Micromechanics: Overall Properties of Heterogeneous Materials* (Amsterdam: North-Holland)

Nemat-Nasser, S. and Shokoon, A., 1980, On finite plastic flows of compressible materials with internal friction. *International Journal of Solids and Structures*, **16**, pp. 495–514.

Nemat-Nasser, S. and Obata, M., 1988, A microcrack model of dilatancy in brittle material. *Journal of Applied Mechanics ASM*E, **55**(B10), pp. 24–35.

Nixon, F.E., 1965, *Handbook of Laplace Transform: Fundamentals, Applications, Tables, and Examples,* 2nd ed. (Englewood Cliffs: Prentice-Hall).

Niumpradit, B. and Karasudhi, P., 1981, Load transfer from an elastic pile to a saturated porous elastic soil. *International Journal of Solids and Structures,* **5**, pp. 115–138.

Noble, B. and Daniel, J.W., 1988, *Applied Linear Algebra* (Englewood Cliffs: Prentice-Hall).

Nowacki, W.K., 1978, *Stress Waves in Non-elastic Solids* (Oxford: Pergamon Press).

Oka, F. and Kimoto, S., 2012, *Computational Modelling of Multi-Phase Geomaterials* (London: Spon).

Osinov, V.A. and Wu, W., 2009, Wave speeds, shear bands and the second-order work for incrementally nonlinear constitutive models. *Acta Mechanica*, **202**, pp. 145–151.

Palmer, A.C. and Rice, J.R., 1973, The growth of slip surfaces in the progressive failure of over-consolidated clay. *Proceedings of the Royal Society of London Series A*, **332**, pp. 527–548.

Pan, E., 1999, Green's functions in layered poroelastic half-spaces. *International Journal of Numerical and Analytical Methods in Geomechanics,* **23**, pp. 1631–1653.

Pan, J. and Rice, J.R., 1983, Rate sensitivity of plastic flow and implications for yield-surface vertices. *International Journal of Solids and Structures,* **19**(11), pp. 973–987.

Pao, Y.-H., 1998, Applied mechanics in science and engineering. *Applied Mechanics Reviews,* **51**(2), pp. 141–153.

Perzyna, P., 1963, The constitutive equations for rate sensitive plastic materials. *Quarterly of Applied Mathematics,* **20**, pp. 321–332.

Phillips, A. and Gray, G.A., 1961, Experimental investigation of corners in the yield surface. *Journal of Basic Engineering ASME,* **83**, pp. 275–288.

Pietruszczak, S., 2010, *Fundamentals of Plasticity in Geomechanics* (Boca Raton: CRC Press).

Poulos, H.G. and Davis, E.H., 1974, *Elastic Solutions for Soil and Rock Mechanics* (New York: Wiley).

Press, W.H., Flannery, B.P., Teukolsky, S.A., and Vetterling, W.T., 1992, *Numerical Recipes: The Art of Scientific Computing,* 2nd ed. (New York: Cambridge University Press).

Puswewala, U.G.A. and Rajapakse, R.K.N.D., 1988, Axisymmetric fundamental solutions for a completely saturated porous elastic solid. *International Journal of Engineering Science,* **26**, pp. 419–436.

Qian, W. and Sun, C.T., 1998, A frictional interfacial crack under combined shear and compression. *Composites Science and Technology,* **58**, pp. 1753–1761.

Raniecki, B. and Bruhns, O.T., 1981, Bounds to bifurcation stress in solids with non-associated plastic flow law at finite strain. *Journal of the Mechanics and Physics of Solids*, **29**, pp. 153–172.

Ravi-Chandar, K., 2004, *Dynamic Fracture* (Amsterdam: Elsevier).

Rajapakse, R.K.N.D. and Senjuntichai, T., 1993, Fundamental solutions for a poroelastic half-space with compressible constituents. *Journal of Applied Mechanics ASME,* **60**, pp. 847–856.

Renner, J., Hettkamp, T., and Rummel, F., 2000, Rock mechanical characterization of an argillaceous host rock of a potential radioactive waste repository. *Rock Mechanics and Rock Engineering,* **33**(3), pp. 153–178.

Rice, J.R., 1968a, A path independent integral and the approximate analysis of strain concentration by notches and cracks. *Journal of Applied Mechanics ASME,* **35**, pp. 379–386.

Rice, J.R., 1968b, Mathematical analysis in mechanics of fracture. In *Fracture: An Advanced Treatise,* Vol. 2, edited by H. Liebowitz (New York: Academic Press), pp. 191–311.

Rice, J.R., 1971, Inelastic constitutive relations for solids: An internal-variable theory and its application to metal plasticity. *Journal of the Mechanics and Physics of Solids*, **19**, pp. 433–455.

Rice, J.R., 1973, The initiation and growth of shear bands. In *Proceedings of the Role of Plasticity and Soil Mechanics,* edited by A.C. Palmer (Cambridge: Cambridge University Engineering Department), pp. 263–274.

Rice, J.R., 1975, Continuum mechanics and thermodynamics of plasticity in relation to microscale deformation mechanisms. In *Constitutive Equations in Plasticity*, edited by A.S. Argon, Chapter 2 (Cambridge: MIT Press), pp. 23–79.

Rice, J.R., 1976, The localization of plastic deformation. In *Proceedings of the 14th International Congress of Theoretical and Applied Mechanics*, edited by W.T. Koiter, Delft, Amsterdam, Vol. 1, 1976, pp. 207–220.

Rice, J.R., 1980, The mechanics of earthquake rupture. In *Physics of the Earth's Interior, Proceedings of the International School of Physics "Enrico Fermi,"* Course 78, 1979, Italian Physical Society (Amsterdam: North-Holland) pp. 555–649.

Rice, J.R. and Cleary, M.P., 1976, Some basic stress diffusion solutions for fluid-saturated elastic porous media with compressible constituents. *Reviews of Geophysics and Space Physics*, **14**, pp. 227–241.

Rice, J.R. and Sih, G.C., 1965, Plane problems of cracks in dissimilar media. *Journal of Applied Mechanics ASME,* **32**, pp. 418–423.

Robert, M. and Keer, L.M., 1987a, An elastic circular cylinder with displacement prescribed at the ends–Axially symmetric case. *Quarterly Journal of Mechanics and Applied Mathematics*, **40**(3), pp. 339–361.

Robert, M. and Keer, L.M., 1987b, An elastic circular cylinder with prescribed displacements at the ends–Asymmetric case. *Quarterly Journal of Mechanics and Applied Mathematics,* **40**(3), pp. 365–381.

Rongved, L., 1955, Force interior to one of two joined semi-infinite solids. *Proceedings, Second Midwestern Conference on Solid Mechanics,* pp. 1–13.

Rooke, D.P. and Cartwright, D.J., 1976. *Compendium of Stress Intensity Factors* (London: Her Majesty's Stationery Office).

Roscoe, K.H., Schofield, A.N., and Worth, C.P., 1958, On the yielding of soils. *Geotechnique*, **8**, pp. 22–53.

Roscoe, K.H. and Burland, J.B., 1968, On the generalized stress-strain behavior of "wet" clay. In *Engineering Plasticity,* edited by J. Humane and F.A. Leckie (Cambridge: Cambridge University Press), pp. 535–609.

Rudnicki, J.W., 1977, The effect of stress-induced anisotropy on a model of brittle rock failure as localization of deformation. In *Energy Resources and Excavation Technology, Proceedings of the 18th U.S. Symposium on Rock Mechanics*, Keystone, Colorado, June 22–24, 1977, pp. 3B4-1–3B4-8.

Rudnicki, J.W., 1980, Fracture mechanics applied to the earth's crust. *Annual Review of Earth and Planetary Sciences,* **8**, pp. 489–525.

Rudnicki, J.W., 1981, On fundamental solutions for a fluid-saturated porous solid by M.P. Cleary. *International Journal of Solids and Structures*, **17**, pp. 855–857.

Rudnicki, J.W., 1982, Discussion of "On Finite Plastic Flows of Compressible Materials with Internal Friction" by S. Nemat-Nasser and A. Shokooh. *International Journal of Solids and Structures*, **18**, pp. 357–360.

Rudnicki, J.W., 1984, A class of elastic-plastic constitutive laws for brittle rocks. *Journal of Rheology*, **28**, pp. 759–778.

Rudnicki, J.W., 1985, Effect of pore fluid diffusion on deformation and failure of rock. In *Mechanics of Geomaterials*, edited by Z.P. Bažant (New York: John Wiley & Sons), pp. 315–347.

Rudnicki, J.W., 1986, Fluid mass sources and point forces in linear elastic diffusive solid. *Mechanics of Materials*, **5**, pp. 383–393.

Rudnicki, J.W., 1987, Plane strain dislocations in linear elastic diffusive solids. *Journal of Applied Mechanics ASME,* **54**, pp. 545–552.

Rudnicki, J.W., 1988, Class Notes of Mechanics of Earthquakes, Department of Civil Engineering, Spring Quarter, 1988, Northwestern University.

Rudnicki, J.W., 1991, Boundary layer analysis of plane strain shear cracks propagating steadily on an impermeable plane in an elastic diffusive solid. *Journal of the Mechanics and Physics of Solids,* **39**, pp. 201–221.

Rudnicki, J.W., 1996, Moving and stationary dislocations in poroelastic solids and applications to aseismic slip in the Earth's crust. In *Mechanics of Poroelastic Media,* edited by A.P.S. Selvadurai (Dordrecht: Kluwer), pp. 3–22.

Rudnicki, J.W., 2000, Geomechanics. *International Journal of Solids and Structures* **37**, pp. 349–358.

Rudnicki, J.W., 2002, Conditions for compaction and shear bands in a transversely isotropic material. *International Journal of Solids and Structures,* **39**(13–14), pp. 3741–3756.

Rudnicki, J.W. and Chau, K.T., 1996, Multiaxial response of a microcrack constitutive model for brittle rock. In *Tools and Techniques in Rock Mechanics*, Proceedings of NARMS'96, 2nd North American Rock Mechanics Symposium, ISRM Regional Conference, June 19–21, 1996, edited by M. Aubertin, F. Hassani, and H. Mitri (Rotterdam: Balkema), pp. 1707–1714.

Rudnicki, J.W. and Hsu, T.C., 1988, Pore pressure changes induced by slip on permeable and impermeable faults. *Journal of Geophysical Research,* **93**(B4), pp. 3275–3285.

Rudnicki, J.W. and Rice, J.R., 1975, Conditions for the localization of deformation in pressure-sensitive dilatant materials. *Journal of the Mechanics and Physics of Solids*, **23**, pp. 371–394.

Rudnicki, J.W., Yin, J., and Roeloffs, E.A., 1993, Analysis of water level changes induced by fault creep at Parkfield, California. *Journal of Geophysical Research,* **98**, pp. 8143–8152.

Sabbagh, K., 2003, *The Riemann Hypothesis: The Great Unsolved Problem in Mathematics* (New York: Farrar, Straus & Giroux).

Sadeh, W.Z., 1967, A note on the general solution of the two dimensional linear elasticity problem in polar coordinates. *AIAA Journal*, **5**, p. 354.

Sadowsky, M.A. and Sternberg, E., 1949, Stress concentration around a triaxial ellipsoidal cavity. *Journal of Applied Mechanics ASME*, **16**, pp. 149–157.

Salencon, J., 1974, *Applications of the Theory of Plasticity in Soil Mechanics* (Chichester: John Wiley).

Salganik, R.L., 1973, Mechanics of bodies with many cracks. *Mechanics of Solids* (English translation from Russian), **8**, pp. 135–143.

Salman, A.D., Ghadiri, M., and Hounslow, M.J., (Eds.), 2007, *Particle Breakage, Handbook of Powder Technology,* Vol. 12 (Amsterdam: Elsevier).

Sanders, J.L. Jr., 1954, Plastic stress-strain relations based on linear loading functions. *Proceedings of the 2nd U.S. National Congress of Applied Mechanics*, pp. 455–460.

Savage, J.C., 1980, Dislocation in seismology. In *Dislocations in Solids,* edited by F.R.N. Nabarro, Vol. 3, Chapter 12 (Amsterdam: North-Holland), pp. 251–353.

Savin, G.N., 1961, *Stress Concentration around Holes,* translated from the Russian by W. Johnson (Oxford: Pergamon).

Schapery, R.A., 1962, Approximate methods of transform inverse for viscoelastic stress analysis. *Proceedings of the 4th U.S. National Congress of Applied Mechanics,* pp. 1075–1085.

Schapery, R.A., 1967, Stress analysis of viscoelastic composite materials. *Journal of Composite Materials,* **1**(3), pp. 228–267.

Schapery, R.A. and Park, S.W., 1999, Methods of interconversion between linear viscoelastic material functions. Part II–an approximate analytical method. *International Journal of Solids and Structures,* **36**(11), pp. 1677–1699.

Schiffman, R.L. and Fungaroli, A.A., 1965, Consolidation due to tangential loads. *Proceedings 6th International Conference on Soil Mechanics and Foundation Engineering*, Montréal, Vol. 2 (Toronto: University of Toronto Press), pp. 188–192.

Schofield, A. and Wroth, P., 1968, *Critical State Soil Mechanics* (New York: McGraw-Hill).

Scholz, C.H., 1990, *The Mechanics of Earthquakes and Faulting* (Cambridge: Cambridge University Press).

Scholz, C.H. and Kranz, R., 1974, Notes on dilatancy recovery. *Journal of Geophysical Research,* **79**(B14), pp. 2132–2135.

Scholte, J.S., 1947, The range of existence of Rayleigh and Stoneley waves. *Geophysical Journal International*, **5**, Suppl. s5, pp. 120–126.

Segel, L.A., 1987, *Mathematics Applied to Continuum Mechanics* (New York: Dover).

Selvadurai, A.P.S., 2000, *Partial Differential Equations in Mechanics*, Vol. 2, *The Biharmonic Equation, Poisson's Equation* (Berlin: Springer-Verlag).

Selvadurai, A.P.S., 2007, The analytical methods in geomechanics. *Applied Mechanics Reviews,* **60**, pp. 87–106.

Senjuntichai, T., 1994a, Dynamic Green's functions for homogeneous poroelastic half-planes. *Journal of Engineering Mechanics ASCE,* **120**, pp. 2381–2404.

Senjuntichai, T., 1994b, Green's Functions for Multi-Layered Poroelastic Media and an Indirect Boundary Element Method, Ph.D. thesis submitted to University of Manitoba, Canada.

Senseny, P.E., Fossum, A.F., and Pfeifle, T.W., 1983, Non-associative constitutive laws for low porosity rocks. *International Journal of Numerical and Analytical Methods in Geomechanics*, **7**, pp. 101–115.

Sewell, M.J., 1974, A plastic flow rule at a yield vertex. *Journal of the Mechanics and Physics of Solids*, **22**, pp. 469–490.

Shah, S.P., 1995, *Fracture Mechanics of Concrete: Applications of Fracture Mechanics to Concrete, Rock and Other Quasi-brittle Materials* (New York: Wiley).

Sih, G.C., 1973, *Handbook of Stress Intensity Factors* (Bethlehem, PA: Institute of Fracture and Solid Mechanics).

Silverman, R.A., 1974, *Complex Analysis with Applications* (New York: Dover).

Simons, D.A., 1977, Boundary-layer analysis of propagating mode II cracks in porous elastic media. *Journal of the Mechanics and Physics of Solids,* **25**, pp. 99–115.

Skempton, A.W., 1954, The pore-pressure coefficients A and B. *Geotechnique*, **4**, pp. 143–147.

Slepyan, L.I., 2002, *Models and Phenomena in Fracture Mechanics* (Berlin: Springer).

Sneddon, I.N., 1951, *Fourier Transforms* (New York: McGraw-Hill).

Sokolnikoff I.S., 1956, *Mathematical Theory of Elasticity*, 2nd ed. (New York: McGraw-Hill).

Southwell, R.V. and Gough, H.J., 1926, On the concentration of stress in the neighborhood of a small spherical flaw. *Philosophical Magazine*, **1**, pp. 71–87.

Spiegel, M.R., 1963, *Schaum's Outline Series: Theory and Problems of Advanced Calculus* (New York: McGraw-Hill).

Spiegel, M.R., 1964, *Schaum's Outline Series: Theory and Problems of Complex Variables* (New York: McGraw-Hill).

Spiegel, M.R., 1965, *Schaum's Outline Series: Theory and Problems of Laplace Transforms* (New York: McGraw-Hill).

Spiegel, M.R., 1968, *Schaum's Outline Series: Mathematical Handbook* (New York: McGraw-Hill).

Srinivasan, T.P. and Nigam, S.D., 1969, Invariant elastic constants for crystals. *Journal of Mathematics and Mechanics,* **19**(5), pp. 411–420.

Sternberg, E., 1960, On the integration of the equation of motion in the classical theory of elasticity. *Archive of Rational Mechanics and Analysis*, **6**, pp. 34–50.

Stoneley, R., 1949, The seismological implications of aeolotropy in continental structure. *Monthly Notice of the Royal Astronomy Society of Geophysics,* Suppl. **5**, pp. 343–353.

Struik, D.J., 1987, *A Concise History of Mathematics,* 4th rev. ed. (New York: Dover).

Sulem, J. and Vardoulakis, I., 1990, Bifurcation analysis of the triaxial test on rock specimens: A theoretical model for shape and size effect. *Acta Mechanica,* **83**, pp. 195–212.

Sun, C.T. and Qian, W., 1998, A treatment of interfacial cracks in the presence of friction. *International Journal of Fracture*, **94**, pp. 371–382.

Sun, C.T. and Yin, Z.H., 2012, *Fracture Mechanics* (Waltham: Academic Press).

Tada, H., Paris, P.C., and Irwin, G.R., 1973, *The Stress Analysis of Cracks Handbook* (Hellerton, PA: Del Research Corporation).

Taguchi, I. and Kurashige, M. 2002, Fundamental solutions for a fluid-saturated, transversely isotropic, poroelastic solid. *International Journal of Numerical and Analytical Methods in Geomechanics,* **26**, pp. 299–321.

Talebi, S., Nechtschein, S., and Boone, T.J., 1998, Seismicity and casing failures due to steam stimulation in oil sands. *Pure and Applied Geophysics*, **153**(1), pp. 219–233.

Tang, C.A., Lin, P., Wong, R.H.C. and Chau, K.T., 2001, Analysis of crack coalescence in rock-like materials containing three flaws–Part II: Numerical approach. *International Journal of Rock Mechanics and Mining Science,* **38**(7), pp. 925–939

Tapponnier, P. and Brace, W.F., 1976, Development of stress-induced microcracks in Westerly granite. *International Journal of Rock Mechanics and Mining Science & Geomechanics Abstract,* **13**, pp. 103–112.

Tarn, J.-Q. and Lu, C.-C., 1991, Analysis of subsidence due to a point sink in an anisotropic porous elastic half space. *International Journal of Numerical and Analytical Methods in Geomechanics,* **5**, pp. 573–592.

Taya, M., 1981, On stiffness and strength of an aligned short-fiber reinforced composite containing penny-shaped cracks in the matrix. *Journal of Composite Materials,* **15**, pp. 198–210.

Terzaghi, K., 1943, *Theoretical Soil Mechanics* (New York: Wiley).

Terzaghi, K. and Richart, F.E., 1952, Stresses in rock about cavities. *Geotechnique,* **3**, pp. 57–90.

Timoshenko, S.P., 1953, *History of Strength of Materials: With a Brief Account of the History of Theory of Elasticity and Theory of Structures* (New York: McGraw-Hill).

Timoshenko, S.P. and Goodier J.N., 1982, *Theory of Elasticity*, 3rd ed. (New York: McGraw-Hill).

Vallée, M., Landès, M., Shapiro, N.M., and Klinger, Y., 2008, The 14 November 2001 Kokoxili (Tibet) earthquake: High-frequency seismic radiation originating from the transitions between sub-Rayleigh and supershear rupture velocity regimes. *Journal of Geophysical Research,* **113B**, B07305, doi:10.1029/ 2007JB005520, pp. 1–14.

Van, P. and Vasarhelyi, B., 2010, Centenary of the first triaxial test-recalculation of the results of Karman. In *Rock Mechanics in Civil and Environmental Engineering,* edited by J. Zhao, J.-P. Labiouse, J. Dudt and J.-F.Mathier (London: Taylor & Francis), pp 59–62.

Vardoulakis, I., 1979, Bifurcation analysis of the triaxial test on sand samples. *Acta Mechanica,* **32**, pp. 35–54.

Vardoulakis, I., 1983, Rigid granular plasticity model and bifurcation in the triaxial test. *Acta Mechanica,* **49**, pp. 57–79.

Vardoulakis, I., 1994, Potentials and limitations of softening models in geomechanics: The role of second order work. *European Journal of Mechanics A/Solids,* **13**(4), pp. 195–226.

Vardoulakis, I. and Harnpattanapanich, T., 1986, Numerical Laplace-Fourier transform inversion technique for layered-soil consolidation problems: I. Fundamental solutions and validation. *International Journal of Numerical and Analytical Methods in Geomechanics,* **10**, pp. 347–365.

Vardoulakis, I. and Sulem, J., 1996, *Bifurcation Analysis in Geomechanics* (Glasgow: Blackie Academic & Professional).

Verruijt, A., 1969, The completeness of Biot's solution of the coupled thermoelastic problem. *Quarterly Applied Mathematics,* **26**, pp. 485–490.

Verruijt, A., 1971, Displacement functions in the theory of consolidation or in thermoelasticity. *Journal of Applied Mathematics and Physics (ZAMP),* **22**, pp. 891–898.

Vijayakumar, S. and Cormack, D.E., 1987, Green's functions for the biharmonic equation: Bonded elastic media. *SIAM Journal of Applied Mathematics,* **47**(5), pp. 982–997.

Viktorov, I.A., 1967, *Rayleigh and Lamb Waves: Physical Theory and Applications* (New York: Plenum Press).

Wan, F.Y.M., 1968, Comments on "A note on the general solution of the two-dimensional linear elasticity problem in polar coordinates". *AIAA Journal,* **6**, p. 569.

Wang, H.F., 2000, *Theory of Linear Poroelasticity: With applications to Geomechanics and Hydrology* (Princeton: Princeton University Press).

Wang, K.A. and Chong, K.P., 1989, Diametrical compression of transversely isotropic disks. *21st U.S. Symposium on Rock Mechanics,* pp. 243–248.

Wang, M.Z., 2002, *Advanced Theory of Elasticity* (Beijing: Peking University Press) (In Chinese).

Wang, M.Z., Xu, B.X., and Gao, C.F., 2008, Recent general solutions in linear elasticity and their applications. *Applied Mechanics Reviews,* **61**, pp. 1–20.

Wang, W.M., Sluys, L.J., and de Borst, R., 1997, Viscoplasticity for instabilities due to strain softening and strain-rate softening. *International Journal of Numerical Method for Engineering*, **40**, pp. 3839–3864.

Wang, Y.B. and Chau, K.T., 1998, A new boundary element method for plane elastic problems involving cracks and holes. *International Journal of Fracture,* **87**, pp.1–20.

Wang, Y.B. and Chau, K.T., 2001, A new boundary element method for mixed boundary value problems involving cracks and holes: Interactions between rigid inclusions and cracks. *International Journal of Fracture,* **110**(4), pp. 387–406.

Washizu, K., 1958, A note on the conditions of compatibility. *Journal of Mathematical Physics,* **36**, pp. 306–312.

Watson, G.N., 1952, *A Treatise on the Theory of Bessel Functions* (Cambridge: Cambridge University Press).

Watanabe, S., 1996, Elastic analysis of axi-symmetric finite cylinder constrained radial displacement on the loading end. *Structural Engineering/Earthquake Engineering JSCE,* **13**(2), pp. 175s–185s.

Wawersik, W.R., Rudnicki, J.W., Olsson, W.A., Holcomb, D.J., and Chau, K.T., 1990, Localization of deformation in brittle rock: Theoretical and laboratory investigations. In *Micromechanics of Failure of Quasi-Brittle Materials,* edited by S.P. Shah, S.E. Swartz, and M.L. Wang (Amsterdam: Elsevier), pp. 115–124.

Weertman, J., 1964, Continuum distribution of dislocations on faults with finite friction. *Bulletin of the Seismological Society of America,* **54**, pp. 1035–1058.

Weertman, J., 1996, *Dislocation Based Fracture Mechanics* (Singapore: World Scientific).

Weertman, J. and Weertman, J.R., 1964, *Elementary Dislocation Theory* (London: Macmillian).

Wei, X.X. and Chau K.T., 1998, Spherically isotropic spheres subject to diametral point load test: analytic solutions. *International Journal of Rock Mechanics and Mining Science*, **35**(4–5), pp. 623–624, Paper No. 006.

Wei, X.X. and Chau, K.T., 2000, Finite solid circular cylinders subjected to arbitrary surface load: Part II. Application to double-punch test. *International Journal of Solids and Structures* **37**, pp. 5733–5744.

Wei, X.X. and Chau, K.T., 2002, Analytic solution for transversely isotropic cylinders under the axial point load test. *Journal of Engineering Mechanics ASCE*, **128**, pp. 209–219.

Wei, X.X. and Chau, K.T., 2009, Finite transversely isotropic linear elastic cylinders under compression with end constraint induced by friction. *International Journal of Solids and Structures*, **46**, pp. 1953–1965.

Wei, X.X., Chau, K.T., and Wong R.H.C., 1999, Analytic solution for the axial point load strength test on solid circular cylinders. *Journal of Engineering Mechanics ASCE*, **125**, pp. 1349–1357.

Westergaard, H.M., 1939, Bearing pressures and cracks. *Journal of Applied Mechanics ASME*, **61**, pp. A49–A53.

Westergaard, H.M., 1952, *Theory of Elasticity and Plasticity* (Cambridge: Harvard University Press).

Westmann, R.A., 1965a, Asymmetric mixed boundary-value problems of the elastic half-space. *Journal of Applied Mechanics ASME*, **32**, 411–417.

Westmann, R.A., 1965b, Simultaneous pairs of dual integral equations. *SIAM Review*, **7**, pp. 341–348.

Widder, D.V., 1946, *The Laplace Transform* (Princeton: Princeton University Press).

Williams, C.D. and Burk, C.F., 1970, Upper cretaceous. In *Geological History of Western Canada,* edited by R.G. McCrossan and R.P. Glaister (Calgary: Alberta Society of Petroleum Geologists), pp. 169–188 (Chapter 12).

Williams, M.L., 1957, On the stress distribution at the base of a stationary crack. *Journal of Applied Mechanics ASME*, **24**, pp. 109–114.

Williams, M.L., 1959, The stresses around a fault or crack in dissimilar media. *Bulletin of the Seismological Society of America*, **49**(2), pp. 199–204.

Willis, J.R., 1970, Stress fields produced by dislocations in anisotropic media. *Philosophical Magazine*, **21**, pp. 931–949.

Willis, J.R., 1980, A polarization approach to the scattering of elastic waves II. Multiple scattering from inclusions. *Journal of the Mechanics and Physics of Solids*, **28**, pp. 307–327.

Wineman, A.S. and Pipkin, A.C., 1964, Material symmetry restrictions on constitutive equations. *Archive for Rational Mechanics and Analysis*, **17**, pp. 184–214.

Wong, C.W., 1991, *Introduction to Mathematical Physics: Methods and Concepts* (New York: Oxford University Press).

Wong, R.C.K., 1998, Swelling and softening behaviour of Labiche shale. *Canadian Geotechnical Journal*, 35, pp. 206–221.

Wong, R.H.C. and Chau K.T., 1998, Crack coalescence in a rock-like material containing two cracks. *International Journal of Rock Mechanics and Mining Science*, **35**(2), pp. 147–164.

Wong, R.H.C., Chau, K.T., Tang, C.A., and Lin, P., 2001, Analysis of crack coalescence in rock-like materials containing three flaws-Part I: Experimental approach. *International Journal of Rock Mechanics and Mining Science*, **38**(7), pp. 909–924.

Wong, R.H.C., Chau, K.T. and Wang, P., 1996, Microcracking and grain size effect in Yuen Long marbles. *International Journal of Rock Mechanics and Mining Science & Geomechanics Abstract*, **33**(5), pp. 479–485.

Wong R.H.C., Tang C.A., Chau K.T., and Lin P., 2002, Splitting failure in brittle rocks containing pre-existing flaws under uniaxial compression. *Engineering Fracture Mechanics,* **69**(17), pp. 1853–1871.

Wong, T.-f., 1990, A note on the propagation behavior of a crack nucleated by a dislocation pileup. *Journal of Geophysical Research*, **95**(B6), pp. 8639–8646.

Wong, T.-f. and Baud, P., 1999, Mechanical compaction of porous sandstone. *Oil Gas Science and Technology,* **54**(6), pp. 715–727.

Wong, T.-f., Baud, P. and Klein, E., 2001, Localized failure modes in a compactant porous rock. *Geophysical Research Letters*, **28**, pp. 2521–2524.

Wong, T.-f., Wong, R.H.C., Chau, K.T., and Tang, C.A., 2006, Microcrack statistics: Weibull distribution and micromechanical modeling of compressive failure in rock. *Mechanics of Materials*, **38**, pp. 664–681.

Wu, S.Z. and Chau, K.T., 2006, Dynamic response of an elastic sphere under diametral impacts. *Mechanics of Materials*, **38**, pp. 1039–1060.

Wu, S.Z., Chau, K.T., and Yu, T.X., 2004, Crushing and fragmentation of brittle spheres under double impact test. *Powder Technology*, **143–144**, pp. 41–55.

Xu, Z.L., 1982, *Elasticity*, Vols. 1 and 2 (Beijing: People Education Press) (In Chinese).

Yatomi, C., Yashima, A., Uzuka, A., and Sano, I., 1989, General theory of shear bands formation by a non-coaxial Cam-clay model. *Soils and Foundations*, **29**(3), pp. 41–53.

Yu, H.S., 2006, *Plasticity and Geotechnics* (New York: Springer).

Author Index

A

Abramowitz, M., 137, 141, 287, 319, 324, 326, 375, 403
Achenbach, J.D., 343, 350, 359, 366–368, 377, 385, 390, 403
Adachi, T., 186, 189, 403, 406
Airy, G.B., 26, 36, 39–40, 43, 49, 60–61, 63–66, 69–70, 89, 103–104, 206–207, 385
Aki, K., 235, 366, 403
Apirathvorakij, V., 309, 403
Ashby, M.F., 197, 235, 403
Atkinson, B.K., 218–219, 249, 403, 413
Atkinson, J.H., 175, 403

B

Baladi, G.Y., 191, 407
Barber, J.R., 60, 136, 403
Bardet, J.P., 148, 403
Barenblatt, G.I., 316, 403
Baud, P., 191, 424
Bažant, Z.P., 26, 36, 197, 237–240, 385, 403, 418
Beltrami, E., 33, 93, 101–104, 106, 147, 386, 397
Ben-Menahem, A., 149, 366, 403
Bert, C.W., 40, 403
Bertsch, P.K., 166, 404
Bessel, F.W., 141–142, 386, 392–393
Besuelle, P., 191, 404
Bigoni, D., 18, 404
Biot, M.A., 36, 295–296, 298, 299, 301–302, 304–306, 310, 323, 330–331, 367, 386, 391, 399–400, 404
Bland, D.R., 268, 404
Bodner, S.R., 186, 404
Boland, J.N., 413
Boley, B.A., 330, 404
Boltzmann, L.E., 258, 289, 386, 392, 396
Booker, J.R., 315, 404
Boone, T.J., 421
Borcherdt, R.D., 290, 359, 404
Boresi, A.P., 60, 404
Boussinesq, J.V., 93-94, 113, 115, 117–118, 122, 133–134, 140–141, 144–145, 147, 156, 271, 273–274, 279, 289, 292, 366, 386
Brace, W.F., 197, 235, 386, 342, 404, 421
Bransby, P.L., 175, 403
Britton, J.R., 82, 404
Broberg, K.B., 88, 249, 360, 367, 404
Budiansky, B., 160, 165–167, 242, 244, 387, 402, 404
Burgers, J.M., 49–50, 55–56, 58, 290, 387
Burk, C.F., 284, 423
Burland, J.B., 178, 417

C

Cagniard, L., 352, 366, 369, 377, 405
Campbell, H.G., 162, 405
Carrier, G.F., 63–64, 72, 315, 405
Carter, J.P., 315, 404
Cartwright, D.J., 214, 249, 417
Carslaw, H.S., 295, 324, 330, 375, 393, 405
Carvalho, J.L., 284, 315, 405
Cauchy, A.L., 18–21, 23–24, 35, 52, 55, 57, 62, 64, 72–73, 80–81, 89, 181, 192, 199, 207, 242, 387, 391, 399
Cerruti, V., 93–94, 118–119, 122, 141, 147, 150, 156–157, 387, 405
Chan, K.S., 148, 311, 405
Chatterjee, A.K., 245, 405
Chau, K.T., 17–18, 29–30, 39, 48, 148–149, 163, 184, 191, 211, 227, 235, 242, 247, 249–250, 280–282, 284–285, 287–289, 295, 309–311, 315, 365–367, 405–407, 410–412, 415, 418, 421, 420–424
Chen, W.F., 173–176, 179–180, 191, 407

Cheng, A.H.-D., 295, 298, 305, 306, 331, 408
Cherepanov, G.P., 219, 407
Chinnery, M.A., 53, 407
Choi, S.K., 17, 413
Chong, K.P., 48, 218, 404, 413, 422
Chou, P.C., 13, 60, 94, 102, 148, 407
Christensen, R.M., 258–259, 265, 290, 407
Christoffersen, J., 166, 407
Chuang, T.-j., 398, 408
Clapeyron, B.F.E., 27, 387
Cleary, M.P., 88, 166, 250, 296, 298–304, 315–316, 320–319, 321–323, 327–328, 330–333, 399, 407, 417–418
Cook, N.G.W., 48, 60, 237, 393, 412
Cormack, D.E., 285, 422
Costin, L.A., 167, 235, 247, 249, 408, 411
Coulomb, C.A.de, 159, 168–169, 171, 174–175, 195–196, 388, 397
Cristescu, N., 191, 408
Curran, J.H., 284, 315, 405

D

Daniel, J.W., 162, 416
Davis, E.H., 45–46, 60, 416
Davis, R.O., 60, 177, 191, 408
Davis, S.H., 373, 408
de Boer, R., 295. 331, 408
de Borst, R., 410
Deresiewicz, H., 331, 367, 396, 408, 413
Derski, W., 295, 408
Desai, C.S., 187, 408
Detournay, E., 298, 305–306, 331, 408
Dirac, P.A.M., 58, 271, 277–278, 311, 313, 323, 337, 375, 377–378, 380, 388
Drnevich, V.P., 364, 410
Drucker, D.C., 159, 163–164, 166–167, 169, 176-177, 258, 388, 394, 398–400, 408
Duncan, J.M., 174, 413
Dundurs, J., 53, 132, 135–136, 154–155, 288, 388–389

E

Edison, L., 412
Einstein, A., 2–4, 388, 391–392, 396, 399
England, A.H., 88, 409
Erdelyi, A., 52, 287, 375, 409
Eringen, A.C., 366, 387, 409
Eshelby, J.D., 58, 219, 389, 397, 409
Ewing, W.M., 352, 366, 409

F

Fairbairn, E.M.R., 46, 409
Filon, L.N.G., 29, 54, 63, 148, 212, 389, 409, 412, 414
Filonenko-Borodich, M., 40, 409
Findley, W.N., 166, 404
Flannery, B.P., 416
Flugge, W., 261, 263, 281, 290, 409
Fossum, A.F., 419
Freund, L.B., 347, 360–362, 367, 389–390, 409
Fung, Y.C., 60, 97, 102, 111, 138–139, 257-259, 263, 265, 280–281, 290, 390, 409
Fungaroli, A.A., 295, 307, 306–310, 315, 330, 400–401, 419

G

Galerkin, B.G., 26, 94, 97–100, 111, 118, 138–140, 147–148, 390
Gao, C.F., 422
Gaylord, E.W., 13, 411
Ghadiri, M., 419
Gharpuray, V.M., 42, 409
Gibbs, J.W., 5, 382, 391–392
Gibson, R.E., 295, 298, 306–308, 310–311, 315, 330, 332, 391, 401, 408, 414
Gladwell, G.M.L., 148, 409
Goodier, J.N., 11, 23, 25, 28, 37, 39, 60, 88, 93, 421
Goodman, L.E., 99, 147–148, 398, 409
Goodman, R.E., 263, 409

Goursat, E., 63
Gradshteyn, I.S., 146, 287, 319, 324, 375, 409
Gray, G.A., 166, 416
Green, A.E., 60, 88, 410
Green, G., 24, 26, 35, 36, 58, 59, 60, 63, 93, 109, 110, 111, 310, 331, 338, 366, 391, 394
Griffith, A.A., 52–53, 197, 204, 210–212, 214, 217, 219, 224, 237, 391, 410
Gurtin, M.E., 258, 260, 290, 391, 410

H

Hadamard, J., 353, 354, 365, 392
Haddad, Y.M., 263, 290, 410
Hallam, S.D., 197, 235, 403
Hankel, H., 94, 140–142, 144–148, 156, 295, 309, 312, 315, 330, 366, 386, 392, 397
Hardin, B.O., 364, 410
Harnpattanapanich, T., 331, 421
Hashin, Z., 244–245, 410
Hearmon, R.F.S., 27, 60, 410
Heaviside, O. 271–272, 277–278, 285, 292, 310–311, 313, 325, 337, 373, 375, 392
Heeres, O.M., 189, 410
Hellan, K., 207, 249, 410
Helmholtz, H. von, 26, 94–99, 139–140, 147, 183, 258, 303, 340–341, 365, 382, 392, 397
Henkel, D.J., 408
Henyey, F.S., 245, 410
Hetnarski, R.B., 40, 60, 373, 410
Hetenyi, M., 136, 154–155, 409
Hettkamp, T., 417
Higo, Y., 412
Hill, D., 53, 410
Hill, R., 26, 163-164, 166, 171, 183–184, 191, 393, 410
Hirth, J.P., 60, 410
Ho, K.C., 39, 410
Hohenemser, K., 186, 411
Hori, M., 60, 242, 244, 415
Horii, H., 235, 245, 397, 411
Hounslow, M.J., 419
Hu, H.C., 30, 60, 393, 411
Hughes, W.F., 13, 411
Hutchinson, J.W., 166, 407
Hyman, B.I., 40, 411

I

Iai, S., 148, 403
Idriss, I.M., 364, 369, 411
Ignaczak, J., 40, 60, 373, 410
Inglis, C.E., 197, 411
Irwin, G.R., 197, 214, 217, 362, 393, 411, 420
Issen, K.A., 191, 411

J

Jaeger, J.C., 48, 60–61, 237, 295, 323-324, 330, 375, 393, 405, 412
James, I., 395, 412
Jardetzky, W.S., 409
Jeffery, G.B., 387, 412
Jeffrey, H., 401, 412
Jenkins-Jones, S., 385, 412
Jeyakumaran, M., 53, 235, 412
Johnston, I.W., 413

K

Kachanov, L.M., 242, 412
Kalandiya, A.I., 88, 412
Kannien, M.F., 214, 232, 249, 363, 412
Kanok-Nukulchai, W., 309, 311, 315, 412
Karasudhi, P., 37, 40, 60, 136, 309, 403, 405, 412, 416
Karman, von T., 175, 385, 387–388, 393–394, 396–397, 412
Keer, L.M., 148, 394, 409, 412, 417
Kelvin, Lord, 9, 59, 93-94, 109, 111–115, 122–123, 132, 138–139, 147, 262–263, 265, 268, 273–274, 276–277, 281, 289–292, 315, 329, 387, 391, 394, 401
Kimoto, S., 186, 189–191, 331, 412, 416
Kirchhoff, G.R., 19–21, 26, 35–36, 60, 161, 366, 371, 392, 394, 397, 398, 400

Klinger, Y., 421
Knopoff, L., 405
Koehler, J.S., 51, 412
Kolosov, G.V., 63, 101, 394, 397
Koo, K.K., 365, 412
Krajcinovic, D., 242, 412
Kranz, R.L., 197, 235, 412, 419
Kriegh, R.B., 404
Kreyszig, E., 8, 412
Krook, M., 405
Kuhn, G., 397, 412
Kurrer, K.-E., 396, 413
Kuruppu, M.D., 218, 413

L

Lade, P.V., 174, 413
Lam, S.S., 412
Lama, R.D., 263, 281, 413
Lamb, H., 147-148, 338, 342, 366, 395, 411
Lamé, G., 29, 96, 98–99, 106, 114, 117–118, 132, 147, 266, 296, 301, 329, 341, 395
Landès, M., 360
Laplace, P.S., 64, 98, 118, 124, 129, 137, 140, 142, 185, 207, 265–272, 274, 279–281, 285, 287, 289-290, 295, 309–310, 312, 315, 332, 357, 360, 366, 373–377, 387, 389, 392–393, 395, 397
Lapusta, N., 413
Lee, E.H., 270, 285, 395, 413
Lee, J.D., 404
Lee, S.L., 405
Legendre, A.M., 34, 137, 146, 184, 215, 303, 382–383, 395–396
Lekhnitskii, S.G., 27, 30, 60, 243, 245, 413
Lemaître, J., 242, 413
Lewis, J.L., 409
Li, V.C., , 218–219, 249, 413
Lim, I.L., 217–218, 413
Lin, P., 421, 423
Liu, X., 191, 407
Little, R.W., 60, 88, 136–137, 147–148
Love, A.E.H., 32, 60, 98–99, 109, 111, 114, 117, 138–139, 141, 144, 147–148, 325, 329, 337–338, 339, 342, 346–351, 354, 359, 365, 368–369, 390, 396, 413
Lode, W., 169–172, 174
Lothe, J., 60, 410
Lu, C.-C., 315, 421
Lu, X., 362, 413
Lubliner, J., 161, 164-165, 184, 186, 191, 260, 353, 356, 413

M

Mahmoud, M.S., 331, 413
Mal, A.K., 8, 60, 366, 414
Malvern, L.E., 8, 13, 21, 26, 31, 34, 36, 60, 96, 102, 185, 414
Mandel, J., 353
Mann, E.H., 40, 414
Maor, E., 325, 414
Markov, M.G., 367, 414
Mase, G.E., 290, 414
Mase, G.T., 290, 414
Matsuoka, T., 174, 414
Maxwell, J.C., 94, 102–104, 106, 147, 185, 187, 262–263, 265, 267, 271–272, 274–276, 289–290, 292, 387, 391, 396, 398
McNamee, J., 295, 298, 306–308, 310–311, 315, 330, 332, 401, 414
McClintock, F.A., 237, 414
Meleshko, V.V., 39–40, 63, 139, 389, 414
Michell, J.H., 33, 39–40, 43, 93, 101–102, 396, 414
Millar, D., 385, 391, 394, 414
Millar, I., 385, 391, 394, 414
Millar, J., 385, 391, 394, 414
Millar, M., 385, 391, 394, 414
Mindlin, R.D., 93–94, 99–100, 122, 126, 128, 135, 147, 150–152, 164, 310, 388, 394, 396, 400, 408–409, 415
Milne-Thomson, L.M., 53, 60, 63, 88, 212, 414
Mizuno, E., 173, 175–176, 179–180, 191, 407
Mohr, O., 168–169, 171, 174, 195–196, 396–397

Morera, G., 94, 102, 104, 106, 147, 397
Muhlhaus, H.B., 17, 23, 235, 237, 415
Muki, R., 94, 139–142, 148, 156, 309, 397, 415
Mura, T., 52, 58–60, 183, 397, 411, 415
Murakami, Y., 214, 249, 287, 415
Muskhelishvili, N., 52, 57, 63, 75, 81, 88-89, 101, 198, 202, 205, 223, 395, 397, 415

N

Nakai, T., 174, 191, 414–415
Nayak, G.C., 170, 415
Nechtschein, S., 421
Newmark, N.M., 93, 113, 402, 415
Nemat-Nasser, S., 60, 163, 183, 197, 235, 242, 244, 397, 403, 411, 415, 418
Neuber, H., 94, 99–101, 109, 115, 122–123, 126, 128–132, 135, 138, 147–156, 304–305, 307, 397
Niumpradit, B., 309, 416
Nixon, F.E., 287, 375, 415
Noble, B., 162, 416
Nowacki, W.K., 366, 416

O

Obata, M., 197, 235, 415
Oka, F., 186, 189, 331, 403, 406, 412, 416
Ord, A., 415
Osinov, V.A., 354, 416

P

Pagano, N.J., 13, 60 ,94, 102, 148, 407
Pan, E., 331, 416
Pan, J., 166, 416
Palmer, A.C., 227–234, 416–417
Pao, Y.H., 416
Papkovitch, P.F., 94, 99–101, 109, 115, 122–123, 126, 128–132, 135, 138, 147–156, 304–305, 307, 397, 415
Paris, P.C., 420
Park, S.W., 290, 419
Pearson, C.E., 405
Perzyna, P., 186–189, 410, 416
Pfeifle, T.W., 419
Phillips, A., 166, 416
Pietruszczak, S., 159, 191, 416
Piola, G., 19–21, 26, 60, 371, 394, 398
Pipkin, A.C., 315, 423
Planas, J., 237, 239, 401
Poisson, S.D., 27, 29, 30, 43, 86, 95–97, 107–108, 117–118, 140, 242, 266, 282–283, 286, 295–296, 298, 300, 340, 342–343, 345–346, 365, 367–368, 370, 398, 411, 419
Pomphery, N., 245, 410
Popelar, C.H., 214, 232, 249, 363, 412
Poulos, H.G., 45–46, 60, 416
Prager, W., 29, 159, 163–164, 169, 176, 186, 387–388, 398, 408, 411
Press, F., 409
Press, W.H., 283, 369, 416
Puswewala, U.G.A., 309, 416

Q

Qian, W., 250, 288, 416, 420

R

Rahardjo, H., 331, 409
Rajapakse, R.K.N.D., 148, 309, 331, 416, 418
Rajapakse, Y.D.S., 237, 403
Rayleigh, Lord, 337–339, 341–346, 351–352, 359, 361–362, 365, 367, 370, 398, 400-401, 419, 421–422
Renner, J., 281–283, 289, 416
Rice, J.R., 17, 60, 88, 159, 162–163, 165–168, 183–184, 189, 191–192, 219, 221, 227–234, 242–243, 249–250, 287, 296, 298–302, 304, 330–332, 354, 360, 389, 398–399, 407–408, 410, 416–418
Richart, F.E., 43, 86, 421
Richards, P.G., 235, 366, 403
Riemann, G.F.B., 2, 55, 64, 89, 207, 392, 399, 418

Robert, M., 148, 417
Roeloffs, E.A., 418
Rooke, D.P., 214, 249, 417
Rosakis, A.J., 413
Roscoe, K.H., 177–178, 399, 417
Rudnicki, J.W., 17, 53, 159, 162–163, 165–168, 183, 189, 191–192, 214, 218, 233–235, 247, 249–250, 296, 298, 301-304, 307, 315, 322–323, 327–333, 354, 399, 404, 406–407, 411-412, 417–418, 422
Rummel, F., 416
Rutland, L.W., 404
Ryzhik, I.M., 146, 287, 319, 324, 375, 409

S

Sabbagh, K., 399, 418
Sadeh, W.Z., 40, 418
Sadowsky, M.A., 148, 418
Sahasakmontri, K., 245, 411
Saleeb, A.F., 174, 191, 407
Salencon, J., 191, 419
Salganik, R.L., 243, 245, 419
Salman, A.D., 367, 419
Sanders, J.L. Jr., 165–166, 419
Sano, I., 424
Savin, G.N., 81–82, 88, 198, 419
Schiffman, R.L., 295, 307, 306–310, 315, 330, 400–401, 419
Schofield, A.N., 175, 417, 419
Scholte, J.S., 352, 366, 419
Scholz, C.H., 214, 235, 419
Seed, H.B., 364, 369, 400, 411
Segel, L.A., 8, 13, 342, 345, 400, 419
Selvadurai, A.P.S., 60, 139, 177, 191, 389, 408, 414, 418–419
Senjuntichai, T., 148, 331, 416, 419
Senseny, P.E., 163, 168, 419
Sewell, M.J., 166, 420
Shapiro, N.M., 421
Shokoon, A., 163, 415
Sih, G.C., 214, 249, 287, 399, 417, 420
Simons, D.A., 250, 331, 420
Singh, S.J., 8, 60, 149, 366, 403, 414
Slepyan, L.I., 249, 362, 367, 420
Sluys, L.J., 422
Sneddon, I.N., 142, 144, 148, 266, 415, 420
Sokolnikoff, I.S., 60, 420
Spiegel, M.R., 3, 72, 88, 146, 182, 207, 287, 310, 318, 375, 420
Stegun, I.A., 137, 141, 287, 319, 324, 326, 375, 403
Sternberg, E., 100, 148, 258, 260, 290, 391, 397, 400, 410, 415, 418, 420
Stieltjes, T.J., 260–261
Stokes, G.G., 8–9, 59, 338, 394, 400–4001
Stoneley, R., 337, 350–352, 365–369, 401, 412, 414, 419–420
Struik, D.J., 392, 420
Suhubi, E.S., 366, 409
Suiker, A.S.J., 410
Sulem, J., 18, 23, 148, 191, 420–421
Sun, C.T., 250, 288, 416, 420
Symonds, P.S., 186, 404

T

Tada, H., 214, 226, 249, 420
Talebi, S., 284, 421
Tang, C.A., 250, 421
Tapponnier, P., 197, 235, 421
Tarn, J.-Q., 315, 421
Taya, M., 244, 421
Terzaghi, K., 43, 86, 295, 297, 421
Timoshenko, S.P., 11, 23, 25, 28, 37, 39, 60, 88, 93, 163, 270, 385–401, 421
Teukolsky, S.A., 416

U

Ulm, F.-J., 46, 409
Uzuka, A., 424

V

Vallée, M., 362, 421
Van, P., 175, 421
Vardoulakis, I., 18, 23, 148, 191, 331, 354, 401, 415, 420–421
Vasarhelyi, B., 175, 421
Verruijt, A., 304–308, 330, 401, 421–422

Vetterling, W.T., 416
Vijayakumar, S., 285, 422
Viktorov, I.A., 343, 422
Voigt, W., 2, 262–263, 265, 268, 273–274, 276–277, 289–292, 387, 401
Volterra, V., 49–50, 59, 258, 402
Vutukuri, V.S., 263, 281, 413

W

Walsh, J.B., 237, 386, 414
Wan, F.Y.M., 40, 422
Wang, G.S., 235, 406
Wang, H.F., 331, 422
Wang, K.A., 48, 422
Wang, M.Z., 102–104, 106, 122, 128, 148, 330, 422
Wang, P., 424
Wang, W.M., 188, 422
Wang, Y.B., 211, 250, 406, 422
Washizu, K., 26, 382, 393, 422
Watanabe, S., 148, 422
Weertman, J., 50, 52–53, 55, 60–61, 422
Weertman, J.R., 50, 53, 55, 60, 422
Wei, X.X., 30, 39, 48, 148–149, 406, 422–423
Westergaard, H.M., 60, 97, 128, 148, 171–173, 197, 223–227, 249–250, 402, 415, 423
Westmann, R.A., 148, 423
Widder, D.V., 377–379, 423
Williams, C.D., 284, 423
Williams, M.L., 197, 206–207, 423
Willis, J.R., 59, 244, 423
Wineman, A.S., 315, 423
Wong, C.W., 13, 423
Wong, R.C.K., 250, 280–282, 284–285, 287, 289, 407, 423
Wong, R.H.C., 48, 184, 235, 242, 250, 404, 407, 421, 423-424
Wong, T.-f., 53, 191, 250, 402, 424
Wong, Y.L., 412
Wroth, P., 175, 419
Wu, S.Z., 149, 367, 424
Wu, W., 354, 416

X

Xu, B.X., 422
Xu, Z.L., 78, 88, 198, 424

Y

Yang, X., 407, 412
Yashima, A., 424
Yatomi, C., 181–183, 192, 424
Yin, J., 418
Yin, Z.H., 238, 420
Young, T., 27, 29–30, 185, 246, 282, 405–406
Yu, H.S., 168, 191, 424
Yu, T.X., 406, 424

Z

Zerna, W., 60, 88, 410
Zhang, D., 187, 408
Zienkiewicz, O.C., 170, 415

Subject Index

A

Adachi–Oka model, 189
Airy stress function, 26, 36, 39–40, 43, 49, 60–61, 63–66, 69–70, 89, 103–104, 206–207, 385
Acceleration waves, 338, 353
Acoustic tensor, 354–357
Alamani's strain tensor, 24
Analytic functions, 63, 65–72, 75, 79, 85, 89, 91, 202, 205, 223, 392

B

Bažant size effect, 240
Beltrami–Michell compatibility, 33, 93, 101–102
Beltrami stress function, 102, 104, 106, 147
Bessel function, 142, 386, 393, 422
Biharmonic functions, 39, 61, 65, 89, 94–95, 97–99, 103, 111–112, 118, 135, 138–139, 147, 156, 207, 414, 419, 422
Biot's function, 304
Biot's theory, 298
Boltzmann integral, 258
Boltzmann solid, 258
Boussinesq problem, 113, 141, 271–274, 366
Brazilian test, 46, 48
Burgers circuit, 58
Burgers material, 290, 387
Burgers vector, 49–51, 58

C

Cagniard–de–Hoop method, 366, 377
Cam-clay model, 160, 177, 181, 399, 424
Cap models, 160, 175
Cartesian tensor, 2, 7, 22, 30
Cauchy–Riemann relation, 55, 64, 89, 207, 399
Cauchy singular integral, 57, 62, 88
Cauchy strain tensor, 24
Cauchy stress tensor, 18, 23, 181, 242
Center of dilatation, 227, 284–286, 288–289, 296, 325, 407
Cerruti problem, 93–94, 118-119, 122, 141, 147, 150, 156–157, 387
Christoffel stiffness, 355
Christoffel symbols, 8
Circular hole, 40–42, 68, 85, 86
Clapeyron formula, 27
Compatibility, 11, 13, 26, 32, 33, 38, 54, 60, 94, 101–103, 263, 300, 354, 385-386, 392, 396, 422
Complementary energy, 34–35, 184, 215, 382, 384
Complex variable method
  boundary condition, 70
  coordinate transformation, 76
  multi-connected body, 72–75
  single-valued condition, 71
  uniqueness, 69
Conformal mapping, 76, 78, 81–82, 88, 198
Constitutive laws
  isotropic solids, 27
  transversely isotropic solids, 29
Continuum damage mechanics, 240
Correspondence principle, 270
Cosserate continuum, 23
Cracks, 197
Crack tip singularity
  mode I, 207
  mode II, 211
  mode III, 212
Creeping tests, 275
Cylindrical coordinate, 10, 15, 39, 51, 98, 113, 115–116, 137–140, 142, 150, 172, 298, 308, 386
Cylindrical transverse isotropy, 245, 254-255
Cyclic steam injection, 284

D

Deformation theory, 160, 163, 166–167, 355

Differential scheme, 245–247, 249, 254–255, 410
Dimensional scaling, 315
Dirac delta function, 58, 271, 277–278, 323, 337, 375, 377–378, 380, 388
Dislocation
  Burgers vector, 49–51, 58
  curved, 58, 397
  edge, 49
  glide, 61
  Mura formula, 58, 397
  pile-up, 51, 53, 62
  screw, 53
  wedge crack, 53
Dispersion of waves, 351
Divergence theorem, 9
Dot product, 3
Drucker's postulate, 163
Drucker–Prager model, 163, 170, 176
Duhamel integral, 258
Dundurs parameter, 288
Dyadic tensor, 5
Dynamic fracture
  asymptotic field at tip, 361
  energy release rate, 362
  fragmentation, 367
  moving cracks, 361
  stationary cracks, 360

E

Earthquakes
  seismic waves, 338, 366
  energy release rate, 233
Edge dislocation, 49
Effective compliance, 243
Einstein notation, 3
Elasticity
  hyperelastic, 35
  hypoelastic, 35
  three-dimensional, 93
Elastic body with holes, 78
Elliptical hole, 198
Energy
  strain energy, 33
  complementary energy, 35
Energy release rate, 214
Equations of motion, 31
Eulerian strain, 24

F

Fading memory hypothesis, 259
Faulting, 53–54, 56–57, 197, 214, 233–234, 284, 296, 338, 32
First Piola–Kirchhoff stress, 19
Fluid mass dipole, 328
Fluid mass source, 323, 325
Flow theory, 160
Fracture mechanics
  crack-tip singularity, 205–212
  energy release rate, 214–217
  J-integral, 219
  shear mode, 205
  slip in slopes, 227
  tear mode, 212, 250, 252, 290, 360
  tensile mode, 202, 207
  Westergaard function, 223
  wing crack model, 235
Fracture toughness, 217–219, 288–289, 363, 367, 393
  Bažant size effect, 240
  dynamic arrest toughness, 363
  dynamic growth toughness, 363
  dynamic initiation toughness, 363
  in rocks, 217
Fracture speed
  intersonic, 362
  sub-Rayleigh, 362
  subsonic, 362
  super-Rayleigh, 362
  supersonic, 362
  transonic, 362
Fragmentation, 367

G

Galerkin vector, 97
Generalized Kelvin model, 263, 265
Generalized Maxwell model, 262, 265

H

Hadamard compatibility condition, 353–355
Half-space

Boussinesq problem, 113, 141, 271–274, 366
Cerruti problem, 93–94, 118-119, 122, 141, 147, 150, 156–157, 387
Mindlin problem, 93, 94, 122, 126, 128, 147, 150, 151, 152, 308
Hankel transform, 94, 140–142, 144–148, 156, 309, 312, 366, 386, 392, 397
Hardening
isotropic, 165–166
kinematic, 165–166
Harmonic functions, 132, 136, 137
Heaviside step function, 271–272, 277–278, 285, 292, 310–311, 313, 325, 337, 373, 375, 392
Helmholtz theorem, 94, 340
Heredity solid, 256
Hollow Sphere, 106–107
Hohenemser-Prager model, 186
Hooke's law, 27, 30, 37, 62, 65, 90, 91, 96, 210, 259

I

Isotropic solids, 18, 27–29
Il'iushin's postulate, 163, 165
Indirect tensile test, 46–47
Internal variables, 160, 183, 240

J

J-integral, 219
Jacobian, 21, 369
Jaeger's modified Brazilian test, 48
Jaumann stress rate, 36

K

Kelvin solution, 109
Kelvin–Voigt model, 263
Kirchhoff stress tensor, 26
Kirsch solution
by Airy stress function, 41
by complex variable method, 68
Kolosov–Muskhelishvili formalism, 63
Kronecker delta, 3

L

Lamé constants, 29
Lamé strain potential, 96
Laplace transform
Schapery inversion method, 379
inverse Laplace transform, 375
Left Cauchy–Green tensor, 24
Legendre transformation, 184, 215, 301, 380–381, 394
Love's displacement potential, 98, 111, 117, 141, 147
Love waves
dispersion, 348
nonexistence in half-space, 345
layer on half-space, 345
Lode angle, 169
Lorentz's fundamental solution, 128

M

Maculey bracket, 353
Malvern viscoplastic model, 185
Matsuoka–Nakai model, 174
Maxwell stress function, 103
McNamee–Gibson–Verruijt potential, 304
Melan's fundamental solution, 130
Microcracks, 241
Micropolar elasticity, 23, 32, 396
Mindlin solution, 93, 94, 122, 126, 128, 147, 150, 151, 152, 310
Modified Cam-clay model, 178
Mohr–Coulomb model, 168
Mooney–Rivlin material, 35
Morera stress function, 104
Moving cracks, 359
Muki approach, 139–140
Muki vector potentials, 140
Mura formula, 58
Muskhelishvili formalism, 63

N

Nanson formula, 369
Neo-Hookean material, 35
Noninteracting model, 243, 245–249

P

Papkovich–Neuber potentials, 99
Permutation tensor, 4
Perzyna model, 186
Piola–Kirchhoff stress
  first PK stress, 20
  second PK stress, 21
Planar transverse isotropy
Plasticity
  Cam-clay, 177, 181
  cap models, 175
  deformation theory, 160
  Drucker's postulate, 163
  elasto-plastic, 162
  finite strain, 181
  flow theory, 160
  internal variables, 183
  Il'iushin's postulate, 163
  Lade–Duncan model, 174
  Lode angle, 169
  Matsuoka–Nakai model, 174
  modified Cam-clay model, 178
  Mohr-Coulomb model, 168
  π-plane, 171
  PMPR, 163
  plastic potentials, 161
  Rudnicki–Rice, 163
  yield function, 161
  yield vertex, 165–166
  viscoplasticity, 184
Poisson condition, 342
Poroelasticity
  anisotropic form, 302
  Biot's theory, 298
  Biot–Verruijt function, 304
  Cleary solution, 315
  fluid mass dipole, 328
  fluid mass source, 323, 325
  McNamee–Gibson–Verruijt potential, 306
  point force in full space, 315, 323
  point force in half-space, 310
  Rice–Cleary constitutive form, 298
  Rudnicki constitutive form, 301–302
  Rudnicki solution, 323
  Schiffman–Fungaroli–Verruijt potential, 307
Principle of superposition, 27, 45, 389

R

Rate of deformation, 25
Rayleigh wave
  Characteristics equation, 341
  Particle motions, 342, 345
Relaxation tests, 277
Rice–Cleary constitutive form, 298
Riemann geometry, 2
Rivlin–Saunders material, 35
Rudnicki constitutive form, 301–302
Rudnicki–Rice model, 163

S

Schwarz–Christoffel integral, 82
Schwarz Inequality, 3
Screw dislocation, 53
Second Piola–Kirchhoff stress, 21
Seismic waves, 336
Self-consistent method, 244
Shear band, 354
Size effect, 237
Soil dynamics, 364
Slip in slopes, 227
Slip line theory, 58
Spherical coordinate, 12
Square hole, 82
Standard linear solid, 263, 268
St.-Venant's compatibility, 26
Stieltjes integral, 258
Stokes theorem, 9
Stoneley waves, 351
Strain energy, 6
Strain Energy Density, 33
Strain localization, 227, 354
Stress concentration
  circular hole, 41, 68
  elliptical hole, 198
  square hole, 82
Stress tensor, 18
Supershear, 362
Surface breaking fault, 56
Surface loads, 45
Surface waves, 341–342

T

Tensor
cross product, 4
curl, 7
derivatives, 7
divergence, 8
dot product, 6
dyadic, 5
e-δ identity, 5
free index, 1
grad, 7
rank, 2
order, 2
Thermoelasticity, 329
Three-dimensional elasticity
Beltrami stress functions, 101, 104
Beltrami–Schaefer stress function, 101
displacement formulation, 94
Galerkin Vector, 97
Helmholtz theorem, 94
Kelvin's Fundamental Solution, 109
Lamé strain potential, 96
Love's Displacement Potential, 98
Maxwell stress functions, 103
Mindlin solution, 93, 94, 122, 126, 128, 147, 150, 151, 152, 310
Morera stress function, 104
Muki potential, 139
Papkovitch–Neuber Potential, 109
Stress formulation, 101
Transversely isotropic solids, 30

V

Vector, 3
Viscoelasticity
Burgers model, 290
Boussinesq problem, 271–274, 290
complex moduli, 357
corresponding principle, 270
crack problems, 284
creeping tests, 277
differential form, 261
fading memory hypothesis, 259
Kelvin–Voigt model, 263
generalized Kelvin model, 265
generalized Maxwell model, 265
heredity solid, 258
Maxwell model, 262
model calibration, 281
relaxation compression tests, 281
relaxation tests, 276, 281
shales, 281–284
standard linear solid, 278
Stieltjes convolution, 260
three parameter models, 263
with elastic bulk modulus, 267
Viscoplasticity
Adachi–Oka model, 189
consistency model, 188
Hohenemser–Prager model, 184
Perzyna model, 186–187

W

Waves
in elastic-plastic solids, 353
in isotropic solids, 338
in viscoelastic solids, 357
Wedge, 43
Wedge crack, 53
Westergaard stress function, 223
Wing crack model, 235

Y

Yield function, 161
Yield vertex, 165–166

Z

Zerner–Stroh crack, 53

Kelvin–Voigt model, 261
generalized Kelvin model, 265
generalized Maxwell model, 265
hereditary solid, 268
Maxwell model, 262
model calibration, 284
relaxation compression tests, 281
relaxation tests, 276, 281
shakes, 281, 284
standard linear solid, 278
Stieltjes convolution, 260
three-parameter models, 265
with elastic–plastic model, 185
Viscoplasticity
Anand's model, 189
Bodner's model, 185
Chaboche–Perzyna model, 184
Perzyna model, 184, 187

W

Waves
in elastic–plastic solids, 358
in isotropic solids, 335
in viscoelastic solids, 357
Wedge, 423
Wedge crack, 538
Westergaard stress function, 523
Wriggers model, 235

Y

Yield function, 161
Yield surface, 158–161

Z

Zener–Stroh crack, 531

T

Tensor
cross product, 7
curl, 7
derivatives, 8
divergence, 8
Three-dimensional elasticity
Beltrami stress equations, 101, 103
Beltrami–Schaefer stress functions, 103
displacement formulation, 97
Galerkin vector, 97
Helmholtz theorem, 98
Kelvin's fundamental solution, 104
Lamé strain potential, 96
Love's Displacement Potential, 98
Maxwell stress functions, 103
Navier equations, 95, 97, 126, 128, 148, 152, 310
Morera stress function, 104
Stokes potential, 99
Papkovich–Neuber Potential, 100
Stress formulation, 101
Transversely isotropic solids, 80